6th International Conference on Advances in Materials, Machinery, Electronics (AMME 2023)

MATEC Web of Conferences Volume 382 (2023)

Wuhan, China
27 - 28 May 2023

Editors:

Lin Liu
Jiaru Wu

ISBN: 978-1-7138-8336-4

Printed from e-media with permission by:

Curran Associates, Inc.
57 Morehouse Lane
Red Hook, NY 12571

Some format issues inherent in the e-media version may also appear in this print version.

For additional information, please contact EDP Sciences – Web of Conferences
at the address below.

EDP Sciences – Web of Conferences
17, Avenue du Hoggar
Parc d'Activité de Courtabœuf
BP 112
F-91944 Les Ulis Cedex A
France

Phone: +33 (0) 1 69 18 75 75
Fax: +33 (0) 1 69 28 84 91

contact-edps@webofconferences.org

Additional copies of this publication are available from:

Curran Associates, Inc.
57 Morehouse Lane
Red Hook, NY 12571 USA
Phone: 845-758-0400
Fax: 845-758-2633
Email: curran@proceedings.com
Web: www.proceedings.com

TABLE OF CONTENTS

Study on the Sealing Test Method of Hydraulic Solenoid Valve in Nuclear Power Station 1
Wang Haibao, Tan Xudong, Xiao Qunxiong, Chang Xincai

Ni3S2/Zn0.76Co0.24S for High Performance Hybrid Supercapacitor 5
Sun Mengxuan, Yan Yijun, Li Zhijie, Jia Chunyang

Electrochromic Properties of Monosubstituted Benzylboronic Acid Viologen 9
Tang Shengwei, Niu Junlong, Jia Chunyang

Health Status Assessment of Scraper Conveyor Conveying Equipment Based on Cross-Entropy Theory and Immune Principle 14
Wang Chunsheng, Wu Dangdang, Qi Yuhao, Liu Hongfei, Yu Tongzhu

High Catalyst Performance of Two-Dimensional Ti3B3N3S6 for Hydrogen Evolution Reaction 20
Sun Yongxiu, Li Zhijie

Research on the Development of Sensor Tracking Technology Under the Guidance of Artificial Intelligence 24
Ge Shan

Research on Dynamic Path Exploration and Planning Method of Quadrotor Unmanned Aerial Vehicle in Complex Environment 28
Huang Hanting

Experimental Study on Solidification/stabilization of Heavy Metal Lead and Chromium Sludge Solidified by Cement-Based Materials 32
Zheng Chao, Xiong Kangwei, Yang Kun, Feng Deluan

Research on Properties of Polypropylene/Flake Graphite Molded Composite Materials 44
Zhou Haijun, Zhou Mengmeng, Zhang Xiaolei, Bai Lu, Chen Xiaoqi, Zhang Fen, Li ChengYu, Li Yantao

Optimization of Asynchronous Motor with Taguchi Method Based on Finite Element Analysis 49
Zhang Lixi

Kinematics Analysis of a Medication Delivery Robot Based on a 3-DOF Manipulator 55
Huang Shanyue

Research Status of Cathode Materials for Lithium Ion Batteries 61
Lin Zihao

Finite Element Simulation Calculation and Optimal Design of Permanent Magnet Synchronous Motor 69
Yuan Peizhi

Optimizations of the Internal Structure of the Reel of a Double Rope Winder 75
Fu Jia

Research Status and Optimization Methods of Zinc Ion Battery 82
Wang Shurui

Mechanism and Control Strategy of Large-Area Concrete Shrinkage Cracks in the Underground Railway Section 93
Yu Meilin

Research Progress in Nanotechnology for Enhanced Silicon-Based and Iron-Based Anodes for Lithium-Ion Batteries 104
Yao Yaqi

Research and Improvement of Electromechanical Transient Performance of Power System Based on SIMULINK 111
Wang Duo

Research Progress of High Entropy Materials as Water Electrolysis Catalysts in Recent Years 117
Wu Baizhen

Highly Dispersed Pt with CoFeP Hollow Nanocages for Efficient Oxygen Evolution Reaction 126
Xie Jingyi, Dong Bin

Ultrahigh Density Defects of Mn3O4 Boosting Oxygen Evolution Reaction 130
Zhou Yanan, Chai Yongming, Dong Bin

Sulphur Doping of Nickel Ferrite Nanosheet Array for Enhanced Water Oxidation 134
Liu Haijun, Dong Bin

A Coupled Radial-Torsional Lumped Parameter Model of Rubber Bushing for Multibody Dynamic Simulations 138
Cao Chang, Feng Xingxing, Liu Min

Time Response and Natural Frequency Analysis for Structures with Interval Parameters 151
Feng Xingxing, Cao Chang, Liu Min

Research on Distance Measurement Method of Field Driving Test System 159
Yao Yao, Chao Han, Yang Wu, Jinzhou Dai, Peng Xu

Intelligent Mechanical Manufacturing Technology Based on Intelligent Manufacturing Technology 163
Zheng Kang

Preparation of MXene and Its Application and Research Progress in Supercapacitors 167
Gao Zhenkui

Nonlinear Effects-Based 1.7 ìm Fiber Lasers: A Review and Prospect 173
Ma Xinning

Experimental Study on the Factors Affecting the Performance of Wire-Plate Biefeld-Brown Thrust Unit 183
Li Jinbao, Lai Lin

Research on Anti-Loosening Technology of Transmission Line Bolts 189
Li Qinghua, Li Maohua, Zhu Binrong, Hu Xiaoguang

Experimental Study on Bearing Capacity of Corroded Angle Steel Members of In-Service Transmission Towers 195
Zhu Binrong, Huang Yao, Li Zheng, Chen Bin, Liao Hailin

Study on Looseness Characteristics of Bolted Joints of Transmission Towers Under Lateral Vibration 199
Zhu Binrong, Huang Guo, Su Zhigang, Chen Bin, Liao Hailin

Chemical Composition and Identification of Glass Products Based on Generalized Linear Model 203
Pan Zhuo

Review of Recent Distillation Studies 209
Gao Minghong

Influence of Material Characteristics of Hoppers on Particle Flow Behavior 217
Zhang Panpan, Huang Guoqing, Chen Qingwei

Research on Privacy Protection of Intelligent Construction 221
Shang Jingshi, Li Xiuwen, Wang Qingyong

Study on the Measuring and Reading Mode of High Precision Thermometer in Hydraulic Engineering 225
Zhen Shisen, Wu Hao

Key Influencing Factors Research of Domestic and Import Aviation Lubricating Oil Foam Characteristics 229
Geng Yuxiong, Wang Xiaoai, Wang Yongtao, Zheng Yitian, Ma Jun

Vibrational Analysis of Stiffened Double-Layered Plate in Vacuum 234
Wang Peng, Li Zhao, Zhang Hao, Chen Jiandong, Li Wei, Liu Min

Experimental Platform Design for Particle Behavior Investigation 238
Zhang Panpan, Zhou Siyang, Chen Qingwei

Simulation Analysis of Internal and External Faults in Transformer Area 242
Bai Peixian, Jia Yixuan

A Method for Monitoring Operation Status of Oil-Immersed Transformer Based on the Baffle Angle Characteristic of Gas Relay 251
Bai Peixian, Xie Yiling, Dou Longjiang

Author Index

Preface: 6th International Conference on Advances in Materials, Machinery, Electronics (AMME 2023)

This conference proceedings includes original and peer-reviewed research papers from the 6th International Conference on Advances in Materials, Machinery, Electronics (AMME 2023), and it was held on May 27-28, 2023 in Wuhan, China.

The aims of the conference are to provide an interactive platform for the academicians, R & D organizations, industries, innovators, entrepreneurs, government agencies and policy-makers etc., from both China and abroad to exchange ideas on recent researches and advances in the domain of materials, machinery, electronics with its prospective application in the various interdisciplinary domains of engineering. This two-day conference is being organized in technical association with Research Center of Engineering and Science and include invited keynote talks and oral paper presentations from both academia and industry to ignite our young minds, to initiate them in the field of meaningful research and thereby enrich their existing knowledge.

All full paper submissions to the AMME 2023 must be written in English and were sent to at least two reviewers and evaluated based on originality, technical or research content, correctness, relevance to conference, contributions, and readability. The full paper submissions were chosen based on technical merit, interest, applicability, and how well they fit a coherent and balanced technical program. AMME 2023 has received 71 manuscripts. And 42 submissions have been accepted by our reviewers. The Committees of AMME 2023 invest great efforts in reviewing the papers submitted to the conference and organizing the sessions to enable the participants to gain maximum benefit.

Hopefully, all participants and other interested readers benefit scientifically from the proceedings and find it stimulating in the process.

With our warmest regards,

Lin Liu, Jiaru Wu
Conference Organizing Committee
Wuhan, China

Statement of Peer review

In submitting conference proceedings to *Web of Conferences*, the editors of the proceedings certify to the Publisher that

1. They adhere to its **Policy on Publishing Integrity** in order to safeguard good scientific practice in publishing.
2. All articles have been subjected to peer review administered by the proceedings editors.
3. Reviews have been conducted by expert referees, who have been requested to provide unbiased and constructive comments aimed, whenever possible, at improving the work.
4. Proceedings editors have taken all reasonable steps to ensure the quality of the materials they publish and their decision to accept or reject a paper for publication has been based only on the merits of the work and the relevance to the journal.

Title, date and place of the conference

6th International Conference on Advances in Materials, Machinery, Electronics (AMME 2023)
May 27-28, 2023
Wuhan City, China

Proceedings editor(s):

Lin Liu
Jiayu Wu

Date and editor's signature June 19, 2023

 https://doi.org/10.1051/matecconf/202338201001

Study on the sealing test method of hydraulic solenoid valve in nuclear power station

Haibao Wang [1], Xudong Tan[1], Qunxiong Xiao[1] and Xincai Chang [2,*]

[1] Yangjiang Nuclear Power Co.,ltd. China
[2] Shanghai Engineering Science &Technology Co., Ltd Cgnpc, Beijing, 100086, China

Abstract. In order to improve the reliability of tripping solenoid valve and switching solenoid valve of steam turbine valve in nuclear power plant and reduce abnormal valve closing. According to the working principle of valves in nuclear power plant, this paper designs a kind of electromagnetic valve sealing test system, including test base, measurement and control system and hydraulic system. The detection system is an off-line system. By simulating the real load condition of the steam turbine system, the tightness of the tripping solenoid valve and the switching solenoid valve is tested to improve the reliability of the solenoid valve. The actual application results show that the system operates stably and reliably, is simple to operate, has high detection efficiency and accuracy, and has good use effect.

1. Introduction

The steam turbine of nuclear power plant is composed of high-pressure cylinder, low-pressure cylinder and several air inlet pipelines. Each intake pipeline is equipped with one GSE stop valve and one GRE regulating valve. Among them, GSE cut-off valve has two tripping solenoid valves and one switching solenoid valve. During normal operation, the tripping solenoid valve is energized and energized, and the switching solenoid valve is energized to control the opening of the cut-off valve through high-pressure hydraulic oil. In case of emergency tripping of the turbine, the tripping solenoid valve loses power and makes the valve close quickly. Therefore, the main function of the solenoid valve on the GSE valve is to protect the safety of the steam turbine. GRE control valve and GSE stop valve have repeatedly failed to trip solenoid valve, which directly affects the normal function of the valve. If the solenoid valve leaks, the pressure cannot be maintained, the valve closes abnormally, and the oil pump acts frequently, which seriously affects the safe operation of the unit and even causes certain accidents.

Therefore, during the overhaul, the nuclear power station will inspect the solenoid valve on the inlet valve of the GSE / GRE system turbine. On the one hand, it will carry out routine inspection, including testing the insulation value and continuity resistance value of the solenoid valve. Based on experience, it will formulate the fault standard of the solenoid valve. Once the tested solenoid valve has deviation, it will be replaced; On the other hand, the tripping solenoid valve shall be replaced regularly. Although the failure of the solenoid valve will be avoided, the maintenance cost will be increased and the resource utilization rate is not high. At the same time, there is no effective means for the detection and cause analysis of equipment failure. The solenoid valve realizes the on-off of the liquid path or the change of the liquid flow direction in the control system, and plays different roles in cooperation with different control systems. The sealing of the solenoid valve is the key factor that affects the reliability of the control system under the real load condition. Therefore, it is necessary to develop a set of special hydraulic solenoid valve sealing test system.

2. System composition and working principle

The solenoid valve on the inlet valve of GSE / GRE system is a two position three-way DC solenoid valve. The test system of hydraulic solenoid valve developed and designed for it mainly includes test base, measurement and control system and hydraulic system. Used to test the tightness of the solenoid valve. According to the actual working environment pressure of the system, the pressure of the test system is set to 150bar. The servo motor drives the oil pump to extract the hydraulic oil in the oil tank, and supplies the hydraulic pressure to the whole test system through the filter, one-way valve and overflow valve. The test system controls the hydraulic pressure by adjusting the output voltage of the servo motor. Check the digital pressure gauge. When the pressure is stable at 150bar, the test solenoid valve can be tested for tightness. The upper computer software of the measurement and control system controls the power on and power off of the test solenoid valve according to the test requirements, tests the inlet pressure in real time, calculates the leakage rate and other parameters of the solenoid valve, and judges the performance of the solenoid valve. And print the standard report.

* Corresponding author: changxincai@cgnpc.com.cn

2.1 Test base

Including the main body of the test platform, the mounting base plate of the test solenoid valve, the oil pan and parts. The main structure of the designed test platform is compact, and the requirements of the field test are fully considered to ensure that the test equipment is not damaged.

2.2 Measurement and control system

The measurement and control unit and data processing unit are the centers of signal processing, data communication, motor control and data acquisition, including pressure sensor, temperature sensor, AI input module and do output module. This design can test both tripping solenoid valve and switching solenoid valve

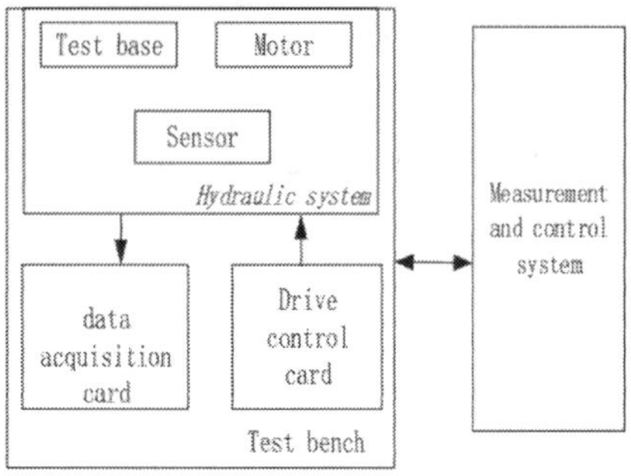

Figure 1. Schematic diagram of measurement and control system.

1) Data processing unit

The acquisition chassis with Ethernet communication is adopted, which is equipped with digital input module, digital output module, voltage input module and voltage output module. It communicates with the upper computer through Ethernet to realize the control, acquisition and signal processing functions of the measurement and control system.

2) Measurement and control unit

The measurement and control unit is divided into acquisition and control.

The acquisition unit includes analog signals such as pressure, temperature, liquid level and current, and switching signals such as filter alarm and oil tank low liquid level alarm.

The control unit includes a voltage output and a switching value output. The output voltage drives the motor, and adjusts the output hydraulic pressure of the oil pump by changing the motor speed; Control the output of the program-controlled power supply and adjust the power supply voltage of the test solenoid valve; The output control of the switching value mainly controls the relay to complete the power supply switch of the test solenoid valve, motor enable, emergency stop control, etc.

3) Upper computer software

The system runs in the windows environment, and the client computer can install the running program. The main functions of the upper computer software include: human-computer interaction interface, information management of test solenoid valve, pressure test of solenoid valve, collection and display of real-time curve, data analysis, report output, data storage, data playback and other functions.

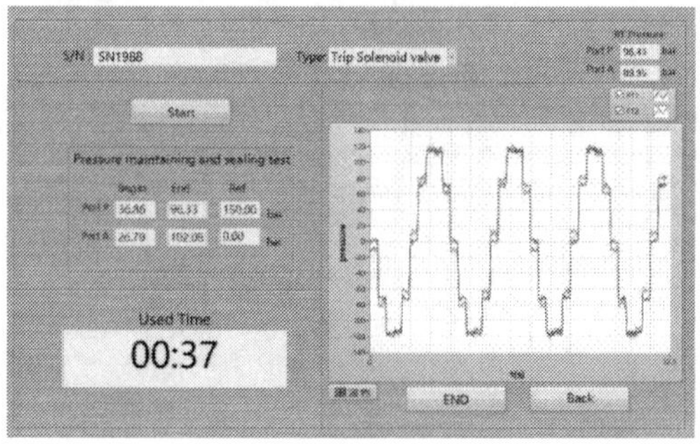

Figure 2. Schematic diagram of sealing test interface

2.3 Hydraulic system

The hydraulic system consists of motor, oil pump, pressure gauge, ball valve, hydraulic hose, test solenoid valve, oil tank, etc. The hydraulic system provides a pressure environment for the entire test system.

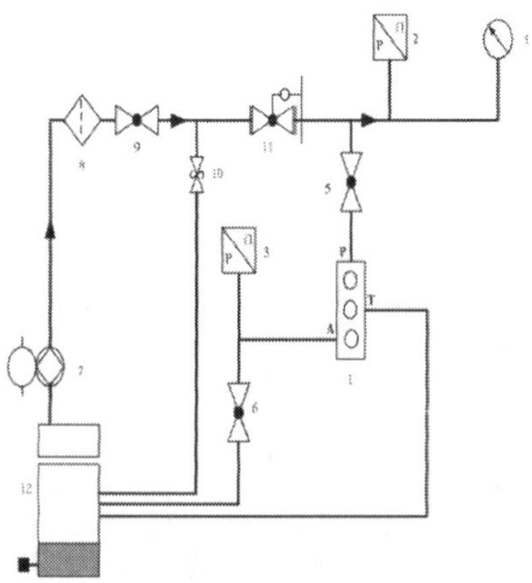

Figure 3. Working principle of hydraulic system

https://doi.org/10.1051/matecconf/202338201001

3. Sealing test

The pressure measurement method is used to test the leakage rate of the solenoid valve when it loses excitation / excitation. During the test, the pressure at the inlet side of the solenoid valve can be kept basically unchanged. The back pressure reduction value of the solenoid valve can be measured within the time t, and the actual leakage rate can be calculated through the pressure change value and other parameters.

Specifically, first of all, make preparations before pressure:

1) Run the upper computer software and fill in the parameters of the test solenoid valve: SN code, valve type, state, oil pressure reference value and other parameters

2) Use an Ohmmeter to test the continuity resistance value of the test solenoid valve and the resistance value of the positive and negative electrodes to the ground, and record it in the corresponding position of the upper computer software.

3) Keep the ball valve at port a of the test solenoid valve closed and open the ball valve at port P (Port P is the oil inlet)

Start pressure test:

1) The program controls the motor to start. After 2S, open the oil circuit solenoid valve, observe the pressure gauge in real time, and test that the oil inlet pressure of the solenoid valve reaches the set value.

2) Check the upper computer software interface. After the pressure at port P is stable, manually close the ball valve at port p to maintain the pressure at port P.

Start the pressure maintaining test:

1) Click the upper computer interface to execute the command of "start pressure maintaining", and the system will automatically close the motor and oil circuit solenoid valve.

2) The pressure curves of port P and port a start to be drawn, and the timer starts to count. At this time, the system automatically records the initial oil pressure;

3) After a certain time t, click "end pressure maintaining", the curve will stop drawing and the timing will stop;

4) The upper computer sends the "pressure relief" command to completely relieve the pressure of the test solenoid valve;

5) The system automatically calculates the leakage rate and records it in the database for subsequent data query and playback.

Comparison of test results

It can be clearly seen from the following figure that the pressure retaining capacity of the solenoid valve a in Fig. 4 is obviously insufficient, and the sealing performance is decreased, which needs to be replaced.

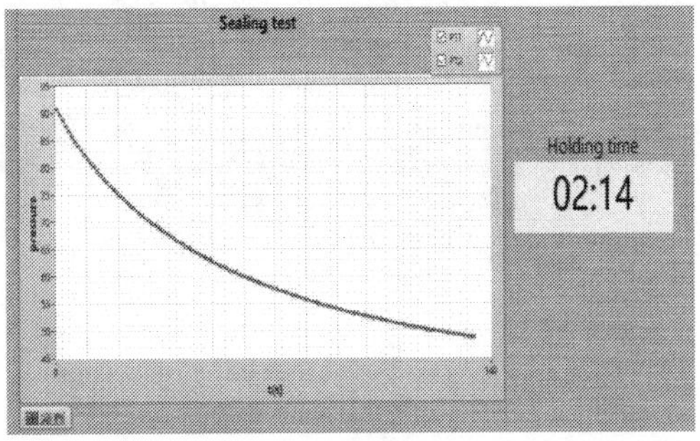

Figure 4. Replay interface of sealing test of solenoid valve A

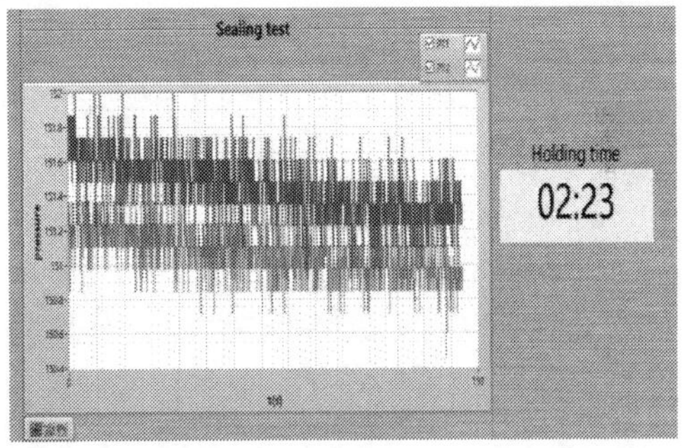

Figure 5. Replay interface of sealing test of solenoid valve B

4. Conclusion

Through this method, the sealing state performance of the solenoid valve on the inlet valve of the GSE / GRE system can be accurately detected, the fault can be found in time, and the decision-making basis can be provided for the adjustment or maintenance of the solenoid valve. In addition, there are clear and quantifiable judgment standards for the performance of solenoid valves, which no longer rely on the experience of test personnel. The test report is also in a unified version, which effectively avoids human error and provides guarantee for the safe operation of the steam turbine inlet valve of the nuclear power plant.

References

1. Jian-Wei Yang; Jin-Hai Wang; Qiang Huang; Ming Zhou; Reliability Assessment for the Solenoid Valve of a High-Speed Train Braking System under Small Sample Size [J]. Chinese Journal of Mechanical Engineering. 2018 (3)
2. Angadi S V , Jackson R L , Choe S Y , et al. Reliability and life study of hydraulic solenoid valve. Part 2: Experimental study[J]. Engineering Failure Analysis, 2009, 16(3):944-963.

3. Zai-Qing, Ren , and W. Jian-Hui . "Test-system design for hydraulic solenoid valve of a certain aircraft." IOP Conference Series: Materials Science and Engineering 1043.4(2021):042023 (6pp).
4. KalinowskiMarcinmarcin@o.plKamińskiZbigniewz.kaminski@pb.edu.plFaculty of Mechanical Engineering,Bialystok University of Technology,ul. Wiejska C, -Biaystok,Poland. (2021). Measurement and evaluation of functional and operational coefficients of hydraulic solenoid valve prototypes used for variable valve timing control in combustion engines. Acta Mechanica et Automatica, 15(2), 74-81.

$Ni_3S_2/Zn_{0.76}Co_{0.24}S$ for high performance hybrid supercapacitor

Mengxuan Sun [1], Yijun Yan [1], Zhijie Li [1,*] and Chunyang Jia [2]

[1]School of Physics, University of Electronic Science and Technology of China, Chengdu, 610054, P. R. China.;
[2]School of Integrate Circuit Science and Engineering, University of Electronic Science and Technology of China, Chengdu 610054, P. R. China.

Abstract. Supercapacitors have become a research hotspot in the field of energy storage due to their high-power density, fast charging/discharging ability, long-term stability and safety. Nevertheless, relatively low energy density hindered their application. Herein, $Ni_3S_2/Zn_{0.76}Co_{0.24}S$ (NZCS) microsphere were synthesized using a facile two-step hydrothermal process. The polymetallic synergies can improve conductivity and shorten ion transport path. The uniform particle distribution provided numerous active sites for faradaic reactions. The NZCS mircosphere showed a large capacity of 571.5 C g^{-1} at 1 A g^{-1} and 87% rate rentention when the current increases by 10 times. A hybrid supercapacitor assmebled by NZCS cathode and active carbon anode demostrate a high energy density of 44.1 Wh kg^{-1} (407.0 W kg^{-1}) and a stable cycliability of 15,000 cycles with 15% loss. Thus, NZCS is a promising electrode material for high performance supercapacitor.

Keywords: Supercapacitor; $Zn_{0.76}Co_{0.24}S$; Ni_3S_2; hydrothermal.

1. Introduction

Environmental problems and energy shortage have become increasingly serious due to the rapid economic development has led to the huge consumption of coal, oil and other resources. High efficiency, clean energy and related advanced technologies have been put forward higher requirements for energy storage device (EES).[1, 2] Supercapacitors (SCs) has attracted much attention due to its high-power density, ultra-long cycle life and reliable security. However, the relatively low energy density of SCs compared with that of lithium-ion batteries and fuel cells has hindered its further application. Hybrid supercapacitors (HSCs) use a dual charge storage mechanism, both capacitive-controlled and difussion-controlled, to enable higher operating voltage and higher energy density.[3]

Electrode material is an important factor affecting the supercapacitor, which directly determines the charge storage capacity of the capacitor. Metal sulfides, such as Ni_3S_2 and Co_9S_8, have attracted more attention as electrode materials for SCs due to their excellent electrical conductivity and electrochemical activity. However, metal sulfides stored energy by Faraday reaction occurring through the surface of the material with electrolyte ions, and tend to show a poor long-term stable performance in the process of charge and discharge.

Designing polymetallic sulfide composites is an effective strategy to improve electrochemical stability due to the high theoretical capacity compared with monometal and more abundant redox reaction.[4] Thus, we prepared $Ni_3S_2/Zn_{0.76}Co_{0.24}S$ microsphere by hydrothermal method and exhibit excellent electrochemical properties. The uniform distribution microsphere structure can provide effective active area and has a short ion transport path during energy storage, which can enhance the Faraday reaction between electrolyte ions and the active materials. Moreover, the polymetallic synergies help to enhance electrical conductivity. At the current density of 1 A g^{-1}, $Ni_3S_2/Zn_{0.76}Co_{0.24}S$ exhibit an large capacity of 571.5 C g^{-1}. The NZCS//AC HSC can achieve a high energy density of 44 Wh kg^{-1} at a power density of 407 W kg^{-1} and a long-term cycling stability of 15,000 with 75% capacity retention, which exhibits great application prospect.

* Corresponding author: zhijieli@uestc.edu.cn

 https://doi.org/10.1051/matecconf/202338201002

2. Experimental section

2.1 Materials

All the chemical regents were purchased from Aladdin Biochemical Technology Co., Ltd. and Tianjin iweixin Huagong Technology Co., Ltd. All the chemical regents were of analytical purity grade, and used without any further purification.

2.2 The synthesis of NZCS

The synthesis of NZCS was displayed in **Fig. 1**. Typically, 0.49 g $Zn(CH_3COO)_2 \cdot 2H_2O$, 0.49 g $Co(CH_3COO)_2 \cdot 4H_2O$ and 0.662 g $Ni(CH_3COO)_2 \cdot 4H_2O$ were dissolved in 75 mL deionized water and stirring for 15 min. Then, 0.48 g urea, 0.11 g KOH and 0.38 g $C_6H_8O_7$ were added in turn to the above solution and stirring until uniform. Furthermore, the above mixture was transferred into a Teflon-lined stainless-steel autoclave and reacted at 160°C for 6 h to obtain NZCO. For prepared NZCS, NZCO (0.16 g) and $Na_2S \cdot 9H_2O$ (2.8 g) were further processed by hydrothermal method for 8 hours at 120°C.

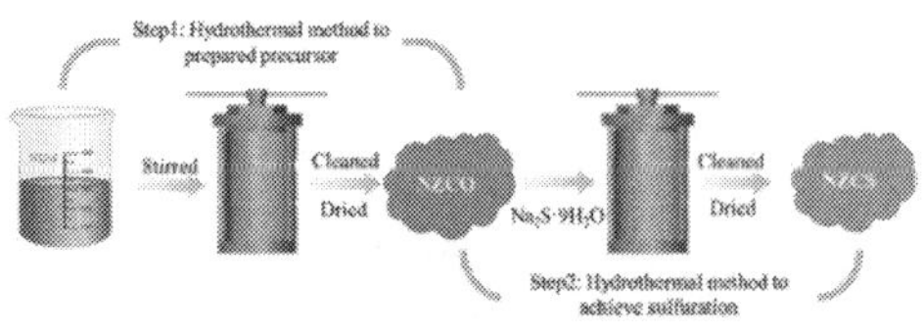

Fig. 1 Schematic diagram of preparation process of the NZCS

2.3 Materials and electrochemical characterization

X-ray diffraction (XRD, Rigaku D/max-2400) was used for quantitative analysis and macroscopic stress analysis. The molecular structure and valence states were analyzed by XPS. Scanning electron microscope (SEM, Inspect F50) was used to observe the surface morphology of materials.

Cyclic voltammetry (CV), galvanostatic chargedischarge (GCD), and electrochemical impedance spectroscopy (EIS), were characterized by using electrochemical workstation (CHI660E).

The Specific capacity (C, C g^{-1}) can be calculated based on GCD curves, as follow:

$$C = \frac{I \times \Delta t}{m} \quad (1)$$

Where I (A), Δt (s) and m (g) are current, discharge time and mass of the active material.

The energy density (E) and power density (P) can be calculated as the following equations:

$$E = \frac{C \times \Delta V}{2 \times 3600} \quad (2)$$

$$P = \frac{E \times 3600}{\Delta t} \quad (3)$$

Where C (C g^{-1}) is the specific capacity of the HSC device, Δt (s) is the discharge time and ΔV (V) is the potential window.

3. Results and discussion

NZCS were characterized by XRD, and the results were shown in Fig. 2a. The main diffraction peaks are indexed to (110), (003), (202) of Ni_3S_2 (JCPDS card No. 44-1418) and (111), (220), (311) of $Zn_{0.76}Co_{0.24}S$ (JCPDS card No. 47-1656), respectively. There is no spare other phases, which illustrate the product is composed of Ni_3S_2 and $Zn_{0.76}Co_{0.24}S$. Fig. 2b and Fig. 2c showed the SEM image and the corresponding element distribution of NZCS composite, respectively. It can be seen that NZCS composite is microsphere structure, and the diameter is about 1.6 μm. The uniform distribution of such particles can increase the contact area between the electrode material and the electrolyte during the electrochemical reaction process, providing a good way for electron transfer and ion diffusion.

Fig. 2 (a) XRD pattern, (b) SEM images, and (c) Mapping of NZCS

As shown in Fig. 3a, CV curves of NZCS and NZCO composite electrode materials was measured in a three-electrode system with 2 M KOH as for electrolyte at the scan rate of 10 mV s^{-1}. There are two obvious peaks can be observed, which correspond to the redox process in the charging and discharging process. Compared with NZCO, NZCS has a significantly larger potential window and absolute area, which indicate a higher specific capacity. CV curves of NZCS at different scanning rates was measured in a voltage window of 0 ~ 0.7V as shown in Fig. 3b. The anodic and cathodic peaks can be clearly observed in each curve, which can be interpreted as the following reactions occurring.

$$Zn_{0.76}Co_{0.24}+OH^- \leftrightarrow 0.76ZnSOH+0.24CoSOH+e^- \quad (4)$$

$$Ni_3S_2+3OH^- \leftrightarrow Ni_3S_2(OH)_3+3e^- \quad (5)$$

With the scan rate increasing, the peak current increases gradually and the positions of cathodic and anodic peaks are shifted to negative and positive potentials, respectively, which due to the rapid electronic/ionic transports and good reversibility.

Moreover, in order to further analyze the electrochemical behavior, the capacitance contribution is evaluated according to the Eq. (6), and the results were shown in Fig. 3c.

$$i(v)=k_1v+k_2v^{0.5} \quad (6)$$

Where k_1 and k_2 are constants, v (mV s^{-1}) is the scanning rate, and i (A) is the current. When the scanning rate is 5, 10, 20, 30 and 50 mV s^{-1}, the percentage of capacitance contribution (k_1v) is 40.0%, 52.0%, 64.0%, 72.2% and 87.2% respectively, which illustrate that the surface capacitive-type behavior and battery-type diffusion-control behavior exist in the process of charge and discharge and the surface-controlled capacitive processes start to be dominant with the increase of scanning rate.

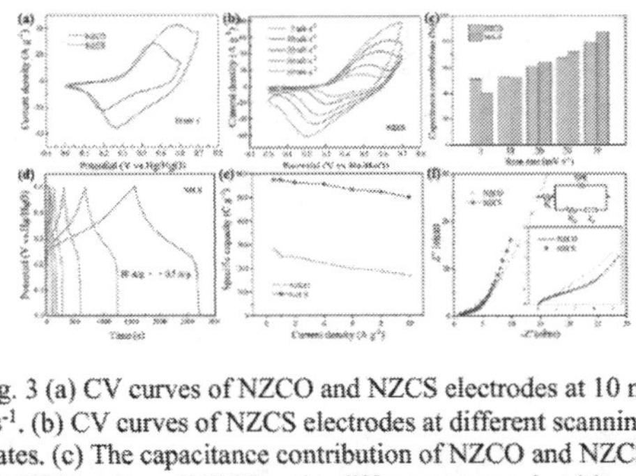

Fig. 3 (a) CV curves of NZCO and NZCS electrodes at 10 mV s^{-1}. (b) CV curves of NZCS electrodes at different scanning rates. (c) The capacitance contribution of NZCO and NZCS. (d) GCD curves of NZCS under different current densities. (e) Specific capacity of NZCO and NZCS electrodes. (f) EIS curves of NZCO and NZCS.

The GCD curves of NZCS at different current densities within a potential range of 0 ~ 0.5 V as shown in Fig. 3d. There is an obvious charging and discharging platform can be observed, which is caused by the reversible redox reactions. The specific capacity of the two electrodes can be calculated by Eq. (1) and the results were showed in Fig. 3e. Compared with NZCO, the NZCS electrode has a larger capacity. When the current density is 0.5, 1, 2, 4, 6, 8 and 10 A g^{-1}, the capacity of NZCS is 572.8, 571.7. 561.35, 554.4, 531.9, 522.9 and 498.9, respectively, which shows an excellent rate capability of 87% from 1 A g^{-1} to 10 A g^{-1}. The Nyquist plots of NZCS and NZCO were shown in Fig. 3f. According to the fitting model, the series resistance of NZCO (R_s=1.04Ω) and NZCS (R_s=1.02Ω) is close, and the charge transfer resistance of NZCS (R_{ct}=0.75Ω) is smaller than those of NZCO (R_{ct}=1.08Ω), which indicates a relatively fast electron transfer rate of NZCS. In addition, the slope of NZCS curve in the low frequency region is higher than that of NZCO composite electrode, indicating that NZCS composite electrode has a faster ion diffusivity during the energy storage process.

In order to further evaluate the electrochemical performance of NZCS electrodes, the HSC device was assembled with NZCS nanocomposite electrode as positive electrode and AC electrode as negative electrode (NZCS//AC). CV curves of NZCS//AC HCS at different scanning rates was shown in Fig. 4a. When the scanning rate increases gradually, the CV curve remains consistent and the overall appearance is rectangular-like, indicating that NZCS//AC HCS has good capacitance characteristics. The energy density and power density of NZCS//AC HSC can be calculated according to the Eq. (2-3). It shows a maximum power density of 7250.0 W kg^{-1} (20.7 Wh kg^{-1}) and a maximum energy density of 44.1 Wh kg^{-1} (407.0 W kg^{-1}), respectively, which is higher than that of many supercapacitor devices in literatures as shown in Fig. 4b. When it is packaged into a button battery, it can light up the timer, indicating that NZCS//AC HSC has a good application prospect. The cycling stability of NZCS//AC HSC devices was tested at 4 Ag^{-1}, and the results were shown in Fig. 4c. The specific capacitance of NZCS//AC can maintain 75% of the initial after 15,000 charge-discharge cycles, and the coulomb efficiency can always maintain above 98%, which indicates an excellent long-term stability.

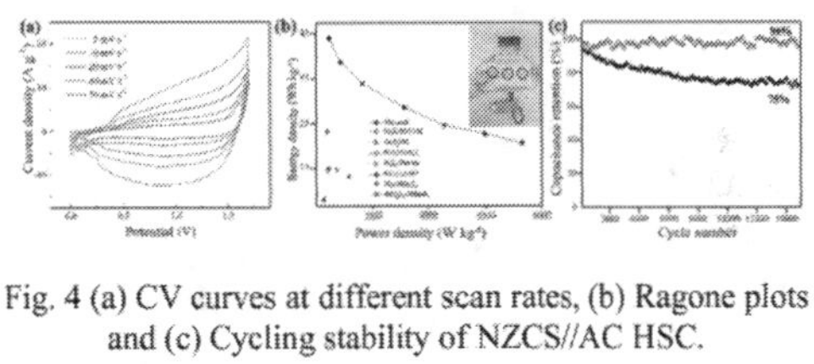

Fig. 4 (a) CV curves at different scan rates, (b) Ragone plots and (c) Cycling stability of NZCS//AC HSC.

4. Summary

$Ni_3S_2/Zn_{0.76}Co_{0.24}S$ composite was successfully prepared by a two-step hydrothermal method with a following sulfidation method. NZCS electrode exhibits a high specific capacity of 571.5 C g^{-1} at 1 A g^{-1}, due to the lower charge-transfer resistance and the synergistic of multiple metals, which is more than twice that of NZCO. Moreover, the NZCS electrode was assembled with an AC electrode to assemble an HSC with an energy density of 44.1 W h kg^{-1} at a power density of 407 W kg^{-1}. After 15,000 charge and discharge cycles, it can still maintain 75% of the initial capacitance value. Therefore, the good electrochemical performance of NZCS composite makes it have a wide application prospect.

References

1. Y. S. Luo, D. S. Tsai, et al. Lithium-ion hybrid capacitor with a scaffold electrode of Tin sulfide and Tin metal and its electrolyte issue. Journal of Physical Chemistry C. Vol. 124 (2020) No. 40, p.21909-21918.
2. A. Eftekhari. The mechanism of ultrafast supercapacitors. Journal of Materials Chemistry A. Vol. 6 (2018) p.2866-2876.
3. C. Liu, Y. Xia, Y. Zhang, et al. Pseudocapacitive crystalline $MnCo_2O_{4.5}$ and amorphous $MnCo_2S_4$ Core/Shell heterostructure with graphene for high-Performance K-ion hybrid capacitors. ACS applied materials & interfaces. Vol. 12 (2020) No. 49, p54773-54781.
4. X. Ren, M. Sun, Z. Gan, et al. Hierarchically nanostructured $Zn_{0.76}Co_{0.24}S@Co(OH)_2$ for high-performance hybrid supercapacitor. Journal of Colloid Interface Science. Vol. 618 (2022) No. 15, p.88-97.

Electrochromic properties of monosubstituted benzylboronic acid viologen

Shengwei Tang, Junlong Niu, Chunyang Jia

State Key Laboratory of Electronic Thin Films and Integrated Devices, School of Integrated Circuit Science and Engineering, University of Electronic Science and Technology of China, Chengdu 610054, P. R. China.

Abstract. Synthesis of monosubstituted benzyl boronic acid viologen (mBBV) has been achieved. The boronic acid group in mBBV can form boronic ester bonds with the hydroxyl groups in polyvinyl alcohol (PVA), resulting in an all-in-one electrochromic gel. The immobilized viologen by boronic ester bonds can slow down the diffusion rate and effectively mitigate viologen dimerization, which may address the issues caused by dimerization. The optical contrasts of mBBV and mBBV-PVA reached 49.02% and 45.01%, respectively. The electrochemical and electrochromic properties obtained from the tests suggest the potential application of mBBV in the field of electrochromics.

Keywords: Electrochromic; all-in-one; boronic ester linkages; viologen.

1. Introduction

The phenomenon of reversible color change of materials under an applied potential is referred to as electrochromism, and the materials exhibiting this property are known as electrochromic materials. Using materials with this property, various electrochromic devices such as smart windows, anti-glare rearview mirrors, low-power displays, and bionic devices can be fabricated. The key component of electrochromic devices is the electrochromic material, and the most important feature of these materials is that they should remain stable under electrochemical action and exhibit significant color changes between their oxidized and reduced states. Based on their type, electrochromic materials can be classified into two main categories: inorganic and organic electrochromic materials. Common types of inorganic electrochromic materials include transition metal oxides, Prussian blue and its analogues, etc. Organic electrochromic materials include small molecules such as viologens and anthraquinones, as well as conductive polymers such as polyaniline and polypyrrole.

Among these materials, viologens have been extensively studied due to their low driving voltage, fast response, high contrast, and ease of design. Viologens are organic materials composed of conjugated pyridine groups. This substance can usually be directly dissolved in liquid or gel electrolytes to form electrochromic electrolytes. Furthermore, viologens possess excellent redox reversibility, maintaining their structural stability even after repeated redox cycles. During the reduction process, viologens accept electrons and affect their molecular configuration and conjugated structure, resulting in changes in light absorption, leading to different colors.

Fangyuan Sun et al. investigated the electrochromic behavior of a water-soluble monosubstituted viologen derivative, BPYDB-C_6H_{13}, in different electrolyte systems. The gel-state ECD, especially the double network channel CMC-Na-PAM-ECD, exhibited better cycling stability (maintaining initial optical contrast of 99.0% after 12000 seconds compared to 95.5% for solution-state ECD), higher coloring efficiency (168.5 cm^2/C compared to 64.3 cm^2/C), lower power consumption (0.55 mW/cm^2 compared to 2.11 mW/cm^2), and longer memory time than the solution-state ECD [1]. Tae Yong Yun et al. fabricated an electrochromic device based on an ion gel. They used two electrochromic materials (ECs): diheptyl viologen (DHV^{2+}) and monoheptyl viologen (MHV^+). They found that MHV^+ ECD performed better than DHV^{2+} ECD, with higher transmittance contrast, coloring efficiency, color purity, and operational stability (MHV^+ ECD had a coloring efficiency of 94.5 cm^2/C). In addition, MHV^+ ECD exhibited better operational stability and a larger ΔT. The initial ΔT of MHV^+ ECD was about 65%, while that of DHV^{2+} ECD was about 45%, which performed poorly in the magenta state. The ion gel containing MHV^+ will be used as a key electrochromic material for magenta (red) pixel applications in ECDs [2]. Hwan Oh et al. prepared two electrochromic materials, monoheptyl viologen (MHV^+) and diheptyl viologen (DHV^{2+}), which appeared pale yellow in the bleached state and magenta and blue in the colored state, respectively. Both electrochromic devices showed good coloring efficiency, with 87.5 cm^2/C and 91.3 cm^2/C, respectively. The device had multiple

colors adjustable by voltage, operated at a low voltage of less than 1.5 V, and was flexible [3].
Here, we report the synthesis of 1-(3-boronobenzyl)-[4,4'-bipyridine] bromide (mBBV). When mixed with a PVA-containing electrolyte, the boronic acid group of mBBV undergoes esterification with hydroxyl groups on PVA chains, resulting in the formation of an all-in-one electrochromic gel. In this gel, mBBV is immobilized on PVA, which is expected to slow down the diffusion rate of mBBV and alleviate its adverse effects due to dimerization. Various characterization techniques, including electrochemical and spectroscopic methods, were employed to systematically investigate the properties of mBBV.

2. Experimental

2.1 Materials

4,4′-Bipyridine (98%), 3-(bromomethyl)-phenylboronic acid (98%), Ferrocene (Fc, 98%) and Polyvinyl alcohol (PVA, Mw 61 000) were purchased from Shanghai Macklin Biochemical Co., Ltd. Dimethyl sulfoxide (DMSO, 99%), N, N- Dimethylformamide (DMF, 99%) and ethyl acetate (99%) were purchased from Kelong Chemical Co.
Glasses with ITO layer (15 Ω/□) were cut into rectangles (4.5 cm × 2.5 cm). After rinsing with deionized water, ethanol and isopropanol respectively and ultrasonication for 5 min, then dried, and immersed it in ethanol for use.

2.2 Synthesis of monosubstituted benzylboronic acid viologen

4,4 '- bipyridine (1 g, 6.4 mmol) and 3 - (bromomethyl) - phenylboronic acid (0.86 g, 4 mmol) were dissolved in 20 ml DMF and stirred at 70 ° C for 2 h. The solution was to be cooled and poured into 200 ml ethyl acetate, and the off-white precipitate was collected by centrifugation and washed with ethyl acetate several times. The collection was dried at room temperature to obtain mBBV (0.43 g, 72% yield). ^{1}H NMR (400 MHz, Deuterium Oxide) δ 8.79 (d, J = 7.0 Hz, 2H), 8.68 – 8.55 (m, 2H), 8.20 (d, J = 7.0 Hz, 2H), 7.80 – 7.69 (m, 2H), 7.63 (d, J = 7.5 Hz, 1H), 7.51 – 7.46 (m, 1H), 7.44 (d, J = 2.5 Hz, 1H), 7.44 – 7.37 (m, 1H), 5.89 (s, 2H).

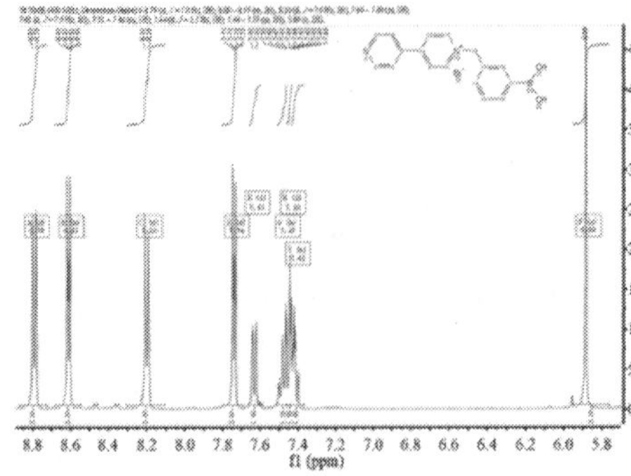

Fig1. The synthetical route of mBBV and 1H-NMR spectra of mBBV in D_2O

2.3 Fabrication of the ECDs

To prepare the DMSO solution containing 5% PVA, 5 g of PVA was mixed with 95 g of DMSO and stirred and heated to 110°C until the polymer completely dissolved. Then, 10 mM mBBV and 10 mM ferrocene were added separately to the DMSO solution and the DMSO solution containing 5% PVA. Afterward, all components were sonicated for 15 minutes. These two formulations were referred to as mBBV and mBBV-PVA, respectively.

To assemble the ECDs, transparent double-sided adhesive tape (200 μm thick) was used as the spacer, and the edges of two ITO glasses (with two small holes reserved) were firmly bonded. Then, the solution was injected into one of the holes using a syringe. Finally, both holes were sealed with glue to prevent solution leakage. The effective area of all ECDs was controlled at 1 cm × 2 cm.

2.4 Characterization

The electrochemical properties of the ECDs were tested using an electrochemical workstation (CHI760E, Chenhua Shanghai). The three-electrode electrochemical system used was comprised of a working electrode (ITO glass), a counter electrode (Pt), and a reference electrode (Ag/Ag+). The transmittance spectroscopy and dynamic response tests were carried out using a UV-visible spectrophotometer (UV6100, MAPADA) equipped with the electrochemical workstation.

3. Results and discussion

3.1 Electrochemical performance of ECDs

The cyclic voltammetry characteristics of two ECDs were tested at eight different scan rates ranging from 5mV/s to 300mV/s, as shown in Fig. 2(a) and (b). A pair of oxidation-reduction peaks corresponding to the oxidation-reduction process of mBBV can be observed at different scan rates for both ECDs.

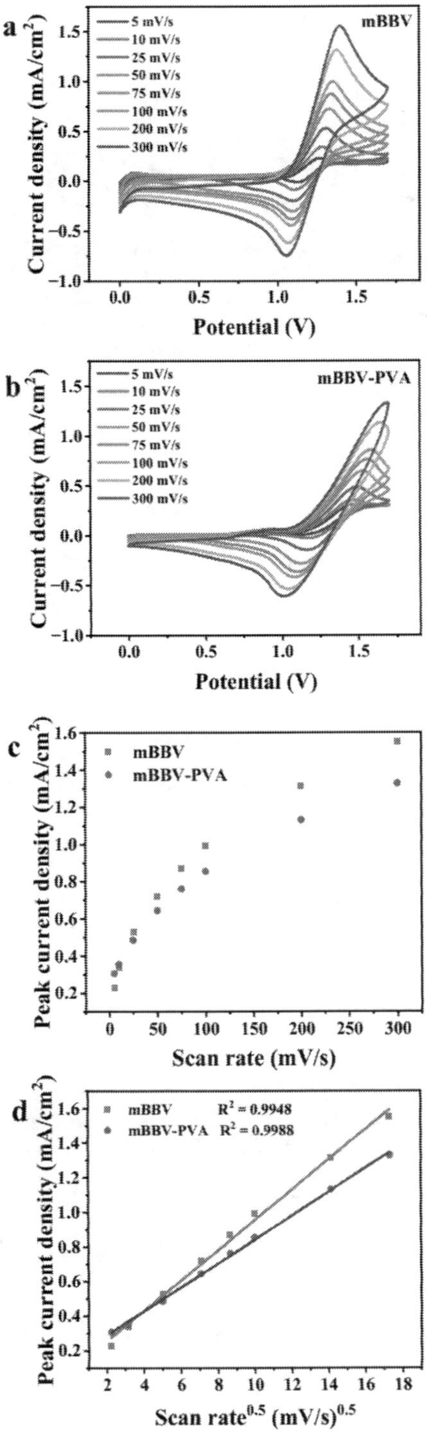

Fig. 2 CVs with scan rates varied from 5 to 300 mV/s for (a) mBBV ECD and (b) mBBV-PVA ECD. Plots of peak current density against (c) scan rate and (d) square root of scan rate for the mBBV ECD and mBBV-PVA ECD.

Based on the peak current density measured at different scan rates, the relationship between peak current density and scan rate or square root of scan rate is obtained, as shown in Fig. 2(c) and (d). In Fig. 2(d), the goodness of fit (R^2) of both curves is close to 1, indicating that the peak current density of both ECDs is linearly related to the square root of the scan rate [4]. When the peak current density is linearly related to the square root of the scan rate, it indicates that the electrochemical reaction process is limited by diffusion. For such ECDs, the apparent diffusion rate can be calculated according to Cottrell's equation (Eq. (1)):

$$i_d = \frac{nFAD_{app}^{0.5}C_0^*}{\Pi^{0.5}t^{0.5}} \quad (1)$$

In Cottrell's equation, A is the effective area and further simplification of the equation yields $j_d=kt^{-0.5}$. By testing two ECDs using chronoamperometry, the relationship between diffusion-limited current density (j_d) and time (t) can be obtained, as shown in Fig. 3(a). Further plotting the relationship between diffusion-limited current density (j_d) and negative square root of time ($t^{-0.5}$), as shown in Fig. 3(b). After fitting the corresponding straight lines for the two curves in Fig. 3(b), their slopes are denoted as K. Then we can get:

$$D_{app} = \frac{K^2\Pi}{(nFC_0^*)^2} \quad (2)$$

According to Eq. (2), D_{app} for mBBV and mBBV-PVA ECDs can be calculated, respectively. Since the apparent diffusion coefficient of mBBV-PVA ECD is smaller than that of mBBV ECD, it can indicate that the possibility of dimerization of mBBV in mBBV-PVA ECD is smaller.

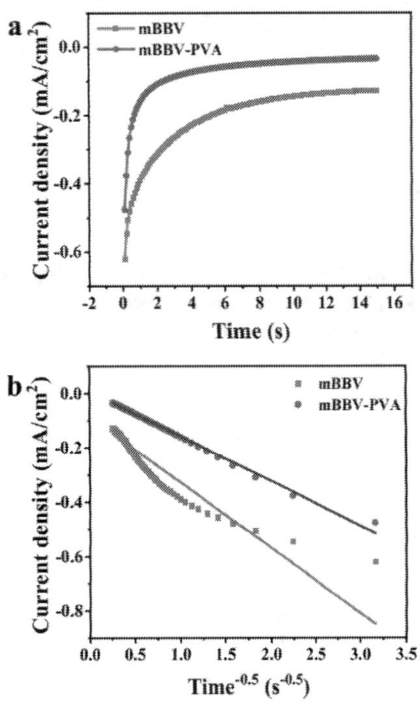

Fig. 3 Plots of diffusion limited current density against (a) time and (b) minus square root of time at 1.5 V for the mBBV ECD and mBBV-PVA ECD

 https://doi.org/10.1051/matecconf/202338201003

In order to further explore the electrochemical behavior of two ECDs, electrochemical impedance spectroscopy (EIS) tests were performed on them. The frequency range was from 100kHz to 0.01Hz and AC amplitude was kept at 5mV. The Nyquist plots of two ECDs in full frequency range and high frequency range are shown in Fig. 4(a) and (b). In Nyquist plot, slope of straight line in low frequency region corresponds to ion diffusion. It can be seen that curve of mBBV-PVA ECD in low frequency region has lower slope, which confirms its slower ion diffusion and further proves its lower dimerization possibility [5].

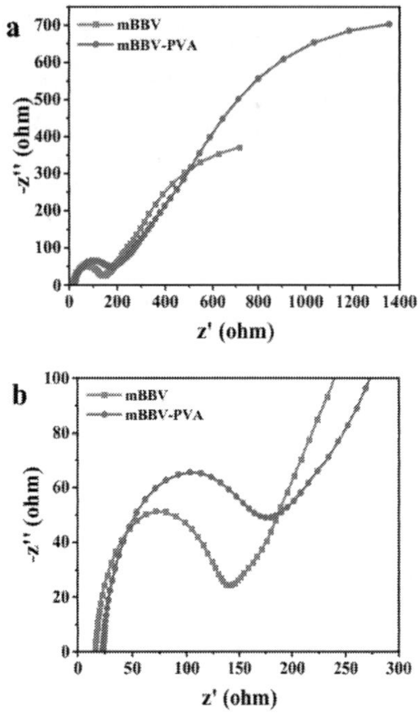

Fig. 4 Nyquist plot of (a) ECDs (full frequency region) and (b) (high frequency region).

3.2 Transmittance spectra and dynamic transmittance responses of ECDs

The transmittance spectra of mBBV and mBBV-PVA ECDs were measured at different reduction potentials, as shown in Fig. 5(a) and (b). The maximum transmittance change was observed at 375nm for mBBV ECD and at 370nm for mBBV-PVA ECD.

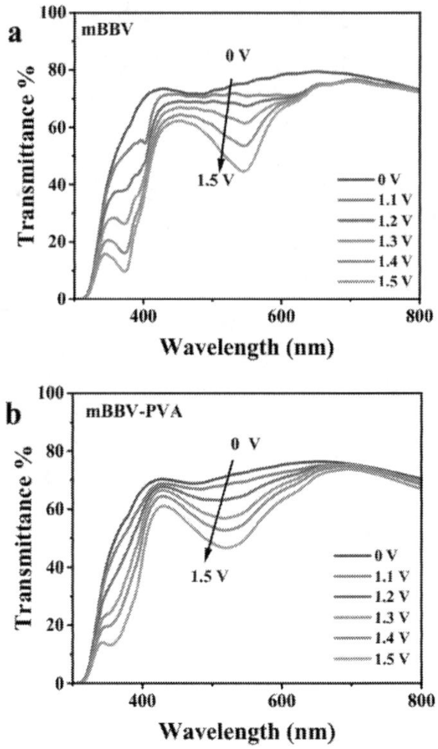

Fig. 5 The transmittance spectrum of (a) mBBV ECD, and (b) mBBV-PVA ECD at various potential biases.

To further understand the performance of these ECDs, their dynamic response performance was tested at the wavelength of maximum transmittance change, as shown in Fig. 6(a) and (b). The specific parameter settings are as follows: each potential step is set to 0 V (15 s) and 1.5 V (30 s). The bleaching time (t_b) and coloring time (t_c) are calculated from the transmittance of the bleaching/coloring state to the time when the change reaches 90% of the maximum transmittance change, as shown in Fig. 6(c) and (d).

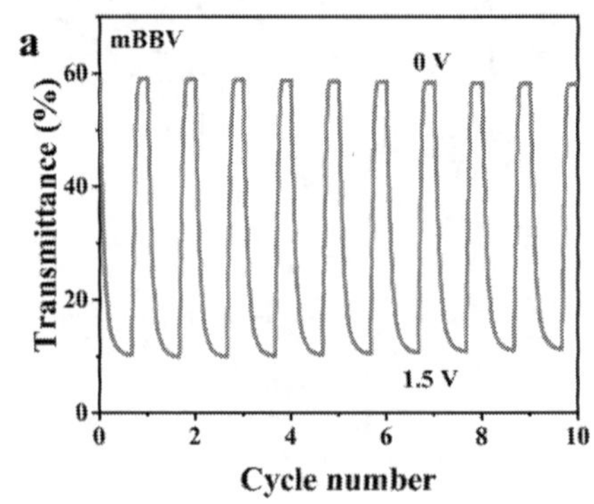

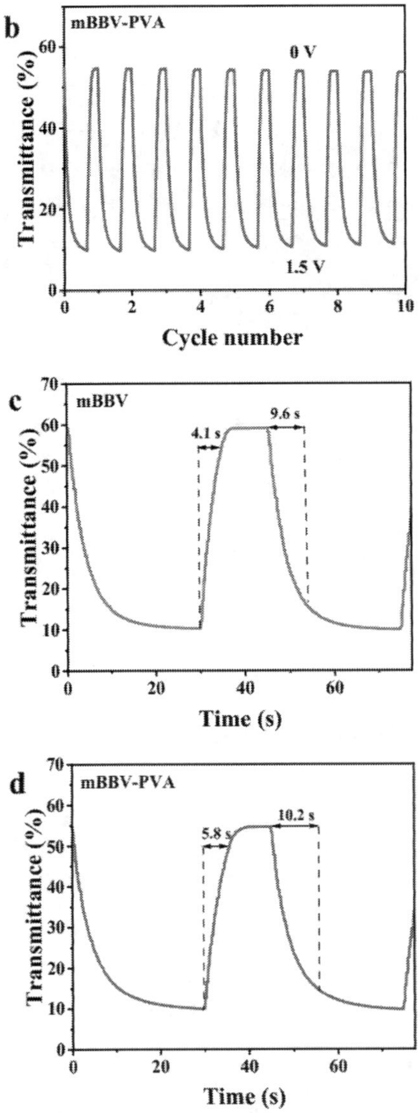

Fig. 6 The dynamic transmittance curves of (a) mBBV ECD, and (b) mBBV-PVA ECD. Bleaching time and coloring time of (c) mBBV ECD, and (d) mBBV-PVA ECD.

Table 1 records relevant information for both ECDs, including: wavelength corresponding to maximum transmittance change (λ_{max}), bleaching potential (V_b), coloring potential (V_c), bleaching time (t_b), coloring time (t_c), transmittance under bleaching state (T_b), transparency under colored state (T_c), maximum transmittance change ($\Delta T_{max}=T_b-T_c$). Both ECDs exhibit large transmittance changes and rapid response, demonstrating good electrochromic performance.

Table 1 The Dynamic transmittance responses for the ECDs.

ECD	λ_{max} (nm)	V_b/V_c (V)	t_b/t_c (s)	T_b/T_c (%)	ΔT_{max} (%)
mBBV	375	0/1.5	4.1/9.6	59.25/10.23	49.02
mBBV-PVA	370	0/1.5	5.8/10.2	54.78/9.77	45.01

4. Summary

In this paper, we successfully synthesized a monosubstituted viologen derivative containing boronic acid group and studied their electrochromic performance for the first time. Due to the presence of boronic ester bond, viologen is immobilized on the polymer mesh of the gel through dynamic chemical bond. By this strategy, it will slow down the diffusion rate of viologen and is an effective means to alleviate viologen dimerization, which is expected to solve problems caused by dimerization. Specifically, our prepared mBBV ECD and mBBV-PVA ECD showed response times less than 11s during switching from 0 to 1.5V, with maximum transmittance changes exceeding 45%, exhibiting fast response speed and high transmittance change, thus having good application prospects.

References

1. Ming-yue Pan, Qi He, Jia-ning Liu, Kui Du et al. A double network hydrogel electrolyte-based electrochromic device for enhanced electrochromic performance and lower power consumption. Dyes and Pigments. Vol.212 (2023), p.111126.
2. Tae Yong Yun, Hong Chul Moon. Highly stable ion gel-based electrochromic devices: Effects of molecular structure and concentration of electrochromic chromophores. Organic Electronics. Vol. 56 (2018), p. 178-185.
3. Hwan Oh, Dong Gyu Seo, Tae Yong Yun, et al. Voltage-Tunable Multicolor, Sub-1.5 V, Flexible Electrochromic Devices Based on Ion Gels. ACS Applied Materials & Interfaces. Vol. 9 (2017) No. 8, p. 7658-7665.
4. A.A. El-Shafei. Electrocatalytic oxidation of methanol at a nickel hydroxide/glassy carbon modified electrode in alkaline medium. Journal of Electroanalytical Chemistry. Vol.471 (1999) , p.89-95.
5. Rongzong Zheng, Yi Wang, Jianbo Pan, et al. Toward Easy-to-Assemble, Large-Area Smart Windows: All-in-One Cross-Linked Electrochromic Material and Device. ACS Applied Materials & Interfaces. Interfaces. Vol. 12 (2020) No. 24, p.27526-27536.

Health status assessment of scraper conveyor conveying equipment based on cross-entropy theory and immune principle

Chunsheng Wang[1], Dangdang Wu [1,*], Yuhao Qi [2], Hongfei Liu [1], Tongzhu Yu[3]

[1]Scraper Research Branch, Ningxia Tiandi Benniu Industrial Group Co.,Ltd. Yin Chuan, 750000, China
[2]ShanDong Energy Group CO.LTD, Ji Nan, 250000, China
[3]Yankuang Eneergy Group Company Limited, Ji Ning, 272000, China.

Abstract. Based on cross-entropy theory and immunity principle, the paper evaluates the health status of scraper conveying equipment. Firstly, the hierarchical analysis model and immune monitoring model of scraper conveyor equipment are established, and then the cross-entropy theory integrating AHP, fuzzy statistical method and entropy weight method is used to calculate the composite weight of scraper conveyor conveyor equipment index, and finally the health status assessment results are obtained. Combined with the case analysis, the proposed health status assessment method can effectively combine the physical characteristics and operating status of each index of scraper conveying equipment to complete the health status assessment of scraper conveying equipment.

Keywords: Scraper conveying equipment; Health status assessment; Cross-entropy theory; Principles of immunity.

1. Introduction

The paper is supported by the Shandong Province Key R&D Plan (Major Science and Technology Innovation Project) "Research and Application of Key Technologies for Intelligent Mining of Complex Geological Conditions in Working Faces" (project number: 2020CXGC011501), and completed the research by collecting relevant data. As an important underground coal and material transportation equipment, the health status of scraper conveying equipment directly affects the economic benefits of mines, so it is particularly important to establish a suitable health assessment method for this equipment. In order to evaluate the health status of an object, it is necessary to select the relevant algorithm to calculate the weight of the object index, and the commonly used algorithms include AHP[1], fuzzy comprehensive evaluation[2], gray correlation method[3], BP neural network method[4], etc. Literature [5] proposes to combine analytic hierarchy method and fuzzy comprehensive evaluation method to evaluate the construction of new energy projects, but new methods can still be found to improve the accuracy of weight calculation. Literature [6] uses analytic hierarchy method and entropy weight method to calculate the weight coefficient of building curtain wall, but this method has limitations for the problem of redundant mining equipment data collected. Literature [7] for three-phase transformers, a cooperative game method combining three methods is proposed for state of health assessment, which further improves the calculation accuracy by calculating the proportional coefficient of the three weights, but the method does not have computer adaptive ability. Literature [8] uses cross-entropy theory to predict photovoltaic power generation, and improves the accuracy of combined prediction by dynamically changing the weights of different prediction methods. In the paper, by reviewing the data, selecting suitable performance indicators to build a hierarchical analysis model and immune model of scraper conveying equipment, and using the cross-entropy theory fusion of AHP, fuzzy statistical method and entropy weight method to calculate the composite weight of indicators, the accuracy of the final evaluation results is further improved.

2. Health assessment model

2.1 Hierarchical analysis model

In the paper, the key component indicators affecting the healthy operation of scraper conveyor equipment are selected to build a hierarchical analysis model, as shown in Figure 1. The figure divides the scraper conveyor equipment into a two-stage system, the first stage is the component layer and the second stage is the indicator layer.

* Corresponding author: 2412209445@qq.com

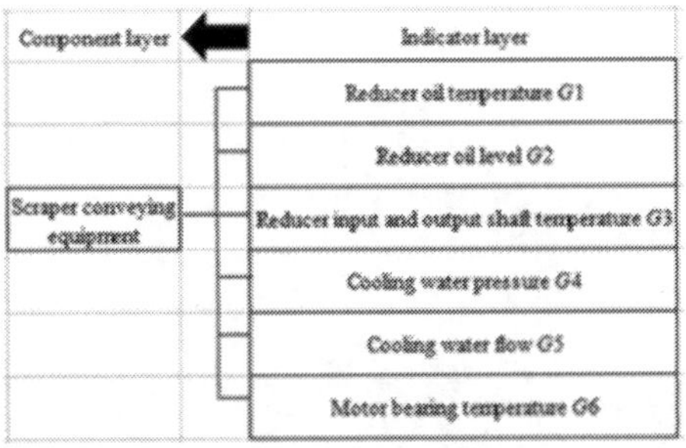

Fig. 1 Hierarchical analysis model

2.2 Immune models

Immunizing a target device goes through three steps: recognition, immunization, and management. Immunological flow diagram shown in Figure 2 and immunity model of scraper delivery device shown in Figure 3.

According to the hierarchical analysis model, the sensor cells corresponding to each index are set up to collect initial data, if the data is within the normal range, it means that there is no antigen, if the data is abnormal, it means that there are abnormal antigens entering the system and the information needs to be transmitted to immune cells. Immune cells are installed at the element layer and are responsible for collecting and aggregating abnormal antigen information for calculation. That is, the cross-entropy theory is used to determine the composite weight, the health value is calculated by the immune principle, the results are uploaded to the management cells, and the backup fault information is updated at the same time. The management cell is equivalent to the data processing terminal computer, which receives the status sequence output by the immune cell and formulates various maintenance measures for failure; Antibodies represent specific repair methods, actuator cells represent maintenance workers, and the goal is to repair individual components to achieve good health.

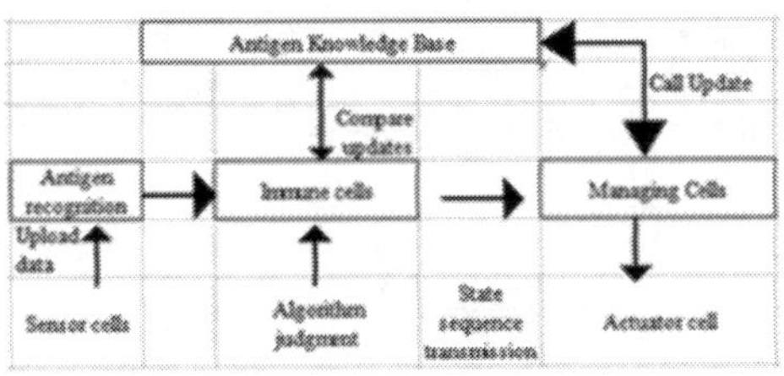

Fig. 2 Immunomonitoring flowchart

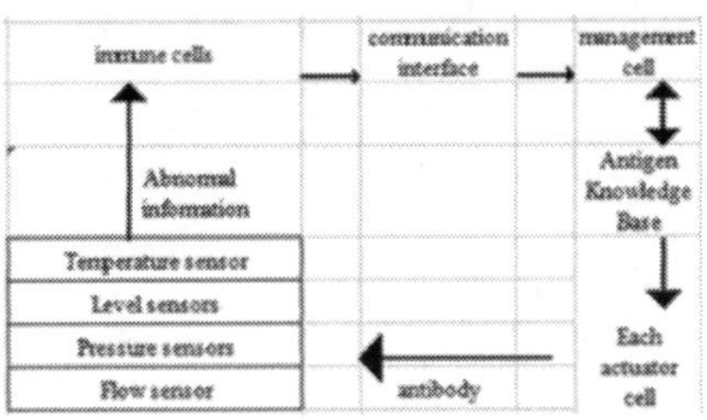

Fig. 3 Immunity model of scraper delivery equipment

3. Algorithmic judgment

3.1 Determine weights based on AHP

The AHP is a subjective method of calculating weights, and preliminary estimates of indicator weights are made based on expert experience. According to the hierarchical analysis model established above, the affiliation between the indicators is determined, and the pair-by-two comparison is carried out, and an n-th order judgment matrix $\boldsymbol{A}$ is obtained for the importance of n different indicators of the same level relative to the corresponding components of the previous layer, as follows.

$$A=(a_{ij})_{n\times n} \tag{1}$$

Formula: $a_{ij}>0$; $a_{ij}=1/a_{ji}$; $a_{ii}=1$. The 1-9 scale method is used to represent the results of pair-by-two comparisons of the importance of different indicators, as shown in Table 1.

Table 1. 1-9 proportional scale method

Scale value	Meaning
1	Consistent importance
3	Slightly more important
5	Important
7	Very important
9	Extremely important
2、4、6、8	Take the middle value

Calculate the maximum eigenvalue of $\boldsymbol{A}$ and its corresponding eigenvector, and perform normalization calculations to obtain the initial weight vector as

$$\vec{W}_i^{(1)}=[w_1^{(1)},w_2^{(1)},\ldots,w_n^{(1)}] \tag{2}$$

Use equations (3) and (4) to test the consistency of the constructed matrix.

$$CI=\frac{\lambda_{\max}-n}{n-1} \tag{3}$$

$$CR=\frac{CI}{RI} \tag{4}$$

where CR is the $\boldsymbol{A}$ random consistency ratio, CI is a conformance indicator of $\boldsymbol{A}$, RI is the average random consistency index of $\boldsymbol{A}$, which can eliminate the influence of matrix order. $\lambda_{\max}$ is the maximum feature value of the judgment matrix, and n is the total number of indicators. Table 2 lists the values of RI. In general, in the case of $n \geqslant 3$, when $0<CR<0.10$, it indicates that each index can be assigned a reasonable weight value. Conversely, the values of the comparison matrix should be readjusted according to Table 2 until the degree of agreement is satisfactory.

https://doi.org/10.1051/matecconf/202338201004

Table 2 Pairs of order 1-9 compare the *RI* values of matrices

n	1	2	3	4	5	6	7	8	9
R	0.0	0.0	0.5	0.9	1.1	1.2	1.3	1.4	1.4
I	0	0	8	0	2	4	2	1	5

3.2 The weights are determined based on the fuzzy statistical method

Fuzzy statistics is an objective method for calculating weights. The health status of scraper conveyor conveying equipment is divided into 5 levels, and the definition evaluation set is $T_s=\{T_1, T_2, T_3, T_4, T_5\}$={health, sub-health, mild pathology, moderate morbidity, severe morbidity}. The fuzzy statistical method used here is to let the n experts participating in the assessment obtain the level of each health index according to the set T_s given earlier, and then count the frequency of each health indicator Q_i belonging to each assessment level δ_{is}, then the degree of membership of the health assessment index belongs to the health level is

$$\lambda_i^{(s)}=\frac{\delta_{is}}{n} \tag{5}$$

In view of the practical significance of each health index in the evaluation system, two fuzzy distribution forms of benefit type and cost type are considered. Set the value in (a, b), and insert three equinoxes in the interval. Assuming that the actual value of any indicator Q_i is z, the expression of the two membership degree functions is given.

3.2.1 *Benefit type*

$$\lambda_i^{(1)}=\begin{cases}1 & z\ge b\\ (z-c_4)/d & c_4\le z<b\\ 0 & z<c_4\end{cases} \tag{6}$$

$$\lambda_i^{(s)}=\begin{cases}0 & z\ge c_{5-s+2}\\ (c_{5-s+2}-z)/d & c_{5-s+1}\le z<c_{5-s+2}\\ (z-c_{5-s})/d & c_{5-s}\le z<c_{5-s+1}\\ 0 & z\le c_{5-s}\end{cases} \tag{7}$$

$$\lambda_i^{(5)}=\begin{cases}0 & z\ge c_2\\ (c_2-z)/d & a\le z<c_2\\ 1 & z<a\end{cases} \tag{8}$$

Formula a is the worst critical value, b is the optimal cut-off value, $d=(b-a)/4$, s=2,3,4. Let $a=c_1$ and $b=c_5$.

3.2.2 *Cost type*

$$\lambda_i^{(1)}=\begin{cases}1 & z<b\\ (c_2-z)/d & a\le z<c_2\\ 0 & z\ge c_2\end{cases} \tag{9}$$

$$\lambda_i^{(s)}=\begin{cases}0 & z<c_{s-1}\\ (z-c_{s-1})/d & c_{s-1}\le z<c_s\\ (c_{s+1}-z)/d & c_s\le z<c_{s+1}\\ 0 & z\ge c_{s+1}\end{cases} \tag{10}$$

$$\lambda_i^{(5)}=\begin{cases}0 & z<c_4\\ (c_5-z)/d & c_4\le z<a\\ 1 & z\ge a\end{cases} \tag{11}$$

Formula a is the optimal cut-off value, b is the worst critical value, $d=(b-a)/4$, s=2,3,4. Let $b=c_1$ and $a=c_5$.

According to the degree of affiliation of each health index in the indicator layer to the five health levels, the index evaluation matrix is constructed

$$D_i=[\lambda_i^{(1)}\quad \lambda_i^{(2)}\quad \lambda_i^{(3)}\quad \lambda_i^{(4)}\quad \lambda_i^{(5)}] \tag{12}$$

Sensor cells can output status monitoring information based on membership values. Make

$$e_i=\max\{\lambda_i^{(s)}\}\qquad s=1,2,\ldots,5 \tag{13}$$

When $e_j=\lambda_j^{(4)}$ *or* $e_j=\lambda_j^{(5)}$, the corresponding sensor cell output alarm information is 1. When $e_j=\lambda_j^{(1)}$、$e_j=\lambda_j^{(2)}$ *or* $e_j=\lambda_j^{(3)}$, the sensor cell outputs a normal message of 0. Get the objective correction weights of each indicator

$$\alpha_i=\frac{1/e_i}{\sum 1/e_i} \tag{14}$$

Formula: $\sum 1/e_i$ is the sum of the reciprocals of the e_i values corresponding to all health indicators of a component. Finally, the comprehensive weight of each indicator in the indicator layer is obtained

$$\beta_i=\frac{\alpha_i w_i^{(1)}}{\sum \alpha_i w_i^{(1)}} \tag{15}$$

The indicator layer synthesizes the weight vector

$$\vec{W}_i^{(2)}=[\beta_1^{(2)},\beta_2^{(2)},\ldots,\beta_n^{(2)}] \tag{16}$$

3.3 The weights are determined based on the entropy weight method

The entropy weight method is also an objective method for determining weights. The main calculation steps are as follows:

1. Normalize the original data. The system indicators collected by the sensor are normalized to the original data matrix with k objects, l evaluation indicators, and the i-th component and j-th index

$$x_{ij}(n)=\frac{x_a-x_{ij}(n)'}{x_a-x_b},(i=1,2,\ldots,k;j=1,2,\ldots,l_i) \tag{17}$$

where $x_{ij}(n)'$ is the n-th detection value of the j-th index of the i-th component, x_a is the worst value of the indicator,

 https://doi.org/10.1051/matecconf/202338201004

x_b is the optimal value of the indicator, $x_{ij}(n)$ is the normalized value of $x_{ij}(n)'$.

2. The information entropy of the i-th object and the j-th indicator in the system is

$$E_{ij} = -\alpha \sum_{n=1}^{n} (p_{ij}(n) \ln p_{ij}(n)) \tag{18}$$

where $p_{ij}(n) = x_{ij}(k) / \sum_{n=1}^{n} x_{ij}(k)$, α=1/Inn.

3. Determine the weight. The weight of the i-th element and the j-th indicator is defined as

$$w_{ij}^{(3)} = \frac{1 - E_{ij}}{l_i - \sum_{j=1}^{l_i} E_{ij}} \tag{19}$$

Indicator weight vector for the i-th object

$$\vec{W}_i^{(3)} = [w_{i1}^{(3)}, w_{i2}^{(3)}, ..., w_{il_i}^{(3)}] \tag{20}$$

3.4 Composite weights are determined based on cross-entropy theory

Cross-entropy theory is used to quantify the difference between two probability distributions. In the uniform probability space, cross-entropy is defined for two different distributions $\boldsymbol{p}$ and $\boldsymbol{q}$, as follows: When discrete, $\boldsymbol{p}$ and $\boldsymbol{q}$ are probability vectors, defined as

$$D(p\|q) = \sum_{j=1}^{n} p_j \ln \frac{p_j}{q_j} \tag{21}$$

When continuous, $\boldsymbol{p}$ and $\boldsymbol{q}$ are probability density functions, defined as

$$D(p\|q) = \int_a^b p(x) \ln \frac{p(x)}{q(x)} dx \tag{22}$$

Cross-entropy quantifies the "distance" between two probabilities, and also quantifies the amount of information required between the source of information from probability $\boldsymbol{p}$ to probability $\boldsymbol{q}$, reflecting the degree of agreement between the two probabilities, with the following theorems and properties.

Theorem: $\boldsymbol{p}$ and $\boldsymbol{q}$ are density functions, then $D(p\| q) \geq 0$。

Properties: $D(p\| q) \geq 0$, equal sign holds: if and only if $\boldsymbol{p} = \boldsymbol{q}$ exists almost everywhere.

In the process of combining weight calculation, cross-entropy theory completes data fusion by calculating the proportion coefficient of each single method weight in the combined weight, and obtains a more accurate and reasonable combination weight.

3.5 Calculate health values

The sensor cell uploads the collected data information to the immune cell and outputs a state sequence vector

$$\vec{K}_I = [K_1 \quad K_2 \quad ... \quad K_n] \tag{23}$$

where n is the total number of sensor cells, $K_i(i=1,2,...,n)$ outputs the status monitoring information for each sensor cell, with a value of 0 or 1.

The influence factor of the indicator layer on the component layer is

$$\gamma_E = \vec{K}_I \vec{w}_i^T \tag{24}$$

The health value of the component can be obtained as

$$H = 1 - \gamma_E \tag{25}$$

where the health value $H \in [0,1]$, the larger H, the better the health state of the element. By calculating the health value of each element, the target health status assessment is achieved. Table 3 shows the breakdown of health value and health status.

Table 3 The division of health values and health states

health values	1-0.9600	0.9599-0.9	0.8999-0.8	0.7999-06	0.5999-0
health states	Health	Sub-health	Mildly morbid	Moderately morbid	Severely morbid

4. Case analysis

Table 4 Indicator layer parameters

Health indicators	Reducer oil temperature G_1	Reducer oil level G_2	Reducer input and output shaft temperature G_3	Cooling water pressure G_4	Cooling water flow G_5	motor bearing temperature G_6
Critical value	0℃≤G_1≤200℃	0≤G_2≤ 450mm	0℃≤G_3≤200℃	0≤G_4≤5MPa	0≤G_5≤60L/min	0≤G_6≤200℃
Distribution form	Cost type	Benfit type	Cost type	Cost type	Benfit type	Cost type

Taking a coal mine underground scraper conveying equipment as an example. The sensor cells were installed separately to collect the data of reducer oil temperature G_1, reducer oil level G_2, reducer input and output shaft temperature G_3, cooling water pressure G_4, cooling water flow G_5 and motor bearing temperature G_6. These six indicators are classified in a fuzzy distribution form, as shown in Table 4. The critical value is based on the equipment factory parameters, historical operation records and historical maintenance records.

By collecting the results of each sensor at a certain time, the initial data of each indicator is $G^{(1)}=\{G_1,G_2,G_3,G_4,G_5,G_6\}=\{160,448,131,1.5,59,113\}$. Construct a pairwise comparison matrix A

$$A=\begin{bmatrix} 1 & 1/2 & 3/2 & 1/3 & 1/2 & 1/4 \\ 2 & 1 & 3 & 2/3 & 1 & 1/2 \\ 2/3 & 1/3 & 1 & 2/9 & 1/3 & 1/6 \\ 3 & 3/2 & 9/2 & 1 & 3/2 & 3/4 \\ 2 & 1 & 3 & 2/3 & 1 & 1/2 \\ 4 & 2 & 6 & 4/3 & 2 & 1 \end{bmatrix}$$

The maximum characteristic value is obtained as λ_{max}=6.0, so CI=0, and the consistency check CR=0<0.1 is performed to meet the requirements and subsequent calculations are performed

1) The feature vector corresponding to λ_{max} is found and normalized to obtain the weight vector $\vec{W}^{(1)}=[0.0789,0.1579,0.0526,0.2368,0.1579,0.3158]$ based on analytic hierarchy.

2) The cost-type fuzzy membership function is used for the index G_1, G_3, G_4 and G_6, and the benefit-type fuzzy membership function is used for the index G_2 and G_5, and the weight vector $\vec{W}^{(2)}=[0.0769,0.1231,0.1062,0.2313,0.1336,0.329]$ based on the fuzzy statistical method is obtained according to equations (6)-(16).

3) The second set of data of each index was collected as a reference $G^{(2)}=\{G_1,G_2,G_3,G_4,G_5,G_6\}=\{161, 448,132,1.5,58,114\}$. Firstly, the entropy of each index information $E_j=[E_1, E_2, E_3, E_4, E_5, E_6]$=[0.9999, 0.9999, 0.9997, 0.9999, 0.9264,0.9999] is calculated, and then the weight vector $\vec{W}^{(3)}=[0.0013,0.0013,0.0004,0.0013,0.9906,0.0013]$ based on the entropy weight method is determined.

4) The calculation process of the scale factor for each single method is as follows

1. Define the fusion method. Let $\boldsymbol{\omega}$ be the combined weight vector after fusing the three methods, $\boldsymbol{\omega}^{(r)}$ is the weight vector of the r method, and $\boldsymbol{\omega}^{(r)}$ is known, the dimensions of $\boldsymbol{\omega}$ and $\boldsymbol{\omega}^{(r)}$ are equal to the number of indicators, that is, 6 dimensions, and the fusion is carried out according to the following formula

$$\boldsymbol{\omega}=\sum_{r=1}^{3}k_r\omega^{(r)} \tag{26}$$

where k_r is the scale coefficient of each single method and satisfies the following formula.

$$\sum_{r=1}^{3}k_r=1 \tag{27}$$

2. Define the cross-entropy calculation formula.

$$\begin{aligned}\sum_{r=1}^{3}D\left[\omega\|\omega^{(r)}\right]&=\sum_{r=1}^{3}\sum_{j=1}^{6}\omega_j\ln\left[\omega_j/\omega_j^{(r)}\right]\\&=\sum_{r=1}^{3}\sum_{j=1}^{6}(k_1\omega_j^{(1)}+k_2\omega_j^{(2)}+k_3\omega_j^{(3)})\ln\left[(k_1\omega_j^{(1)}+k_2\omega_j^{(2)}+k_3\omega_j^{(3)})/\omega_j^{(r)}\right]\end{aligned} \tag{28}$$

where ω_j is the weight value of the j-th index after fusion, and $\omega_j^{(r)}$ is the weight value of the j-th index of the r method.

3. Construct the objective function. Given $\boldsymbol{\omega}^{(r)}$, the independent variables in equation (28) are k_1, k_2, k_3, and from equation (27) we get $k_3=1-k_1-k_2$, so the independent variable is defined as $\boldsymbol{x}=[k_1,k_2]$, find the minimum value of equation (28), and construct the objective function $F(\boldsymbol{x})$.

$$F(x)=\min(\sum_{r=1}^{3}D\left[\omega\|\omega^{(r)}\right]) \tag{29}$$

4. Solving the optimization problem $F(\boldsymbol{x})$.

① Make a second-order approximation Taylor expansion for $F(\boldsymbol{x})$.

$$F(x)=F(x_0)+(x-x_0)^T\frac{\partial F}{\partial x}+\frac{1}{2}(x-x_0)^T A(x-x_0) \tag{30}$$

where x_0 is the initial value of the matrix and $\boldsymbol{A}$ is the Heisen matrix.

② Find the $F(\boldsymbol{x})$ differential.

$$G(x)=\frac{\partial F(x)}{\partial x}+\frac{1}{2}(A-A^T)(x-x_0) \tag{31}$$

③ Set the initial value x_0 and the accuracy threshold ε so that H_0 is a second-order identity matrix, i=0.

https://doi.org/10.1051/matecconf/202338201004

④ Solve the search direction D_i.

$$\boldsymbol{D}_i = -\boldsymbol{H}_i \boldsymbol{G}_i \tag{32}$$

⑤ The direction of each iteration still uses $\boldsymbol{D}_i$, but each iteration needs to do a one-dimensional search in this direction to find the optimal step size t_i.

$$t_i = \arg\min_{t \in R} F(\boldsymbol{x}_i + t\boldsymbol{D}_i) \tag{33}$$

⑥ Solve $\boldsymbol{x}_{i+1}$

$$\boldsymbol{x}_{i+1} = \boldsymbol{x}_i + t_i \boldsymbol{D}_i \tag{34}$$

⑦ To determine whether to continue the iteration, bring equation (34) into equation (31) to get $\boldsymbol{G}_{i+1}$, if $\| \boldsymbol{G}_{i+1} \| < \varepsilon$, the iteration ends, otherwise the iteration continues.

⑧ To determine $\boldsymbol{H}_{i+1}$, the formula is as follows

$$\boldsymbol{z}_i = \boldsymbol{G}_{i+1} - \boldsymbol{G}_i \tag{35}$$

$$\boldsymbol{\delta}_i = \boldsymbol{x}_{i+1} - \boldsymbol{x}_i \tag{36}$$

$$\boldsymbol{B}_i = \frac{\boldsymbol{\delta}_i \boldsymbol{\delta}_i^T}{\boldsymbol{\delta}_i^T \boldsymbol{z}_i} \tag{37}$$

$$\boldsymbol{C}_i = \frac{-(\boldsymbol{H}_i \boldsymbol{z}_i)(\boldsymbol{H}_i \boldsymbol{z}_i)^T}{\boldsymbol{z}_i^T \boldsymbol{H}_i \boldsymbol{z}_i} \tag{38}$$

$$\boldsymbol{H}_{i+1} = \boldsymbol{H}_i + \boldsymbol{B}_i + \boldsymbol{C}_i \tag{39}$$

Continue the iterative solution so that $i=i+1$ goes to step ④. Combined with the above can obtain the $\boldsymbol{x}$ value at the end of the iteration, that is, the proportional coefficients k_1, k_2, $k_3=1-k_1-k_2$ of each method are obtained, and the combined weight $\boldsymbol{\omega}$ is obtained from equation (26). By calculation $k_1=0.3851$, $k_2=0.4176$, $k_3=1-k_1-k_2=0.1973$, the combined weight $\vec{W}=[0.0628,0.1122,0.0647,0.1881,0.312,0.2593]$ can be obtained according to equation (26).

5. According to equation (23), the state sequence vector $\vec{K}_I=(1,0,0,0,0,0)$ of the output of each index immune cell was determined, $\gamma_E=0.0628$ was obtained from equation (24), and finally the health value of scraper delivery equipment $H=1-0.0628=0.9372$ was obtained from equation (25).

It can be seen that by random value, the health value of the scraper conveying equipment is 0.9372, which is in a sub-healthy state, and the equipment can still operate normally, but according to the immune cell output state sequence, maintenance personnel need to pay attention to the lubricating oil temperature of the reducer. The equipment has been in operation underground for one month since the complete overhaul, and the parts are in good operating condition, which is consistent with the calculation results.

5. Summary

In the paper, the hierarchical analysis model and immune model of the scraper conveying equipment are first constructed, and then the initial weight is obtained by the analytic hierarchy method for the initial data collected, the index membership degree is calculated by the fuzzy statistical method, the initial weight is corrected by the entropy weight method, and finally the three algorithms are fused by cross-entropy theory to calculate the health value of the scraper conveying equipment by combining the immune principle. Taking a coal mine underground scraper conveying equipment as an example, the health status assessment was completed to lay the foundation for the subsequent construction of a health management system.

References

1. Jiyong Guo. Comprehensive evaluation of the safety of a dam based on analytic hierarchy method. Heilongjiang Hydraulic Science and Technology, 2023, 51(02): 58-61.
2. Kai Yang, Yongjun Qin. Quality evaluation of rockfill concrete based on fuzzy comprehensive evaluation method. Journal of Engineering Management, 2023, 37(01): 108-113.
3. Zhongguang Fu, Yunxi Yang, Binghan Liu. Comprehensive evaluation of entropy weight grey correlation degree of gas-fired power plant by combining principal component analysis. Thermal Power Generation, 2020, 49(02): 71-76.
4. Zhiyong Zhang, Meng Zhang, Jingui Lu. Fatigue life prediction of hydraulic support front connecting rod based on improved BP neural network. Coal Mine Machinery, 2023, 44(02): 177-179.
5. Yu Liu. Research on safety risk evaluation of new energy construction projects based on AHP-fuzzy comprehensive evaluation. Project Management Technology, 2022, 20(10): 123-127.
6. Jiabin Du, Yingjie Chen, Cheng Luo. Cost analysis and countermeasures of building curtain wall based on AHP-entropy weight method. Construction Economy, 2022, 43(S1): 228-234.
7. Yuxuan Liu, Feng Zhao. Urban rail vehicle auxiliary three-phase transformer health assessment. Electric Drive, 2020, 50(11): 89-94.
8. Lixia Chen. Photovoltaic power generation combination prediction method based on cross-entropy theory. Electrotechnics Electric, 2022(04): 17-20+48.
9. Fengan Wang. Discussion on fault diagnosis technology of mechatronics equipment. China Plant Engineering, 2023(05): 178-180.
10. Ruiyou Liang, Zhihu Huang, Zhiguang Chen. Analysis of key technologies of self-healing system of smart distribution network based on immunity principle. Automation & Instrumentation, 2020(06): 201-204.

High catalyst performance of two-dimensional Ti3B3N3S6 for hydrogen evolution reaction

Yongxiu Sun, Zhijie Li *

School of Physics, University of Electronic Science and Technology of China, Chengdu, 610054, China

Abstract. Single-atom catalysts (SACs) for the hydrogen evolution reaction (HER) is an efficient electrochemical pathway to produce the green production. However, the development of HER process is hampered by the lack of high-performance electrocatalysts. In this work, we proposed a new π-d conjugated structure of the $Ti_3B_3N_3S_6$ monolayer as the single-atom catalysts for the HER process by using the density functional theory (DFT) calculations. The calculated results show that the Ti atom is active site of the $Ti_3B_3N_3S_6$ monolayer with the high catalytic activity ($\Delta G_H = -0.14$ eV) for HER. The electronic properties of the $Ti_3B_3N_3S_6$ monolayer were explored by the electron localization function (ELF), Bader charge analysis and the polarized density of states (PDOS) density analysis. The $Ti_3B_3N_3S_6$ monolayer can promote the electronic transfer during the HER process, which indicates taht the $Ti_3B_3N_3S_6$ monolayer can is considered to investigate the catalytic activity for HER. The Gibbs free energy of H atoms adsorption on the $Ti_3B_3N_3S_6$ monolayer is –0.14 eV. Furthermore, the origin of high catalytic activity for the $Ti_3B_3N_3S_6$ monolayer was explored by the analysis the PDOS of the H adsorption on the $Ti_3B_3N_3S_6$ monolayer. Therefore, our work propose a new and high catalytic single-atom catalyst for the HER.

Keywords: hydrogen evolution reaction; Two-dimensional monolayer; $Ti_3B_3N_3S_6$ monolayer.

1. Introduction

Due to increased environmental pollution and the global scarcity of resources, the pollution-free and high energy density fuel gains widespread attention. Hydrogen (H_2) is as an ideal energy carrier of sustainable energy source for its high energy density and environmental friendly. The electro-chemical water splitting for the production of H_2 through hydrogen evolution reaction (HER) process plays a critical role as promising and sustainable method for large-scale hydrogen production. So far, platinum-based catalysts are regarded as the best catalysts for HER. However, the limited earth resources of the noble metals hamper their large-scale application. Exploring new catalysis for HER with resource-rich and non-noble metal elements is one promising pathway for the mass production of hydrogen

In recent years, embedding single transition metal (TM) atoms into 2D substrates have been widely explored as effective electrocatalysts for HER process. Owing to their tunable electronic structures and low coordinated configurations, these single-atom catalysts (SACs) are developed to improve the electrocatalytic performance of catalysts for the HER process. SACs can effectively prevent the occurrence of clusters in the reaction process due to the interaction between the metal atom and the substrate, such as Ti, V, Mn and Co–silience [1], Sc, Ti and V–$CuPS_3$[2] and Ni–P_4[3]. At the same time, the non-noble transition metals are cheap and well-resourced on the earth. However, so far many catalysts cannot achieve high metal atom loadings. This becomes a serious problem as the high loading of metal atoms is crucial to increase the number of exposed active sites, thus significantly improving catalytic efficiency of the catalysts for HER. Therefore, the design of a rational, highly atomically loaded and cheap catalyst is the subject of our work.

Moreover, as an analogue of benzene (C_6H_6), borazine ($B_3N_3H_6$) is electronically and isostructurally similar to benzene. Borazine exhibits delocalized π ring currents localized on the nitrogen atoms, while it has a lower level of aromaticity. Recently, there a new π conjugated structure of the B_3N_3 ring was explored the dynamical and mechanical stabilities [4]. Therefore, the transition metal atom and $B_3N_3S_6$ could form a new π-d conjugated structure with high stability.

Herein, we proposed a new 2D structure of the $Ti_3B_3N_3S_6$ monolayer to study the catalytic activity for the HER process and investigate the origin of high catalytic activity by using the density functional theory (DFT) simulations. The calculated results show that the Ti atom is the active center of the $Ti_3B_3N_3S_6$ monolayer with ΔG_H of –0.14 eV, which shows that the $Ti_3B_3N_3S_6$ monolayer is a high electrocatalyst for the HER. Furthermore, the electronic structure (including PDOS, ELF and Bader

* Corresponding author: zhijieli@uestc.edu.cn

charge) of $Ti_3B_3N_3S_6$ monolayer was explored for the HER process.

2. Computational Details

The Vienna ab initio Simulation Package (VASP) code was used to all energy calculations[5]. The electronic exchange correlation was performed by the Perdew – Burke – Ernzerhof (PBE) function based on the generalized gradient approximation (GGA)[6]. The Brillouin zones was performed by the Monkhorst – Pack mesh of 3×3×1. The z-direction was set to 15 Å to avoid the weak interactions of van der Waals interaction. The energy cutoffs was set to 520 eV. The energy and force convergences were set to 10^{-5} eV and 0.02 eV/Å, respectively.

The Gibbs free energy of H adsorption (ΔG_H) on the $Ti_3B_3N_3S_6$ monolayer is used to evaluate the HER catalytic performance, which can be calculated by the following equation:

$$\Delta G_H = \Delta E_H + \Delta ZPE - T\Delta S, \quad (1)$$

Where the ΔE_H is the total energy of H adsorption on the $Ti_3B_3N_3S_6$ monolayer. ΔZPE is zero-point energy difference. T is set to 300 K. ΔS is the entropy difference. The calculated value of $\Delta ZPE - T\Delta S$ is 0.24 eV.

3. Results and discussion

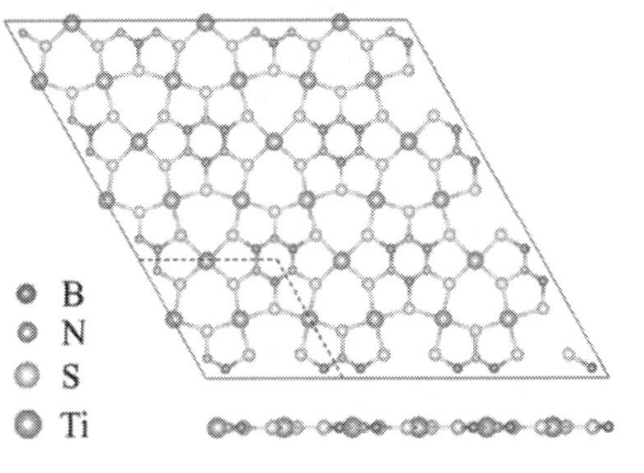

Figure 1. The top and side views of 3×3×1 supercell of $Ti_3B_3N_3S_6$ monolayer. The green, sliver, yellow and blue cells represent B, N, S and Ti atoms. The dotted plot represents a unit cell.

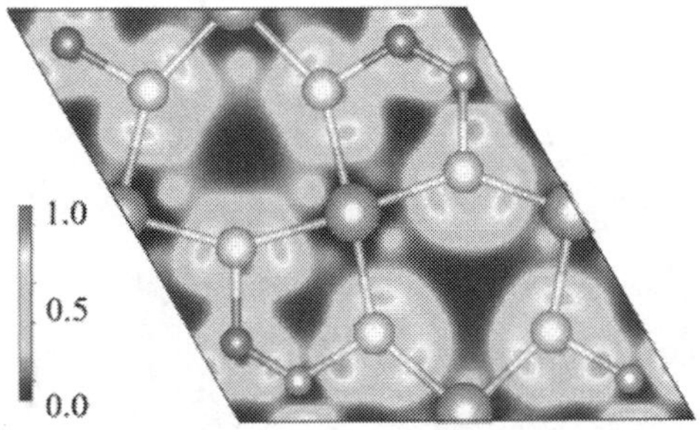

Figure 2. Electron localization function (ELF) map at isosurface value of 0.50 a.u. The values of ELF are between 0.0 and 1.0.

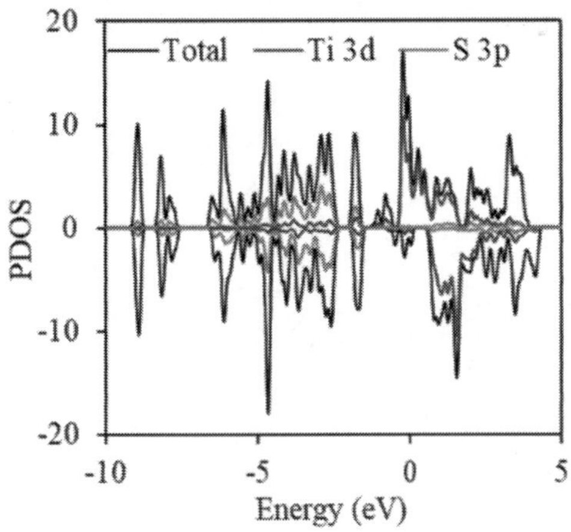

Figure 3. The PDOS of $Ti_3B_3N_3S_6$ monolayer. The black, blue and red lines are DOS of the total, 3d orbital of Ti atom and 3p orbital of S atom. The Fermi energy is located at 0 eV.

Figure 1 shows the top and side views of 3×3×1 supercell of $Ti_3B_3N_3S_6$ monolayer, where the dotted box indicates the unit cell. The green, sliver, yellow and blue cells represent B, N, S and Ti atoms. Each $B_3N_3S_6$ ring is connected to six S atoms, and a Ti atom is connected to four S atoms. After structural optimization, all atoms of $Ti_3B_3N_3S_6$ monolayer are in the same plane, and the lattice constants are a = b = 9.18 Å, c = 15 Å, $\alpha = \beta = 60°$, γ = 120°. It is essential to know the electronic properties of the $Ti_3B_3N_3S_6$ monolayer for catalytic HER performance. To investigate electronic properties of the $Ti_3B_3N_3S_6$ monolayer, the electron localization function (ELF) and the polarized density of states (PDOS) were investigated, and the calculated results are shown in Figures 2 and 3. For the ELF of the $Ti_3B_3N_3S_6$ monolayer as shown in Figure 2, the ELF values of Ti-S bonds are lower than 0.5, which indicated that the bond of Ti-S is ionic bond. At the same time, the B-S and B-N bonds are covalent due to the ELF values are greater than 0.5.

At the same time, the Bader charge of the $Ti_3B_3N_3S_6$ monolayer are calculated, and the calculated results show that the Ti atom transfer about 0.23 electrons to the S atom, which indicates that the transition atom can transfer electrons efficiently in the catalytic process. Therefore, the $Ti_3B_3N_3S_6$ monolayer can generate effective electrons transfer during the HER process.

Moreover, the partial density of states (PDOS) of the $Ti_3B_3N_3S_6$ monolayer are investigated, and the calculated result was shown in Figure. 3. The he black, blue and red lines are DOS of the total, 3d orbital of Ti atom and 3p orbital of S atom. The results show that the Fermi energy level (E_F) is crossed by the electronic states, and the Ti atoms contribute the majority of the electronic state, which indicated that the $Ti_3B_3N_3S_6$ monolayer is metal conductive. The above results indicated that the $Ti_3B_3N_3S_6$ monolayer can promote the electronic transfer during the HER process. Therefore, the $Ti_3B_3N_3S_6$ monolayer is considered to investigate the catalytic activity for HER process.

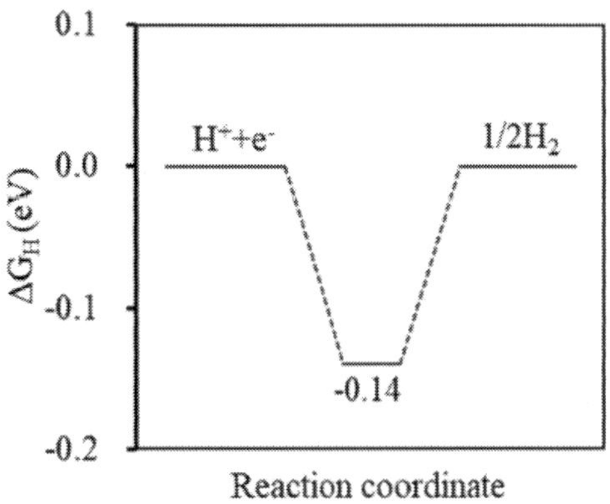

Figure 4. The diagram of Gibbs free energy for HER on the $Ti_3B_3N_3S_6$ monolayer.

According to the previous studies, the sites of the transition metal atom would be the active catalytic center of the HER process. Therefore, the H atom adsorption site on surface of the $Ti_3B_3N_3S_6$ monolayer is the Ti atom in our work. Therefore, the Gibbs free energy diagram was investigated in Figure 4. The Gibbs free energy of adsorbed H atoms (ΔG_H) to evaluate HER performance has been widely accepted. At the same time, the closer the value of ΔG_H is to 0 eV, the better the catalytic effect, i.e., the catalyst can effectively adsorb H atoms without being too tightly adsorbed to allow the H atoms to be released. Fig. 4 shows the diagram of Gibbs free energy for HER on the $Ti_3B_3N_3S_6$ monolayer. The value of ΔG_H is -0.14 eV for the $Ti_3B_3N_3S_6$ monolayer, which indicates that the Ti atom can effectively adsorb H atoms and can easily release H_2 molecules during the HER catalytic process. Therefore, based on the above, the $Ti_3B_3N_3S_6$ monolayer can be act as a novel and efficient electrocatalyst for the HER process.

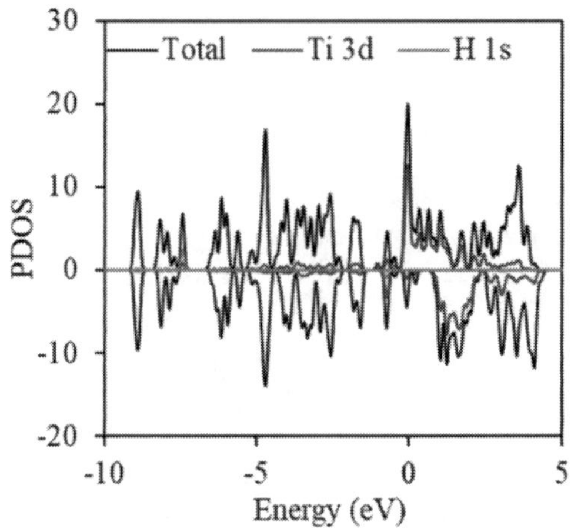

Figure 5. The partial density of states of H adsorption on the $Ti_3B_3N_3S_6$ monolayer. The black, blue and red lines are density of states of the total, 3d orbital of Ti atom and 1s orbital of H atom. The Fermi energy level is located at 0 eV.

To understand the high catalytic mechanism for the HER process on the surface of the $Ti_3B_3N_3S_6$ monolayer, the PDOS of the H adsorption on the $Ti_3B_3N_3S_6$ monolayer was investigated in Figure 5, and the calculated results show that the 1s orbital of H atom with 3d orbital of Ti atoms generate hybridization from the coupling of electronic states, which indicates that the H atom can be adsorbed on the Ti atom of the $Ti_3B_3N_3S_6$ monolayer on the HER process. The density states are below the E_F, which can be full filled, which indicates that the Ti atoms achieve effective charge transfer on the HER process. Therefore, the electronic analysis of the H adsorption on the $Ti_3B_3N_3S_6$ monolayer proved the H atoms can be stable adsorbed during the HER process, and the $Ti_3B_3N_3S_6$ monolayer has the great catalytic performance of the HER.

4. Conclusion

In this paper, we investigated the catalytic activity of $Ti_3B_3N_3S_6$ monolayer for HER by using the DFT calculation. The calculated results implied that the Ti atom is the active center with the ΔG_H value of –0.14 eV, which indicates that the $Ti_3B_3N_3S_6$ monolayer can be acted as a good sing-atom catalyst for the HER process. The electronic properties of the $Ti_3B_3N_3S_6$ monolayer, i.e., the partial density of states, Bader charge and electron localization function, are invaestagated to get the potensial of the perforemance of the HER. At the same time, the origin of excellent catalytic performance was studied by a series of electronic structure analysis, which shows taht the s orbital of H atom with d orbital of Ti atoms generate hybridization from the coupling of electronic states. The electronic analysis of the H adsorption on the $Ti_3B_3N_3S_6$ monolayer proved the H atoms can be stable adsorbed during the HER process, and the $Ti_3B_3N_3S_6$ monolayer has the great catalytic performance of the HER. Therefore, our work proposed a new and excellent two-dimensional electrocatalyst for the HER process.

References

1. Sun, Y. X., Huang, A. J. Wang, Z. G, et al. Transition metal atom (Ti, V, Mn, Fe, and Co) anchored silicene for hydrogen evolution reaction. Rsc Adv. 9. (2019) 26321-26326.
2. Sun, Y. X., Huang, A. J., Li, Z. J., et al. Transition Metal Atoms Anchored on CuPS3 Monolayer for Enhancing Catalytic Performance of Hydrogen Evolution Reactions. Electrocatalysis-Us. 13. (2022) 494-501.
3. Yang, Y. et al. Insights into the Origin of High Activity of Ni5P4(0001) for Hydrogen Evolution Reaction. J Phys Chem C. 127. (2023) 5385-5394.
4. Abdullahi, Y. Z. Antiferromagnetic semiconductor in porous boron nitride (B6N6) sheet: First-principles investigation. Computational and Theoretical Chemistry. 1197. (2021) 113155.
5. Kresse, G. & Furthmuller, J. Efficient iterative schemes for ab initio total-energy calculations using a plane-wave basis set. Phys Rev B. 54. (1996) 11169-11186.
6. Perdew, J. P. et al. Atoms, Molecules, Solids, and Surfaces - Applications of the Generalized Gradient Approximation for Exchange and Correlation. Phys Rev B. 46. (1992) 6671-6687.

Research on the Development of Sensor Tracking Technology under the Guidance of Artificial Intelligence

Shan Ge

Liaoning Communication University, 110136, China

Abstract: The development and application of artificial intelligence from before heavy calculation to today's parallel processing, has a qualitative leap, for the experience of human experts can predict the conclusion, the integration of chip development has been artificial intelligence to create economic benefits of a big research topic, and combined with the analysis of all kinds of indicators to sensor data technology also more directional in artificial intelligence, modern research and development of all kinds of tracking equipment and drone equipment to a certain extent, integrated the sensor detection technology, and on the development of programming language, have the updated research field.

Keywords: Smart sensor; visual VR tracking; evolutionary algorithm; sensor data acquisition

1. Introduction

Due to the extensive promotion of the expert system, Experienced experts to analyze the diversity of AI, And the algorithm evolution of artificial intelligence and brain-like thinking under the condition of stable programming structure, This is of great benefit to the application and development of artificial intelligence, Algorithm generators are also the mainstream development direction in recent years, Combining the evolutionary algorithm of artificial intelligence, The results generated by the algorithm and the adversarial comparison algorithm have a game effect, This allows humans to build on the inference of intelligent algorithms, With the new lines of thinking, From a developmental perspective, The social structure of human-computer interaction will become one of the increasingly mature topics, The use of sensing tracking technology will be greatly improved technology[1].

2. Research on VR in visual sensing technology

2.1 Preparation of VR technology

In the computer vision generation technology, the visual VR transmission of sensors is a way to have an immersive experience of high-quality visual sensing information[2], Back in the 1980s, The concept of virtual reality has been advocated by many scientists, Generating vision technology for computers, In the effect of stereo vision to meet the multi-dimensional needs of users, During the transformation of the visual field, VR's visual sensing technology has also been improved somewhat, Virtual reality devices that meet the needs of their users, Performed the information collection of the sensors, Experiencers use a variety of output devices to complete the digital information experience, And get the multi-dimensional sense of space, The research and development of closed head display has a more authoritative guiding significance in this field, Since the location of the sensor deployment of the head display is somewhat related to the neural network distribution designed in brain-like studies, So the head collected by the sensor has a corresponding relationship with the brain module controlling the movement process[3], The sensors on the virtual reality devices analyze the data, transmit it to the computer, and generate real-time constructed multi-dimensional visual images, so as to improve the pleasure of users' experience content under the action of the visual sensor.

2.2 Development process research of VR application

When applying sensors to collect VR information, it is usually necessary to prepare models, video resources, audio output and other resources. In the actual development process, the sensor needs to be combined with the Unity program compiler to integrate and compare. In the whole process of VR interactive development, the software produced also changes with the collection information content of the sensor[4], To a certain extent, this makes the sensor information has the basis of multi-dimensional information collection.

2.3 Development process of multi-dimensional vision sensor model establishment

Among the many software that establish multi-dimensional model, the application of Blender is the most mainstream. Under the 3D content that provides open source information, the production process is relatively recognized by the studio. In terms of the modeling parameters of sensor acquisition, the design of bone binding and animation is the 3D characteristics of the software in the production process, rendering the physical simulation process around the design demonstration, so that the user's visual experience can rise. In the later synthesis process, the software has the synchronization function of sensor information.

3. Acquisition of sensing and tracking data

3.1 Introduction of the acquisition and tracking controller

In the actual background, the controller of virtual works binds objects in the real world according to the parameter requirements of the tracking system, which completes the fundamental information import in the function of tracking objects[5]. In the mainstream products, V IVE Tracker as V IVE VR accessories, with stronger control tracking sense, it has a strong with chase function, in many experiments, its control parameters is more like the embedded controller handle operation terminal, but no entity button failure, in the case of no design button, implement 6 DOF tracking function, V IVE T racker volume of short lean, sensor data only by binding it to the human body any joint sensor data acquisition[6], Thus, it accurately realizes the expected effect of motion capture and tracking in the virtual world.

3.2 Rendering design after tracking data collection

Using Su bstance Designer can be material rendering design, using the software's advantage of the traditional material on the material provide open source, and in the late rendering, the exported file and related parameters set more valuable, file is smaller, conform to the requirements of the import file, and for exposure parameters, often can set necessary parameters to complete the node parameters design, node parameters exposed output, analyzed by the engine, and make a VR content limited content and attributes[7], The output and display of the engine should be checked after rendering. Under the control of the Inspector panel, the engine can complete the primary automatic modification state, but the relatively large revision requires further benchmarking to perform. In addition, the software can also change the effect of the user experience requirements. After the computer data collection, by replacing multiple material material to complete the comprehensive design, in the design process, there is undoubtedly a lot of program parallel call, the software parallel file processing ability put forward new requirements, rendering design also support for network files, in the case of tracking data existing, rendering design is to enhance software beautification effect, the best way to improve the user experience.

3.3 Setting of the parameter module of the tracker

(1) In the experiment, the whole matching mode is used for position tracking, and the switching mode of the equipment should be the same as the V IVE controller;
(2) Sensor module, the total sensor needs to exchange information with the base station to receive the base station signal and feedback to the equipment. The subsensor needs to be connected to the object to generate a synchronous signal and send it to the equipment;
(3) Standard camera bracket interface, set the same parameters as the SLR camera, with a 1 / 4 inch screw hole to complete the fixation, generally using the standard tripod cloud head for access;
(4) Fixed needle slot, produce concussion in the experiment, use fixed needle slot to complete and strengthen the stability of the equipment and bracket;
(5) Electrical connect the spring pin. During the experiment, when the electrical signal is transmitted, the spring contact interface is used to complete the electrical transmission;
(6) anti-skid equipment, used for the connection between V IVE tracker and accessories, to prevent the two skid phenomenon, providing greater friction;
(7) U SB Port, for the transmission of sensor signals, and provide serial links for the open source update of devices[8].

4. Algorithm fusion in sensing technology

4.1 Overall design

Since the sphere is the most representative of the multiple stereo devices, assume that in the scene, the 3D sphere is used as the research object, for the rendering design, the sensing data of the sphere to the image is the sensor probe project content.
$O = (ka, kb, kc)$ Set the center of the sphere and the radius of the sphere r, there is a tracing design equation:

$$(a-ka)^2+(b-kb)^2+(c-kc)^2=r^2 \quad (1)$$

When tracing, limiting one point to a random point T, the above equation (1) can be rewritten as:

$$(T-K)*(T-k)=r^2 \quad (2)$$

When considering the multidimensional tracking system, set in a sphere in three-dimensional scene, the sensor needs through the form of photoelectric induction signal acquisition, let the acquisition of random points with the movement of the light signal feedback sensing signal, the acquisition output many times, can render the outer boundary of the sphere, so as to determine its position for tracking. As shown in Figure 1.

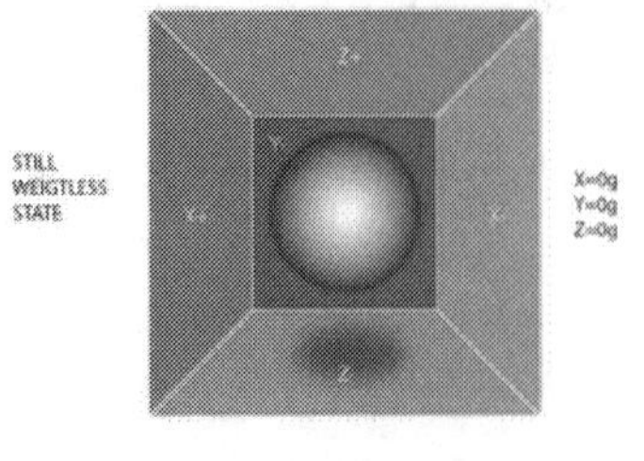

Figure 1. Sphere simulation in 3 D scenes

After determining the position of the sphere is determined, the comprehensive scene carries out diversified feedback. If there are other spheres somewhere in the scene, that is, the unknown amount is not unique, then the emitted light of the collected sensor will cross the feedback with the feedback light source of the known spheres. This is not unique to the number of feedback objects on the tracking results. This is all related to the movement state of the sphere. Assuming that the sphere movement is left and right swing, it is shown in Figure Figure 2.

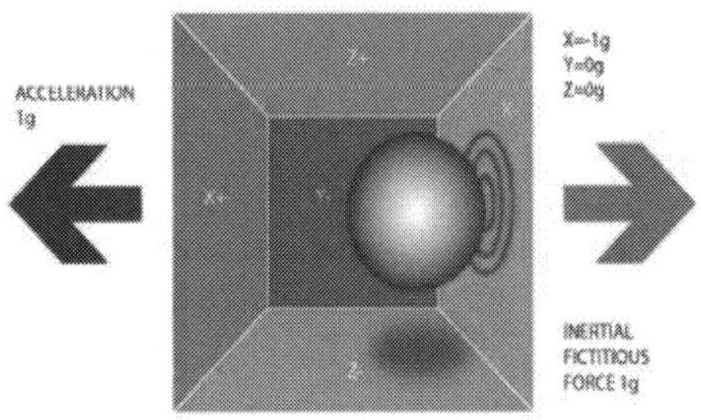

Figure 2. Analysis of the sphere movement state

Then the following question is how to use the sensor to collect information to determine whether there is a cross of the set light with a sphere in a specific scene, and whether the cross beam is unique. Set the contact point where the light meets the sphere as, and the determination formula can be derived from the following formula:$T(u)$

$$\begin{cases} (T(u) - K) * (T(u) - k) = r^2 \\ (J + uL - K) * (J + uL - k) = r^2 \end{cases} \quad (3)$$

4.2 Hardware design

In the hardware module, there is a data transmission interface at the cover plate designed at the top of the head, which opens the interface for data transmission. The internal interface device is mainly composed of USB, which increases the possibility of users to the external device[9]In the current device, the input of gestures and foot sense can be completed by the hardware. In addition, the parameters of the waist can be collected and transmitted, and the hardware interface set as open source interface also has a certain market prospect. In the scene of certain circumstances, the user can apply for open light Angle visual mode, generally adopts the field of view positioning value in around 110, in accordance with the escort of the guidance, as far as possible to use the front-facing camera to complete the image acquisition of the work, sensors, mainly dominated by Steam VR tracking technology, and equipped with G-sensor correction system for auxiliary work, the gyroscope gyroscope configuration is the necessary objects in high-end stability equipment, from a certain extent, solve the problem of position instability.

5. Application of sensing and tracking technology guided by artificial intelligence

In view of the diverse applicability of sensing tracking technology[10]Under the guidance of artificial intelligence, it is also widely used, which is mostly related to the artificial intelligence measurement of tracking data.

5.1 Medical treatment and rehabilitation work

In medical work, medical staff can use sensor tracking means to complete the process of simulation operation, in order to improve the success rate of surgery, using virtual tracking technology with sensor can complete the transmission of 3D imaging technology, when tracking pathogens and target, using simulation surgery can be completed for fear of heights, autism treatment, in addition, tracking sensor for the potential parameters of the nervous system also has certain contribution, this has a new breakthrough in remote medical subject[11].

5.2 Social networking platform

In the context of complex social networks, virtual reality experience with sensors can also complete the basic needs of human beings on mobile platforms. In the process of human-computer interaction, VR experience is mostly received by the new generation of social experts.

6. Challenges for the future development of intelligent sensing and artificial intelligence

6.1 Price war competition

The hardware cost of tracking sensors is currently relatively expensive on the market[12], The implantation of mobile devices makes the market price of hardware not accepted by consumers, thus reducing the desire to consume, which to a certain extent inhibits the development of intelligent sensing tracking technology.

6.2 Limitations of computing technical capabilities

For the high requirements of computing capacity and artificial intelligence parameters, there is a widespread mismatch between effect and technical capabilities. A large number of technologies are supported by the research and development requirements of high pixel and high precision positioning tracking equipment, which is

https://doi.org/10.1051/matecconf/202338201006

undoubtedly a huge test for the computing speed and ability of the internal chip of the equipment.

6.3 Future development

In today's world, with the rapid development of artificial intelligence, the trend of human-computer interaction is imperative. Under the premise of artificial intelligence oriented, providing simulation scenes and using intelligent sensing tracking technology to carry out necessary human actions, which will become increasingly intense in the future of the technology development.

References

1. Hong Wenjin, Huali Xia, Miao Yu, Jin Hongmei, Tang Ying, Shen Lei Design and implementation of intelligent micro sensing interactive system for sports underwear [J]. Shanghai Textile Technology, 2021,49 (09): 16-19 + 36.
2. Umang Singh, Ajith Abraham, Arturas Kaklauskas , Tzung -Pei Hong.Smart Sensor Networks[M].:2021-09-03.
3. Hu Rui, Rui Zhong, Dong Xingfa, Peng Lili, Design and application of intelligent power system with multi-sensor fusion [J]. Modern Electronic Technology, 2021,44 (17): 178-181.
4. Smart City construction in Rome: the trial of artificial intelligence sensor [J]. Industrial Design, 2021, (08): 22.
5. Zhao Jie, Huang Huibao, Li Nuodong, Zhang Hanke, Research and development and application of new intelligent measurement and control equipment based on monitoring sensors [J]. Sichuan Hydropower Power Generation, 2021,40 (04): 5-8 + 12.
6. Wang Yi, Wang Kai, Zhang Yitian, Liu Dun, Liu Yu based on the intelligent following system design based on ultrasonic sensors [J]. Sensors and Microsystems, 2021,40 (08): 92-95.
7. Wang Zhaofeng, Li Boxiao, Plateau, Lu Zejian, an intelligent assisted decision system design for heterogeneous sensor network [J]. Electronic Technology and Software Engineering, 2021, (15): 33-37.
8. Lin Haiqiang motion action intelligent recognition based on laser sensor [J]. Laser Magazine, 2021,42 (07): 84-89.
9. Li Xin, Xu PBM, Cui Jianhua, Li Yang, Tang Zhiguo Research on the evaluation method of local discharge intelligent perception terminal based on gray model [J]. Electrical measurement and instrument: 1-7.
10. Jia Wenxian , Liu Menghan , Zhou Jie, Kaiwartya Omprakash.Adaptive Chaotic Ant Colony Optimization for Energy Optimization in Smart Sensor Networks[J].Journal of Sensors,2021.
11. .Research on the weaving method of Gao Qian intelligent wearable photoelectric sensor [D]. Tianjin University of Technology, 2021.
12. Liu Yilong, Kou Yuanjin development and application of echatronics technology in intelligent manufacturing [J]. Technology wind, 2021, (17): 6-7.

Research on Dynamic Path Exploration and Planning Method of Quadrotor Unmanned Aerial Vehicle in Complex Environment

Hanting Huang

LORS LAND, Shanghai 201400, China

Abstract. As a research hotspot in the field of mobile robots in recent years, UAV(Unmanned Aerial Vehicle) has been improving its autonomous performance. Compared with traditional UAV, UAV has many advantages, such as vertical lifting, strong maneuverability and convenient operation. However, the dynamic path planning of quadrotor UAV, as a typical autonomous ability of mobile robot, has always been the research focus in the field of UAV. In this regard, this paper will try to study the dynamic path exploration planning method of quadrotor UAV in complex environment. In the complex environment model, the comparative simulation from the aspects of calculation time, calculation complexity and success rate shows that the selection of high-order polynomials can ensure high-order continuity, but it will also increase the calculation time and complexity. Compared with CF algorithm, the calculation time in this paper is 57% lower on average and the success rate is 17.6% higher on average, which shows that this algorithm is more suitable for online flight path planning of quadrotor. Comparatively speaking, the success rate of the other two methods decreases with the increase of environmental complexity, and the overall calculation time is longer.

Keywords: Complex environment, Four-rotor UAV, Dynamic path, Explore planning methods

1. Introduction

With the continuous enhancement of human's ability to explore the living environment, mobile robots with autonomous behavior will provide more abundant and comprehensive information for human beings. Compared with the traditional UAV, the four-rotor UAV has the advantages of vertical lifting, strong mobility and easy operation, which makes it play an important role in many fields such as aerospace launch and recovery, military reconnaissance, police search, fire rescue, agricultural monitoring, film and television shooting [1]. Path planning has been widely concerned by researchers in the field of artificial intelligence because of its strong autonomous behavior performance. Path planning is to search for the best path from the starting point to the destination on the basis of meeting the requirements of the planned object's motion trajectory. As a research hotspot in the field of mobile robots in recent years, the four-rotor UAV has continuously improved its autonomous performance. As a typical autonomous capability of mobile robots, path planning has always been the focus of research in the field of UAV [2-3]. The research of four-rotor UAV path planning is usually focused on the outdoor open environment. The indoor environment is more time-varying and unpredictable. It is of great significance to carry out UAV precise path planning in the indoor complex time-varying environment for UAV autonomous cooperation and precise operation [4].

In the complex environment, refer to the research findings of domestic and foreign scholars on local dynamic path planning in recent years. It is particularly important that how to maintain the original flight attitude and independently plan the flight route according to the original waypoint, safely and stably fly to the destination and land, so as to avoid sudden accidents of aircraft damage and personal injury when the four-wheel UAV is out of manual control or other emergencies [5]. The superiority of the search path depends on the accuracy of the environment model and the real-time and effectiveness of the algorithm. These search paths limit the form of flight motion. Therefore, we hope that from the perspective of motion learning, we can not only plan a desired path, but also integrate the motion characteristics by learning the existing motion forms. At present, the research of such problems is mainly aimed at the path planning of UAV in the outdoor environment, while the problems faced by UAV in the indoor environment are more complex. In the case of obstacles, it is also possible to plan a feasible obstacle avoidance path through dynamic motion primitives [6]. Such a motion learning mechanism improves the autonomous navigation capability of the micro four-rotor UAV. Obstacles and environment are more uncertain, so the local dynamic programming algorithm proposed for outdoor environment cannot be applied in indoor environment.

2. Indoor local dynamic path planning of quadrotor unmanned aerial vehicle

2.1 Memoryless A* algorithm

On the path planning problem, A* algorithm, as a heuristic algorithm, has high search efficiency and can meet the real-time performance of UAV path planning in complex indoor environment. At the same time, A* algorithm, as a greedy algorithm, can make up for the shortcomings of heuristic search algorithm and find the optimal path. Then, the nodes with the minimum cost are screened out from the open set and put into the closed set, and the nodes put into the closed set continue to carry out extended search for the current node, and the cost is updated in the process of each extended search [7]. The algorithm node expansion is shown in Figure 1.

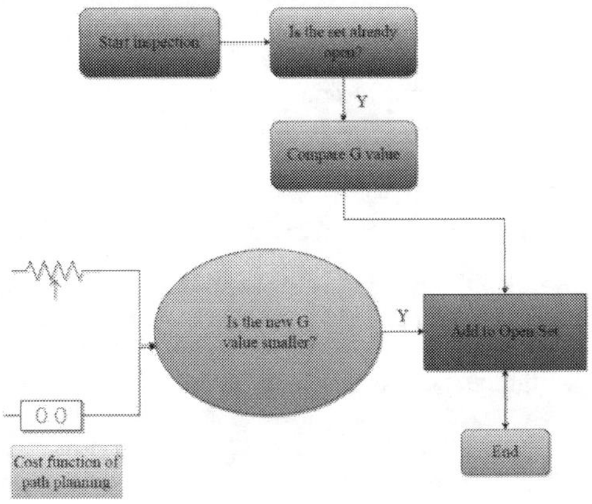

Figure 1 A * algorithm node expansion process diagram

Due to the dynamic obstacles in the environment, it is necessary to continuously rescan the environment information and repeat the environment modeling during the planning process. When the A * algorithm performs path search, its planning efficiency is seriously affected due to frequent environment modeling. The four-rotor UAV adopts a symmetrical "X" shaped rigid body structure, in which two pairs of positive propellers and two pairs of reverse propellers are installed on the same diagonal. The forward rotation of the positive rotor produces clockwise torque, while the reverse propeller produces reverse torque. After the offset between the torques, the UAV obtains vertical upward lift.

This kind of traversal search has great randomness and blindness, resulting in reduced real-time performance of the algorithm, and will generate too many redundant nodes, plan the route twists and turns, increase the four-rotor maneuver, and reduce the track tracking [8]. In a three degree of freedom motion system, a point to point motion model is established in three-dimensional Cartesian space. In formula (1), the coupling term C is added to build a dynamic system with obstacle avoidance function, namely:

$$\tau \upsilon = \alpha_{\upsilon} + C_{i,j} \quad (1)$$

The vector form of motion coupling term with three degrees of freedom is the key to avoid obstacles in $C_{i,j} = \lfloor C_{i,x} \quad C_{i,y} \rfloor$ planning.

A motion perpendicular to the current velocity direction is added to the coupling term, which is a function of distance vector and velocity vector, as shown in the following formula:

$$C_t = kRy\mu e^{-\beta} \quad (2)$$

$$\mu = \arccos \frac{o-y}{o-y(y)} \quad (3)$$

According to the above dynamic model of micro quadrotor UAV, it can be seen that the movement of UAV can be the movement of the center of mass and the rotation of the fuselage, and the movement of the center of mass depends on the lateral acceleration provided by the roll of the fuselage, so the key to control UAV is attitude control. Therefore, in the flight process, the central control unit provides control signals to adjust the motor speed after receiving the feedback information, so that a variety of flight state outputs can be obtained by adjusting the four motor speeds. Next, the stress analysis of the X-shaped quadrotor UAV in several flight states under normal operation will help to explain its flight principle more clearly.

2.2 Memoryless regression A * algorithm

In view of the above ideas about local dynamic path planning solutions, to realize this idea, we must first improve the global planning algorithm and local planning algorithm. Analyze the characteristics of global planning algorithm and local dynamic planning algorithm. Considering the complexity of the indoor environment, this paper takes A * algorithm as the global planning algorithm, and refers to the memoryless A * algorithm, improves and designs the memoryless regression A * algorithm as the local dynamic path planning algorithm [9]. According to the above micro four-rotor UAV dynamics model, it can be seen that the movement of the UAV can be the movement of the center of mass and the rotation of the body. The movement of the center of mass depends on the lateral acceleration provided by the body's roll. Therefore, the key to control the UAV is attitude control. The control structure diagram is established as shown in Figure 2.

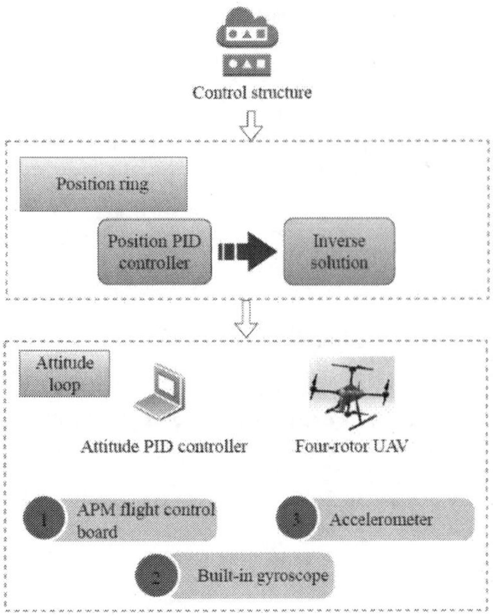

Figure 2 Schematic diagram of control structure

The planned track must meet the dynamic performance constraints of the four-rotor wing, otherwise the four-rotor wing cannot fly according to the planned track, and the planned track is not feasible. After dealing with sudden dynamic environment changes, in order to obtain the local coupling equation, a performance index function that depends on the local adjacency consistency error and coordinated control strategy is designed. The path should return to the originally planned path as soon as possible, so the local dynamic planning algorithm should have the ability to return to the global path.

The reduction of track points can cut most of the tortuous track points and only retain the track points necessary for the UAV to avoid obstacles. This optimization step can significantly reduce the number of UAV maneuvers, not only enhance the safety of UAV flight, but also save fuel and increase the flight distance [10]. The four-rotor UAV conducts global path planning in the indoor environment. Although the global environment information is fully known and can be fully utilized, considering the small indoor environment space, small obstacle avoidance margin, irregular and complex obstacles, and the three-dimensional planning space, a path planning algorithm is required to solve the problem, which can not only ensure the planning efficiency, but also meet the requirements of environmental accuracy.

3. Experiment and analysis

3.1 Experimental data transmission link

Set the IP address of the workstation server running Motive software, click the "Adjust IP" button corresponding to the IP of the client to automatically obtain the IP address of the ground station, and then click the "Enable" button below to realize the communication between the ground station system based on UDP communication protocol and the Motive software workstation, and receive the position, attitude and spatial state environment information of the multi-rotor UAV obtained by the indoor visual positioning system. When the quadrotor UAV passes through the area represented by the Euclidean distance field, the gradient of the Euclidean distance field will suddenly change near these obstacles, making it impossible for the flight path to smoothly bypass these obstacles. Then the data transmission and recovery system can ensure the accurate transmission of remote control instructions, ensure that UAV can feedback flight data in real time, and can fly at a safe altitude and speed according to the instructions.

Under normal circumstances, a complete quadrotor UAV will be loaded with global positioning system, ranging sensor, gas sensor, gyroscope, photoelectric pod and other equipment. By setting the serial number, baud rate and transmission frequency of the wireless data transmission module, click the "Enable" button below to establish the communication interactive link between the ground station system and the multi-rotor UAV. In order to further solve this problem, additional parameter information is introduced, and the penalty terms of distance cost, collision cost, speed and acceleration are added, and the on-line trajectory avoidance optimization model of four-rotor UAV is constructed. Tick the "path planning" option under the position expectation, and the ground station system will carry out environmental modeling and dynamic path planning according to the current space state information and the position and attitude data of the multi-rotor UAV. After the modeling and planning are completed, the planned route trajectory points will be sent to the four-rotor UAV one by one according to the set transmission frequency.

3.2 Dynamic path planning simulation in complex environment

When the four-rotor UAV is flying in an unknown environment, due to the limited range of perception, it is necessary to re-plan the flight path frequently, that is, when the scheduled flight path will collide with the new obstacles, it is necessary to re-plan the flight path online to ensure that once any collision is detected, a new safe flight path will be available. Use the same reference track to initialize all methods, and limit the optimization time to within 18ms. This paper compares CF (Collaborative Filtering) with the algorithm in this paper, and the relevant data statistics are shown in Table 1. The comparison results show that the method in this paper can always maintain a high success rate and obtain a feasible escape route in the unknown environment. In comparison, the success rate of the other two methods decreases with the increase of the complexity of the environment, and the overall calculation time is longer.

Table 1 Comparison of simulation data of two algorithms in different complex environments

Obstacle environment	Algorithm	Time consuming	Success rate
Low density	CF	12.4	86.5
	This text	4.2	100
Medium density	CF	15.3	88.7
	This text	7.3	100
High density	CF	15	90.3
	This text	6.24	100

This chapter combines the actual flight environment of the four-rotor UAV, based on the traditional dynamic window algorithm, adds the improvement of obstacle circle and gives collision detection conditions, further speeds up the trajectory simulation speed, improves the feasibility of the safe flight of the UAV, ensures the stability of the closed-loop system, and ensures the good control performance of the multi-UAV formation system in the case of switching communication topology.

4. Conclusions

In this paper, the memoryless regression A* algorithm is designed and combined with the traditional A* algorithm for dynamic planning, which overcomes the limitations and blindness of the algorithm in the process of local dynamic path planning. Four-rotor UAV carries out global path planning in indoor environment. Although the global environmental information is completely known and can be fully utilized, considering the small indoor environment space, small obstacle avoidance margin, irregular and complex obstacles, and the planning space is three-dimensional, a path planning algorithm is needed to solve this problem, which can not only ensure the planning efficiency, but also meet the requirements of environmental accuracy. In the complex environment model, the comparative simulation from the aspects of calculation time, calculation complexity and success rate shows that the selection of high-order polynomials can ensure high-order continuity, but it will also increase the calculation time and complexity. Compared with CF algorithm, the calculation time in this paper is 57% lower on average and the success rate is 17.6% higher on average, which shows that this algorithm is more suitable for online flight path planning of quadrotor. The comparison results show that the method in this paper can always maintain a high success rate and obtain feasible evasive paths in unknown environments. Comparatively speaking, the success rate of the other two methods decreases with the increase of environmental complexity, and the overall calculation time is longer.

References

1. Liu X, Du X, Zhang X, et al. Evolution-algorithm-base d unmanne d aerial vehicles path planning in complex environment[J]. Computers & Electrical Engineering, 2019, 80(30):106493-106499.
2. Kun W, Bingyin R. A Method on Dynamic Path Planning for Robotic Manipulator Autonomous Obstacle Avoidance Based on an Improved RRT Algorithm[J]. Sensors, 2021, 18(2):571-632.
3. Liang Y. A geometrical path planning method for unmanned aerial vehicle in 2D/3D complex environment[J]. Intelligent Service Robotics, 2020, 11(3):22-30.
4. Zhang J, Li J, Yang H, et al. Complex Environment Path Planning for Unmanned Aerial Vehicles[J]. Sensors, 2021, 21(15):5250-5311.
5. Xiaoyu S, Yuanhang L, Ligang C, et al. Research and Design of UAV Environmental Monitoring System[J]. International Journal of Modelling, Identification and Control, 2021, 48(13):41-57.
6. Yang L, Hu D, Chen J, et al. Research on Obstacle Avoidance Algorithm for Four-rotor UAV[J]. Journal of Physics Conference, 2021, 1061(19):012016-012027.
7. Wang K, Song M, Li M. Cooperative Multi-UAV Conflict Avoidance Planning in a Complex Urban Environment[J]. Sustainability, 2021, 13(5):26-38.
8. Chu Y, Han Z, Yang L. Track planning of multi-rotor unmanned aerial vehicle in the complex environment space[J]. International Journal of Modelling, Identification and Control, 2021, 15(1):37-49.
9. Zammit C, Kampen E. Real-time 3D UAV Path Planning in Dynamic Environments with Uncertainty[J]. Unmanned Systems, 2022, 11(03):203-219.
10. Lv Z, Huo L, Shen M. Four-Rotor UAV Safe Landing Risk Assessment Method[J]. IOP Conference Series: Materials Science and Engineering, 2019, 677(3):032005-032025.

Experimental study on solidification/stabilization of heavy metal lead and chromium sludge solidified by cement-based materials

Chao Zheng [1], Kangwei Xiong [1], Kun Yang [1], Deluan Feng [2,*]

[1] Guangzhou Grandtop Environment Service Co., Ltd., Guangzhou 510006, China
[2] School of Civil and Transportation Engineering, Guangdong University of Technology, Guangzhou 510006, China

Abstract: To explore the strength change law and solidification mechanism of cement solidified heavy metal lead and chromium contaminated soil, ordinary Portland cement (OPC) and magnesium phosphate cement (MPC) were used for solidification/stabilization treatment. Through the unconfined compressive strength (UCS) test, scanning electron microscopy (SEM) and the leaching toxicity test, the microscopic characteristics of the cement type, the cement content, the ion type, the ion concentration, and the age and their influence law on the strength characteristics of heavy metal contaminated soil were obtained. The UCS test shows that the compressive strength increases with the cement content and age. When OPC and MPC were used to solidify lead and chromium heavy metal contaminated soil, the compressive strength of OPC was greater than that of MPC, and when the curing age was 28 days, 20% of OPC and MPC satisfied the landfill requirements. The leaching toxicity test showed that after the solidification/stabilization of heavy metal lead and chromium contaminated soil by OPC and MPC, the leaching content of lead ions in the leaching solution conforms to the national limit. For chromium contaminated soil, the concentration was less than or equal to 5000mg/kg, which conform to the limit standard. When the concentration was 10000mg/kg, the content exceeded the limit. Therefore, heavy metal lead contaminated soil is more easily stabilized than chromium contaminated soil. Microscopic tests show that OPC solidified heavy metal contaminated soil is solidified/stabilized by heavy metal ions through hydration products, namely, ettringite and amorphous cementing materials. MPC is obtained by the hydration reaction of magnesium oxide and ammonium dihydrogen phosphate to form struvite, and heavy metal ions are encapsulated in the lattice to achieve solidification/stabilization.

Keywords: heavy metal; normal Portland cement; magnesium phosphate cement; solidification/stabilization; leaching toxicity

1. Introduction

Although the rapid development of heavy chemical industry can significantly improve the country's economy, it will also have a bad impact on our lives. Among them, heavy metals cause great harm to our ecological environment and health, such as air pollution[1,2], groundwater pollution[3], and soil pollution[4]. Heavy metals not only reduce the yield and quality of crops, indirectly affecting human health, but also change the soil properties, reduce the shear strength and bearing capacity, cause foundation settlement, and cause structural damage. Given the large amount of harmful sludge produced in the industrial wastewater treatment process, including electroplating sludge containing heavy metals lead and chromium, oily sludge, and printing and dyeing sludge, to treat the heavy metals in the sludge, scholars have treated contaminated soil through physical remediation[5,6], chemical remediation [7–9], bioremediation [10–14], and solidification/stabilization remediation [15–19]. Among these treatments, solidification/stabilization technology has the advantages of low cost and simple operation and has become an effective method for treating heavy metal contaminated soil[20]. This method mainly produces precipitates through the chemical reaction between the cement curing agent and the heavy metal contaminated soil and fixes the heavy metals in the cement soil. To prevent the diffusion of heavy metals to the surrounding soil, reduce the leaching toxicity of ions, and achieve the purpose of stabilizing heavy metal contaminated soil[21]. Zha Fusheng et al.[22] used ordinary Portland cement (OPC) to solidify heavy metal Pb and Zn contaminated soil. They observed the obvious solidification effect of OPC on heavy metal contaminated soil and the improved leaching characteristics and strength of the cement soil with the increase of the cement content. Liang Shihua et al.[23] treated a contaminated soil by mixing cement with slag and observed the good effect of the treatment on the stabilization of zinc contaminated sludge with a concentration of 5000 mg/kg and below. Chen Lei et

* Corresponding author: dede_de123@163.com

https://doi.org/10.1051/matecconf/202338201008

al.[24,25] found that the unconfined compressive strength of cement solidified lead contaminated soil increased with the cement content, and the critical concentration was reached, which was related to the cement content and age. Li Xilin et al.[26] found that OPC is an effective curing agent for solidified contaminated soil by studying the contaminated soil of chromium slag yard, and the optimum dosage is 20%. Xiong Lu et al.[27,28] used OPC to solidify chromium contaminated soil. They found that the compressive strength of the contaminated soil increased with the cement content and age and decreased with the increase of the ion concentration and the freeze-thaw cycles. Leaching experiments show that the leaching rate of metal ions decrease gradually with the increase of curing agent content and age, and the curing effect is improved. Luo Yunlong et al.[29] found that the heavy metals treated by magnesium phosphate cement (MPC) mainly include Pb, Ni, Cu, Zn, Cr, Mn and Hg. The main solidification mechanisms include chemical bonding, adsorption and physical encapsulation. Zhang Tingting et al.[30,31] studied the solidification of lead contaminated soil using MPC and found that the water–soil ratio had a certain influence on the strength and permeability coefficient of the solidified body of contaminated soil, and the critical value was 0.45. The compressive strength and solidification effect of the solidified body improved and the permeability coefficient decreased with the increase of the MPC content. Hou Shiwei et al.[32–34] used MPC to study the strength characteristics and freeze–thaw stability of heavy metal contaminated soil. They found that MPC has a significant effect on the solidification of heavy metal contaminated soil. However, the compressive strength gradually decreased with the increase of the ion concentration. An extremely high concentration would decrease of the solidification effect, and the compressive strength would decrease with the decrease of the MPC cement content and the increase of the freeze–thaw cycles. Zha Fusheng et al.[35] think that different kinds of heavy metals will affect the compressive strength of solidified soil. Generally, heavy metal ions will reduce the strength of the solidified body, but the compressive strength will continue to increase with the cement content and age. Yin et al.[36] believe that a small amount of Pb^{2+} could improve the compressive strength of contaminated soil.

The above research, revealed that OPC or MPC has a good effect on the stabilization of heavy metal contaminated soil. Therefore, this paper compares and studies heavy metal lead and chromium contaminated soil under different types of cement, cement contents, ion concentrations and ages, and uses macro and micro means to analyze the strength characteristics and curing mechanism providing theoretical and data support for the solidification/stabilization remediation of heavy metal contaminated soil and a basis for the resource utilization of contaminated soil in the future[37].

2. Test Sample Preparation and Scheme

2.1 Test Sample Preparation

The soil used for the test was taken from the muddy soil of the Nansha District of Guangzhou City. The soil properties were as follows: the water content was 52%, the organic matter was 4.49%, and the liquid and plastic limits were 51% and 32.3%, respectively. The basic physical and mechanical properties of the soil samples were measured using the geotechnical test method (GB/T50123-1999), as shown in Table 1. The Shijing OPC and MPC produced in Guizhou were used as the cement materials. MPC is made by mixing magnesium oxide (MgO, represented by M) and ammonium dihydrogen phosphate ($NH_4H_2PO_4$, represented by P) at 1:1 ratio. The heavy metal pollutants came from Aladdin brand lead nitrate (analytically pure) and McLean brand chromium nitrate (analytically pure). Tap water was used in the testing.

Contaminated soil samples were prepared as follows. (1) The retrieved soil was placed in an oven at a temperature of 105°C for drying to constant weight, crushed with a ball mill, passed through a 2 mm sieve, and then set aside. (2) Tap water with 50% dry soil mass, was weighed, and pollutants were added at different concentrations ($Pb(NO_3)_2$ and $Cr(NO_3)_3$); the solution was stirred evenly, added to the soil for mixing, and set aside; (3) Water and cement at 0.5 ratio and a cement curing agent were added into the prepared contaminated soil, and the mixture was place it a mortar mixer for slow mixing for 2 min and fast mixing for 2 min; (4) The cylindrical mold with a diameter of 39.1 mm and a height of 80 mm was adopted, and Vaseline was uniformly smeared on the inner wall of the cylindrical mold; (5) The mixed slurry was evenly poured into the cylindrical mold in three layers, and a vibration table was used to vibrate and compact each layer of mixed slurry, vibrate out the bubbles, and smoothen the surface; the sample was sealed and place in a standard curing box for curing. The curing temperature was 20 °C, and the humidity exceeded 95%.

Table 1. Basic physical and mechanical properties of muddy soil

Water Content/%	Density /(g/cm3)	Liquid Limit/%	Plastic Limit/%	Liquidity Index/-	Plasticity Limit Index/%	Maximum dry density/(g/cm3)	Organic matter/%
52	1.74	51	32.3	0.813	18.7	0.482	4.49

2.2 Test Scheme and Method

In this paper, different cement types and admixtures, heavy metal pollutants, ion concentrations and ages were used to carry out solidification tests on the contaminated soil samples. Through unconfined compressive strength and leaching toxicity tests, the influence of different factors on the strength of heavy metal contaminated soil was analyzed. The details are presented in Table 2:

Table 2. Test plan for solidification of heavy metal lead and chromium contaminated soil with ordinary Portland cement and magnesium phosphate cement

Types of heavy metals/-	Cement type/-	Cement content /%	Ion concentration/mg/kg	Age/d
Cr	OPC MPC	20% 30% 40%	0 1000 3000 5000 10000	7 14 28
Pd	OPC MPC	20% 30% 40%	0 1000 3000 5000 10000	

Unconfined compressive strength (UCS): A yaw-50d microcomputer controlled constant pressure tester is selected, the loading rate is set at 1.6 mm/min, and the average value of three parallel samples is taken as the result.

Leaching test: The toxic leaching method is used for testing. The toxic leaching test is carried out according to HJT300-2007 "Solid waste - Extraction procedure for leaching toxicity-acetic acid buffer solution method." The specific process determines the pH value of the test sample, selects a suitable leaching solution according to the pH value, passes the air-dried soil samples through 9.5 mm sieve, places dry soil and the leaching solution a bottle at a solid–liquid ratio of 1:20 in a constant temperature oscillator, at the rotation speed of 30 ± 2 r/min, shakes the samples at 23 ± 2 °C for 18 ± 2 h, and passes the leaching solution through 0.45 μ. The concentration of Pb^{2+} and Cr^{3+} in the leaching solution is measured by inductively coupled plasma mass spectrometry (ICAP RQ) to determine whether it is within the leaching toxicity limit.

Scanning electron microscopy (SEM) test: After the UCS test, the samples are air dried at low temperature (60 ± 3 °C) for 24 h, and the representative samples are sprayed with gold and photographed with corresponding multiples. Energy dispersive spectrometry (EDS) is often used in conjunction with SEM to scan the point, line and surface morphology and analyze the composition of target parts.

X - ray Diffractometer (XRD) test: The aged samples are placed in an oven for drying treatment, ground and passed through a 0.075 mm sieve, and placed in the sample box for testing with a diffractometer.

3. Test Results and Discussion

3.1 Unconfined Compression Strength Test Results and Analysis

The strength regularity results of OPC and MPC solidified heavy metal contaminated soil are shown in Figures 1 to 4.

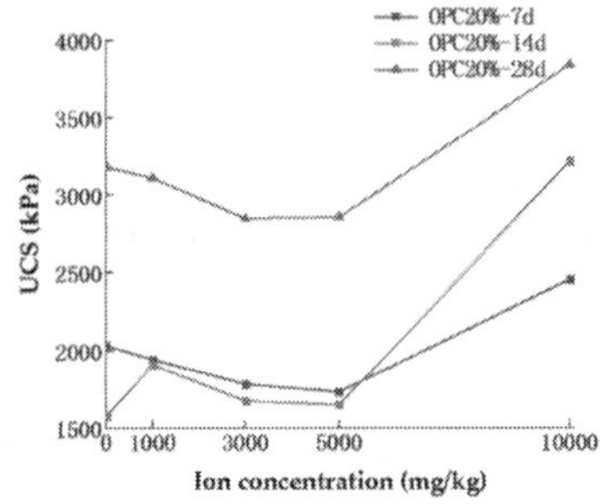

(a) OPC content: 20%

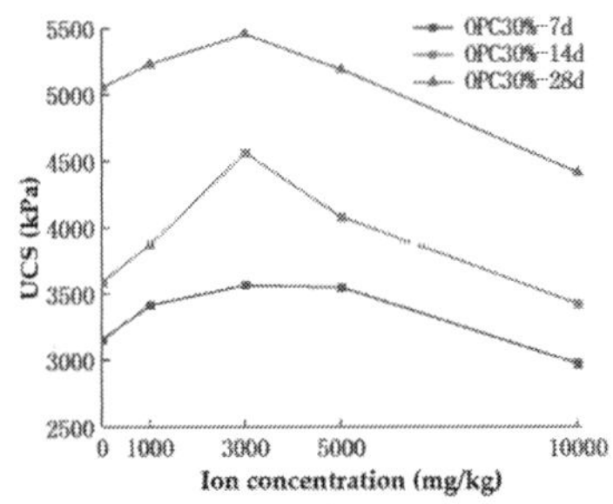

(b) OPC content: 30%

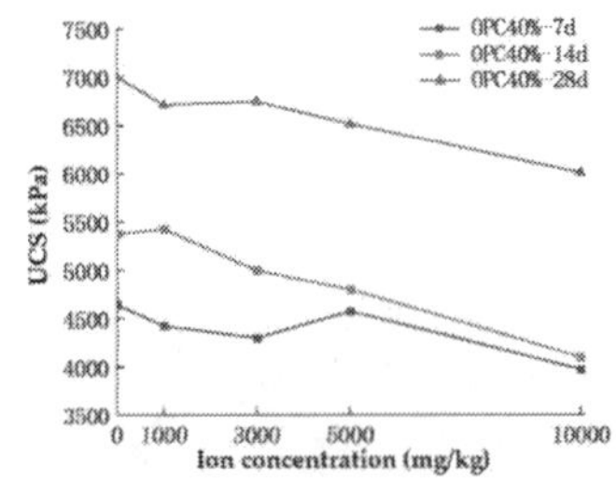

(c) OPC content: 40%

Figure 1. Compressive strength regularity of OPC solidified chromium contaminated soil

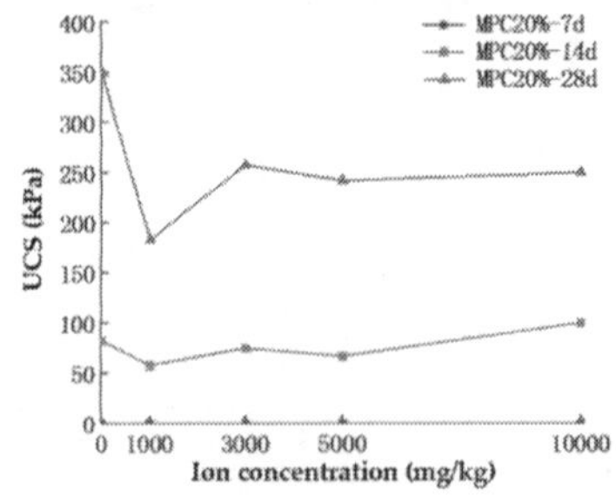

(a) MPC content: 20%

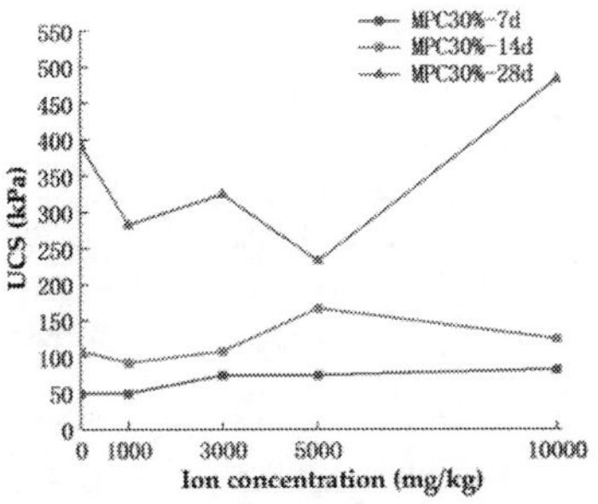

(b) MPC content: 30%

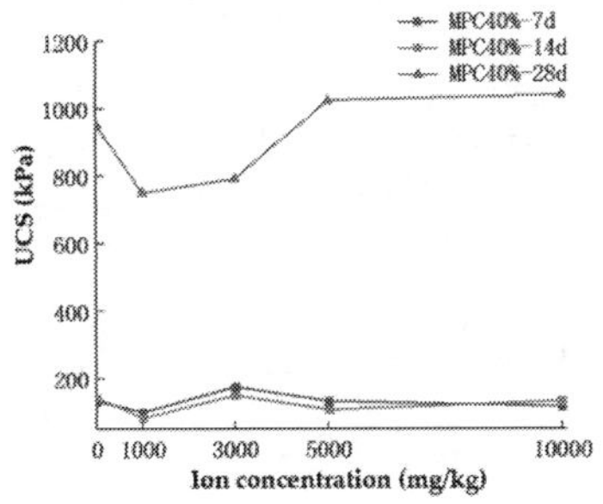

(c) MPC content: 40%

Figure 2. Compressive strength regularity of MPC solidified chromium contaminated soil

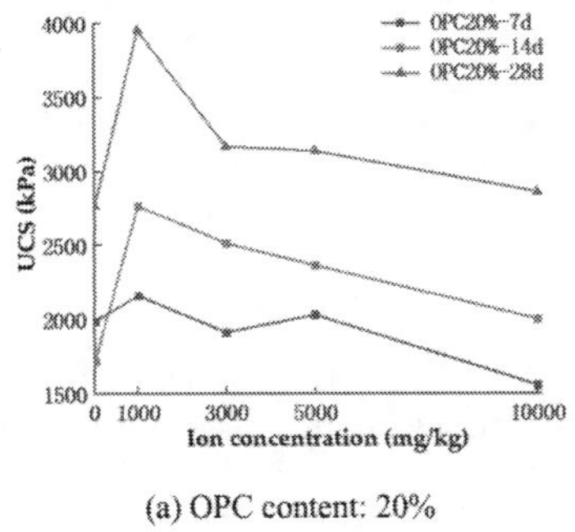

(a) OPC content: 20%

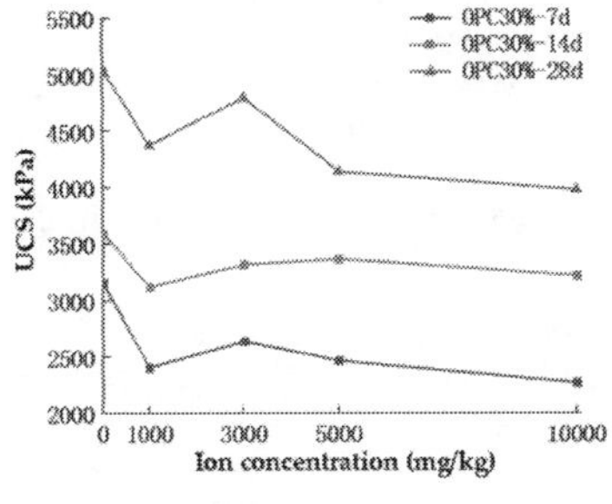

(b) OPC content: 30%

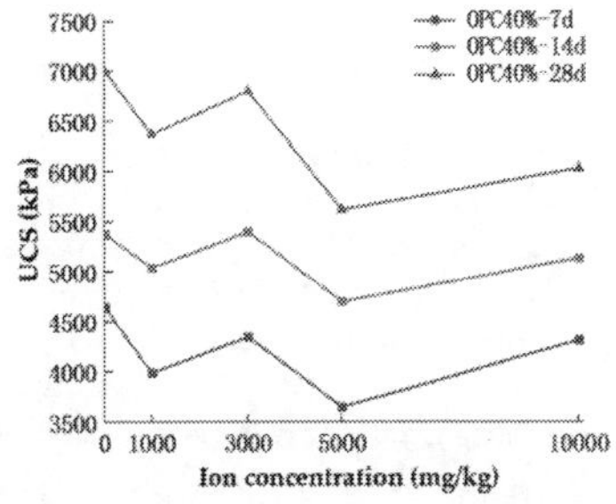

(c) OPC content: 40%

Figure 3. Compressive strength regularity of OPC solidified lead contaminated soil

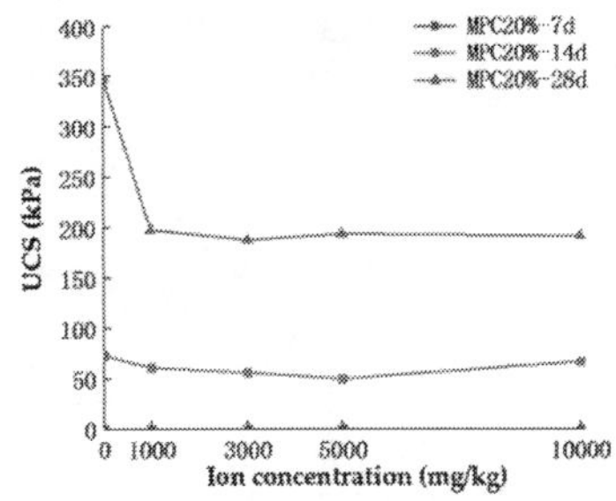

(a) MPC content: 20%

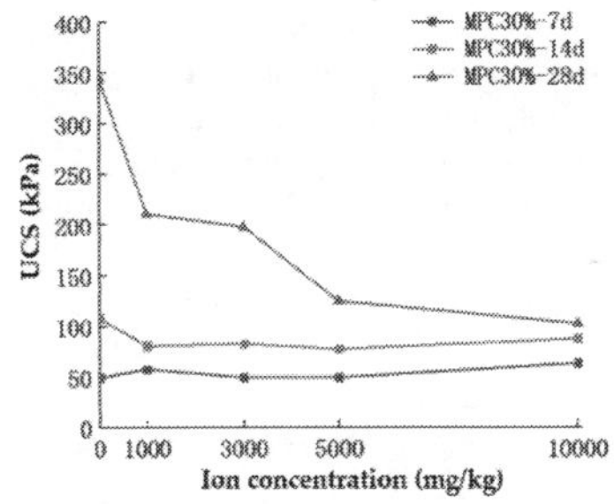

(b) MPC content: 30%

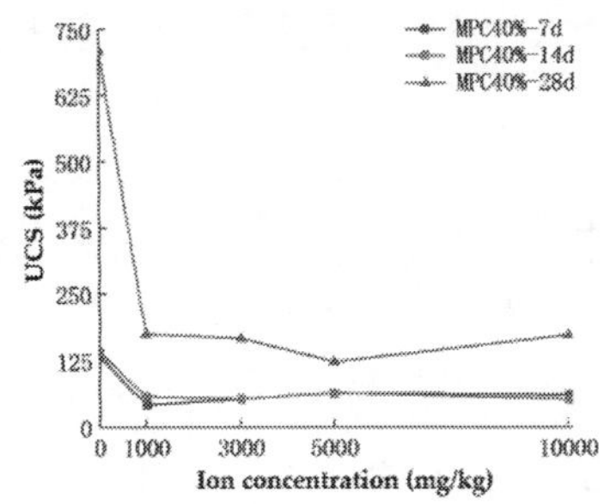

(c) MPC content: 40%

Figure 4. Compressive strength regularity of MPC solidified lead contaminated soil

3.1.1 Effect of Curing Age on Compressive Strength of Heavy Metal Contaminated Soil

The influence of OPC and MPC on the compressive strength of heavy metal contaminated soil at different ages is shown in Figures 1 to 4. The compressive strength of OPC and MPC solidified heavy metal lead and chromium contaminated soil is positively related to the curing age. The growth rate of OPC solidified contaminated soil is fast in the early stage, and the relative growth rate is slow in the later stage[32]. When the OPC content is 30% and the ion concentration is 3000 mg/kg, the compressive strength of Cr contaminated soil at 14 days is 1.28 times that at 7 days, and that at 28 days is 1.53 times that at 7 days. The compressive strength of lead contaminated soil at 14 days is 1.26 times that at 7 days, and that at 28 days is 1.81 times that at 7 days. The MPC content is also 30%. When the ion concentration is 3000 mg/kg, the compressive strength of chromium contaminated soil at 14 days is 1.44 times that at 7 days, and that at 28 days is 4.33 times of that at 7 days. The compressive strength of lead contaminated soil at 14 days is 1.66 times that at 7 days, and that at 28 days is 3.96 times that at 7 days. The test shows that age is an important factor affecting the compressive strength of heavy metal contaminated soil. Under the same conditions, the compressive strength of OPC solidified heavy metal contaminated soil is higher than that of MPC. Although MPC has a good early strength and fast hardening, when MPC is used to solidify heavy metal soil, it does not give full play to its own characteristics, because the hydration and hydrolysis reactions of OPC and MPC take time. When the curing agent is mixed into the heavy metal contaminated soil, the mineral material reacts with the water in the soil to generate a small amount of hydration products in a relatively short time, resulting in the poor cohesion of the soil particles and low strength in the early stage. Over time the hydration and hydrolysis reactions of the cement continue, the products gradually increase, and the hardening and ion exchange reactions occur. The products of these reactions can fill the pores of contaminated soil, reduce the porosity of soil, and increase the cohesion between soil particles, to form a tight whole to improve the strength of the solidified body[38,39].

3.1.2 Influence of Cement Content on Compressive Strength of Heavy Metal Contaminated Soil

The effect of the cement content on the compressive strength of the OPC and MPC cured heavy metal contaminated soil is shown in Figures 1 to 4. The compressive strength of the OPC and MPC cured heavy metal lead and chromium contaminated soil increases with the cement content. At 28 days of curing and 3000 mg/kg ion concentration, when OPC was used to solidify the chromium contaminated soil, the compressive strength of 30% cement content is 1.92 times that of 20%, and that of 40% is 2.37 times that of 20%. When MPC solidified the chromium contaminated soil, the compressive strength of 30% cement content is 1.26 times that of 20%, and that of 40% cement is 3.07 times that of 20%. When OPC solidified the lead contaminated soil, the compressive strength of 30% cement content is 1.51 times that of 20%, and that of 40% is 2.15 times that of 20%. When MPC solidified the lead contaminated soil, the compressive strength of 30% cement content is 1.05 times that of 20%, and that of 40% cement is 0.89 times that of 20%. The results show that OPC and MPC have obvious effects on the solidification of heavy metal lead and chromium contaminated soil, and the compressive strength of heavy metal contaminated soil gradually increases with the OPC content. When MPC solidified the chromium contaminated soil, the compressive strength increased with the cement content, and when MPC solidified the lead contaminated soil, the strength decreased with the increase of the cement content, indicating that the type of heavy metal also affects the strength of contaminated soil. The phenomenon probably occurred because the alkalinity of contaminated soil was continuously enhanced and the solubility of the silicon and aluminum in the cement and the contaminated soil was greatly increased with the increase of the OPC content at the same curing age, generating more hydrated calcium silicate and aluminate, which improved the strength of contaminated soil[40]. With the increase of MPC content, large numbers of hydration products produced by MPC improve the cohesion between contaminated soil particles, and the hydration products fill the pores between soil particles, making the contaminated soil a dense whole. Zhang Tingting[31,38] believes that when MPC solidifies lead contaminated soil, the strength will also increase with the cement content, because the more intense the physical and chemical reactions between MPC and the contaminated soil, the greater the amount of hydrated products that can be generated by the hardening, ion exchange, and hydration and hydrolysis reactions, filling the pores of contaminated soil, reducing the pore ratio, and improving the strength of contaminated soil. However, the critical value is 500mg/kg for the hydration reaction of the MPC solidification of lead contaminated soil. When the Pd content is low, it can promote the hydration reaction of cement and generate an amorphous colloidal precipitate with cementation ability. This precipitate can enhance the cementation between soil particles and has a compact structure, which improves the compressive strength of contaminated soil. However, the increase of the Pb content, does not promote the hydration reaction, but hinders the cement hydration reaction and reduces the generation of hydration products and the curing effect. Pb reacts with OH^- to form lead sulfite $[Pb(OH)_4]^{2-}$, weakening the cementation between soil particles and reducing the strength of solidified soil.

3.1.3 Influence of Ion Species and Concentration on Compressive Strength of Heavy Metal Contaminated Soil

The effects of the ion type and concentration on the compressive strength of OPC and MPC solidified heavy metal contaminated soil are shown in Figures 1 to 4. The relationship between the unconfined compressive strength of contaminated soil and the cement content is non-linear but related to the ion concentration. Under the curing age of 28 days, Figure 1 shows that when the OPC content is 20%, the strength of solidified chromium contaminated soil tends to decrease first and then increase with the increase of the ion concentration, and the minimum value appears at the concentration of 5000 mg/kg. Figure 3 reveals that when 20% OPC is used to solidify lead contaminated soil, the strength tends to increase first and then decrease with the increase of the ion concentration, and the maximum value appears at the concentration of 1000mg/kg. Figures 2 and 4 indicate that when MPC with 20% content solidifies chromium and lead contaminated soils, the strength decreases first and then tends to flatten with the increase of the ion concentration, and the minimum value is 1000mg/kg. The compressive strength of lead contaminated soil at this point is higher than that of chromium contaminated soil. The results show that the ion species and concentration affect the compressive strength of heavy metal contaminated soil. Luo Yunlong[29] et al. found that when MPC solidifies Pb, Cu, Cd, Ni, and other heavy metals, different kinds of heavy metals present different curing effects. Under the same conditions, the presence of Zn reduces the compressive strength of MPC, while Cu, Cd, and Ni increase the compressive strength of MPC, but all of them extend the setting time. Pd has the best effect when MPC solidifies Pb and Cd. In addition, interaction exists between different kinds of heavy metals. Under the same M/P, heavy metals are cured by MPC, and metal ions will react with phosphate to form the corresponding phosphate during the curing process. When the concentration of heavy metals treated by MPC increases, the addition of heavy metals will hinder the formation of struvite, decrease the setting rate of materials, and extend the setting time. However, the compressive strength of heavy metal contaminated soil will decrease, and the toxic leaching concentration will increase.

3.2 Leaching Characteristics of Cement Solidified Lead and Chromium Contaminated Soil

The leaching test results of OPC and MPC for curing different ion types and concentrations are shown in Table 3. The test shows that when OPC and MPC solidify heavy metal chromium contaminated soil, the effect of MPC is better than that of OPC when the original ion concentration is small (1000 mg/kg). As the original ion concentration gradually increases, the leaching concentration of OPC is lower than that of MPC. When OPC and MPC solidify chromium contaminated soil with the original ion concentration less than or equal to 5000 mg/kg, the leaching values all satisfy the national limits (the concentration limit of the total chromium leaching solution is 15 mg/L), but when the original concentration is 10000 mg/kg, the specified limit is not satisfied. OPC and MPC solidification of lead contaminated soil satisfies the national limit (the concentration limit of lead leaching solution is 5 mg/L) when the original ion concentration is less than or equal to 10000 mg/kg. The solidification/stabilization effect of MPC is better than that of OPC. The limits specified by the state shall be obtained in accordance with the identification standard for hazardous waste leaching toxicity (GB 5085.3-2007). To sum up, the types of heavy metals have a great impact on the leaching effect of cement solidified contaminated soil, while the types of cement have a small impact on the leaching effect. Furthermore, lead contaminated soil is more stable than chromium contaminated soil.

Table 3. Leaching toxicity results of the types and concentrations of cement solidified heavy metal ions

Cement type	Types of heavy metals	Original ion concentration/mg/kg	Leaching concentration/mg/L	Types of heavy metals	Original ion concentration/mg/kg	Leaching concentration/mg/L
OPC	Cr	1000	3.167	Pd	1000	0.288
		3000	7.286		3000	0.269
		5000	10.312		5000	0.282
		10000	20.513		10000	0.303
MPC	Cr	1000	3.031	Pd	1000	0.161
		3000	7.996		3000	0.162
		5000	11.229		5000	0.177
		10000	22.422		10000	0.224

https://doi.org/10.1051/matecconf/202338201008

3.3 Micro Test Results and Analysis

3.3.1 Scanning Electron Microscopy Test Analysis

The micro test on heavy metal lead and chromium contaminated soil was carried out through SEM, and the results are shown in Figure 5.

(a)

(b)

(c)

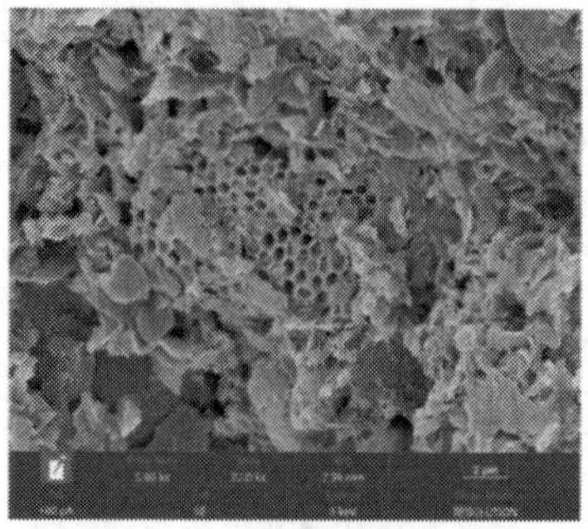

(d)

(e)

(f)

Figure 5. SEM of OPC and MPC solidified heavy metal contaminated soil

(a) OPC solidified contaminated soil (concentration of 0); (b) OPC solidified chromium contaminated soil (concentration: 5000 mg/kg); (c) OPC solidified lead contaminated soil (concentration 10000 mg/kg); (d) MPC solidified chromium contaminated soil (concentration 1000 mg/kg); (e) MPC solidified chromium contaminated soil (concentration: 5000 mg/kg); (f) MPC solidified lead contaminated soil (concentration: 5000 mg/kg).

Figure 5(a) shows that when OPC solidified unpolluted soil (ion concentration is 0), large amounts of needle rod crystal ettringite (AFt) and amorphous calcium silicate hydrate (C-S-H) were generated, filling the pore and forming a dense whole. Figures (5) b – c show that with the addition of heavy metals, chromium contaminated soil (with a concentration of 5000 mg/kg) and lead contaminated soil (with a concentration of 10000 mg/kg) will also produce a small amount of AFt with the increase of the ion concentration, but the amount is not enough to fill the pores. Thus, many pores will remain in the contaminated soil, resulting in a decrease in the strength of the solidified body. In addition, when OPC solidifies the uncontaminated soil, cementitious substances can be generated between minerals and soil particles, making cure soil cohesive and reducing the generation of pores. When OPC solidifies heavy metal chromium and lead contaminated soils, heavy metal ions will hinder the production of cement, resulting in the absence of cohesion between soil particles and the loose state of the soil. The unconfined compressive strength test results also show that the strength of the OPC solidified unpolluted soil is higher than that of the heavy metal contaminated soil. Figures (5) d – f show that the MPC solidification of heavy metal contaminated soil can be explained from the following four aspects[25]:

(1) Chemical bonding mode[41–44]: Mg^{2+} reacts with NH_4^+ and HPO_4^{2-}, $H_2PO_4^-$ and PO_4^{3-} to form amorphous or crystalline phosphate hydration products ($MgNH_4PO_4 \cdot 6H_2O$), which is commonly known as struvite. The crystals are usually equiaxed or wedge-shaped, short column shaped, and thick plate-shaped. The specific reaction equation is shown in Formula 1:

$$Mg^{2+} + PO_4^{3-} + 6H_2O + NH_4^+ \rightarrow MgNH_4PO_4 \cdot 6H_2O\downarrow \quad (1)$$

During the hydration of MPC, heavy metal ions participate in chemical bonding reactions to generate insoluble phosphates such as $Pb_3(PO_4)_2 \cdot nH_2O$.

(2) The chemical interaction between phosphate and heavy metal ions is interpreted as the adsorption mechanism of heavy metal ions[45]. The chemical action includes the ion exchange on the surface[46]. The product recovered by chemical adsorption forms a phosphate precipitate containing metal[47], during the recrystallization process, other metal substitution [48] and the combination of functional groups and heavy metals on its surface[49].

(3) Physical encapsulation: the main hydrated product of MPC is struvite. The crystal is plate-shaped, the surface is relatively flat, and the connected with each other and form a dense structure[50]. Struvite combines with a large amount of unreacted MgO and exhibits a good encapsulation performance.

(4) Synergistic action mode: The solidification mechanism of the MPC solidification/stabilization of heavy metals is chemical bonding, adsorption, and physical encapsulation. The process is that free metal ions (Pb^{2+}) react first and form low dissolved $Pb_3(PO_4)_2$, and then fix heavy metal ions in the soil through physical adsorption and encapsulation through $MgNH_4PO_4 \cdot 6H_2O$ and the excess MgO.

3.3.2 X - ray Diffractometer and Energy Dispersive Spectrometer Test Analysis

The XRD of OPC and MPC solidified heavy metal contaminated soils is shown in Figure 6. Through EDS, we can obtain: the elemental composition of OPC and MPC solidified heavy metal contaminated soils as shown in Table 4. As shown in Figure 6 (a): SEM reveals that the OPC solidified heavy metal contaminated soil produces aft and amorphous gel. The EDS energy spectrum is used for spot scanning. The main elements are O, C, Si, Al, Ca, and Fe. The heavy metal ions and the hydration products are fixed in the crystal together, preventing leaching into the environment. As shown in Figure 6 (b): SEM also reveals that struvite is produced in MPC solidified heavy metal contaminated soil. XRD and EDS, show that in addition to struvite, a hydrated product, MgO has not completed the hydration reaction, indicating that the hydration of cement is incomplete. The main elements of the energy spectrum analysis are: O, Si, Al, C, Mg, Fe, K, and P. Figure 5(f) shows that the crystal shape of the sheet-like struvite is poor due to the incorporation of heavy metals. Ions will also hinder the production of hydration products, loosening the structure[41,51,52].

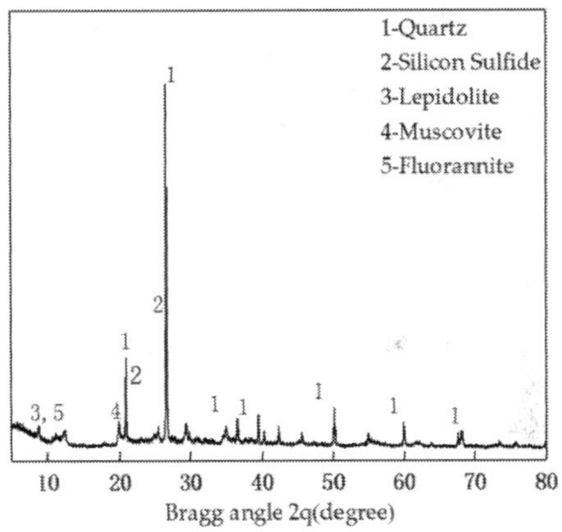

(a) OPC solidification of contaminated soil

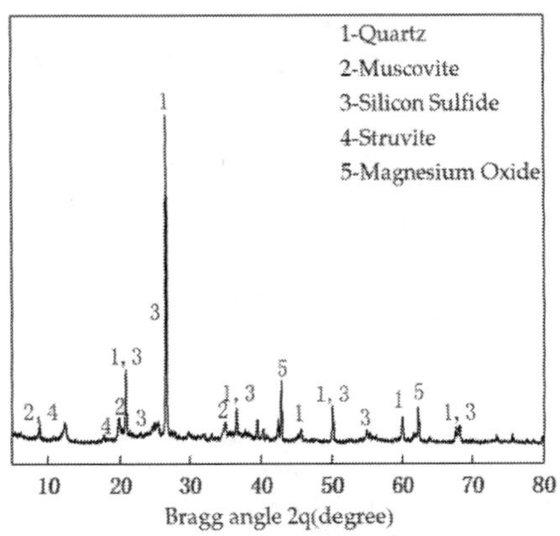

(b) MPC solidification of contaminated soil

Figure 6. XRD pattern of OPC and MPC solidified heavy metal contaminated soil

Table 4. Element composition of OPC and MPC solidified heavy metal contaminated soil

Cement type	element	W_t/%	A_t/%	Cement type	element	W_t/%	A_t/%
OPC	C	9.58	15.77	MPC	C	5.17	9.25
	O	44.16	54.57		O	35.28	47.39
	Na	0.51	0.44		Na	0.36	0.34
	Mg	0.88	0.72		Mg	8.99	7.95
	Al	12.04	8.82		Al	14.17	11.28
	Si	18.56	13.06		Si	21.14	16.17
	K	1.85	0.93		P	2.71	1.88
	Ca	9.06	4.47		S	0.72	0.48
	Ti	0.31	0.13		Cl	0.44	0.27
	Fe	3.06	1.08		K	3.50	1.92
					Ca	0.78	0.42
					Ti	0.85	0.38
					Fe	5.88	2.26
Total/%		100%	100%	Total/%		100%	100%

Where:
Wt/%: Weight percentage
At/%: Atomic percentage

4. Conclusions

(1) When OPC and MPC solidify different kinds of heavy metal contaminated soil, and the compressive strength will increase with the age and cement content. The UCS test shows that the strength of OPC solidified heavy metal contaminated soil is higher than that of MPC solidified heavy metal contaminated soil.

(2) When the same cement solidifies different ion concentration of heavy metal, different laws of change will be observed. Some ions will promote the hydration reaction and improve the strength of the contaminated soil when the concentration is low. The compressive strength continues to decrease with the increase of the ion concentration. Many studies have found that lead ions can promote the formation of hydration reactions at low concentrations.

(3) Under the same cement content, different cements were used to solidify the heavy metal lead and chromium contaminated soils. The heavy metal ions in the lead contaminated soil were more likely to be solidified by the hydration products of cement. When different cements are used to solidify the same heavy metal, the compressive strength of OPC is better than that of MPC for chromium contaminated soil, for lead contaminated soil, MPC has better stability than OPC. When the ion concentration is below 5000mg/kg, the leaching solution concentration of heavy metal lead and chromium contaminated soil meets the national standard.

(4) SEM, XRD, EDS, and other microscopic tests reveal that the solidification of heavy metal contaminated soil by OPC mainly uses needle-like ettringite and amorphous cementitious materials to solidify and wrap up heavy metal ions. MPC is used to solidify contaminated soil by hydrating magnesium oxide and ammonium dihydrogen phosphate to produce flaky struvite, which solidifies heavy metal ions and reduces the ion content in the leaching solution.

In this paper, OPC and MPC were used to solidify heavy metal lead and chromium contaminated soil. The strength and toxic leaching characteristics of heavy metal contaminated soil were analyzed and studied through UCS test, SEM test and leaching toxicity test. However, there is no systematic study on the follow-up tests of contaminated soil, such as the influence of the combined action of two heavy metals on the strength of contaminated soil, and the water stability and dry wet cycle tests are also lacking. In addition, the influence degree of each factor can be calculated through orthogonal test and range analysis. The solidified heavy metal contaminated soil shall be evaluated whether it can meet the strength requirements and be used as construction soil, such as subgrade and pipeline landfill.

Acknowledgements

This research was funded by the National Natural Science Foundation of China, grant number No. 52078142, Natural Science Foundation of Guangdong Province, grant number No. 2022A1515011047, Science and Technology Program of Guangzhou, China, grant number No. 202002030194.

References

1. Bing, H.; Wu, Y.; Zhou, J.; Ming, L.; Sun, S.; Li, X. Atmospheric Deposition of Lead in Remote High Mountain of Eastern Tibetan Plateau, China. Atmospheric Environment 2014, 99, 425–435, doi:10.1016/j.atmosenv.2014.10.014.

2. Guan, P.; Zhou, Y. Cheng, S.; Duan, W.; Yao, S. Li, J.; Yue, L. Characteristics of heavy pollution process and source appointment in typical heavy industry cities, China. China Environmental Science 2020, 40, 31–40.

3. Miao, Y.; Kong, X.; Li, C. Distribution and Sources of Polycyclic Aromatic Hydrocarbons in a Karst Groundwater System in a Strongly Industrial City, China. Environmental Science 2019, 40, 239–247, doi:10.13227/j.hjkx.201805212.
4. Guo, C.; Zhang, H.; Yang, G. Mercury Pollution and Its Control of Agriculture Eco-System in Heavy Industry City, China. Journal of Agro-Environment Science 2000, 245–247, 4, doi:10.3321/j.issn:1672-2043.2000.04.016.
5. Zhao, C.; Dong, Y.; Feng, Y.; Li, Y.; Dong, Y. Thermal Desorption for Remediation of Contaminated Soil: A Review. Chemosphere 2019, 221, 841–855, doi:10.1016/j.chemosphere.2019.01.079.
6. Wen, D.; Fu, R.; Li, Q. Removal of Inorganic Contaminants in Soil by Electrokinetic Remediation Technologies: A Review. Journal of Hazardous Materials 2021, 401, 123345, doi:10.1016/j.jhazmat.2020.123345.
7. Hu, W.; Niu, Y.; Zhu, H.; Dong, K.; Wang, D.; Liu, F. Remediation of Zinc-Contaminated Soils by Using the Two-Step Washing with Citric Acid and Water-Soluble Chitosan. Chemosphere 2021, 282, 131092, doi:10.1016/j.chemosphere.2021.131092.
8. Cheng, S.; Lin, Q.; Wang, Y.; Luo, H.; Huang, Z.; Fu, H.; Chen, H.; Xiao, R. The Removal of Cu, Ni, and Zn in Industrial Soil by Washing with EDTA-Organic Acids. Arabian Journal of Chemistry 2020, 13, 5160–5170, doi:10.1016/j.arabjc.2020.02.015.
9. Song, P.; Xu, D.; Yue, J.; Ma, Y.; Dong, S.; Feng, J. Recent Advances in Soil Remediation Technology for Heavy Metal Contaminated Sites: A Critical Review. Science of The Total Environment 2022, 838, 156417, doi:10.1016/j.scitotenv.2022.156417.
10. Azhar, U.; Ahmad, H.; Shafqat, H.; Babar, M.; Shahzad Munir, H.M.; Sagir, M.; Arif, M.; Hassan, A.; Rachmadona, N.; Rajendran, S.; et al. Remediation Techniques for Elimination of Heavy Metal Pollutants from Soil: A Review. Environmental Research 2022, 214, 113918, doi:10.1016/j.envres.2022.113918.
11. Irfan, A.; Amtul, M. Halophytes for Phytoremediation of Hazardous Metal (Loid)s: A Terse Review on Metal Tolerance, Bio-Indication and Hyperaccumulation. Journal of Hazardous Materials 2022, 424, 127309, doi:10.1016/j.jhazmat.2021.127309
12. Shah, V.; Daverey, A. Phytoremediation: A Multidisciplinary Approach to Clean up Heavy Metal Contaminated Soil. Environmental Technology & Innovation 2020, 18, 100774, doi:10.1016/j.eti.2020.100774.
13. Li, C.; Tian, L.; Dong, C.; Zhang, Y.; Wang, Y. Experimental study on zinc-lead composite contaminated soil solidified/stabilized by MICP technology combined with porous silicon adsorption materials. Rock and Soil Mechanics 2022, 43, 307–316, doi:10.16285/j.rsm.2021.1403.
14. Shan, Y.; Liang, J.; Tong, H.; Yuan, J.; Zhao, J. Effect of Different Fibers on Small-Strain Dynamic Properties of Microbially Induced Calcite Precipitation-Fiber Combined Reinforced Calcareous Sand. Constr. Build. Mater. 2022, 322, 126343, doi:10.1016/j.conbuildmat.2022.126343.
15. Mu, B.; Liu, X.; Nian, L.; Li, L.; Yang, Y. A Bibliometric Analysis of Research Trends in Soil Remediation from 2011 to 2020. Chinese Agricultural Science Bulletin 2022, 38, 143–151.
16. Huang, Z.; Zhao, P.; Wang, Y.; Ma, Y.; Liu, X. Progress in basic research & development and its application on solidification and stabilization materials of heavy metals in soil. Journal of Agricultural Resources and Environment 2022, 39, 435–445.
17. 17. Li, W.; Ni, P.; Yi, Y. Comparison of Reactive Magnesia, Quick Lime, and Ordinary Portland Cement for Stabilization/Solidification of Heavy Metal-Contaminated Soils. Science of the Total Environment 2019, 671, 741–753, doi:10.1016/j.scitotenv.2019.03.270
18. 18. Shan, Y.; Wang, X.; Cui, J.; Mo, H.; Li, Y. Effects of Clay Mineral Composition on the Dynamic Properties and Fabric of Artificial Marine Clay. J. Mar. Sci. Eng. 2021, 9, 1216, doi:10.3390/jmse9111216.
19. Li, Y.; Li, J.; Cui, J.; Shan, Y.; Niu, Y. Experimental Study on Calcium Carbide Residue as a Combined Activator for Coal Gangue Geopolymer and Feasibility for Soil Stabilization. Constr. Build. Mater. 2021, 312, 125465, doi:10.1016/j.conbuildmat.2021.125465.
20. Chang, C.; Cao, H.; Tao, L.; Lv, Y.; Dong, M. Advances on Heavy Metal Stability and Reactivation for Soil After Solidification/Stabilization Remediation. Soils 2021, 53, 682–691, doi:10.13758/j.cnki.tr.2021.04.003.
21. Liu, X.; Luo, S. Study on heavy metals in MSWI fly ash by solidification stabilization technology. Applied Chemical Industry 2022, 51, 816–820, doi:10.16581/j.cnki.issn1671-3206.20220128.019.
22. Zha, F.; Liu, J.; Xu, L.; Cui, K. Cyclic wetting and drying tests on heavy metal contaminated soils solidified/stabilized by cement Chinese Journal of Geotechnical Engineering 2013, 35, 1246–1252.
23. Liang, S.; Niu, J.; Wang, M.; Liu, Y.; Yi, Y. Experimental study of zinc contaminated sludge solidification with cement and slag. Industrial Construction 2017, 47, 89–94.
24. Chen, L.; Du, Y.; Liu, S.; Jin, F. Experimental study of stress-strain properties of cement treated lead-contaminated soils. Rock and Soil Mechanics 2011, 32, 715–721, doi:10.16285/j.rsm.2011.03.017.
25. Chen, L.; Liu, S.; Du, Y.; Jin, F. Unconfined compressive strength properties of cement solidified/stabilized lead-contaminated soil. Chinese Journal of Geotechnical Engineering 2010, 32, 1898-1903.

26. Li, X.; Zhang, J.; Chen, D.; Sun, J.; Hou, Z.; Sun, B.; Wang, C. Strength and Leaching for Remediation of Chromium Contaminated Soil Using Cement. Bulletin of the Chinese Ceramic Society 2017, 36, 979-983+990.
27. Xiong, L. Research on Meachinacs and Frost Resistance of Cement Solidified Hexavalent Chromium Contaminated Soils. Inner Mongolia Agricultural University 2018.
28. Xiong, L.; Shen, X.; Xue, H.; Li, G.; Yuan, Q.; Zhou, Y. Experimental Research of Mechanical Properties and Microstructure of Cement Solidified Hexavalent Chromium Contaminated Soils. Journal of Ecology and Rural Environment 2018, 34, 946–954. doi:10.11934/j.issn.1673-4831.2018.10.012.
29. Luo, Y.; Mu, W.; Zhou, X.; Huang, J.; Luo, Z.; Ma, Y.; Wang, L.; Shao, Z. Research progress on solidification/stabilization mechanism and properties of heavy metals using magnesium phosphate cement. Environmental Chemistry 2021, 40, 3875–3886.
30. Zhang, T.; Li, J.; Wang, P.; Huang, Q.; Xue, Q. Experimental study of mechanical and microstructure properties of magnesium phosphate cement treated lead contaminated soils. Rock and Soil Mechanics 2016, 279–286, doi:10.16285/j.rsm.2016.S2.034.
31. Zhang, T.; Li, J.; Wang, P.; Li, Z. Experimental study of stress-strain properties of lead-contaminated soils treated by magnesium phosphate cement. Rock and Soil Mechanics 2016, 215–225, doi:10.16285/j.rsm.2016.S1.028.
32. Hou, S.; Zhang, F.; Gao, G.; Liu, X.; Ding, Z. Engineering Characteristics of Magnesium Phosphate Cement Solidified Copper Contaminated Soil. Science Technology and Engineering 2021, 21, 2105–2111.
33. Hou, S.; Zhang, Y.; Li, H.; Zhang, Y.; Zhang, Y.; Ding, Z. Experimental Study on Leaching Characteristics of Copper Cu-contaminated Soil Cured by Magnesium Phosphate Cement under Freeze-thaw Cycles. Science Technology and Engineering 2020, 20, 1993–1999.
34. Hou, S.; Zhang, H.; Yang, Z.; Zhang, Y.; Zhang, Y.; Ding, Z. Study on freeze-thaw stability of Cu-contaminated soil solidified by magnesium phosphate cement. Chinese Journal of Rock Mechanics and Engineering 2020, 3123–3129.
35. Zha, F.; Xu, L.; Cui, K. Strength characteristics of heavy metal contaminated soils stabilized/solidified by cement. Rock and Soil Mechanics 2012, 33, 652-656+664, doi:10.16285/j.rsm.2012.03.003.
36. Yin, C.-Y.; Mahmud, H.B.; Shaaban, M.G. Stabilization/Solidification of Lead-Contaminated Soil Using Cement and Rice Husk Ash. Journal of Hazardous Materials 2006, 137, 1758–1764, doi:10.1016/j.jhazmat.2006.05.013.
37. Agency, E. Review of Scientific Literature on the Use of Stabilisation/Solidification for the Treatment of Contaminated Soil, Solid Waste and Sludges. 2004.
38. Zhang, T.; Wang, P.; Li, J.; Wang, Y.; Xue, Q.; Wang, S. Effect of curing time and lead concentration on mechanical properties of lead-contaminated soils stabilized by magnesium phosphate cement. Rock and Soil Mechanics 2018, 39, 2115–2123, doi:10.16285/j.rsm.2016.1844.
39. Zhang, Y. Study on engineering characteristics of solidified/stabilized chromium-contaminated soil by soil stabilizer and biochar. Northwest A&F University, 2018.
40. Zhu, C.; Pan, L.; Yu, C.; Wang, H.; Jiang, J.; Liao, R. Mechanical properties and microstructure characteristics of cement solidification pollution soil from Wenzhou. Journal of Civil and Environmental Engineering 2018, 40, 117–123.
41. Du, Y.-J.; Wei, M.-L.; Reddy, K.R.; Jin, F.; Wu, H.-L.; Liu, Z.-B. New Phosphate-Based Binder for Stabilization of Soils Contaminated with Heavy Metals: Leaching, Strength and Microstructure Characterization. Journal of Environmental Management 2014, 146, 179–188, doi:10.1016/j.jenvman.2014.07.035.
42. Sanderson, P.; Naidu, R.; Bolan, N.; Lim, J.E.; Ok, Y.S. Chemical Stabilisation of Lead in Shooting Range Soils with Phosphate and Magnesium Oxide: Synchrotron Investigation. Journal of Hazardous Materials 2015, 299, 395–403, doi:10.1016/j.jhazmat.2015.06.056.
43. Feng, Y.-S.; Du, Y.-J.; Reddy, K.R.; Xia, W.-Y. Performance of Two Novel Binders to Stabilize Field Soil with Zinc and Chloride: Mechanical Properties, Leachability and Mechanisms Assessment. Construction and Building Materials 2018, 189, 1191–1199, doi:10.1016/j.conbuildmat.2018.09.072.
44. Wang, L.; Chen, L.; Guo, B.; Tsang, D.C.W.; Huang, L.; Ok, Y.S.; Mechtcherine, V. Red Mud-Enhanced Magnesium Phosphate Cement for Remediation of Pb and As Contaminated Soil. Journal of Hazardous Materials 2020, 400, 123317, doi:10.1016/j.jhazmat.2020.123317.
45. Dima, J.B.; Sequeiros, C.; Zaritzky, N.E. Hexavalent Chromium Removal in Contaminated Water Using Reticulated Chitosan Micro/Nanoparticles from Seafood Processing Wastes. Chemosphere 2015, 141, 100–111, doi:10.1016/j.chemosphere.2015.06.030.
46. Moghimi, F.; Jafari, A.H.; Yoozbashizadeh, H.; Askari, M. Adsorption Behavior of Sb(III) in Single and Binary Sb(III)—Fe(II) Systems on Cationic Ion Exchange Resin: Adsorption Equilibrium, Kinetic and Thermodynamic Aspects. Transactions of Nonferrous Metals Society of China 2020, 30, 236–248, doi:10.1016/S1003-6326(19)65195-2.
47. Angkawijaya, A.E.; Santoso, S.P.; Bundjaja, V.; Soetaredjo, F.E.; Gunarto, C.; Ayucitra, A.; Ju, Y.-H.; Go, A.W.; Ismadji, S. Studies on the Performance of Bentonite and Its Composite as Phosphate Adsorbent and Phosphate Supplementation for Plant. Journal of Hazardous Materials 2020, 399, 123130, doi:10.1016/j.jhazmat.2020.123130.

48. Gorski, C.A.; Fantle, M.S. Stable Mineral Recrystallization in Low Temperature Aqueous Systems: A Critical Review. Geochimica et Cosmochimica Acta 2017, 198, 439–465, doi:10.1016/j.gca.2016.11.013.
49. Wang, H.; Wang, X.; Ma, J.; Xia, P.; Zhao, J. Removal of Cadmium (II) from Aqueous Solution: A Comparative Study of Raw Attapulgite Clay and a Reusable Waste–Struvite/Attapulgite Obtained from Nutrient-Rich Wastewater. Journal of Hazardous Materials 2017, 329, 66–76, doi:10.1016/j.jhazmat.2017.01.025.
50. Lai, Z.; Lai, X.; Shi, J.; Lu, Z. Effect of Zn2+ on the Early Hydration Behavior of Potassium Phosphate Based Magnesium Phosphate Cement. Construction and Building Materials 2016, 129, 70–78, doi:10.1016/j.conbuildmat.2016.11.002.
51. Li, Y.; Fang, J. Study on the Solidification Performance and Mechanism of Magnesium Phosphate Cement with Volatile Heavy Metal Pb2 + and Cd2 +. Bulletin of the Chinese Ceramic Society 2019, 38, 901-904+917.
52. Wang, J.; Ma, B.; Li, X.; Li, H.; Tian, Z. The solidification and hydration products of magnesium phosphate cement with Pd2+, Zn2+ and Cu2+. Journal of Functional Materials 2014, 45, 5060–5064.

Research on Properties of Polypropylene/Flake Graphite Molded Composite Materials

Haijun Zhou [1], Mengmeng Zhou [2], Xiaolei Zhang [3,*], Lu Bai [1], Xiaoqi Chen [1], Fen Zhang [1], ChengYu Li [3], Yantao Li [1]

[1].Institute of Energy Source, Hebei Academy of Sciences, Shijiazhuang, 050081, China
[2]. School of Chemical Engineering, Shijiazhuang University, Shijiazhuang, 050018, China
[3]. Hebei University of Science and Technology, Shijiazhuang, 050018, China

Abstract: PP/FG composite materials were prepared by compressing moulding under high temperature through high-speed mixer, mixing mill and vulcanization machine taking PP as the matrix resin, FG as the thermal conductive filler, and NDZ-201 as the coupling agent. Effects of PP proportion, FG content and particle size on the thermal conductivity and mechanical properties of composite materials were studied. Results showed that PP proportion has little effect on the thermal conductivity of PP/FG composite materials, but it greatly affects the mechanical properties of the composite materials. Overall, the quality ratio of PP230 to K-4038 at 50:50 is more suitable. With the increasing of FG content, the thermal conductivity increases, and a significant critical value appears when the FG content is 50 wt%; The strength of the composite material first decreases and then increases, with an increase in modulus and a decrease in strain. As the FG particle size increases, the thermal conductivity of the composite material increases and the mechanical properties decrease.

Keyword: Polypropylene; Flake graphite; Compression molding; Composite materials; Thermal conductivity

1. Introduction

Polymer based thermal conductive composite materials can replace metal materials for manufacturing equipment such as condensers and heat exchangers due to their excellent mechanical properties, corrosion resistance, scale inhibition, easy processing, light weight, and low maintenance costs. They are widely used in fields such as chemical, pharmaceutical, construction, electronics, and seawater desalination[1-4]. At present, polymer based thermal conductive composites are usually prepared by adding fillers with high thermal conductivity to the matrix resin, such as nitrides, CFs, flake graphite, graphene, carbon nanotubes, etc[5-9]. Pan Shihao et al.[10] prepared AlN/PA6 thermal conductive composites by melt blending method and found that the thermal conductivity of the composite gradually increased with the increasing of AlN filling content. F Khodayaril et al.[11] prepared acrylic based nanocomposites with different metal oxide nanoparticles and multi walled carbon nanotubes (MWCNTs) content using a solvent mixing method. They found that the thermal conductivity of the composite material significantly increased with the addition of thermal conductive fillers. However, some high thermal conductivity fillers such as graphene and carbon nanotubes are expensive and prone to agglomeration, which limits the application of polymer based thermal conductivity composites. Therefore, the development of polymer based thermal conductivity composites with low cost and excellent performance has become a research hotspot.

With the rapid development of research and development technology for polymer based thermal conductive composites, high-performance polypropylene (PP)/flake graphite (FG) thermal conductive composites are receiving more attention. Wu et al.[12] prepared PP/graphite thermal conductive composite material with the graphite size of 20 μm and 2 μm. It was found that when the graphite filling amount was 40 wt%, the interlayer and interlayer thermal conductivity of the composite material were 1.125 W·m-1·K-1 and 2.897 W·m-1·K-1, respectively, also had certain mechanical properties. Xu et al.[13] prepared PA6/PP/FG composite materials through extrusion and injection methods, and found that the thermal conductivity reached 1.88 W·m-1·K-1 when FG content was 20 wt%. PP is a cost-effective matrix resin with low price, corrosion resistance, excellent mechanical properties, and good processing performance; Meanwhile, FG has a wide range of sources and high thermal conductivity, making it an ideal thermal conductive filler. However, FG and PP have poor compatibility and are difficult to mix evenly, making it difficult to increase the filling amount, resulting in low thermal conductivity of the composite materials.

* Corresponding author: zhangxl_2021@hebust.edu.cn

To attack these problems, high thermal conductivity PP/FG composite materials were prepared by compression molding method in this study. The compatibility between FG and PP was improved by adding coupling agents. The dispersion and filling amount of FG in PP were improved by utilizing the shear force of the open mill. In this research, the influence of PP ratios, FG content and particle size on the thermal conductivity and mechanical properties of composite materials were studied. This study provides a basis for the further preparation of composite materials with better comprehensive performance.

2. Experimental

2.1 Materials

The raw materials are as follows: PP230(powder form) and PPK4038 (granular form) were purchased from Maoming Shihua Dongcheng Chemical Co., Ltd and Taiwan Chemical Fiber Co., Ltd, respectively; FG with a carbon content of 99% (particle sizes of 17 μm、27 μm、37 μm、74 μm or 148 μm) was supplied by Hebei Aoteng Trading Co., Ltd; Titanate coupling agent of NDZ-201 and Anhydrous ethanol were obtained from Dinghai Plastic Chemical Co., Ltd and Tianjin Yongda Chemical Reagent Co., Ltd, respectively; Antioxidant 1010 and Antioxidant PLM-168 were purchased from Aladdin Reagent Company; Zinc stearate was got from Aladdin Reagent Company.

2.2 Instruments

Double-roll open mill (ZG-180, Dongguan Zhenggong Electromechanical Equipment Technology Co., Ltd); Electrothermal drum drying box (101-3AB, Tianjin Taisite Instrument Co., Ltd); Plate vulcanizing press machin (ZG-200T, Dongguan Zhenggong Electromechanical Equipment Technology Co., Ltd); Universal prototype (WZY-240, Chengde Hengtong Test Instrument Co., Ltd); Thermal constant analysing(TPS-2500S was produced by Sweden Hot disk Co., Ltd); Electronic universal testing machine (104C,Shenzhen Wance Testing Machine Co., Ltd.).

2.3 Characterization

Thermal conductivity was test according to GB/T 32064-2015, the probe radius is 3.189mm, the test temperature is 25°C, and the specimen size is 40×40×3±0.2mm。The tensile properties of the FG/PP composites were tested according to GB/T 1040.1-2006. Sample thickness: 3±0.2mm, 1B; The bending properties of the composites were tested according to GB/T 9341-2008. Sample thickness: 3±0.2mm, with testing speed of 2mm/min.

2.4 Preparation of PP/FG composite materials

NDZ-201 was dissolved with anhydrous ethanol and mixed evenly with FG. System was dried in a 120°C oven. A certain amount of PP, antioxidant, and zinc stearate were mixed in a high-speed mixer for 3 times, each time for 1 minute. During the mixer mixing process, the roller gap multiple times was adjusted to ensure uniform mixing of the materials. The temperature of the mixer is 180°C for the front roller and 175°C for the rear roller. Place the mixed material slices in a heated mold for preheating, melting, and molding to obtain composite materials. The molding conditions are: the upper template of 193°C, the lower template of 190°C, exhaustion of 3 times. Then the formed composite material was obtained after cooling to room temperature under a pressure of 12 MPa.

3. Results and discussions

3.1 Effect of PP proportion on the properties of PP/FG composite materials

Two different PPs were mixed to prepare composite materials under the condition of FG content of 70 wt%, a graphite particle size of 37 μm and coupling agent FG content dosage of 0.7 wt%. And thermal conductivity and mechanical properties were tested. The results are shown in Figure 1 and Table 1.

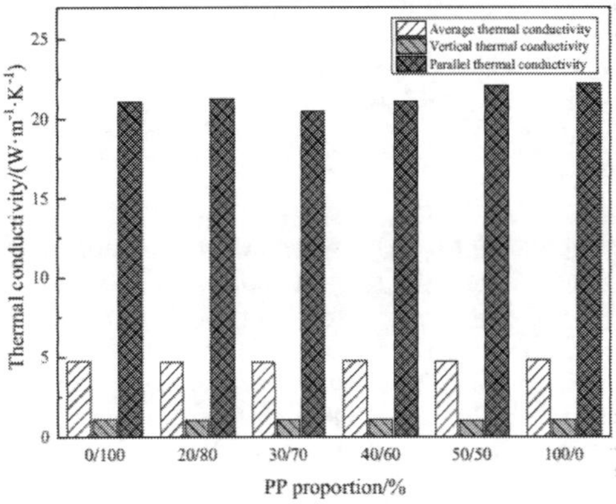

Figure 1 Effect of PP proportion on the thermal conductivity of PP/FG composite materials

As shown in Figure 1, there was no significant change in the thermal conductivity of composite materials prepared with different PP proportion. When the matrix is all PP-230, the average, vertical, and parallel thermal conductivity of the composite material are 4.74 $W \cdot m^{-1} \cdot K^{-1}$, 1.09 $W \cdot m^{-1} \cdot K^{-1}$ and 21.1 $W \cdot m^{-1} \cdot K^{-1}$, respectively. As the PP K-4038 amount increases, the thermal conductivity of the material does not change significantly, indicating that PP proportion have no obvious effect on the construction of thermal conductivity pathways in the composite material.

Table 1 Effect of PP proportion on the mechanical properties of PP/FG composite materials

PP proportion 230/K-4038	Tensile strength /MPa	Elastic tensile modulus /GPa	Breaking tensile strain /%	Bending strength /MPa	Elastic bending modulus /GPa	Fracture bending strain /%
0/100	23.4	6.34	0.48	39.9	8.36	0.68
20/80	24.3	7.86	0.39	41.2	9.59	0.53
30/70	26.3	8.38	0.38	42.1	10.4	0.51
40/60	27.1	8.88	0.36	45.0	10.6	0.49
50/50	29.6	9.10	0.34	50.8	11.7	0.48
100/0	34.8	10.3	0.30	63.0	17.8	0.32

As shown in Table 1, the mechanical properties of composite materials with different PP ratios are significantly different. When the matrix is all PP-230, the tensile and bending strengths of the composite materials are 23.4 MPa and 39.9 MPa, the tensile and bending elastic modulus are 6.34 GPa and 8.36 GPa, and the tensile and bending strains at break are 0.48% and 0.68%, respectively; With the increasing of PP K-4038 dosage, the strength and modulus of the composite material increase, while the breaking tensile strain and fracture bending strain decrease. Considering the comprehensive mechanical properties, the ratio of the two types of PP is optimal at 50: 50.

3.2 2.2 Effect of FG content on the properties of PP/FG composite materials

Composite materials with different FG contents were prepared taking FG with particle size of 17 μm and coupling agent NDZ-201 was 1 wt% of FG content. And thermal conductivity and mechanical properties were tested. The results are shown in Figure 2 and Table 2.

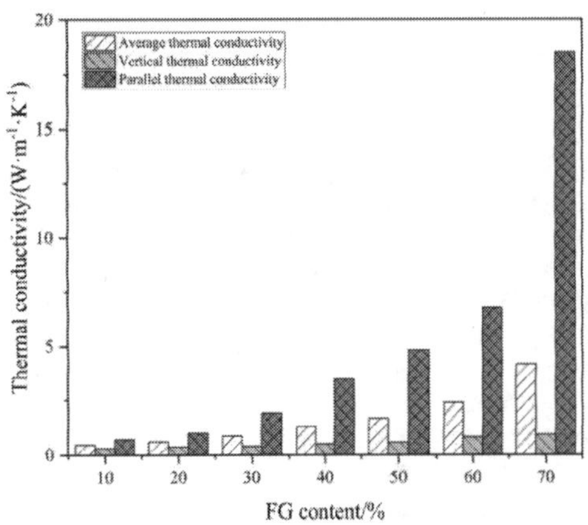

Figure 2 Effect of FG content on the properties of PP/FG composite materials

As shown in Figure 2, with the increase of FG content, the thermal conductivity of the composite material also increases. When the FG content is 10 wt%, the average, vertical, and parallel thermal conductivity of the composite material are 0.475 $W \cdot m^{-1} \cdot K^{-1}$、0.318 $W \cdot m^{-1} \cdot K^{-1}$ and 0.710 $W \cdot m^{-1} \cdot K^{-1}$, respectively. When the FG content increased to 70 wt%, the average, vertical, and parallel thermal conductivity were 4.16 $W \cdot m^{-1} \cdot K^{-1}$、0.938 $W \cdot m^{-1} \cdot K^{-1}$ and 18.4 $W \cdot m^{-1} \cdot K^{-1}$, respectively, corresponding to increasing of 7.76 times, 1.95 times, and 24.9 times, respectively. Through comparison, the parallel thermal conductivity of composite materials has significantly improved, which is closely related to molding. The graphite layer in the vertical state is prone to tilt and collapse when pressure is applied in the vertical direction; Meanwhile, the molten material flows towards fewer parts and promote the orientation of FG along its flow direction under the action of pressure. Therefore, with the increase of FG content, the parallel thermal conductivity of the composite material increases much more than the vertical thermal conductivity. In addition, when the FG content is below 50 wt%, the thermal conductivity of the composite material increases slowly with the increase of FG content, because of the continuous phase in the composite material is mainly PP, making it difficult to form a relatively complete thermal conductivity path; When the FG content exceeds 50 wt%, FG is the mainly continuous phase in the composite material, and the opportunity for FG to overlap with each other greatly increases. It is easy to construct more complete thermal conductivity paths in the composite material, resulting in a sharp increase in thermal conductivity.

Table 2 Effect of FG content on the mechanical properties of PP/FG composite materials

FG particle size/μm	Tensile strength /MPa	Elastic tensile modulus /GPa	Breaking tensile strain /%	Bending strength /MPa	Elastic bending modulus /GPa	Fracture bending strain /%
10	31.2	3.12	1.92	52.8	3.47	2.15
20	26.7	3.24	1.42	47.0	3.95	1.69
30	30.7	4.27	1.19	49.9	5.45	1.33
40	33.7	5.91	1.01	53.0	7.57	0.96
50	34.4	6.02	0.66	51.5	9.10	0.73
60	37.1	10.5	0.47	61.2	11.8	0.64
70	36.1	10.7	0.32	59.0	13.7	0.49

As shown in Table 2, with the increasing of FG content, the tensile strength and bending strength of the composite material first decrease and then increase. When the FG content is 20 wt%, the tensile strength and bending strength of the composite material decrease to the minimum of 26.7 MPa and 47.0 MPa, respectively. As the FG content further increases, the tensile strength and bending strength increase. When the FG content is 60 wt%, the tensile strength and bending strength reach 37.1 MPa and 61.2 MPa, respectively. The tensile and bending elastic modulus of composite materials increase with the

https://doi.org/10.1051/matecconf/202338201009

increase of FG content. The tensile and bending elastic modulus are 3.12 GPa and 3.47 GPa with 10 wt% FG, respectively; When the FG content increases to 70 wt%, the tensile modulus and elastic bending modulus of elasticity increase to 10.7 GPa and 13.7 GPa, respectively. The breaking tensile strain and fracture bending strain of composite materials decrease with the increase of FG content. The breaking tensile strain and fracture bending strain are 1.92% and 2.15% when FG content was 10 wt%. And the breaking tensile strain and fracture bending strain decreased to 0.32% and 0.49%, with the FG content increased to 70 wt%. The reason was that the dispersion concentration of FG in the composite material is too low under lower FG content. Meanwhile, PP matrix bears most of the stress, and the plastic deformation energy that the composite material can absorb is very small; However, with the increase of FG content, according to the reinforcement effect of rigid particles [20], FG, as a rigid inorganic particle, has a reinforcement effect on the material, resulting in higher mechanical properties of the composite material.

3.3 Effect of FG particle size on the properties of PP/FG composite materials

Under the condition of a fixed FG content of 50 wt% and a coupling agent NDZ-201 dosage of 1 wt% of FG content, FG particle size of 17μm、27 μm、37 μm、74 μm、148 μm were selected to prepared composite materials, and the thermal conductivity and mechanical properties were tested. The results are shown in Figure 3 and Table 3.

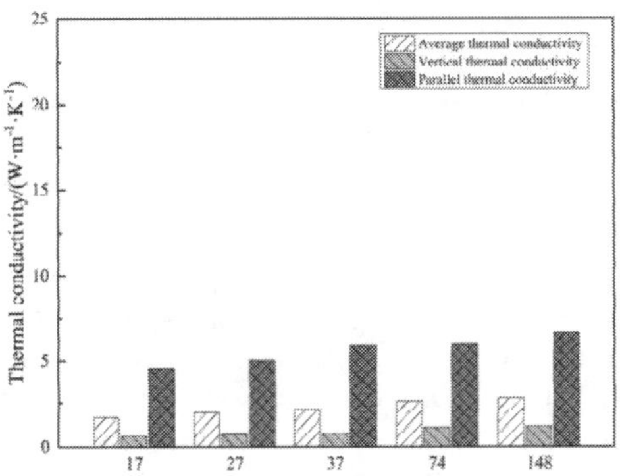

Figure 3 Effect of different particle sizes of FG on the thermal conductivity of PP/FG composite materials

It can be seen From Figure 3 the thermal conductivity of the composite material increases accordingly with the FG particle size increases. With FG particle size of 17 μm, the average, vertical, and parallel thermal conductivity of composite materials are1.67 $W \cdot m^{-1} \cdot K^{-1}$、0.618 $W \cdot m^{-1} \cdot K^{-1}$and 4.51 $W \cdot m^{-1} \cdot K^{-1}$, respectively. while the average, vertical, and parallel thermal conductivity of composite materials with FG particle size of 148 μm reached 2.77 $W \cdot m^{-1} \cdot K^{-1}$、1.16 $W \cdot m^{-1} \cdot K^{-1}$and 6.63 $W \cdot m^{-1} \cdot K^{-1}$, which were 65.9%, 87.7%, and 47.0% higher than the former. The larger FG particle size, the smaller specific surface area, and the PP melt film between the FG particles is thick and uneven, FG particles are easily separated by squeezing and shearing resulting in contact with each other under compressive and shear stress, forming a more complete thermal conductivity network. When FG particle size is larger, the number of gaps between FG particles in the same space is relatively small, which is more conducive to the overlapping of uncoated FG particles to form a relatively complete thermal conductivity path, thereby improving thermal conductivity.

Table 3 Effect of FG particle size on the mechanical properties of PP/FG composite materials

FG particle size/μm	Tensile strength /MPa	Elastic tensile modulus /GPa	Breaking tensile strain /%	Bending strength /MPa	Elastic bending modulus /GPa	Fracture bending strain /%
17	34.4	7.85	0.70	53.0	9.87	0.69
27	32.5	6.74	0.91	50.4	7.79	0.93
37	30.5	6.53	0.77	49.0	8.01	0.86
74	27.9	6.48	0.80	44.3	7.35	0.87
148	25.1	6.83	0.54	43.0	7.12	0.86

Table 3 showed that the strength and modulus of the composite material gradually decrease with the increase of FG particle size. When the particle size of FG is 17 μm, the tensile strength of the composite material is 34.4 MPa, the tensile modulus of elasticity is 7.85 GPa, the bending strength is 53 MPa, and the bending modulus of elasticity is 9.87 GPa; The particle size of FG is 148 μm, its tensile strength decreased to 25.1 MPa, tensile modulus of elasticity decreased to 6.83 GPa, flexural strength decreased to 43.0 MPa, and flexural modulus of elasticity decreased to 7.12 GPa. It is relatively difficult for FG and PP with large particle sizes to mix evenly, and defects such as voids and cracks are prone to occur. This material is prone to cracking when subjected to stress, leading to material fracture, thereby reducing the mechanical properties. In addition, due to the different expansion coefficients of PP and FG, the interface between the two is easy to separate during the cooling process of the material. The larger the FG particle size, the more obvious the separation, and the greater the internal stress generated, thereby reducing the mechanical properties of the material.

4. Conclusion

PP/FG composite materials were prepared through molding process. And the compatibility between FG and PP was improved through the coupling agent NDZ-201, also increase the dispersion and filling amount of FG in PP with the shear force of the open mill. Effects of PP ratios, FG content and particle size on the thermal conductivity and mechanical properties of composite materials are as follows:

(1) Different PP proportion have little effect on the thermal conductivity of PP/FG composite materials, but have a significant impact on the mechanical properties of the composite materials. Overall, the ratio of 50:50 between PP230 and K-4038 is more suitable.

(2) The thermal conductivity of the composite material increases with the increasing of FG content, and a significant critical value appears with FG content of 50

wt%; The tensile strength and bending strength of the material first decrease and then increase, and the tensile modulus and bending modulus also increase with the increase of FG content. The breaking tensile strain and fracture bending strain at break decrease. When the FG content with particle size of 17 μm is 70 wt%, the average, vertical, and parallel thermal conductivity of the material are 4.16 $W{\cdot}m^{-1}{\cdot}K^{-1}$、0.938 $W{\cdot}m^{-1}{\cdot}K^{-1}$and 18.4 $W{\cdot}m^{-1}{\cdot}K^{-1}$, respectively. The tensile strength and bending strength are 36.1 MPa and 59.0 MPa, and the tensile and bending modulus of elasticity are 10.7 GPa and 13.7 GPa; The breaking tensile strain and fracture bending strain are 0.32% and 0.49%, respectively.

(3) As the FG particle size increases, the thermal conductivity of the composite material significantly increases, and the mechanical properties correspondingly decrease. Set FG content is 50 wt% in composite materials, And comparing composite material prepared by FG with a particle size of 148 μm with the materials containing FG particle size of 17 μm, the average, vertical, and parallel thermal conductivity increased by 65.9%, 87.7%, and 47.0%. The tensile strength and bending strength decreased from 34.4 MPa and 53.0 MPa to 25.1 MPa and 43.0 MPa, respectively.

Acknowledgments

This work was financially supported by the Two-institute Cooperation Project of Hebei Academy of Sciences (Project No: 22713) and t he Fundamental Research Funds Pilot Project of the Hebei Academy of Sciences (Project No: 2023PF03-1).

References

1. Hauser R A, King J A, Pagel R M, et al. Effects of carbon fillers on the thermal conductivity of highly filled liquid-crystal polymer based resins[J]. Journal of Applied Polymer Science, 2008, 109(4): 2145-2155.
2. Panwar V, Park J O, Park S H, et al. Electrical, dielectric, and electromagnetic shielding properties of polypropylene-graphite composites[J]. Journal of Applied Polymer Science, 2010, 115(3): 1306-1314.
3. Liu Z H. A study on the graphite/polymer composite bipolar plates in PEMFC[D]. Tianjin University, 2005.
4. Zhang L W. Manufacturing, property and application of thermal conductive polymer composites[D]. East China University of Science and Technology, 2010.
5. Donnay M, Tzavalas S, Logakis E. Boron nitride filled epoxy with improved thermal conductivity and dielectric breakdown strength[J]. Composites Science and Technology, 2015, 110: 152-158.
6. Li J, Ou B Q, Jiao L C, et al. Effects of compound filler contents on properties of polyamide 6 thermally conductive composites[J]. Engineering Plastics Application, 2015, 43(9): 134-137.
7. Zhao J, Ting O Y, Yang Y, et al. Thermal conductivity enhancement of phase change materials with form-stable carbon bonded carbon fiber network[J]. Materials & Design, 2018, 143: 177-184.
8. Hong Y Z, Zhu J, Feng Z M, et al. Preparation and research of thermally conductive PBT/CF composites[J]. Engineering Plastics Application, 2015, 43(4): 92-97.
9. Ma L, Chen K X, Mo D C, et al. Preparation, properties and mechanism of thermal conductive graphene/polymer composites[J]. CIESC Journal, 2017, 68(S1): 18-25.
10. Pan S H, Dai P G, Chen Y J, et al. Research on thermal conduction performance of AlN/PA6 composite material[J]. 2015, 43(11): 83-84, 108.
11. Khodayari F, Jamshidi M. Influence of Mwcts and nanometal oxide/Mwcts hybrids on the thermal conduction of their acrylic-based nanocomposites[J]. Iranian Polymer Journal, 2016, 25(3): 1-8.
12. Wu J W, Ji X Y, Yang J D, et al. Preparation and properties of polypropylene/graphite composites with high thermal conductivity[J]. Engineering Plastics Application, 2016, 44(8): 36-40.
13. Xu H G, Peng X T, Zhou J J, et al. Research of thermally conductive composites based on flake graphite filled in polyamide6/polypropylene[J]. Engineering Plastics Application, 2015, 43(9): 1-6.
14. Zhao M. New modified polypropylene materials[M]. Chemical Industry Press, 2021.
15. S N, Cao G Z, Wang X, et.al, Research of the Polypropylene Modification Methods[J]. 2011.39(4):57-59.
16. Hu M, Yu D, Wei J. Thermal conductivity determination of small polymer samples by differential scanning calorimetry[J]. Polymer Testing, 2007, 26(3): 333-337.
17. Yuan J Y. Preparation and properties of thermally conductive and electrically insulating polypropylene composites[D]. South China University of Science and Engineering, 2014.
18. Yan S. Preparation and properties of polypropylene-based thermal conductive composites[D]. Beijing University of Chemical Technology, 2012.
19. Mhike W, Focke W W, Mofokeng J P, et al. Thermally conductive phase-change materials for energy storage based on low density polyethylene, soft Fischer–Tropsch wax and graphite[J]. Thermochimica Acta, 2012, 527: 75–82.
20. Meng J R, Liang G Z, Qin H Y, et al. Toughening and reinforcing effect of rigid fillers for polymers[J]. Plastics, 2002, 31(2): 47-50.

Optimization of asynchronous motor with Taguchi method based on Finite Element Analysis

Lixi Zhang*

School of Electrical and Information Engineering, Jiangsu University, Zhenjiang, China.

Abstract. In order to study the operation process of the motor and improve the operating performance of the motor, this paper performs finite element simulation of the three-phase asynchronous motor and optimizes the initial design by using the Taguchi method. Firstly, a three-phase asynchronous motor model is established with the voltage, current, torque, flux linkage, distribution of the flux lines and magnetic induction intensity of the motor under rated conditions is vertified by Finite Element Analysis (FEA), and the total accuracy reaches 92%. Then, the parametric simulation of the stator and rotor cogging parameters and stator and rotor parameters is carried and the sensibility of different parameters on the motor torque is obtained. The rotor cogging parameter Hs01 and stator cogging parameter Hs0 had the highest sensitivity, which were 75% and 16%, respectively. Finally, the Taguchi method was applied to optimize the torque of the motor, and the torque was increased by 9.5%.

1. Introduction

Energy is a significant material foundation and driving force for the progress of human civilization, which is related to the national economy and people's livelihood and national security. At present, the following national energy demand and the wide application of motors in equipment manufacturing, national defense, industrial production and daily life in the new era, and with the rapid growth of the demand for high-performance motors, how to carry out efficient motor simulation analysis and optimization design is particularly important.

Traditional motor optimization is mainly carried out by parameter sweeping method, a limited number of optimization variables are analyzed by grid analysis of distance length, and the most suitable candidate design is artificially selected. However, this method is inefficient and difficult to obtain the optimal solution. With the development of optimization algorithms and modeling technology, motor design has been able to achieve its functional requirements through model combined with optimization goals, and achieve goals such as reducing costs, reducing losses, and improving efficiency.

In the simulation and optimization design of motors, scholars at home and abroad have done a lot of research. Literature [1] The preliminary design scheme of the motor is obtained by the magnetic circuit method, and the electromagnetic field analysis of the motor is analyzed by the finite element method to verify the rationality of the scheme, and then the permanent magnet inclined pole is used and the design parameters of the motor are optimized by combining genetic algorithm. Literature [2] Firstly, the development status of high-speed motors is analyzed, and the structure and design characteristics of different types of high-speed motors are summarized, the stator and rotor design are analyzed in detail according to the motor indicators, and the protection measures for the loss of motor operation are considered by analyzing the number of stator poles, grooves, core materials, stator windings, rotor geometry, rotor copper bar materials and other parameters. The literature [3] summarizes the research work carried out by domestic and foreign scholars in the rotor pole optimization technology of permanent magnet synchronous motor in recent years, classifies different rotor pole optimization technologies, compares their advantages and disadvantages. The literature [4] uses the finite element method to simulate the operation of the motor, and introduces different numerical techniques for calculating inductance and magnetic force based on the three axisymmetric open boundary problems, so as to better explain and analyze the phenomena of magnetic saturation and magnetized magnetic inductance. Paper [5] Based on the motor parameters, a numerical model considering the magneto strictive effect of silicon steel sheet is established, and the Multiphysics coupled vibration analysis is carried out. Literature [6] uses finite element analysis to evaluate the influence of rotor and stator diameter, tooth width and number of windings on the optimal design of motors, and studies the feasibility of magneto resistive network limiting the parameter domain of finite element simulation, and finally performs experimental verification. Paper[7] analyzes the sensitivity analysis of the performance index of the rotor geometric parameters of permanent magnet synchronous reluctance motor, including inductance difference, torque, torque ripple and permanent magnet flux linkage. The sensitivity of performance indicators to geometric

* Corresponding author: leosxjm@163.com

parameters such as flux barrier angle, width, length and position is obtained experimentally.
However, there are still additions to the above literature for the study of variables for specific motor parameters. However, there are still more related study about the optimization of motor besides above . In this paper, the finite element simulation and optimization design of three-phase asynchronous motor will be promoted to analyze the sensitivity of motor parameters and optimize the experimental design, and the influence of cogging parameters, stator parameters and rotor parameters on the working performance of three-phase asynchronous motor will be explored and optimized.

2. Simulation analysis of three-phase asynchronous motor

This paper will take the three-phase asynchronous motor as the research object, and use MAXWELL software to do the simulation analysis. Figure 1 is the structure diagram of the three-phase asynchronous motor, and the basic parameters are shown in Table 1.

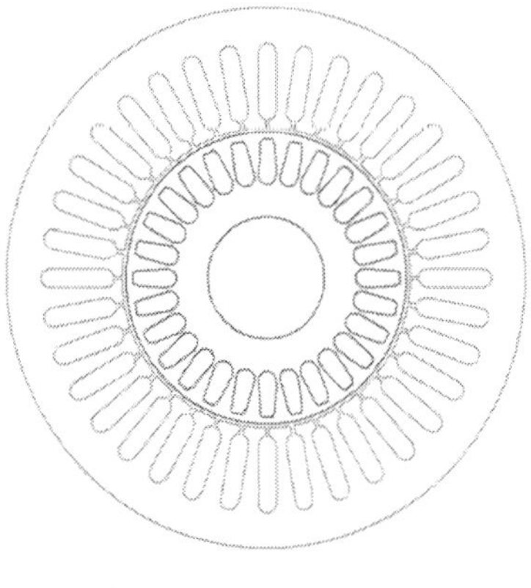

Figure 1. Schematic diagram of three-phase asynchronous motor

Table 1. Basic parameters of the motor

Parameter	Value
Rated power/W	1100
Rated voltage/V	380
Rated speed/rpm	1450
Rated operating temperature/℃	75
Stator outer diameter/mm	140
Stator inner diameter/mm	78
Axial length/mm	250
Core material	Style D21_50
Number of stator slots	36
Rotor outer diameter/mm	76
Rotor inner diameter/mm	32
Roter material	Style D21_50
Rotor guide strips	Style copper

Through the parametric simulation in software, we obtain some parameter images of the motor in the rated operating state, as shown in Figure2, Figure3, Figure4, Figure5, Figure6, Figure7.

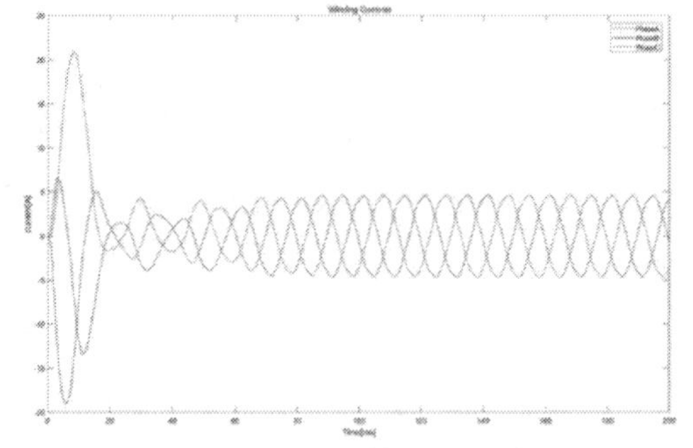

Figure 2. Asynchronous motor simulation of Current

As above figure, the maximum current value reaches 20.96A when the current is in the transient state, which is 4.6 times that of the steady state . Steady state is reached at 24ms, and then it changes in a sinusoidal curve. The rms value is 3.23A, which is 0.34A different from the rated parameters, and the simulation accuracy is 88.4%. According to the above conclusions, it can be obtained that the motor simulation results have high accuracy and meet the design requirements.

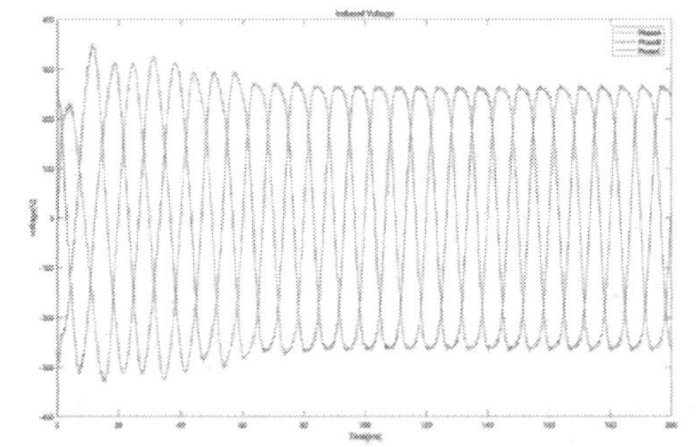

Figure 3. Asynchronous motor simulation of load Voltage

From the above figure, the three-phase voltage waveform is symmetrical, the maximum value is 350.1V, the error from the rated value is 30V, and the simulation accuracy is 92.1%, which meets the simulation accuracy requirements. After reaching the steady-state process in 60ms, the magnetic flux is partially saturated, the voltage is approximately a sine wave change, and the waveform distortion is small.

https://doi.org/10.1051/matecconf/202338201010

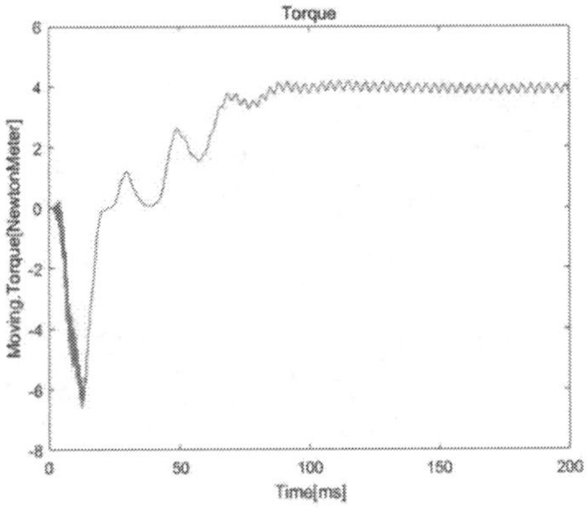

Figure 4. Simulation of Torque

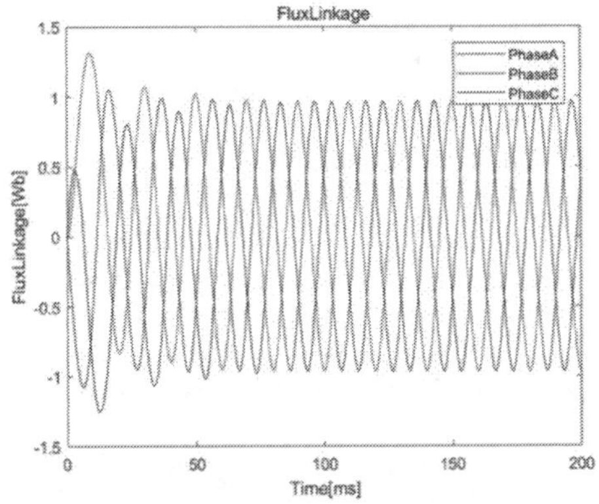

Figure 5. Simulation of FluxLinkage

With the rotation of the motor, the motor reaches a steady state after undergoing a transient process of 90ms, the average value is 4.01Nm, the torque fluctuates in the range of 3.88-4.14, the fluctuation range is 3.14%, and the torque fluctuation is small.
And from the above figure5, it can be seen that the waveform of the flux link has a three-symmetry relationship, the maximum value is 1.31Wb, and after reaching the steady-state process in 60ms, the maximum value of the flux link is stable at 0.95Wb.

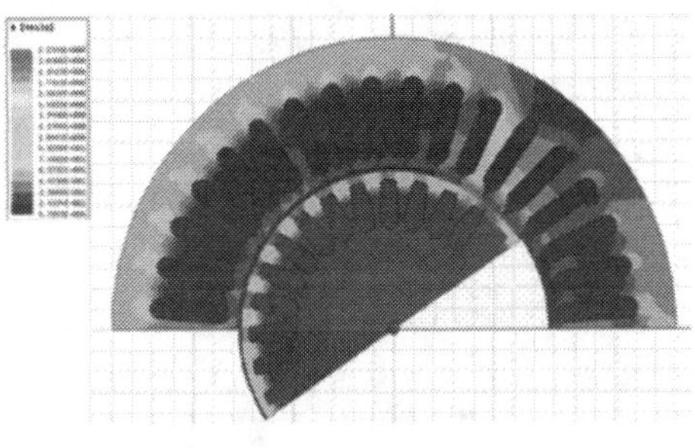

Figure 6. Simulation of Magnetic Density

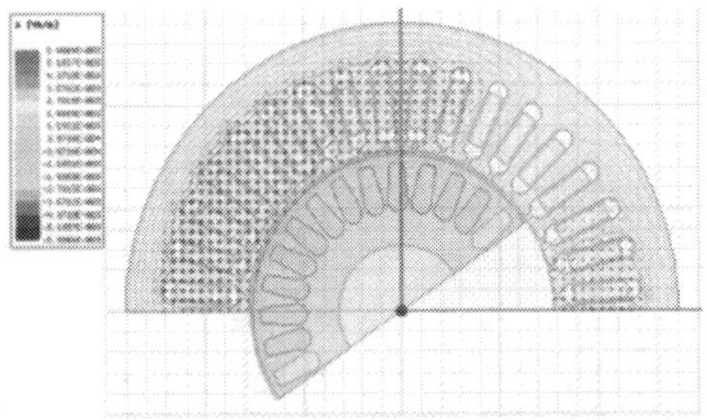

Figure 7. Simulation of Magnetic Induction

As can be seen from the above figure6, the maximum magnetic density of the stator part is 1.74 tesla, which appears in the tooth, and it is necessary to protect and monitor the force condition and thermal condition of the tooth during the operation of the stator. The maximum value of the magnetic density of the rotor part is 3.13tesla, which appears in the rotor tooth at the junction of the stator and rotor, which is located at the junction of the rotor guide bar and the stator copper wire, so the magnetic density is large. The above magnetic density phenomenon is in line with reality and meets the design requirements.
As can be seen from the above figure7, the magnetic induction line closure is small, so the magnetic leakage phenomenon of the motor is relatively small, which meets the design requirements.

3. Finite element simulation of three-phase asynchronous motor

In order to explore the influence of motor parameter changes on motor performance, four parameters of stator, rotor cogging, stator and rotor are selected as independent variables for modification, and finite element simulation is carried out. The change in motor performance is measured using motor torque as a dependent variable, and the results obtained are as follows.

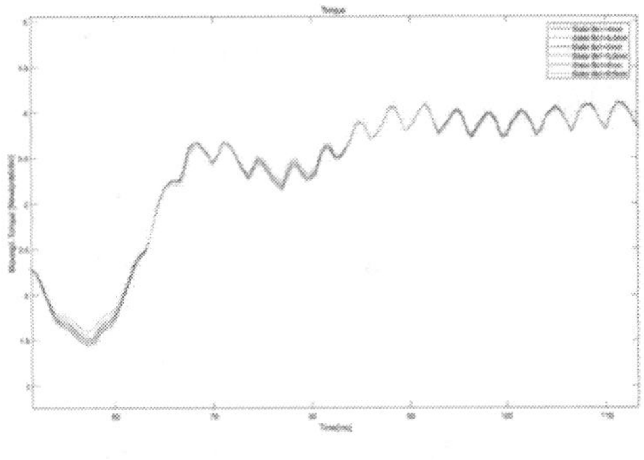

Figure 8. Stator Parametric Simulation

As can be seen from Figure 8, it can obtain that the parameter range of modified stator Bs1 is 4mm~6mm, and its rated parameter is 5mm. When the parameter is smaller, the smaller the torque average, when the parameter increases, the torque average value also increases.

https://doi.org/10.1051/matecconf/202338201010

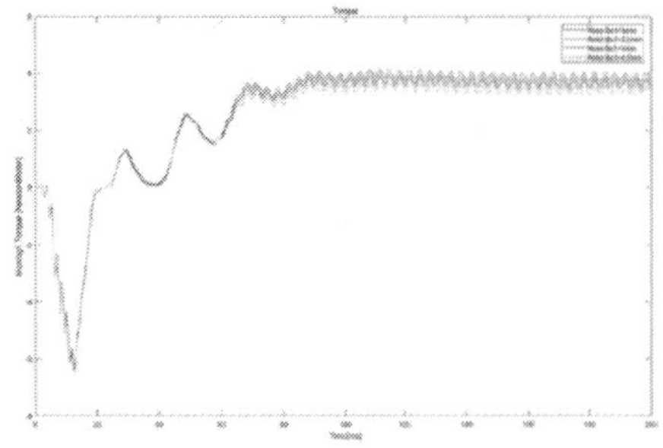

Figure 9. Rotor Parametric Simulation

As can be seen from Figure 9, it can obtain that the parameter range of modified rotor Bs1 is 3mm~4.5mm, and its rated parameter is 4mm. When the parameter is smaller, the larger the torque average, when the parameter increases, the torque average value also decreases.

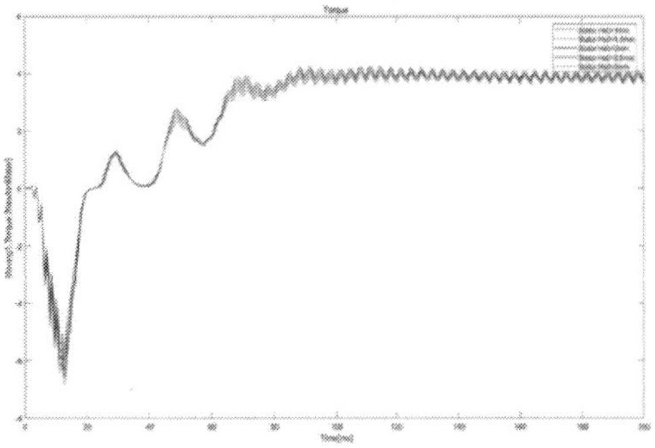

Figure 10. Stator Cogging Parametric Simulation

From Figure 10, it can be obtained that the parameter range of modified stator cogging Hs0 is 1mm~3mm, and its rated parameter is 2mm. When the parameter is smaller, the average torque is larger, and when the parameter increases, the average torque value also decreases.

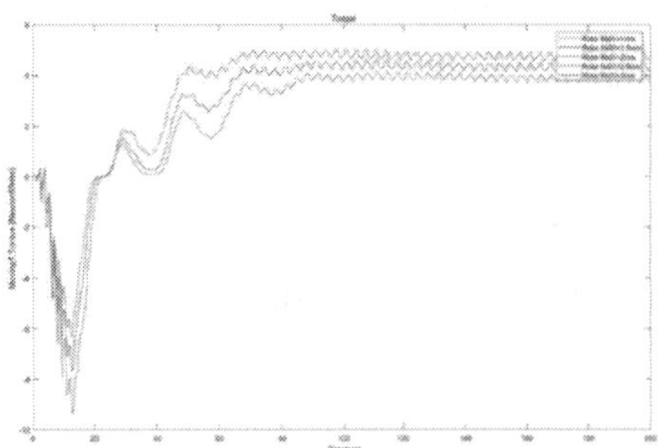

Figure 11. Rotor Cogging Parametric Simulation

From Figure 11, it can be obtained that the parameter range of modifying the rotor cogging Hs01 is 1mm~3mm, and its rated parameter is 2mm. When the parameter is smaller, the average torque is larger, and when the parameter increases, the torque average decreases. There are cases where the torque curves coincide when the parameters change little.

4. Optimized design of three-phase asynchronous motor based on Taguchi method

The Taguchi method was proposed by Dr. Taguchi Genichi in 1952 to find the best combination of parameters to improve product quality, Taguchi method has been rapidly promoted and applied in Japan, achieving huge economic benefits and social impact, so that the method began to be valued and applied by the global industry. The design flowchart of Taguchi's optimization design method is shown in Figure 12.

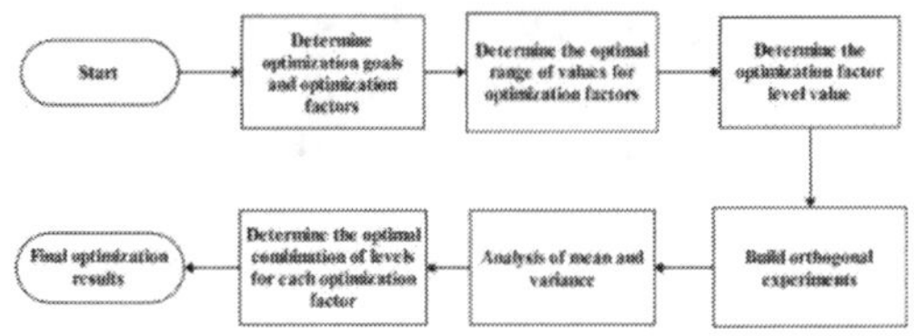

Figure 12. Taguchi method optimization design flowchart

According to the motor design requirements, the stability value of torque should be as high as possible, so the torque Te is selected as the optimization goal, because the motor cogging, stator and rotor have a great influence on the performance and operation of the motor, so this paper selects the stator cogging Hs0, rotor cogging Hs01, stator core inner diameter Bs1 and these four parameter variables are selected as optimization factors.

Maxwell software is used to repeatedly adjust and simulate the optimization factor parameters to obtain the optimal value range, and the value range of each optimization factor is shown in the table. Five influence levels were selected equidistant in the optimal range of each factor, and the five level values were named as level values 1, 2, 3, 4 and 5 according to the principle of order from smallest to largest, and the values of each level were shown in the table 2.

Table 2. The number of optimization factor levels

Optimization factor	LEVEL1	LEVEL2	LEVEL3	LEVEL4	LEVEL5
Stator cogging Hs0/mm	1	1.5	2	2.5	3
Rotor cogging Hs01/mm	1	1.5	2	2.5	3
Stator inner diameter Bs1/mm	4	4.5	5	5.5	6
Rotor inner diameter Bs1/mm	3	3.5	4	4.5	5

According to the four optimization factors (variables) selected in Table 2 and Table 3 and the five horizontal values determined by each optimization factor, the L16(4^5) orthogonal table was established by using the construction principle of orthogonal table and the experimental design principle of Taguchi's algorithm, and a multivariate optimization design can be realized in only 25 experiments, which greatly reduces the degree of calculation and the number of experiments. The finite element method was used to establish 25 motor models, and the parametric simulation analysis was carried out by Maxwell software, and the results of the finite element solution were shown in Table 3.

Table 3. Finite element solution results

Test Number	Experimental Matrix				T/Nm
	Hs0	Hs01	Bs1	Bs1	
1	1	1	1	1	4.872
2	1	2	3	4	4.456
3	1	3	5	2	4.079
4	1	4	2	5	3.903
5	1	5	4	3	4.069
6	2	1	5	4	4.795
7	2	2	2	2	4.417
8	2	3	4	5	3.874
9	2	4	1	3	4.069
10	2	5	3	1	4.005
11	3	1	4	2	4.712
12	3	2	1	5	4.301
13	3	3	3	3	3.974
14	3	4	1	5	3.986
15	3	5	2	4	4.008
16	4	1	3	5	4.582
17	4	2	5	3	4.349
18	4	3	2	1	4.174
19	4	4	4	4	4.483
20	4	5	1	2	4.263
21	5	1	2	3	4.029
22	5	2	4	1	3.979
23	5	3	1	4	3.846
24	5	4	3	2	3.892
25	5	5	5	5	3.750

The statistical analysis of experimental data is carried out from the orthogonal experimental table and finite element analysis results, and in order to obtain the proportion of the influence of all optimization factors on the optimization goal, the average analysis and variance analysis are carried out below.

Firstly, the average value of the finite element simulation results of the optimization target is calculated according to Equation (1) as follows:

$$m = \frac{1}{n}\sum_{i=1}^{n} S_i \quad (1)$$

Where m is the mean of the experiment; n is the number of experiments; S_i is the ith optimization target value.Obtain m=4.195 Nm.

Then, the average value of the simulation analysis of the motor torque performance index corresponding to the change of each optimization factor is analyzed, and the average value of the optimization factor at different levels is first analyzed. According to Equation (2), the average calculation result of torque T is shown in Table 4 when all optimization factors are taken at different levels.

$$m(\eta) = \frac{1}{5}(\eta(1) + \eta(2) + \eta(3) + \eta(4) + \eta(5)) \quad (2)$$

Table 4. Average value of motor indicators at different levels of various optimization factors

Optimization factor	LEVEL	T/Nm
Stator cogging Hs0	1	4.2758
	2	4.232
	3	4.1962
	4	4.3702
	5	3.8272
Rotor cogging Hs01	1	4.598
	2	4.3004
	3	3.9894
	4	4.0048
	5	4.019
Stator inner diameter Bs1	1	4.223
	2	4.1062
	3	4.1818
	4	4.2234
	5	4.2433
Rotor inner diameter Bs1	1	4.2575
	2	4.2726
	3	4.098
	4	4.3176
	5	4.066

From Table 5, the motor optimization factors that can make the highest motor torque are stator cogging Hs0(4), rotor cogging Hs01(1), stator core inner diameter Bs1(5), rotor core inner diameter Bs1(4), respectively, 4.3702, 4.598, 4.2433, 4.3176.

According to the average table of optimization targets analyzed above and the average table of motor performance indicators of various optimization factors at different levels, the variance (SS) of optimization factors under motor torque performance indicators and the proportion of their influence on motor performance can be calculated from equation (3). The calculation results are shown in Table 5.

$$SS = \frac{1}{n}\sum_{i=1}^{n}(m_{Xi}(S) - m(S))^2 \quad (3)$$

Where X is each optimization factor (optimization parameter); S is the optimization goal; m(S) is the average of the optimization target; $m_{xi}(S)$is the average of an optimization target for factor X at level i; SS is the variance value under the motor optimization factor X, and n is the level value of each optimization factor.

Table 5. The influence of each optimization factor on the torque of the motor

Optimization Factor	T/Nm	
	SS	Specific Gravity
Stator cogging Hs0	0.042	16%
Rotor cogging Hs01	0.197	75%
Stator inner diameter Bs1	0.006	2%
Rotor inner diameter Bs1	0.017	7%

 https://doi.org/10.1051/matecconf/202338201010

From Table 5, it can be obtained that the rotor and stator cogging parameters have a relatively large influence on the operating performance of the motor, and the total influence is 91%.

5. Conclusion

In this paper, finite element simulation is carried out for three-phase asynchronous motor, and the optimized design is carried out by the Taguchi method. Firstly, a three-phase asynchronous motor model is established to study the parameters of motor operation under normal conditions, and the total simulation accuracy reaches 92%. Then, the parameters of cogging, stator and rotor are modified for parametric simulation, and the influence of different parameters on the motor torque is obtained. Finally, the Taguchi method was applied to optimize the torque design of the motor, and the torque was increased from 4.01Nm to 4.39Nm, which was increased by 9.5%, and the rotor cogging Hs01 and stator cogging Hs0 had the highest sensitivity, 75% and 16%, respectively.

References

1. WANG Dongmei, BAI Rui, FENG Da, GAN Yuan. Finite element analysis and optimization design of permanent magnet synchronous motor for electric vehicle[J].Control Engineering,2022.1.
2. ZHANG Fengge, DU Guanghui, WANG Tianyu, LIU Guangwei. Review of development and design of high-speed motor[J].Transactions of China Electrotechnical Society,2016.4.
3. WANG Kai, SUN Haiyang, ZHANG Lufeng, LIU Chuang, ZHU Ziqiang. Review of rotor pole optimization technology of permanent magnet synchronous motor[J].Proceedings of the CSEE,2017.12.
4. Joinvile. Simulated Experiments for Teaching Mutually-Coupled Circuits CAD Techniques Using Analytic and Finite Element Solutions[J].Journal of Electromagnetic Analysis and Applications, 2017.
5. Zhao Liu, Lan Lu, Rui Sha. Research on Vibration Reduction Optimization of High Speed Permanent Magnet Synchronous Motor[J].Journal of Electrical Engineering, 2022.
6. Sabrina K, Rechkemmer, Ryosuke I, Wei M Z, Oliver S. Estimation of permanent magnet synchronous machine performance for pre-design using a reluctance network[J]. IFAC Papers OnLine, 2018, 51(31): 75-80.
7. Ibrahim, M.N.F., Sergeant, P. and Rashad, E. (2016) Influence of Rotor Flux-Barrier Geometry on Torque and Torque Ripple of Permanent-Magnet-Assisted Synchronous Reluctance Motors. 2016 XXII International Conference on Electrical Machines (ICEM), Lausanne, 4-7 September 2016.

Kinematics analysis of a medication delivery robot based on a 3-DOF manipulator

Shanyue Huang *

College of Mechanical and Electrical Engineering, Hohai University, Changzhou, China

Abstract. Medication delivery robots, as a type of service robots, are becoming increasingly important in the field of helping the elderly and disabled. Conventional drug delivery robots focus on mobility and often lack the ability to pick and place drugs independently. Based on this phenomenon, a new medication delivery robot structure based on a 3-DOF robot manipulator is designed in this paper. According to the motion process, the 3-DOF manipulator is modeled and simulated. Meanwhile, the influence of different trajectory planning methods on the drug taking and discharging process are compared, which provides theoretical and technical support for the research of robots in related fields.

Keywords: robot manipulator, kinematic analysis, kinematic control, service robot.

1. Introduction

Due to the trend towards an ageing population, service robots are more important than ever to the market, especially for ageing assistance robots, which are emerging to relieve a great deal of caregiving stress. Examples include wearable assisted walking robots [1], robots that can provide medication assistance [2] and companion robots that provide emotional help to the elderly [3].

Medication delivery robots play an important role in these robots. Jonatan et al. team has proposed a care robot that can provide medication and monitor vital signs for elderly people in nursing homes, using ROS as the framework [4]. Dibya et al. also presented an assistance robot that can work at regular intervals [5], and Patel et al. implemented a remote-controlled drug delivery robot using an Arduino Uno and a Bluetooth controller [6].

However, the aforementioned medication delivery robots need to work in conjunction with a caregiver, which do not match the living conditions of most elderly people living alone, so it is crucial to design a robot which can work independently. Multi-degree-of-freedom robotic arms can achieve this requirement [7-10], but multiple degrees of freedom means that the control of the robot arm is more complex. For drug delivery, a reasonably designed 3-DOF robot arm can achieve this. For the functional and structural analysis of three-degree-of-freedom robotic arms, related works include: Noventino et al. performed a DH parametric analysis of a robotic arm with one rotating joint and two pitch joints, established a dynamics model and designed a corresponding PID controller to achieve accurate trajectory tracking of the end-effector [11]. Raj et al. performed a kinematic analysis of a robotic arm with three orthogonal rotating joints and gave joint Jacobi matrices [12]. Liu et al. introduced a new 3-DOF robotic arm, which was simulated using ADAMS after kinematic analysis to verify the accuracy of the kinematic equations [13]. However, none of the service targets in the above literature relate to the field of drug delivery and elderly assistance services.

In this research, a new medication delivery robot based on a 3-DOF robot manipulator is introduced. The rest of this paper is organized as follows. In Section II, the concrete composition of the robot is introduced and the three-dimensional model of the robot is established. Section III presents the kinematic model of the 3-DOF manipulator and deduces the kinematics formulas. In Section IV, MATLAB is used to simulate the motion of the manipulator and two trajectory planning methods are compared [12], [17]. Conclusions are given in Section V. Fig.1 illustrates the main structure and operation of the robot.

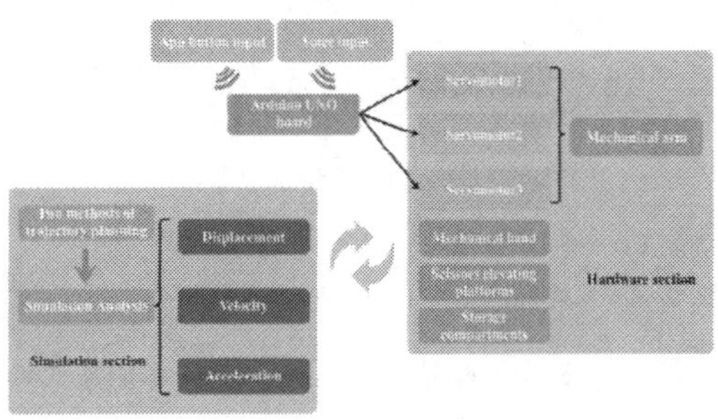

Fig.1 Diagram of the main structure and operation of the robot

* Corresponding author: 2061910129@hhu.edu.cn

2. Three-dimensional model of the medication delivery robot

The structure of the proposed medication delivery robot consists of a 3-DOF robot manipulator, a scissor lift platform [14-16], a cleaning mechanism, and four storage compartments. The three-dimensional mechanism is shown in Fig.2

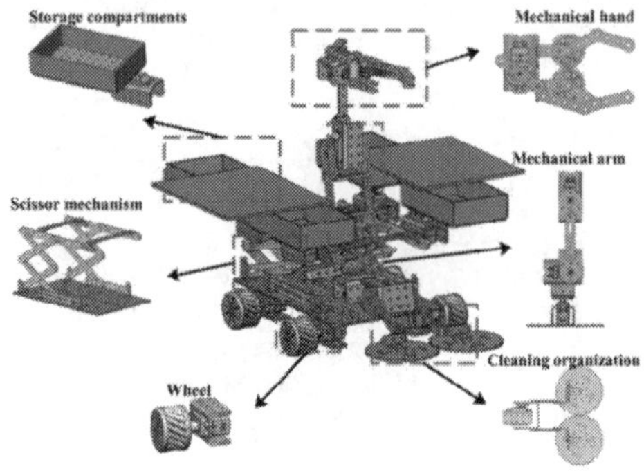

Fig.2 Diagram of the three-dimensional model of the medication delivery robot

The 3-DOF robot manipulator is driven by standard servos and the end-effector is a single-degree-of-freedom mechanical hand. Three linkage lengths from the base to the end-effector are: 37.22 mm, 90.00 mm and 96.54 mm. The main function of the manipulator is to grasp and place the medicine. The scissor lift platform consists of two sets of identical scissor mechanisms. Each scissor mechanism consists of two sets of X-shaped crossed rods, both 112mm in length. The main function of the scissor lift platform is to hold the storage compartments and control the platform for lifting. The storage compartments are driven by servos and connected to both sides of the scissor lift platform. The main function of these compartments is to store different types of medication temporarily. The cleaning mechanism consists of two cleaning wheels connected to the servo, which can be fitted with cleaning cloths and adjusted by the servo. The main function of it is to clean the path during the delivery of the medication. The model has been physically created by sourcing the components and machining. The specific procurement list and the parameters of the physical object are shown in Table 1

Table1 The specific procurement list and the parameters of the physical object

No.	Component	Specifications	No.	Parameters	values
1.	Arduino Uno	R3	7.	Length	385.83mm
2.	Bluetooth Module	HC-06	8.	Width	370.40mm
3.	Voice Module	LD3320	9.	Height	312.96mm
4.	Ultrasonic Distance Sensor	HC-SR04	10.	Weight	1.54kg
5.	Geared Motor and Wheels	100rpm	11.	Lifting distance	132.47mm
6.	Servos	SG90	12.	Chassis height	32.20mm

The built physical model is shown in Fig.3, which is exactly corresponding to the contents of the three-dimensional model shown in Fig.2. The grasping and placing of drugs by the manipulator is the core function of the whole robot. The mathematical model of the 3-DOF robot manipulator will be established for kinematic analysis, and the motion simulation of the robot arm will be carried out in the following sections to verify the accuracy of the mathematical model.

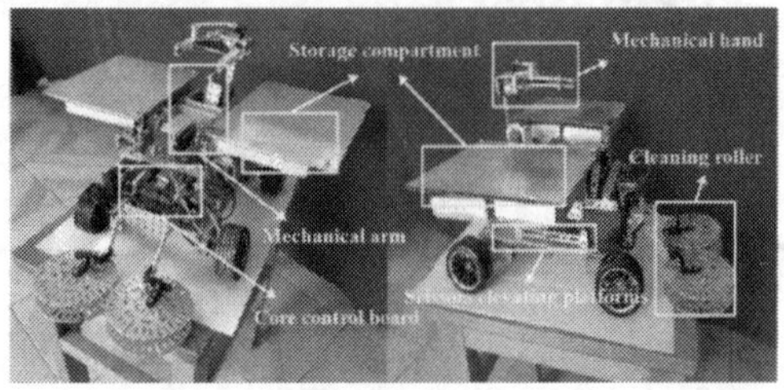

Fig.3 Diagram of the physical model of the medication delivery robot

3. Formula derivation

The DH method is a common method used to describe the kinematics of robotic arms and other robotic systems. It was proposed by Jacques Denavit and Richard S. Hartenberg in 1955 [7]. In this research, the linkage coordinate system of the 3-DOF manipulator is established by using the classic DH method, and the homogeneous coordinate transformation matrix is solved by coordinate transformation. Furthermore, the velocity Jacobian matrix of the manipulator is obtained, and the kinematics analysis of the 3-DOF manipulator is completed.

Table 2 The DH parameter of the 3-DOF manipulator

i	θ_i	d_i	a_i	α_i
i=1	θ_1	67.3mm	-9.8mm	$\pi/2$
i=2	θ_2	0mm	90mm	0
i=3	θ_3	0mm	101.44mm	$-\pi/2$

Fig.4 shows the mechanism model of the 3-DOF manipulator used in this paper, and the DH parameter of the 3-DOF manipulator is shown in Table 2

 https://doi.org/10.1051/matecconf/202338201011

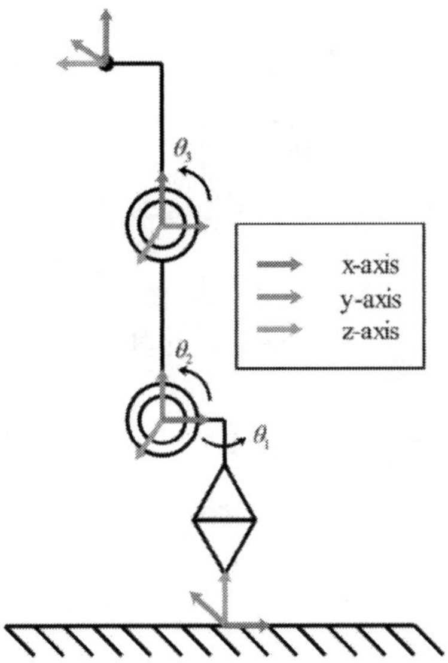

Fig.4 Diagram of the mechanism model of the 3-DOF manipulator

The homogeneous coordinate transformation matrix of the end of the manipulator relative to the base coordinate system can be obtained by the following matrix transformation

$$T_0^3 = T_0^1 T_1^2 T_2^3 \tag{1}$$

where the upper and lower scripts indicate the coordinate system number. For example, T_0^1 represents the homogeneous transformation matrix between the coordinate system x_0-y_0-z_0 and x_1-y_1-z_1.

According to the parameters of DH linkage given in Table 3, the transformation matrix T_0^1, T_1^2, T_2^3 can be obtained by using the following calculation sequence of the classic DH method

$$T_0^1 = Rot(z_0,\theta_1)\cdot Trans\,(-9.8,0,67.3)\cdot Rot(x_1,90^\circ)$$
$$= \begin{bmatrix} c\theta_1 & -s\theta_1 & 0 & 0 \\ s\theta_1 & c\theta_1 & 0 & 0 \\ 0 & 0 & 1 & 0 \\ 0 & 0 & 0 & 1 \end{bmatrix} \begin{bmatrix} 1 & 0 & 0 & -9.8 \\ 0 & 1 & 0 & 0 \\ 0 & 0 & 1 & 67.3 \\ 0 & 0 & 0 & 1 \end{bmatrix} \begin{bmatrix} 1 & 0 & 0 & 0 \\ 0 & 0 & -1 & 0 \\ 0 & 1 & 0 & 0 \\ 0 & 0 & 0 & 1 \end{bmatrix} = \begin{bmatrix} c\theta_1 & 0 & s\theta_1 & -9.8c\theta_1 \\ s\theta_1 & 0 & -c\theta_1 & -9.8s\theta_1 \\ 0 & 1 & 0 & 67.3 \\ 0 & 0 & 0 & 1 \end{bmatrix} \tag{2}$$

$$T_1^2 = Rot(z_1,(\theta_2+90^\circ))\cdot Trans(90,0,0)\cdot Rot(x_2,0^\circ)$$
$$= \begin{bmatrix} c(\theta_2+90^\circ) & -s(\theta_2+90^\circ) & 0 & 0 \\ s(\theta_2+90^\circ) & c(\theta_2+90^\circ) & 0 & 0 \\ 0 & 0 & 1 & 0 \\ 0 & 0 & 0 & 1 \end{bmatrix} \begin{bmatrix} 1 & 0 & 0 & 90 \\ 0 & 1 & 0 & 0 \\ 0 & 0 & 1 & 0 \\ 0 & 0 & 0 & 1 \end{bmatrix} \begin{bmatrix} 1 & 0 & 0 & 0 \\ 0 & 1 & 0 & 0 \\ 0 & 0 & 1 & 0 \\ 0 & 0 & 0 & 1 \end{bmatrix} = \begin{bmatrix} -s\theta_2 & -c\theta_2 & 0 & -90s\theta_2 \\ c\theta_2 & -s\theta_2 & 0 & 90c\theta_2 \\ 0 & 0 & 1 & 0 \\ 0 & 0 & 0 & 1 \end{bmatrix} \tag{3}$$

$$T_2^3 = Rot(z_2,\theta_3)\cdot Trans(101.44,0,0)\cdot Rot(x_3,-90^\circ)$$
$$= \begin{bmatrix} c\theta_3 & -s\theta_3 & 0 & 0 \\ s\theta_3 & c\theta_3 & 0 & 0 \\ 0 & 0 & 1 & 0 \\ 0 & 0 & 0 & 1 \end{bmatrix} \begin{bmatrix} 1 & 0 & 0 & 101.44 \\ 0 & 1 & 0 & 0 \\ 0 & 0 & 1 & 0 \\ 0 & 0 & 0 & 1 \end{bmatrix} \begin{bmatrix} 1 & 0 & 0 & 0 \\ 0 & 0 & 1 & 0 \\ 0 & -1 & 0 & 0 \\ 0 & 0 & 0 & 1 \end{bmatrix} = \begin{bmatrix} c\theta_3 & 0 & -s\theta_3 & 101.44c\theta_3 \\ s\theta_3 & 0 & c\theta_3 & 101.44s\theta_3 \\ 0 & -1 & 0 & 0 \\ 0 & 0 & 0 & 1 \end{bmatrix} \tag{4}$$

where *Rot* represents rotation matrix and *Trans* represents translation matrix.

By substituting Eqs(2)-(4) into Eq(1), the homogeneous transformation matrix T_0^3 from the base coordinate system to the end of the manipulator can be obtained

$$T_0^3 = \begin{bmatrix} -c\theta_1 s\theta_2 c\theta_3 - c\theta_1 c\theta_2 s\theta_3 & -s\theta_1 & c\theta_1 s\theta_2 s\theta_3 - c\theta_1 c\theta_2 c\theta_3 & -101.44(c\theta_1 s\theta_2 c\theta_3 + c\theta_1 c\theta_2 s\theta_3) - 90c\theta_1 s\theta_2 - 9.8c\theta_1 \\ -s\theta_1 s\theta_2 c\theta_3 - s\theta_1 c\theta_2 s\theta_3 & c\theta_1 & s\theta_1 s\theta_2 s\theta_3 - s\theta_1 c\theta_2 c\theta_3 & -101.44(s\theta_1 s\theta_2 c\theta_3 + s\theta_1 c\theta_2 s\theta_3) - 90s\theta_1 s\theta_2 - 9.8s\theta_1 \\ c\theta_2 c\theta_3 - s\theta_2 s\theta_3 & 0 & -c\theta_2 s\theta_3 - s\theta_2 c\theta_3 & 101.44(c\theta_2 c\theta_3 - s\theta_2 s\theta_3) + 90c\theta_2 + 67.3 \\ 0 & 0 & 0 & 1 \end{bmatrix} \tag{5}$$

Three displacement components of the end of the manipulator are the fourth column of the homogeneous transformation matrix T_0^3, which can be obtained by rewriting it into a following system of equations.

$$\begin{cases} P_x = -101.44(c\theta_1 s\theta_2 c\theta_3 + c\theta_1 c\theta_2 s\theta_3) - 90c\theta_1 s\theta_2 - 9.8c\theta_1 \\ P_y = -101.44(s\theta_1 s\theta_2 c\theta_3 + s\theta_1 c\theta_2 s\theta_3) - 90s\theta_1 s\theta_2 - 9.8s\theta_1 \\ P_z = 101.44(c\theta_2 c\theta_3 - s\theta_2 s\theta_3) + 90c\theta_2 + 67.3 \end{cases} \tag{6}$$

where the symbols P_x, P_y, P_z represent displacement components in the x, y, and z directions respectively. So far, we have completed the forward kinematics analysis of the manipulator.

The ultimate goal of kinematic analysis is to make trajectory planning for the 3-DOF manipulator studied in this paper to achieve accurate medication taking. Trajectory planning involves not only the pose planning of path points, but also the planning of velocity and acceleration of path points. Therefore, Jacobian matrix needs to be solved, and the process is as follows

$$\begin{cases} dP_x = \dfrac{\partial P_x}{\partial \theta_1} d\theta_1 + \dfrac{\partial P_x}{\partial \theta_2} d\theta_2 + \dfrac{\partial P_x}{\partial \theta_3} d\theta_3 \\ dP_y = \dfrac{\partial P_y}{\partial \theta_1} d\theta_1 + \dfrac{\partial P_y}{\partial \theta_2} d\theta_2 + \dfrac{\partial P_y}{\partial \theta_3} d\theta_3 \\ dP_z = \dfrac{\partial P_z}{\partial \theta_1} d\theta_1 + \dfrac{\partial P_z}{\partial \theta_2} d\theta_2 + \dfrac{\partial P_z}{\partial \theta_3} d\theta_3 \end{cases} \tag{7}$$

By substituting Eq(6) into Eq(7) to solve the total differential, Eq(8) can be obtained

$$\begin{cases} dP_x = (101.44s\theta_1 s\theta_{23} + 90s\theta_1 s\theta_2 + 9.8s\theta_1)d\theta_1 - (101.44(c\theta_1 c\theta_{23} + 90c\theta_1 c\theta_2))d\theta_2 - (101.44c\theta_1 c\theta_{23})d\theta_3 \\ dP_y = -(101.44c\theta_1 s\theta_{23} + 90c\theta_1 s\theta_2 + 9.8c\theta_1)d\theta_1 - (101.44(s\theta_1 c\theta_{23} + 90s\theta_1 c\theta_2))d\theta_2 - (101.44s\theta_1 c\theta_{23})d\theta \\ dP_z = -(101.44s\theta_{23} + 90s\theta_2)d\theta_2 - (101.44s\theta_{23})d\theta_3 \end{cases} \tag{8}$$

where $s\theta_{23} = \sin(\theta_2 + \theta_3)$ and $c\theta_{23} = \cos(\theta_2 + \theta_3)$.

The velocity Jacobian matrix can reflect the mapping of the tiny transformation of the joint space in the basis coordinate system. We can use the velocity Jacobian matrix to solve for the velocity and acceleration of the end or we can plan the velocity and acceleration in the joint space in reverse. So, solving velocity Jacobian matrix is the core of kinematics analysis of the manipulator. The velocity Jacobian matrix is defined as

$$dP = J(\theta)d\theta \tag{9}$$

By rewriting Eq(8) into a matrix form satisfying the relation of Eq(9), the velocity Jacobian matrix of the manipulator can be written as

$$J(\theta) = \begin{bmatrix} 101.44s\theta_1 s\theta_{23} + 90s\theta_1 s\theta_2 + 9.8s\theta_1 & -(101.44c\theta_1 c\theta_{23} + 90c\theta_1 c\theta_2) & -101.44c\theta_1 c\theta_{23} \\ -(101.44c\theta_1 s\theta_{23} + 90c\theta_1 s\theta_2 + 9.8c\theta_1) & -(101.44s\theta_1 c\theta_{23} + 90s\theta_1 c\theta_2) & -101.44s\theta_1 c\theta_{23} \\ 0 & -(101.44s\theta_{23} + 90s\theta_2) & -101.44s\theta_{23} \end{bmatrix} \tag{10}$$

Since the manipulator designed and studied in this paper has no obstacle within the working range, and the initial position of the end of the manipulator in z direction is much higher than the lifting height of the scissor lift platform, it only needs to design a straight motion trajectory.

Assume that the starting point of the motion is (x_0, y_0, z_0) and the ending point is (x_n, y_n, z_n), both the start point and the end point are the position of the end of the manipulator relative to the base coordinate system. Suppose the number of interpolation steps of the trajectory is N, then the coordinates of each interpolation point $(x_{i+1}, y_{i+1}, z_{i+1})$ can be expressed as

$$\begin{cases} x_{i+1} = x_i + (x_n - x_0)/N \\ y_{i+1} = y_i + (y_n - y_0)/N \quad (i = 0,1,2,\cdots,N-1) \\ z_{i+1} = z_i + (z_n - z_0)/N \end{cases} \tag{11}$$

At the same time, we need to determine the attitude of each interpolation point to list the transformation matrix of the end of the manipulator relative to the base coordinate system. Considering that the manipulator cannot tip over in the process of gripping the medication movement, it should always maintain the attitude as shown in Fig.4, which means the x axis of the end coordinate system should be perpendicular to the x axis of the base coordinate system, so the rotation matrix Rot_{i+1} between the end coordinate system and the base coordinate system is

$$Rot_{i+1} = \begin{bmatrix} 0 & 0 & 1 \\ 0 & 1 & 0 \\ -1 & 0 & 0 \end{bmatrix} \begin{bmatrix} 1 & 0 & 0 \\ 0 & \cos q_{i+1} & -\sin q_{i+1} \\ 0 & \sin q_{i+1} & \cos q_{i+1} \end{bmatrix} \tag{12}$$

where the variable q represents the rotation angle about the x axis during the motion.

By combining the translation vector determined by Eq(12) and the rotation matrix Rot_{i+1} obtained by Eq(11), the homogeneous transformation matrix of each interpolation point relative to the base coordinate system can be written as

$$T_{i+1} = \begin{bmatrix} & & & x_{i+1} \\ & \mathrm{Rot}_{i+1} & & y_{i+1} \\ & & & z_{i+1} \\ 0 & 0 & 0 & 1 \end{bmatrix} \quad (i = 0,1,2,\cdots,N-1) \tag{13}$$

The joint angles $\theta_1, \theta_2, \theta_3$ corresponding to each interpolation point can be solved immediately by combining Eq(11) with the results obtained from Eq(1). The motion control of the 3-DOF manipulator can be completed by writing these angles into the motor driver program and performing smooth interpolation. So far, we have completed the trajectory planning of the 3-DOF manipulator, and then the actual working process simulation will be carried out in the simulation part.

4. Simulation analysis

The main working process of the medication delivery robot studied in this paper is to grab drugs in the storage compartments by 3-DOF robot manipulator. In this section, MATLAB will be used to simulate the process and draw the kinematic parameter curve of the simulation point. Meanwhile, the performance of linear trajectory planning and joint space planning for medication delivery process will be compared. The simulation point selected in this paper is the installation point of the hand, namely the center point of the wrist, as shown in Fig.5

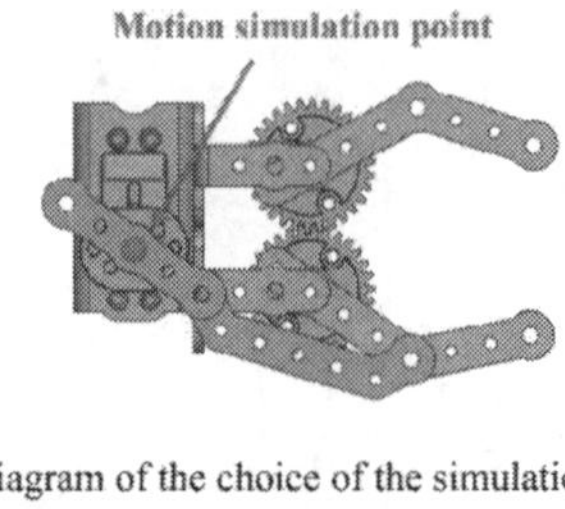

Fig.5 Diagram of the choice of the simulation point

Taking the example of gripping the medicine in the storage compartment on the left side of the manipulator. Firstly, the linear trajectory planning of the simulation point is simulated and its motion characteristics are analyzed. According to the working conditions, the starting point is set to the initial pose as shown in Fig.4, and the ending point is set to be directly above the storage cell and flush with the storage compartment, then the homogeneous transformation matrix T_0 and T_n of the starting point and the ending point are written as follows

$$T_0 = \begin{bmatrix} 0 & 0 & 1 & -0.001 \\ 0 & 1 & 0 & 0 \\ -1 & 0 & 0 & 0.260 \\ 0 & 0 & 0 & 1 \end{bmatrix} \tag{14}$$

$$T_n = \begin{bmatrix} 0.061 & 0.866 & 0.496 & -0.032 \\ 0.106 & -0.500 & 0.860 & -0.056 \\ 0.993 & 0 & -0.122 & 0.143 \\ 0 & 0 & 0 & 1 \end{bmatrix} \tag{15}$$

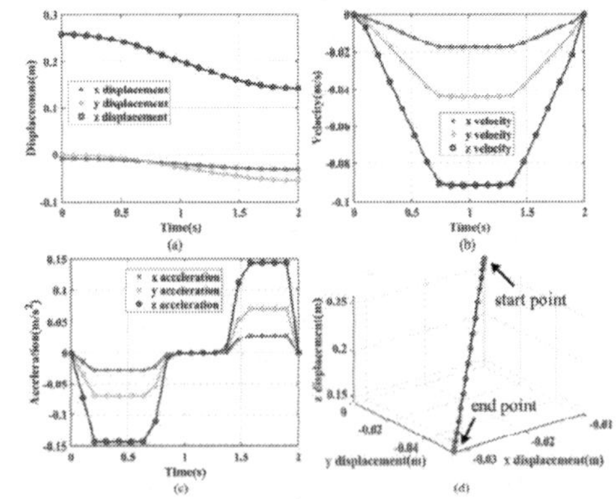

Fig.6 Diagram of the linear trajectory planning of the simulation point. (a)displacement components, (b)velocity components, (c)acceleration components, (d) the linear trajectory

Twenty motion interpolation points are set, and the corresponding homogeneous transformation matrix of each interpolation point can be obtained by using the trajectory planning function *ctraj* provided in MATLAB Robotics Toolbox, so as to obtain the displacement, velocity and acceleration components of the simulation point relative to base coordinate system x0-y0-z0. The curve of its relationship with time is shown in Fig.6(a)-(c). Then the *ikine*, the inverse kinematics solution function, is used to convert the homogeneous transformation matrix of the interpolation points solved by the *ctraj* function into the joint coordinate system, so that we can draw the trajectory of the simulation point. The drawn linear trajectory is shown in Fig.6(d).

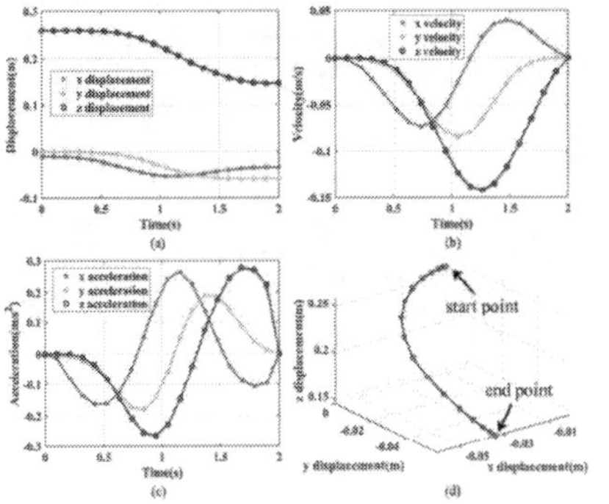

Fig.7 Diagram of the joint space trajectory planning of the simulation point. (a)displacement components, (b)velocity components, (c)acceleration components, (d) the trajectory of the end

Then we simulate the joint space planning of the simulation point. the initial conditions are set to be the same as the linear trajectory planning. Subsequently, the *jtraj* function is used to perform quintic polynomial interpolation for joint motion and the *fkine* function is used to complete the forward kinematics solution of the interpolation results. The generated displacement, velocity and acceleration component curves of the simulation point are shown in Fig.7(a)-(c). The motion trajectory is obtained through the forward kinematics solution, as shown in Fig.7(d).

It can be seen from the linear trajectory shown in Fig.6(d) that the displacement of the manipulator goes through a process from acceleration to deceleration. From Fig.6(c), it can be seen that the transition between interpolation points is not smooth enough. the end-effector suffers a part of flexible impact between 0s-0.5s and 1.5s-2s. Although such impact does not affect the mechanical structure, it seriously affects the positioning accuracy. As a drug taking robot, the motion accuracy of the robot manipulator is one of the problems we focus on. Therefore, smoother trajectory planning is needed to realize. Fig.7(d) shows that the joint space planning trajectory also experienced a process from acceleration to deceleration. But different from the linear trajectory planning, the acceleration curve shown in Fig.7(c) always changed smoothly, indicating that the terminal motion would not be affected by flexible impact in the process of motion and could achieve better motion accuracy than the linear trajectory.

Taking the above two trajectory planning methods into consideration, joint space planning obviously has better motion properties and is competent for the motion process of medication retrieval and delivery.

5. Conclusion

Due to the trend towards an ageing population, medication delivery robots are more important than ever to the market. In this research, a 3-DOF manipulator is used to optimize the structure of traditional drug delivery robot, and the kinematic equations of robotic manipulator is established by DH method. To obtain the velocity and acceleration curves of the robotic manipulator, simulations in MATLAB are carried out. Through the simulation results, we find that joint space planning is more suitable for the motion of the manipulator than linear trajectory planning.

References

1. Z. Wang and G. Chen, "Home intelligent assistance mobile robot for the elderly," 2022 IEEE 4th International Conference on Power, Intelligent Computing and Systems (ICPICS), Shenyang, China, 2022, pp. 873-878, doi: 10.1109/ICPICS55264. 2022.9873607.
2. R. Maan, A. Madiwale and M. Bishnoi, "Design and Analysis of 'Xenia: The Medi-Assist Robot' for Food Delivery and Sanitization in Hospitals," 2021 2nd Global Conference for Advancement in Technology (GCAT), Bangalore, India, 2021, pp. 1-7, doi: 10.1109/GCAT52182.2021.9587776.
3. S. P. A and A. Mohan, "Design and Simulation of an Autonomous Indoor Robot for Elderly Assistance," 2022 IEEE 2nd Mysore Sub Section International Conference (MysuruCon), Mysuru, India, 2022, pp. 1-5, doi: 10.1109/MysuruCon55714.2022.9972508.
4. Jonatan E. Salinas-Avila, Hugo G. Gonzalez-Hernandez, Nashely Martinez-Chan, Carlos M. Bielma-Avendano, Ximena A. Salinas-Molar, and Martin O. Garcia-Garcia. 2022. Assistant Delivery Robot for Nursing Home using ROS: Robotic prototype for medicine delivery and vital signs registration. In The 2022 5th International Conference on Electronics, Communications and Control Engineering (ICECC 2022). Association for Computing Machinery, New York, NY, USA, 141–148. https://doi.org/10.1145/3531028.3531051
5. Dibya Prokash Sarkar. 2018. A Nurse Bot for Elderly People. In Proceedings of the 2018 ACM International Joint Conference and 2018 International Symposium on Pervasive and Ubiquitous Computing and Wearable Computers (UbiComp '18). Association for Computing Machinery, New York, NY, USA, 1798–1799.
https://doi.org/10.1145/3267305.3277821
6. A. Patel, P. Sharma and P. Randhawa, "MedBuddy: The Medicine Delivery Robot," 2021 9th International Conference on Reliability, Infocom Technologies and Optimization (Trends and Future Directions) (ICRITO), Noida, India, 2021, pp. 1-4, doi: 10.1109/ICRITO51393.2021.9596130.

7. O. A. N. Eidsvik, B. O. Arnesen and I. Schjølberg, "SeaArm-A Subsea Multi-Degree of Freedom Manipulator for Small Observation Class Remotely Operated Vehicles," 2018 European Control Conference (ECC), Limassol, Cyprus, 2018, pp. 983-990, doi: 10.23919/ECC.2018.8550326.
8. P. T. Katsiaris, G. Adams, S. Pollard and S. J. Simske, "A kinematic calibration technique for robotic manipulators with multiple Degrees of Freedom," 2017 IEEE International Conference on Advanced Intelligent Mechatronics (AIM), Munich, Germany, 2017, pp. 358-363, doi: 10.1109/AIM.2017.8014043.
9. Nisar, S., Endo, T., and Matsuno, F. (March 23, 2017). "Design and Kinematic Optimization of a Two Degrees-of-Freedom Planar Remote Center of Motion Mechanism for Minimally Invasive Surgery Manipulators." ASME. J. Mechanisms Robotics. June 2017; 9(3): 031013.
10. B. You, S. -Y. Liu, T. -Y. Zhang, Z. -H. Pang and X. Li, "Design of Collaborative Transportation System via Multiple Mobile Manipulators," 2020 39th Chinese Control Conference (CCC), Shenyang, China, 2020, pp. 4765-4770, doi: 10.23919/CCC50068.2020.9189550.
11. T. H. Noventino, M. R. Rosa and A. Z. Fuadi, "PID Control Design and Kinematic Modelling of 3-DoF Robot Manipulator," 2022 International Conference on Electrical Engineering, Computer and Information Technology (ICEECIT), Jember, Indonesia, 2022, pp. 88-94, doi: 10.1109/ICEECIT55908.2022.10030325.
12. N. J. Raj, K. Iyer and A. K. Dash, "Design, fabrication, kinematic analysis and control of a 3-DOF serial manipulator," 2016 International Conference on Next Generation Intelligent Systems (ICNGIS), Kottayam, India, 2016, pp. 1-6, doi: 10.1109/ICNGIS.2016.7854020.
13. F. Liu, G. Gao, L. Shi and Y. Lv, "Kinematic analysis and simulation of a 3-DOF robotic manipulator," 2017 3rd International Conference on Computational Intelligence & Communication Technology (CICT), Ghaziabad, India, 2017, pp. 1-5, doi: 10.1109/CIACT.2017.7977291.
14. M. H. A. Nizami et al., "Proximal Actuation of an Elastically Loaded Scissors Mechanism for the Leg Design of a Quadruped Robot," in IEEE Access, vol. 8, pp. 208240-208252, 2020, doi: 10.1109/ACCESS.2020.3037250.
15. B. Han, D. Zheng, Y. Xu, J. Yao and Y. Zhao, "Kinematic Characteristics and Dynamics Analysis of an Overconstrained Scissors Double-Hoop Truss Deployable Antenna Mechanism Based on Screw Theory," in IEEE Access, vol. 7, pp. 140755-140768, 2019, doi: 10.1109/ACCESS.2019.2930101.
16. Y. Sun, F. Pancheri and T. C. Lueth, "Kinematic Modeling of Scissor-Mechanism-Based Curvilinear Actuator," IECON 2022 – 48th Annual Conference of the IEEE Industrial Electronics Society, Brussels, Belgium, 2022, pp. 1-6, doi: 10.1109/IECON49645.2022.9968802.
17. Z. Li, W. Li and Y. Li, "The kinematics analysis and simulation of the transposition manipulator," Proceedings of 2011 International Conference on Electronic & Mechanical Engineering and Information Technology, Harbin, China, 2011, pp. 2959-2962, doi: 10.1109/EMEIT.2011.6023668.

Research Status of Cathode Materials for Lithium Ion Batteries

Zihao Lin*

School of Materials Science and Engineering, China University of Geosciences Beijing, Bejing, China.

Abstract: Thanks to the promotion of new energy vehicles, the industry of lithium-ion batteries has ushered in its booming period. The current industry of lithium ion batteries is in rapid development with great potential. Therefore, many researchers have turned to focus on lithium ion batteries to obtain better lithium ion batteries. In this paper, the literature review of cathode materials for lithium ion batteries is to be carried out from the following aspects, including the overview of lithium ion batteries, their basic performance indexes, and the classification and preparation methods of cathode materials. Besides, the present situation and modification strategies of cathode materials for lithium ion batteries will be further analyzed, so as to improve their electrochemical performance.

Keywords: Lithium Ion Battery; Cathode Material; Lithium Iron Phosphate; Lithium Cobaltate; Secondary Battery

1. Research Background of Lithium Ion Batteries

1.1 Development of Lithium Ion Batteries

During the development of social civilization, science, and technology, it is precisely the change in energy structure that has led to historic progress and improved people's lives. Since the 1990s, with the rapid development of electronic information technology, many electronic devices need an energy supply to realize the characteristics such as small size, lightweight, high energy density, and long cycle life. Under this background, lithium ion batteries came into being, which not only promoted the upgrading of new electronic equipment, but also directed society towards intelligence and environmental friendliness.

However, the research on lithium ion batteries is not always favorable. In 1913, Gilbert N Lewis, a professor at Massachusetts Institute of Technology, published the paper *The Potential of the Lithium Electrode* in the *Journal of the American Chemical Society*, which expounded and measured the electrochemical potential of lithium metal for the first time. Thus, it was regarded as the earliest systematic study of lithium metal batteries [1]. However, due to lithium being a reactive metal, it has not attracted much attention.

In 1976, Silbernagel et al. [2] thought that lithium can be inserted into $LixTiS_2$ with a small lattice expansion effect, while lithium ions can gradually occupy the octahedral position in the interlayer space. By demonstrating this process, they obtained a patent. The battery has harvested wide attention because of its high energy density. However, it is easy to produce lithium dendrites during the charge-discharge cycle, which leads to short circuits and potential safety hazards.

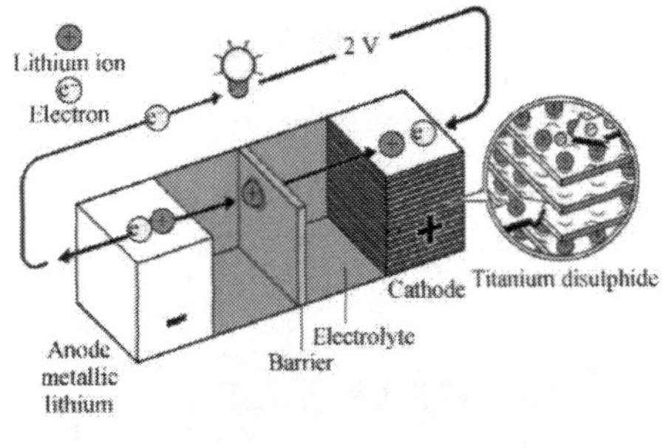

Fig. 1 Lithium-Based Battery with LixTiS2 as Cathode Material [3]

In 1980, the research group of John B. Goodenough first discovered that lithium cobaltate can be used as a cathode material. They believed that $LiCoO_2$ has a large layer spacing in the transition metal oxide $LiXO_2$ (X=V, Cr, Ni, Fe) with a layered structure. Besides, the low spin Co^{4+}/Co^{3+} redox pair has a higher affinity for electrons, which makes oxygen polarize cobalt stronger to obtain 4V voltage and higher energy density. In the following years, $LiNiO_2$ and $LiMn_2O_4$ were discovered one after another [4].

In 1997, Professor Goodenough [5] reported that lithium iron phosphate is a crystal with an ordered olivine structure, in which ferroxide hexahedron is connected with each other on the same side. This structure perfectly balances the redox energy of lithium iron phosphate electrodes with the mobility of lithium ions in the crystal,

* Corresponding author: 2118329060@qq.com

and its characteristics can meet the requirements of power lithium ion batteries, with its capacity, cycle performance, and safety obviously improved. Moreover, this material is cheaper and safer than lithium cobaltate, so it has been greatly popularized in the next 20 years.

1.2 Working Mechanism of Lithium Ion Batteries

The working mechanism of a lithium ion battery is the same as other secondary batteries, that is, the chemical energy of the battery is converted into electric energy and output to the external circuit through the discharge process. Then, the lithium ion battery is restored to its original state through the reverse electrification of the external power supply during the charging. At that time, lithium ions and electrons in the cathode are separated from the lattice, while lithium ions migrate to the anode in the electrolyte, and electrons migrate to the anode through the external circuit. Finally, they are embedded in the anode material by compounding into lithium atoms or alloying there. The discharge is opposite to the charging when electrons drive electronic devices through external circuits [6].

Taking graphite/lithium cobalt oxygen battery as an example [7], lithium ions in cathode $LiCoO_2$ move out during the charging, pass through the electrolyte, and embed between carbon layers of graphite to form lithium-carbon interlayer compounds in the battery. The discharge is just the opposite, that is, lithium ions move out from the interlayer of the graphite anode, pass through the electrolyte, and enter the $LiCoO_2$ cathode.

The reaction of each battery electrode and the battery itself is as follows:

Anode: $6C+xLi^{+}+xe^{-}=Li_xC_6$

Cathode: $LiCoO_2=xLi^{+}+Li_{1-x}CoO_2+xe^{-}$

Total Battery Reaction: $LiCoO_2 + 6C=li_{1-x}CoO_2 + Li_xC_6$

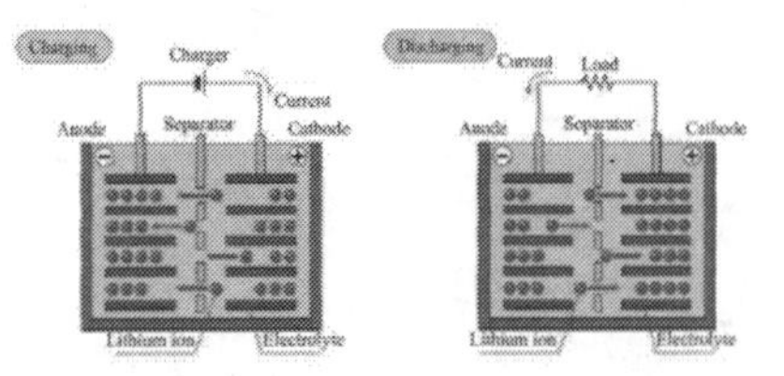

Fig. 2 Schematic Diagram of Working Mechanism of Lithium Ion Battery [8]

2. Basic Performance Indicators of Lithium Ion Batteries

2.1 Capacity

2.1.1 Concepts

The battery capacity represents the amount of electricity discharged by the battery under certain conditions, which is equal to discharge current × discharge time, usually in Ah. As for batteries with different voltages, Wh is their unit. Battery capacity is divided into actual capacity and theoretical capacity. Actual capacity refers to the amount of electricity that can be given under a certain discharge system, while theoretical capacitance refers to the amount of electricity given by all active substances participating in battery reactions.

2.1.2 Influencing Factors

The discharge capacity decreases with the increasing discharge depth [9]. The larger the discharge rate, the faster the battery capacity declines. When the battery discharge rate is 1.5 ~ 3.0, the discharge capacity begins to show signs of recovery. Meanwhile, an excessive discharge rate will lead to battery damage and the degradation of battery performance is affected by temperature. The best working temperature of an ion battery is 20 ℃-40 ℃, with low or high temperatures affecting the battery life [10].

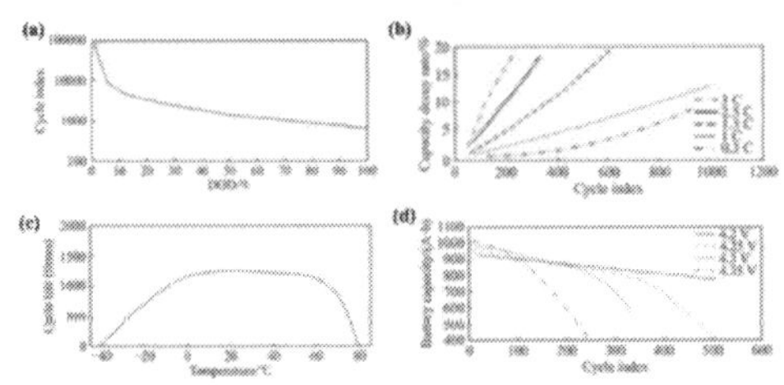

Fig. 3. (a) Relationship Between Battery Cycle Times and Discharge Depth

(b) Battery decay Rate at Different Discharge Rates

(c) Curve of Lithium Battery Life at Different Temperatures

(d) Battery Capacity Decay at Different Charging Cut-Off Voltages [9]

2.2 Specific Capacity

2.2.1 Concepts

Specific capacity has two types. One is the mass-specific capacity, which is what a unit mass battery emits, equal to capacity/mass, with mA·h/g as its unit. The other is volume-specific capacity, which is the amount of electricity that can be released per unit volume of the battery, equal to capacity/volume, with mA·h/L as its unit.

2.2.2 Influencing Factors

With the increasing cycle times and discharge rate, the specific capacity decreases. The specific surface area and porosity of electrode materials will directly affect the mass and diffusion rate of active substances in the electrode, thus affecting the specific capacity of the battery. Moreover, it can be affected by the selection and matching degree of cathode and anode pairs. For example, the combination of cathode and anode with high energy density can improve the specific capacity. The type, concentration, and stability of electrolytes will change the specific capacity by affecting the ion transport rate and electrochemical reaction.

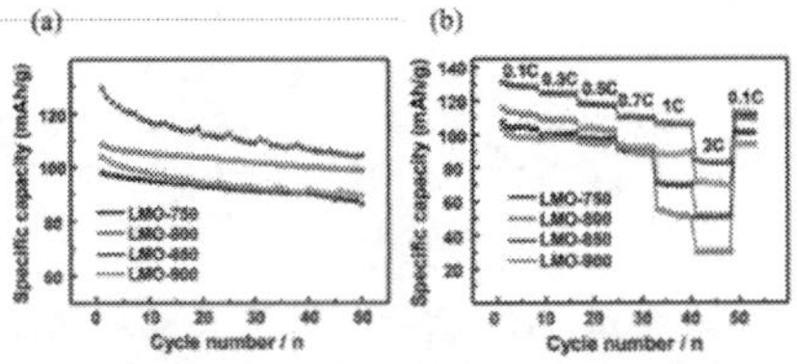

Fig. 4. Cycle Performance of LMO-750, LMO-800, LMO-850 and LMO-900 at 0.3 C

Rate Performance Diagram of LMO-750, LMO-800, LMO-850 and LMO-900 [11]

2.3 Self-Discharge Rate

2.3.1 Concepts

The self-discharge rate refers to the holding ability of the battery's stored electricity under certain conditions when the battery is in an open circuit state. There are always impurities in materials, which inevitably leads to self-discharge.

2.3.2 Influencing Factors

The higher the ambient temperature, the higher the activity of electrochemical materials, the more intense the side reaction, and the more obvious the chemical self-discharge. Ion batteries are easily affected by air pollution and air humidity. High air humidity and conductive particles will lead to an increase in conductivity. The general trend is that the higher the charge of the ion battery, the greater the potential difference, the more intense the side reaction, and the higher the self-discharge rate. Meanwhile, the cycle will cause an internal micro-short circuit and reduce physical self-discharge. Different impurities will affect the self-discharge rate of battery formation. After the battery is charged, Li^+ in the cathode material comes out and is embedded in the anode graphite through the electrolyte. Metal impurities located between the cathode and the separator are oxidized and dissolved into the electrolyte by the cathode to generate metal ions, which are continuously reduced and deposited on the anode surface. Slowly deposited metal dendrites pierce the separator and contact the cathode, forming micro short circuit points and resulting in large self-discharge [12].

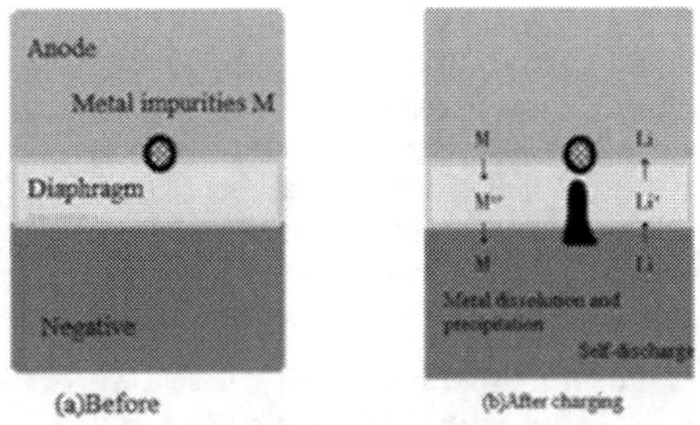

Fig. 5 Principle of Battery Self-Discharge Caused by Metal Impurities Between Cathode and Separator [12]

2.4 Energy Density

2.4.1 Concepts

Energy density refers to the ratio of the energy that can be charged to the mass or volume of the energy storage medium for a given electrochemical energy storage device. Mass energy density is equal to battery capacity × discharge platform/battery weight, with Wh/kg as its unit. Volume energy density is equal to battery capacity × discharge platform/volume, with Wh/L as its unit.

2.4.2 Influencing Factors

The discharge rate of a lithium ion battery refers to the current value required when the battery releases its rated capacity within the specified time. Different discharge platforms have various energy densities. In addition, battery size affects energy density. Increasing the proportion of positive active materials will improve the energy density. More negative active materials are needed to accommodate the swimming lithium ions and store energy. Up against the constant total amount of positive active material, only when as many lithium ions as possible are deintercalated from the cathode, that is, the specific capacity is increased, can the energy density be improved.

2.5 Power Density

2.5.1 Concepts

Power density refers to the ratio of the output power of the battery to its weight, which is usually expressed in W/kg.

2.5.2 Influencing Factors

Energy density refers to the power that a unit mass of a battery can hold. The higher the energy density, the higher the power density. Increasing the proportion of lithium elements and letting many lithium ions carry energy repeatedly is a main measure. For example, lithium hexafluorophosphate has good electrochemical stability and conductivity, but poor thermal stability. With moderate conductivity, lithium perchlorate has battery safety problems as a strong oxidant [13]. Lithium tetrafluoroborate has good low-temperature performance, high conductivity, wide electrochemical window, and good thermal stability. However, it is difficult to form a stable SEI film on the anode surface [14]. Compared with $LiPF_6$, LiFSI has better thermal stability, hydrolysis stability, and higher ionic conductivity, which does not corrode the positive aluminum foil [15].

3. Classification of Cathode Materials for Lithium Ion Batteries

3.1 Layered Oxide $LiMO_2$

3.1.1 $LiCoO_2$ Cathode Material

At present, $LiCoO_2$ with the α-NaFeO2 layered structure is the main cathode material for lithium ion batteries, belonging to the hexagonal system and R-3m space group. $LiCoO_2$ is composed of LiO_6 and CoO_6 octahedrons, in which Li and Co occupy 3b and 3a alternately and respectively, and O occupies 6c, arranged in cubic dense packing [16]. It has the advantages of high voltage, stable discharge, suitable for large current discharge, high specific energy, good cycle performance, and a simple preparation process.

However, as a cathode material for large-scale batteries, $LiCoO_2$ has many problems. With the increasing cycle times, the battery capacity decays greatly, with its poor overcharge resistance and thermal stability. Ohzuku et al. [18] found that when the charging voltage reaches 4.5V, the hexagonal phase begins to change into the monoclinic phase, which is related to the Jahn-Teller distortion of the CoO_6 octahedron. Therefore, in order to reduce irreversible capacity loss and maintain a stable voltage platform, the charging cut-off voltage is usually limited below 4.2V. Aiming to overcome the shortcomings of $LiCoO_2$, the doping coating is used to increase its capacity, improve cycle performance, and promote structure stability.

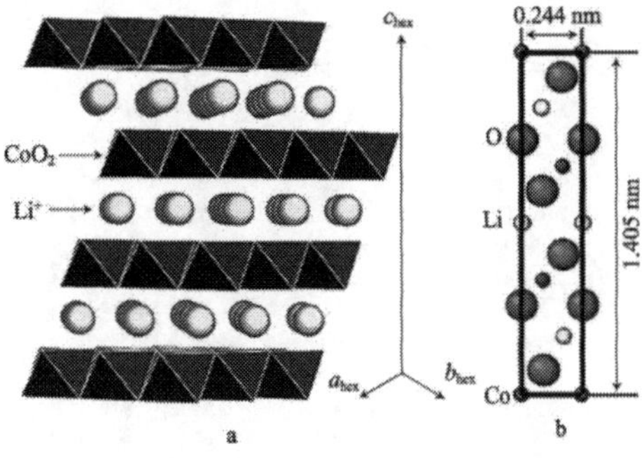

Fig. 6 Crystal Structure Model (a) and Unit Cell Parameter (b) of Layered LiCoO2 with Space Group R-3m [19]

3.1.2 Layered Structure of $LiMnO_2$

The Mn reserves in China are abundant and the application of $LiMnO_2$ can greatly reduce the cost of commercial lithium ion batteries. Therefore, it has become a research hotspot now. The crystal structure of layered LiMnO2 can be divided into a monoclinic phase and an orthogonal phase. The orthorhombic phase is β-$NaMnO_2$ structured and belongs to the Pmnm space group, with a=0. 2805nm, b=0. 5757nm, c=0. 4572nm, and Z=2 as its lattice parameters. LiO_6 and MnO_6 octahedrons are aligned alternately in the lattice. The MnO_6 octahedron skeleton is elongated by 14% due to the Jahn-Teller effect caused by the migration of Mn to the lithium layer. The monoclinic phase is a-$NaFeO_2$ structured, which is similar to that of $LiCoO_2$. Besides, it belongs to the C2/m space group, with a=0. 5439nm, b=0. 2809nm, c=0. 5395nm, and Z=2 as its lattice parameters. MnO_6 octahedron forms an ordered transition metal layer, and the Li layer is distributed in turn. However, the monoclinic phase has poor thermal stability compared with the orthogonal phase, which cannot be synthesized by the traditional method of solid-state reaction, so it is generally prepared by ion exchange with $NaMnO_2$ [20]. $LiMnO_2$ has the following advantages such as high theoretical capacity, overcharge, over-discharge resistance, good thermal stability, abundant resources, no pollution, and low price [21].

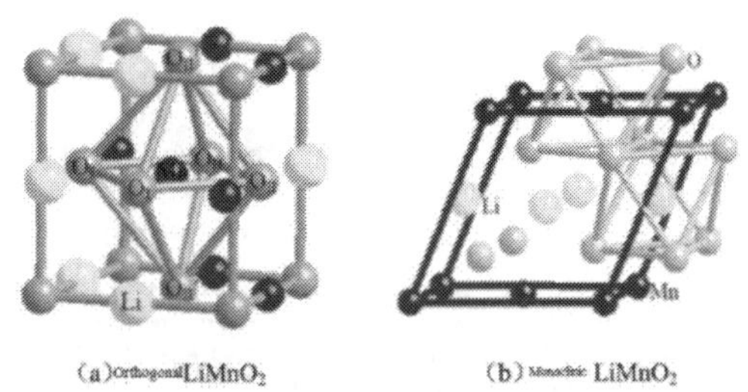

Fig. 7 (a) Orthogonal LiMnO2

(b) Structure of Monoclinic LiMnO2 [22]

However, Mn^{3+} produced during charging and discharging leads to Jahn-Teller distortion [23], which distorts the layered structure into the rhombic structure, seriously damaging the electrochemical performance of layered $LiMnO_2$ material that is uniform orthogonal. Moreover, its preparation conditions are harsh, so it cannot be commercialized up to now.

3.2 Spinel Oxide $LiMn_2O_4$

The crystal structure of $LiMn_2O_4$ belongs to the Fd-3m space group. The oxygen skeleton of $LiMn_2O_4$ is the same as that of $LiMO_2$. O is located at $32e^-$ to form a face-centered cubic structure, and Mn occupies 50% of the oxygen octahedron at 16d, in which trivalent and tetravalent Mn ions account for half each, with Li located at 8a in the oxygen tetrahedron. In this structure, the Mn_2O_4 array is formed by a MnO_6 octahedron with shared edges, and the empty octahedron at 16c provides a diffusion network for lithium ions. Moreover, the rapid migration of lithium ions in this channel makes $LiMn_2O_4$ show excellent rate performance [24].

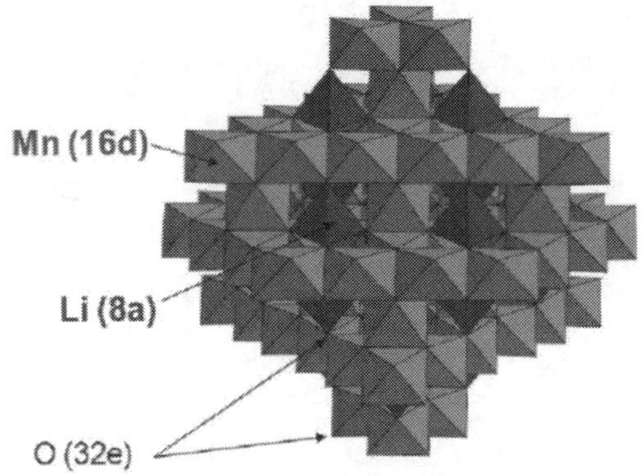

Fig. 8 Space Group Structure of Spinel Lithium Manganese Oxide [25]

It is found that $LiNi_{0.5}Mn_{1.5}O_4$ obtained by replacing part of Mn in $LiMn_2O_4$ doped with Ni has good high voltage stability, and the actual discharge specific capacity even reaches 146mA·h/g[26] [27]. As for $LiNi_{0.5}Mn_{1}$.5O4, its three-dimensional lithium ion transmission channel and high operating voltage ensure the high energy density and power density of lithium ion batteries.

3.3 Olivine Phosphate $LiMPO_4$

3.3.1 *$LiFePO_4$ Cathode Material*

The crystal of lithium iron phosphate has olivine ordered structure and belongs to the orthorhombic crystal system in the Pnmb space group. Concerning the crystal structure of lithium iron phosphate, iron and phosphorus react with oxygen to form FeO_6 octahedron and PO_4 hexahedron respectively. FeO_6 octahedrons and PO_4 tetrahedrons alternately link to form a space skeleton, and Li atoms fill the voids of the skeleton to form LiO_6 octahedrons. P-O covalent bond energy is strong, which makes it high in thermal stability and dynamic stability. Macroscopically, the battery has high safety [28]. In addition, $LiFePO_4$ has a high theoretical capacity, a stable discharge platform, stable structure. It also has the advantages of other cathode materials, such as abundant resource reserves, low prices, and environmental protection [29]. Therefore, the cathode material of lithium iron phosphate has become one of the most potential ones for power lithium ion batteries, which has become a hot topic studied by many researchers [30].

However, $LiFePO_4$ also has some problems. Because there is no continuous octahedral network of $LiFePO_4$, its electronic conductivity is only 10^{-7}~$10^{-9}S{\cdot}cm^{-1}$ at room temperature, which becomes the bottleneck restricting its large-scale utilization. In order to solve these shortcomings, the commonly used methods mainly include nanocrystallization of particles, carbon coating, and ion doping. Nanocrystallization can shorten the transport distance of Li^+ in $LiFePO_4$ particles, and make Li^+ deintercalation faster and more complete, thus improving the electrochemical performance of materials [31].

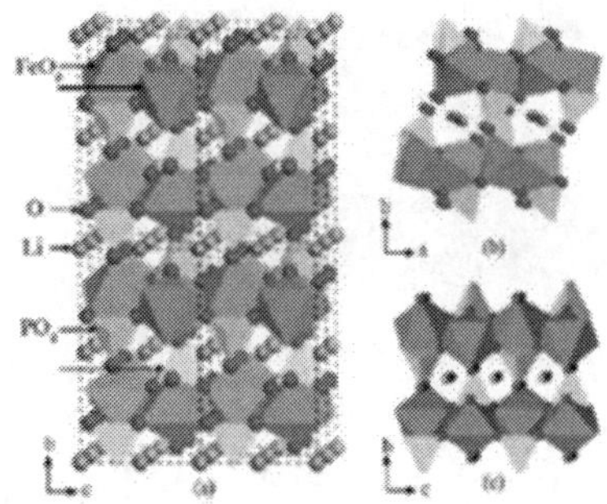

Fig. 9 Schematic Diagram of Crystal Structure of Olivine LiFePO4 [32]

3.3.2 *$LiMnPO_4$ Cathode Material*

The crystal structure of $LiMnPO_4$ is olivine and belongs to the orthorhombic system, with a=0. 0611nm, b=0. 1045nm, and c=0. 0475nm as its unit cell parameters. O is a slightly twisted hexagonal compact packing, while Li and Mn occupy the octahedral O. Meanwhile, Li is in 4a and forms LiO_6 octahedral with O. Mn is at 4c and forms MnO_6 octahedron with O. P and O form PO_4 tetrahedron. In this structure, two LiO_6 octahedrons including one MnO_6 octahedron and one PO_4 tetrahedron share the same side. This alternating structure also makes the material stable in lithium ion migration channels. At the same time, in the tetrahedral structure of PO_4, the P-O bond has strong covalence, which enhances the structural stability of $LiMnPO_4$. During the charging and discharging, O in the lattice is not easy to escape, which makes the material extremely safe [33].

However, problems also occur in $LiMnPO_4$ for the 2eV band gap leading to low electronic conductivity. The disproportionation reaction of Mn^{3+} and the side reaction of electrolytes are also shortcomings [34].Martha et al. [35] used carbon black as the carbon source to prepare carbon-coated $LiMnPO_4/C$ as the cathode material. The synthesized cathode material showed excellent electrochemical performance, and the charge-discharge capacity reached $140mAh{\cdot}g^{-1}$ at 0.10 times.

Carbon material, as the coating material of electrode material, shows electrochemical activity to electrolyte at low potential and does not oxidize at high voltage [36] with good chemical stability, which can be used as a good protective layer to delay the degradation of material and slow down the capacity attenuation during battery charge/discharge cycle.

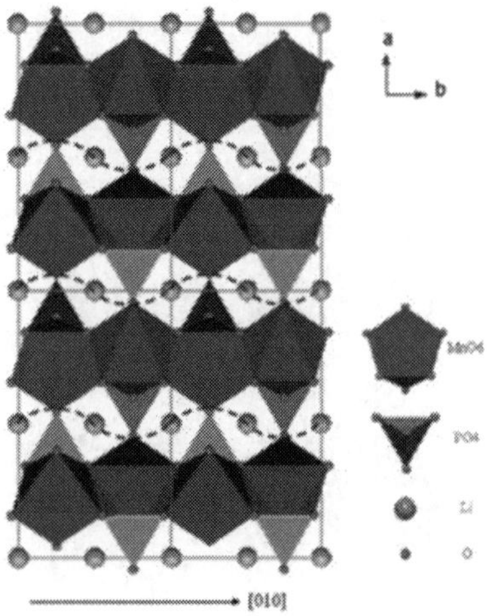

Fig. 10 Schematic Diagram of LiMnPO4 Structure [37]

4. Preparation Method of Cathode Material for Lithium Ion Battery

4.1 Sol-gel Method

The preparation of the sol-gel method is to put the precursor in a specific solvent system and mix it uniformly so that it undergoes hydrolysis and condensation chemical reaction to form a transparent and uniform colloidal system. [38].

For example, Zhang et al. [39] prepared $Li_3V_2(PO_4)_3$ materials by methods of sol-gel and solid-phase. The LVP materials prepared by the solid-phase method have a large and uneven particle size, while those prepared by the sol-gel method have good particle uniformity and small particle size.

4.2 Method of High-Temperature Solid State

At present, a high-temperature solid state is still the mainstream method to prepare cathode materials. Generally, it refers to ball milling and mixing powders evenly. Meanwhile, a large number of composite oxides are generated through contact, reaction, nucleation, and crystal growth reactions between solid interfaces [41]. The main technological processes are as follows. 1. The raw materials of each element are fully physically mixed. The common methods are agate mortar hand grinding and machine ball milling. Compared with the former, the latter is more uniform and smaller in granularity to obtain a powder with a larger specific surface area, which is beneficial to crystal growth during sintering. 2. Put the uniformly mixed powder in the corresponding atmosphere, and adjust the sintering time, temperature, batching ratio, etc. before calcining it. This method is simple to operate and control with short process procedures and easy industrial production. However, it takes a long calcination time with high energy consumption and low efficiency. Besides, it's poor in the uniformity of target materials and electrochemical performance of prepared target products, apart from its difficult formula control [42] [43].

4.3 Coprecipitation Method

The coprecipitation method means that two or more cations exist in the solution in a homogeneous phase. After the precipitation reaction, uniform precipitation of various components can be obtained. It is an important method to prepare ultrafine powder of composite oxide containing two or more metal elements [44] [45].

For example, Lin Zhicheng et al. [46] prepared the black precursor by using methanol as solvent and adding a lithium source at 450 °C for 8h. The lithium, lithium carbonate, and lithium hydroxide are uniformly mixed with lithium acetate as lithium sources. Then, they sintered the red-brown powder after lithium mixing at 800 °C to synthesize different lithium-rich cathode materials. Because of the same morphology of precursors, the lithium-rich cathode materials synthesized by sintering at 800 °C with different lithium sources keep similar morphology to the precursors before sintering, all in elliptical long blocks. It is found that the length of the three samples is from 3.5 μm to 5 μm with relatively uniform size, but Li_2CO_3-800 are more seriously agglomerated, and the morphology of CH_3COOLi-800 and Li_2CO_3-800 remains more complete with dense nanoparticles on the surface. The nanoparticles on the CH_3COOLi-800 surface are smaller with more micropores. However, LiOH-800 is incomplete with a smoother surface.

4.4 Hydrothermal Method

The hydrothermal method is one that dissolves the required raw materials into water (or solvent) and seals them in a container with certain pressure. Then, the powder is dissolved and recrystallized to obtain cathode material powder due to hydrothermal reaction. In order to realize good electrochemical performance, it is necessary to select appropriate reaction conditions, including raw material particle size, water or solvent temperature, pressure, time, pH, heating, stirring rate, solvents, adjuvants, and dispersants, etc [47][48].

The hydrothermal method can be used to prepare materials with fine grains in a simple preparation method, which takes a short time. However, the preparation of materials by hydrothermal method requires higher instruments, which is also the main factor affecting the industrialization of the hydrothermal method [49].

5. Conclusion and Prospect

With the development of science and technology, several technical barriers to lithium ion batteries have been solved and lithium-ion batteries have stepped onto the stage with a large-scale application. Due to the emergence of new energy vehicles and the support of relevant policies, the industry of lithium ion batteries has developed rapidly, with more and more people beginning to pay attention to it.

Cathode materials, as the bottleneck and focus of lithium ion battery research in recent years, are a necessary step for researchers to pursue their higher performance. There are two ways to clear the barriers in the research of cathode materials. One is to explore materials and constantly find new elements, organic or inorganic, layered or 3D. The second is the modification, which can also significantly improve the electrochemical performance of the original cathode material by doping metal ions and conductive agents as well as controlling the particle size. From the recent literature, new topics appear in the cathode materials of lithium ion batteries. Presumably, in the next few years, cathode materials with high energy density, high conductivity, and good recycling will emerge to improve the existing energy structure.

References

1. Lewis, G. N. & Keyes, F. G. (1913). The potential of the lithium electrode. *Journal of the American Chemical Society*, *35*(4), 340.
2. Silbernagel, B. G. & Whittinggham, M. S. (1976). An NMR study of the alkali metal intercalation phase Li_xTiS_2:Relation to structure,thermodynamics, and ionicity. *Journal of Chemical Physics*, *64*(9), 3670-3673.
3. Whittingham, M. S. (2021). Lithium titanium disulfide cathodes. *Nat Energy*. (6), 214.
4. Zhang, J. F. (2020). Development and prospect of lithium-ion batteries from the 2019 Nobel Prize in Chemistry. *Chemical Engineering Design Communications*, *46*(3),2.
5. Padhi, A. K., Goodenough, J. B., Nanjundaswamy, K. S. (1997). Phospho-olivines as positive-electrode materials for rechargeable lithium batteries. *Journal of the Electrochemical Society*, (144), 1188-1194.
6. Zhao, L. Z. (2009). Summaries of lithium-ion battery and progress of its anode materials. *Guangdong Chemical Industry*, *36*(5), 3.
7. Dai, Y. S., Xian, Q. Y. & Huang, Z. M. (2002). Research on lithium-ion battery performance. *Chinese Battery Industry*, *7*(5), 258-261.
8. Cheng, X. Q. (2008). Chemical power source. *Beijing: Chemical Industry Press*.
9. Xie, S. B., Hu, X. S., Zhang, Q. K. et al. (2020). Aging-aware co-optimization of battery size,depth of discharge,and energy management for plug-in hybrid electric vehicles. *Power Sources*, 450:227638.
10. Engbroks, L., Görke, D., Schmiedler, S. et al. (2019). Combined energy and thermal management for plug-in hybrid electric vehicles-analyses based on optimal control theory. *IFAC Papers OnLine*, *52*(5), 610.
11. Li, Z. K. (2022). Preparation and modification of spinel type $LiMn_2O_4$ cathode materials. *Xuzhou: Master Dissertation of China University of Mining and Technology*. DOI:10.27623/d.cnki.gzkyu.2022.001728.
12. Pei, P. C., Chen, J. Y. & Wu, Z. Y. (2019). Self-discharge mechanism and measurement methods for lithium ion batteries. *Journal of Tsinghua University (Science and Technology)*, *59*(1), 53-65.
13. Me, D. Y., Liu, Y. B. & Dai, C. S. (2019). Research progress in liquid organic electrolyte for li-ion battery. Battery Bimonthly, 49(1), 68.
14. Ehteshami, N., Ibing, L., Stolz, L. et al. (2020). Ethylene carbonate-free electrolytes for Li-ion battery:Study of the solid electrolyte interphases formed on graphite anodes. *Power Sources*, 451: 227804.
15. Li, M., Qiu, J. Y. & Yu, Z. B. (2015). New use of conducting salts in electrolytes of high power Li-ion batteries. *Chinese Journal of Power Sources*, *39*(1), 191.
16. Mizushima, K., Jones, P. C., Wiseman, P. J. & Goodenough, J. B. (1980). $Li_xCoO_2(0<x<1)$:A new cathode material for batteries of high energy density. *Mater. Res. Bull*, *15*(6), 783-789.
17. Goodenough, J. B., Mizushima, K. & Takeda, T. (1980). Solid-solution oxides for storage-battery electrodes. *Jpn.J. Appl. Phys*, 19: 305-313.
18. Shao-hao-horn, Y., Croguennec, L., Delmas, C. et al. (2003). Atomic resolution of lithium ions in $LiCoO_2$. *Nature Materials*, *2*(7): 464-467.
19. Myung, S. T., Komaba, S. & Kumagai, N. (2002). Hydrothermal synthesis and electrochemical behavior of orthorhombic LiMnO2.
20. Armstrong, A. R. & Bruce, P. G. (1996). Synthesis of layered $LiMnO_2$ as an electrode for rechargeable lithium batteries. *Nature*, 381, 499-500.
21. Zhao, L. Z., Chen, Y. W. & Wang, G. R. (2010). Raman spectra study of orthorhombic $LiMnO_2$. *Solid State Ionics*, 181: 1399-1402 .
22. Zhao, H. Y., Liu, X. Q. & Zhang, Z. (2013). Research progress of $LiMnO_2$ cathode material for lithium-ion rechargeable batteries. *Electronic Components and Materials*, *32*(6), 1-6.
23. Jolanta, D., Christian, L., Pawel, P. et al. (2014). Hooked on switch:strain-managed cooperative Jahn-Teller effect in $Li_{0.95}Mn_{2.05}O_4$ spinel. *Rsc Advances*, *4*(110), 65205-65212.
24. Thackeray, M. M. (1997). Manganese oxides for lithium batteries. *Progress in Solid State Chemistry*, *25*(1-2), 1-71.
25. Aurbach, D. Levi, M. D., Gamulski, K., Markovsky, B., Salitra, G., Levi, E., Heider, U., Heider, L. & Oesten, R. (1999). Capacity fading of $Li_xMn_2O_4$spinel electrodes studied by XRD and electroanalytical techniques. *J.Power*, 81-82,472-479.
26. Yamada, A., Tanaka, M. & Teller, J. (1995). Structural phase transition around 280K in LiMn2O4. *Materials Research Bulletin*, *30*(6), 715-721.
27. Spence, S. L., Hu, A., Jiang, M. et al. (2022). Mapping lattice distortions in $LiNi_{0.5}Mn_{1.5}O_4$ cathode materials. *ACS Energy Letters*, 7:690-695.
28. Zhang, Y. J., Zhu, Z. Y., Dong, P. et al. (2017). New research progress of the electrochemical reaction mechanism, preparation, and modification for $LiFePO_4$. *Acta Physico-Chemica Sinica*, *33*(6), 1085-1107.

29. Zu, Z. H. (2022). First principle study on modification of $LiFePO_4$ cathode material for lithium-ion battery. *Xi'an: Master Dissertation of Xi'an University of Architecture and Technology*, DOI:10.27393/d.cnki.gxazu.2022.000295.
30. Xu, G. G., Zhong, K. H., Zhang, J. M. et al. (2014). First-principles investigation of the electronic and Liion diffusion properties of $LiFePO_4$ by sulfur surface modification. *Journal of Applied Physics, 116*(6): 063703.
31. Bao, L., Xu, G. & Wang, J. (2015). Hydrothermal synthesis of flower-like $LiMnPO_4$ nanostructrues self-assembled with (010) nanosheets and their application in Li-ion batteries. *CryEngComm, 17*(33): 6399-6405.
32. Yamada, A., Chung, S. C. & Hinokuma, K. (2010). ChemInform abstract: Optimized $LiFePO_4$ for lithium battery cathodes. *ChemInform, 32*(29):17-17.
33. Luo, D. D. (2017). Preparation and modification of $LiMnPO_4$ cathode material by solvothermal method for Li-ion battery. Tianjin: Master Dissertation of Tianjin University.
34. Yamada, A., Hosoya, M., Chung, S. C. et al. (2003). Olivine-type cathodes. *Power Sources*, 119-121, 232-238.
35. Martha, S. K., Markovsky, B., Orinblat, J., Gofer, Y., Hark, O., Zinigrad, E., Aurbach, D., Drezen, T., Wang, D., Deghenghi, Q. & Exnar, I. (2009). LiMnPO4 as an advanced cathode material for rechargeable lithium batteries. *Joumal of The Electrochemical Soeiety, 156*(7):8559-8563.
36. Liu, L. Y., Cao, Z. J., Cui, Y. et al. (2019). Nanocomposites $LiMn_{1-x}M_xPO_4/C$ synthesized via freeze drying assisted sol-gel routine and their magnetic and electrochemical properties. *Journal of Alloys and Compounds*, 779:339-346.
37. Dong, Y., Wang, L., Zhang, S. et al. (2012). Two-phase interface in $LiMnPO_4$.nanoplates. *Jounal of Power Sources, 215*(5):116-121.
38. Wang, J. F. & Zhang, W. G. (2001). Progress on organic/inorganic hybrid materials for sol-gel progress — 1. Organic/inorganic hybrid materials with chemical bond between phases. *Polymer Bulletin, 1*(1), 60-67.
39. Li, Y., Zhen, Z., Ren, M. et al. (2006). Electrochemical performance of nanocrystalline $Li_3V_2(PO_4)_3$/carbon composite material synthesized by a novel sol-gel method. *Electrochimica Acta*, 51(28):6498-6502.
40. Dong, P., Xia, S. B., Zhang, Y. J. et al. (2016). Effect of complexing agent on electrochemical performance of LiNi_(080)Co_(0.15)_Al(0.05)O_2 cathode materials prepared by sol-gel method. *Journal of the Chinese Ceramic Study, 44*(10), 1415-1420.
41. Guo, R., Shi, P. F., Cheng, X. Q. et al. (2007). Synthesis and characterization of $LiNi_{1/3}Mn_{1/3}Co_{1/3}O_2$ by high temperature solid-state method. *Chinese Journal of Inorganic Chemistry, 23*(8), 1387-1392.
42. Cheng, B. (2012). Doped $LiMn_2O_4$ and $LiMn_2O_4$/graphene composites as high power and energy cathode materials for lithium-ion batteries. *Jinan: Master Dissertation of Shandong University*, 14-15.
43. Li, K. (2018). Synthesis of Li_2FeSiS_4 material by high-temperature solid-state method. *Liaoning: Master Dissertation of Northeast University*. DOI:10.27007/d.cnki.gdbeu.2018.002738.
44. Dong, E. H., Liu, B. G., Su, W. T. et al. (2023). Status of cathode material regeneration technology for waste lithium-ion batteries. *Mining and Metallurgy, 32*(1), 91-97, 114.
45. Lu, N. Q., Wang, C. G., Wang, J. Q. et al. (2018). Development of ternary cathode material $LiNi_{1/3}Co_{1/3}Mn_{1/3}O_2$ and its modification for lithium-ion batteries. *Science &* Technology *in Mining and Technology, 26*(6), 58-63.
46. Lin, Z. C. (2020). Preparation of lithium-rich manganese-based cathode materials by oxalate coprecipitation method. *Xiamen: Master Dissertation of Xiamen University*. DOI:10.27424/d.cnki. gxmdu.2020.003064.
47. Ren, Y. M. (2019). Preparation methods for Ni-Co-Al cathode material. *Value Engineering, 38*(22), 151-152.
48. Wang, Z. Y., Wang, B. X., Li, C. G. (2022). Preparation of $NiCo_2O_4$ and Li_2CO_3 from leaching solution of cathode material of spent NCM523 battery. *Mining and Metallurgical Engineering, 42*(6), 157-162.
49. Zhang, X. Y., Di, Y. L., Dong, Q. et al. (2022).Research progrss on preparation of $Li_3V_2(PO_4)_3$ cathode material for lithium-ion batteries. *Inorganic Chemicals Industry, 54*(3), 38-44, 08.

 https://doi.org/10.1051/matecconf/202338201013

Finite Element Simulation Calculation and Optimal Design of Permanent Magnet Synchronous Motor

Peizhi Yuan*

College of Electronic and Information Engineering, Hebei University, Baoding, China

Abstract: In order to promote the development of green energy applications, this paper selects a 4-pole 24-slot surface-mounted permanent magnet synchronous motor and performs finite element simulation and optimization design on it. Firstly, finite element analysis is carried out, and the performance of the motor is verified by finite element simulation. Then, the parameterized simulation and sensitivity analysis of cogging torque and three-phase current are performed by selecting the variables of permanent magnet thickness, stator inner diameter, slot width, and air gap. Finally, simulation experiments provide optimization for each variable for the motor.

Keywords: Permanent Magnet Synchronous Motor(PMSM), Finite Element Simulation, Sensitivity Analysis, Cogging Torque.

1. Introduction

China is actively promoting green energy such as photovoltaic and hydro power generation. Permanent magnet synchronous motors have a great impact in the field of renewable energy with their high efficiency, high power density and low noise[1]. At the same time, the new energy vehicle industry in China is rapidly developing, and as the core component of electric vehicles, permanent magnet synchronous motors have a good development prospect[2]. Permanent magnet synchronous motors have a significant role in China's energy strategy and align with the national development direction.

Due to the excellent characteristics of permanent magnet synchronous motor (PMSM), research on it has become a hot topic in the industry. This paper adopts PMSM for optimization design. Literature [3] uses Taguchi method for multi-objective optimization design, optimizing the efficiency, torque ripple, etc. of the motor. Literature [4] based on analytical method designs, optimizes the loss and torque of PMSM. Literature [5] utilizes the idea of permanent magnet pole offset to weaken the cogging torque. Literature [6] analyzes the influence of armature slot width on the cogging torque of internal PMSM. This paper takes a 4-pole 24-slot surface PMSM as the optimization object, first verifies the motor performance using finite element simulation method, and then parameterizes the rotor diameter, permanent magnet thickness, slot width and air gap size for simulation and sensitivity analysis, providing ideas for optimizing the cogging torque and three-phase current of the motor.

2. Parameters of Motor

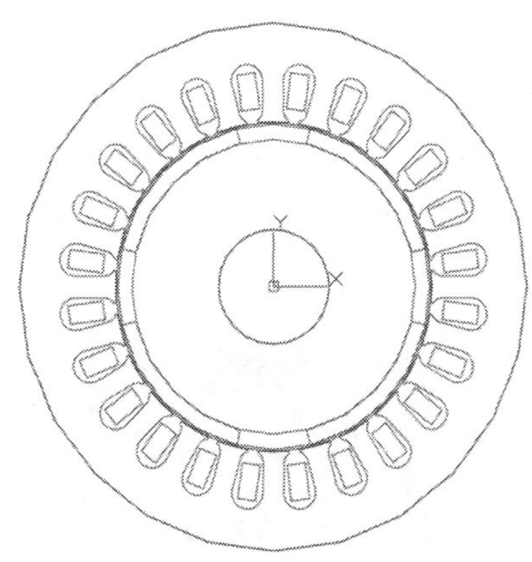

Figure 1: CAD Model of Permanent Magnet Synchronous Motor

Permanent magnet synchronous motors are classified into IPM (interior permanent magnet synchronous motor) and SPM (surface permanent magnet synchronous motor) based on the location of permanent magnets. SPM motors, with low cost and simple production process and maintenance, have been widely used in the motor industry. This paper focuses on the optimization of a 4-pole 24-slot SPM motor, and its CAD rotor lamination diagram is shown in Fig. 1. The model of the motor was established based on the parameters in Table 1, Table 2, and Table 3.

* Corresponding author: yuanpeizhi@stumail.hbu.edu.cn

https://doi.org/10.1051/matecconf/202338201013

Table 1: Main parameters of the simulated motor

Parameters	Value
Number of Poles	2
Rated Voltage / V	220
Rated Power / W	550
Rated Speed / (r/min)	1500
Rated Frequency / Hz	50
Maximum output power/W	1836.05
Rated torque/(N*m)	3.507

Table 2:Dimension Parameters of Stator-mounted Permanent Magnet Synchronous Motor

Parameters	Value
Stator outer diameter/mm	1200
Stator inner diameter/mm	75
Rotor outer diameter/mm	74
Rotor inner diameter/mm	26
Permanent magnet thickness/mm	3.5

Table 3: Materials of Permanent Magnet Synchronous Motor Parts

Synchronous motor component	Material
Stator conductor	Copper
Stator core	DW310_35
Rotor core	DW310_35
Permanent magnet	XG196/96

In order to ensure the speed and accuracy of finite element calculation results, the stator and rotor are relatively sparsely divided, and the permanent magnet is more densely divided. The dividing diagram of the motor is shown in Figure 2. The boundary conditions are set using a symmetrical method.

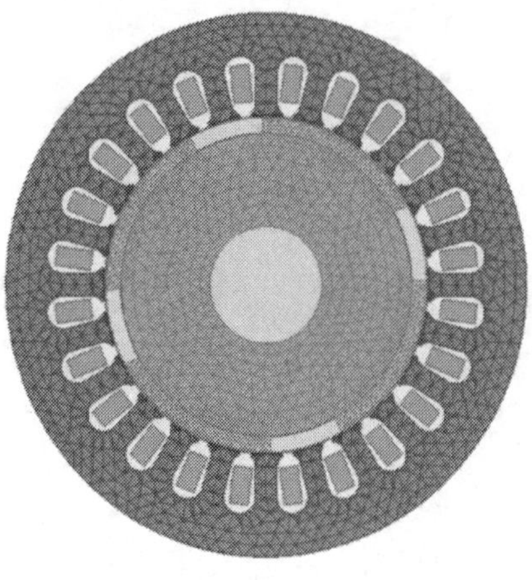

Figure 2: Mesh of the PMSM in Finite Element Analysis

3. Electromagnetic Finite Element Calculation of the Motor

3.1 Finite Element Calculation

This paper will optimize the motor parameters through the finite element algorithm. During the motor optimization process, parameterized simulation is usually performed using analytical methods and finite element methods. The analytical method sacrifices some accuracy for faster parameter calculation, while finite element method is used in this paper for more accurate results. Finite element divides the motor into several small units and solves the partial differential equations for each unit. Through finite element, we can accurately obtain the working condition inside the motor, thus better improving the motor performance. It not only provides help in motor design, but also plays a huge role in motor debugging and fault detection.

3.2 Steps of Finite Element Analysis

This paper uses the finite element method based on Maxwell for simulation. Maxwell simulates the electromagnetism of the motor and calculates the magnetic field distribution, torque, induced electromotive force and other parameters inside the motor through finite element simulation. Through the software, we can simulate the performance of various types of motors under different operating conditions, thus solving problems encountered by the motor in actual use. The modeling process is shown in Figure 3.

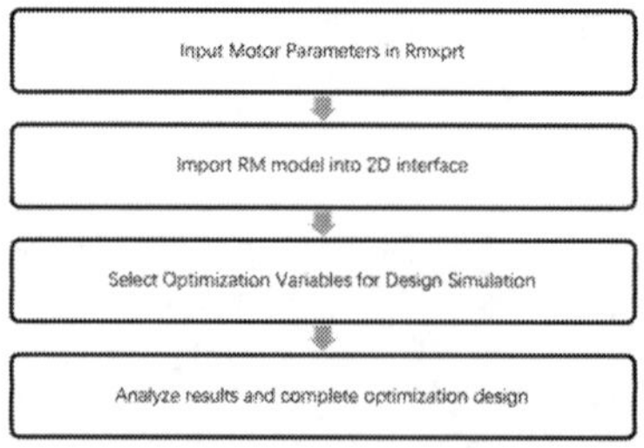

Figure 3. Flowchart of Modeling Process Based on Maxwell

3.3 The finite element analysis results of the motor

3.3.1 No-load simulation

Observing the distribution of magnetic field of the motor can be accurately depicted through Figures 4 and 5 under no-load conditions. The images indicate that the magnetic field strength inside the motor meets design standards, the magnetic lines of force are closed, and leakage is minimal, suggesting that the initial design meets basic design requirements. This is crucial for subsequent motor optimization.

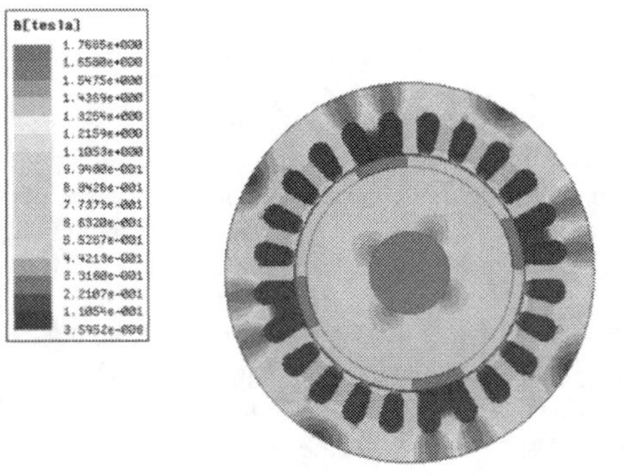

Figure 4: Magnetic Flux Density Cloud of the Motor under No-Load Conditions

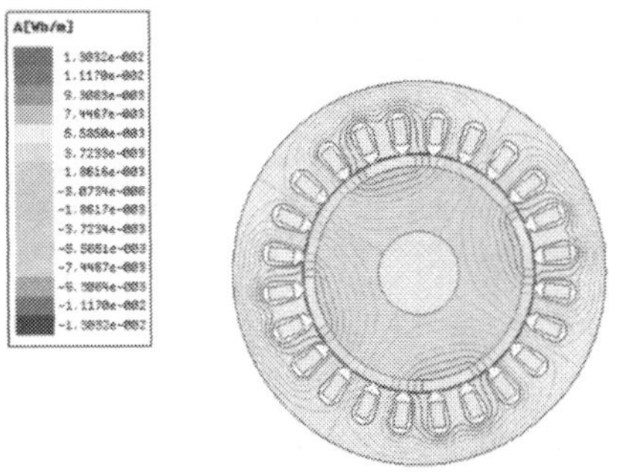

Figure 5: Magnetic Sensitivity Line Diagram of the Motor under No-Load Conditions

The three-phase counter electromotive force is generated by the induction of the permanent magnet cutting the stator three-phase winding, and it has a significant impact on the performance of the motor during operation. The finite element simulation waveform in Figure 6 shows that it is basically a sine waveform. The no-load counter electromotive force calculated from the waveform is 211.42 V, slightly lower than the rated phase voltage, and the simulation error is 3.9%, which meets the motor design standards.

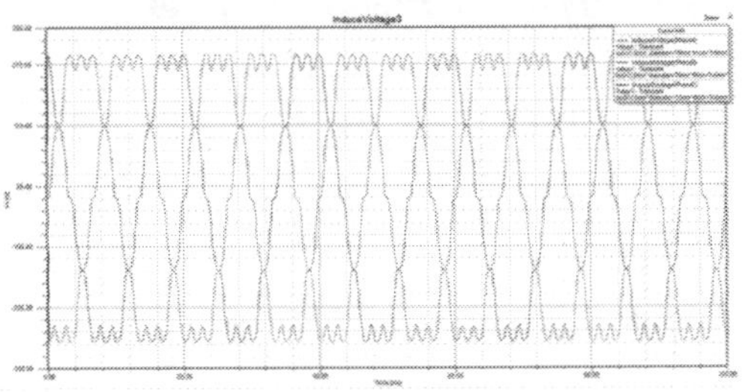

Figure 6: Three-phase counter-electromotive force under no-load condition of the motor

3.3.2 Rated Load Simulation

Under rated load conditions, observation of the motor's magnetic field distribution through magnetic flux cloud and magnetic field line maps can better understand the working state of the motor under load, in order to verify the reliability of the motor and make further optimizations. As can be seen from Figures 7 and 8, the maximum magnetic induction strength of the motor is 1.8 T, the minimum is $4.7*10^{-6}$ T, and the maximum position is on one side of the slot close to the stator. Due to the large magnetic resistance of the slot air, this position is also the most stressed. The recommended numerical value for the motor is 1.6-2.7 T, which complies with the motor standards. The motor's magnetic field line is closed, resulting in minimal leakage. The maximum value of A is 0.01297 Wb/m, which meets the motor design standards.

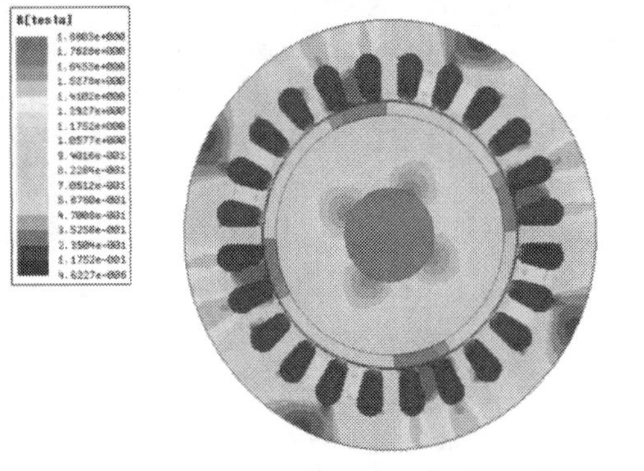

Figure 7: Magnetic Flux Density Cloud Map Under rated Load Conditions

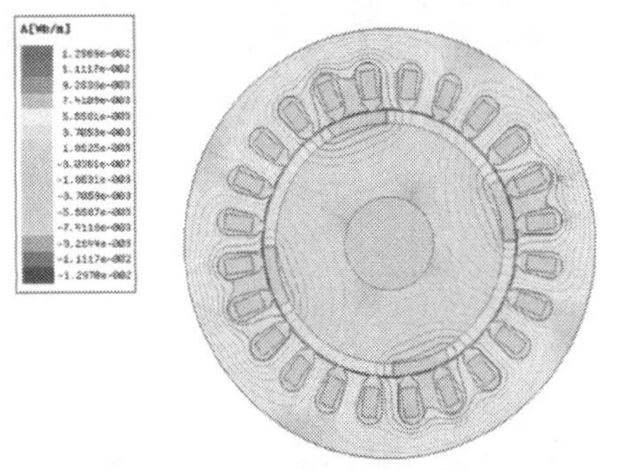

Figure 8: Magnetic field lines of the motor under rated load conditions

In permanent magnet synchronous motors, the three-phase magnetic chain is equal to the product of the number of turns of the energized coil and the magnetic flux. As the motor rotates at a synchronous speed of 1500 r/min, the three-phase magnetic chain is symmetrical, as shown in Figure 9. The three-phase induced current generated by the motor also exhibits approximately three-phase symmetry, as shown in the waveform of Figure 10. Due to the harmonic presence in the openings of the motor, the maximum value is 1.5461 A, the effective value is 1.1637 A, and the rated current is 1.443 A, with a simulation error of 7.14%.

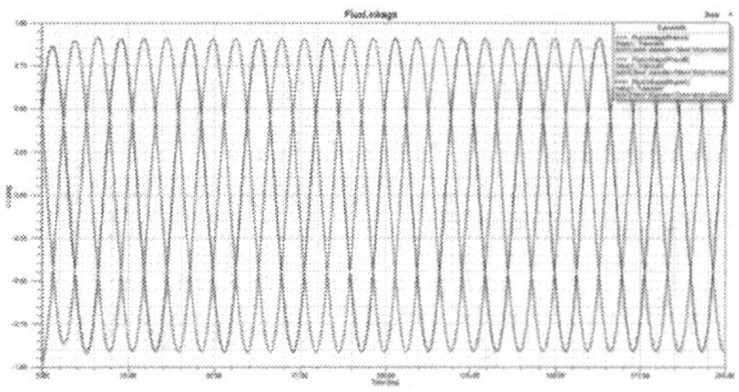

Figure 9 Three-phase magnetic chain of the motor under rated load

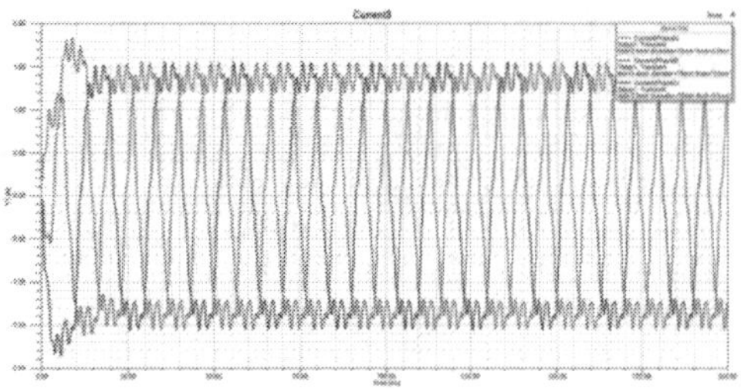

Figure 10 The three-phase current of the motor under rated load.

The cogging torque of the motor will cause vibration and noise of the motor, seriously affecting the normal operation of the motor. The simulation results in the cogging torque waveform of the motor as shown in Figure 11. The cogging torque waveform of the motor is also symmetrical in three phases, and the cogging torque exhibits periodic changes, with a maximum value of 0.5518 N*m, which meets the design requirements.

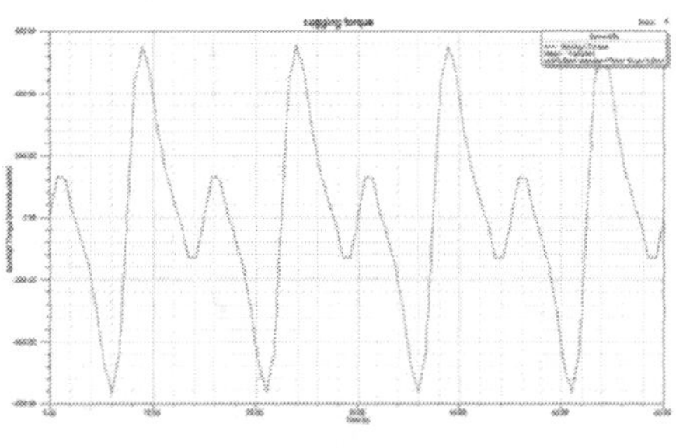

Figure 11 Cogging Torque of the Motor

4. Parameterized calculation of motor

4.1 Parameterized calculation

Parameterized simulation is performed by specifying the type, size, etc. of the target motor through a series of parameters, and obtaining performance indicators of the motor during operation by modifying the parameters. Parameterized simulation can greatly shorten the motor design cycle, reduce design costs, and ultimately obtain the desired motor model through continuous optimization.

4.2 Parameterized calculation with Maxwell

In this paper, parameterized simulation is carried out using Maxwell. The motor is set to no load, with a rated speed of 1°/s. The resulting motor torque is the cogging torque. By applying the control variable method and changing the parameters of permanent magnet thickness, stator inner diameter, slot width, and air gap, the change in the cogging torque can be obtained. Under load conditions, the motor speed is set to 1500r/min, and the reaction current can be obtained through parameterized scanning. By analyzing the images, we can carry out subsequent sensitivity analysis, thereby providing optimization ideas for the motor.

4.3 Cogging torque varies with the change of permanent magnet thickness

The cogging torque of the motor presents periodic changes during rotation. As shown in Figure 12, when the thickness of the permanent magnet is changed, the maximum value of the cogging torque can be observed to change as well. The maximum value is 2.1686 N*m, and the minimum value is 5.5518 N*m. The range of change is from 2.5 to 4.5 mm, with a step size of 0.5 mm. With the increase of the permanent magnet thickness, the cogging torque first decreases and then increases.

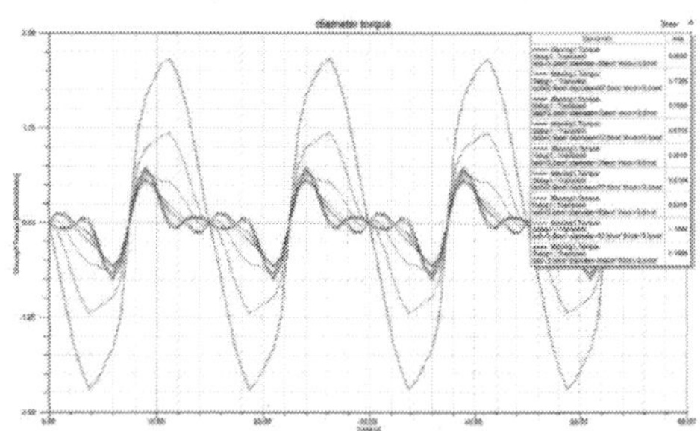

Figure 12: Variation of Cogging Torque with Permanent Magnet Thickness

5. Optimization Design of Motor

In this paper, four variables, permanent magnet thickness, stator inner diameter, slot width, and air gap, are selected. By observing the changes in cogging and induced current, the sensitivity corresponding to each variable is calculated, thereby further optimizing the motor.

5.1 Cogging torque varies with four variables

This paper defines the calculation of cogging torque and induced current through sensitivity analysis, using the following formula:

$$\partial = \frac{\Delta y}{\Delta x} \tag{1}$$

Among them: ∂ ——Sensitivity Factor;

Δx ——Normalized length;

Δy ——Cogging torque variation;

https://doi.org/10.1051/matecconf/202338201013

The purpose of normalization is to bring the data of different parameters into the same unit, so as to calculate the sensitivity of the same objective to them. The normalization formula for length is:

$$\Delta x = \frac{x - \min A}{\max A - \min A} \tag{2}$$

Among them: x ——The length of the parameters;

A ——The set of all parameter lengths;

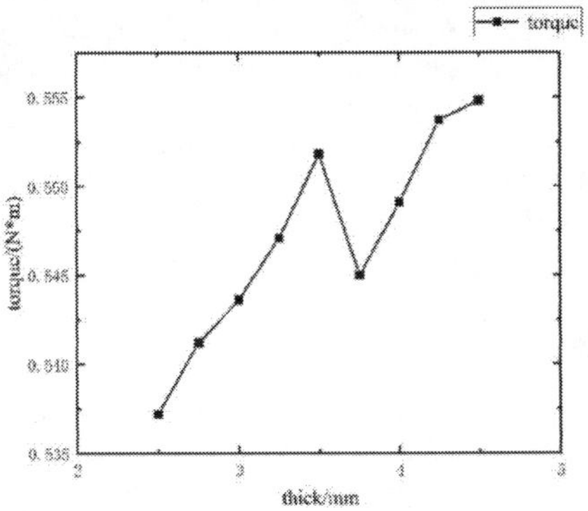

Figure 13: Variation of Cogging torque with Permanent Magnet Thickness

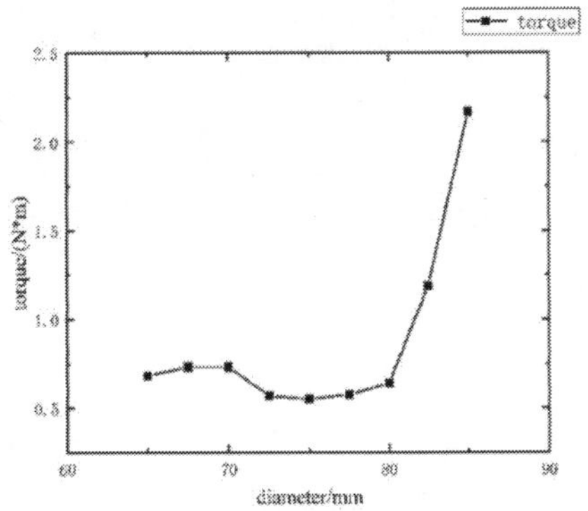

Figure 14: Variation of Cogging torque with Stator Inner Diameter

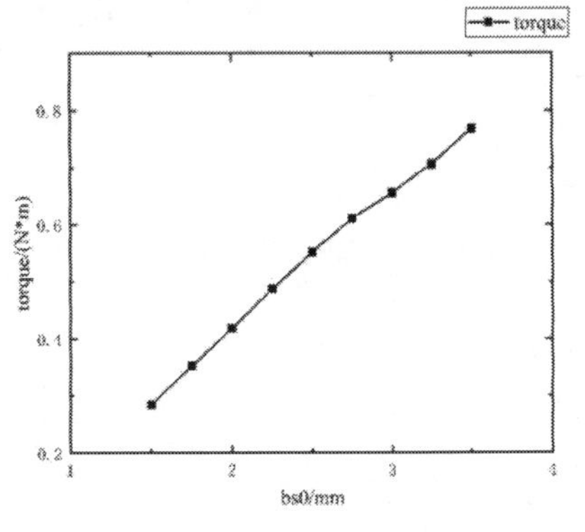

Figure 15: Cogging torque variation with respect to slot width

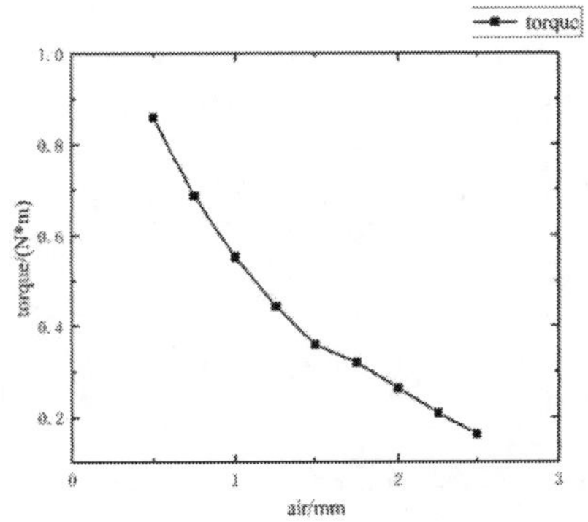

Figure 16 Cogging torque changes with air gap

As shown in Figure 13-15, the cogging torque increases first, then decreases and then increases as the thickness of the permanent magnet increases. As the inner diameter of the nail increases, the cogging torque changes little at first, and then increases significantly. The cogging torque increases gradually with the increase of the slot width. As the air gap increases, the cogging torque decreases gradually. Through the sensitivity analysis of formulas (1) and (2), it can be obtained that the maximum sensitivity of the inner diameter of the stator is 5.8608, and the minimum sensitivity of the slot is 2.8200.

5.2 Three-phase current varies with four variables

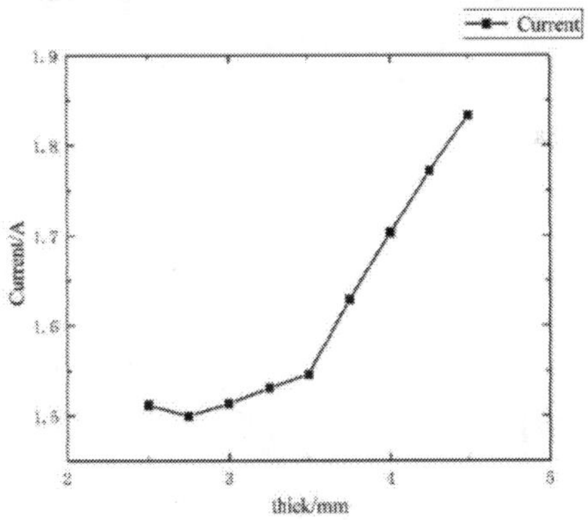

Figure 17 The three-phase current changes with the thickness of the permanent magnet

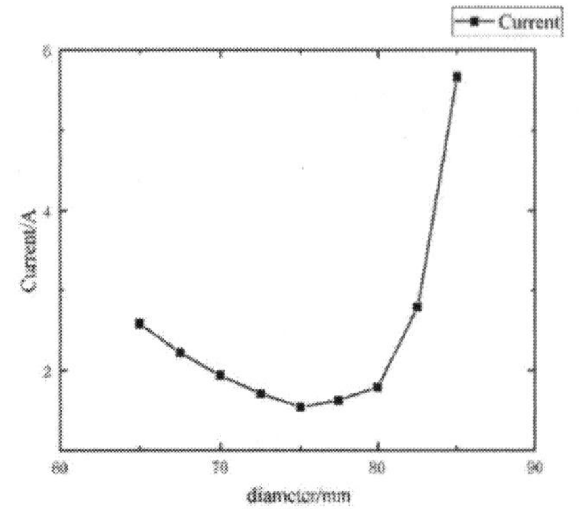

Figure 18 Three-phase current changes with stator inner diameter

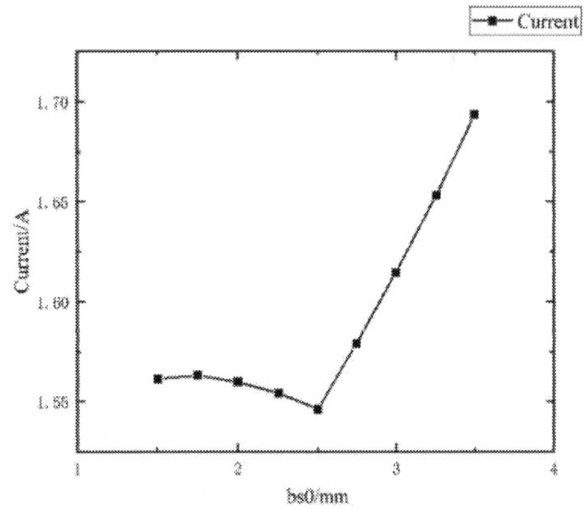

Figure 19 Three-phase current changes with slot width

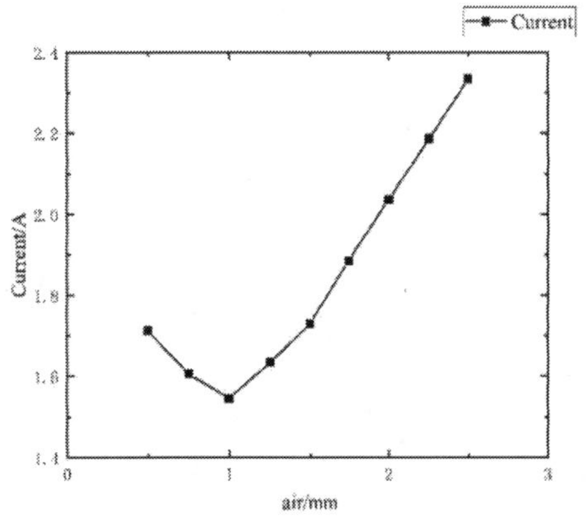

Figure 20 Three-phase current changes with air gap

As shown in Figure 17-20, as the thickness of the permanent magnet increases, the three-phase current increases gradually. The three-phase current first decreases and then increases with the increase of stator inner diameter, slot width and air gap. Through the sensitivity analysis of formulas (1) and (2), it can be obtained that the maximum sensitivity of the inner diameter of the stator is 17.8056, and the minimum sensitivity of the air gap is 12.8488.

6. Conclusion

In this paper, the permanent magnet synchronous motor is designed based on Maxwell, and the finite element simulation of the motor parameters is carried out. By changing the thickness of the permanent magnet, the inner diameter of the stator, the slot width and the air gap, the cogging torque and the induced current of the motor are observed. Using the sensitivity analysis method, the motor optimization ideas are provided for each variable. Through the simulation experiment, it is observed that the sensitivity of the inner diameter of the stator to the cogging torque and the induced current is the largest. Carry out the optimization design of the motor, and select the appropriate optimization algorithm to provide more detailed design ideas for the performance optimization of the motor.

References

1. LI Yifei,JIANG Quan.Research Development of Initial Rotor Position Detection of Permanent Magnet Synchronous Motors [J]. Electronic Science and Technology,2021,34(04):24-33.DOI:10.16180/j.cnki.issn1007-7820.2021.04.005.
2. Ouyang Mingke,Li Jianfei ,et al. Research on Energy-Saving and Emission-Reduction Efficiency of China's Automobile Industry from the Perspective of Technological Innovation [J]. Forum on Science and Technology in China,2021(08):70-81.DOI:10.13580/j.cnki.fstc.2021.08.009.
3. CHEN Shuang, ZHANG Zhi,et al. Multi-objective Optimization Design of a Permanent Magnet Synchronous Motor Based on Taguchi Method [J]. micromotors,2021,54(07):17-22.DOI:10.15934/j.cnki.micromotors.2021.07.004.
4. ZHU Tuo,LI Zheng,et al.Optimal Design of Permanent Magnet Synchronous Machine Based on Analytic Model[J/OL]. Electronic Science and Technology:1-8[2023-01-26].DOI:10.16180/j.cnki.issn1007-7820.2023.03.011.
5. LIU Mei, XIANG Zixuan,et al.Research on Cogging Torque Reduction of Magnetic Gear Motor Based on Permanent Magnet Inter-polar Deflection [J]. micromotors, 2021, 54(12): 10-15. DOI: 10.15934/j.cnki.micromotors.2021.12.002.
6. YANG Yu-bo,WANG Xiu-he,et al.Effect of slot opening on the cogging torque of permanent magnet synchronous motor [J]. Electric Machines and Control,2011,15(07):21-25. DOI: 10.15938 / j.emc. 2011.07.007.

Optimizations of the internal structure of the reel of a double rope winder

Jia Fu*

College of Mechanical Engineering, University of Jinan, Jinan, China

Abstract. Due to the shortage of resources, the deep development of resources is imminent, and the superdeep mine hoist plays a leading role in the development process. According to the structure, the hoists can be divided into winding hoists and friction hoists. Winding hoists can be single-rope winding and multi-rope winding. With the increase of mining depth, friction hoists will have the disadvantage of increasing the mass of the tail rope, while wire rope diameter of single-rope winding hoists will increase with the increase of mining depth. So double-rope winding hoists can be used to balance the above two problems. Therefore, this study will use finite element analysis, and treat hoist reel as a uniformly pressurized shell, through the static and modal analysis of reel. Under the maximum static tension, this study gets the reel the force and deformation distribution. Through the change of the internal structure of the reel, it improves the reel force weakness and reduce the reel deformation. By reducing the deformation of the reel, this study tells how to reduce the tension difference of the wire rope. This study will improve the synchronization and stability of the hoisting system and provide an alternative and effective way for the deep development of the resources with the double-rope winding hoists.

Keywords: Reel construction, Finite element analysis, Optimal design, Poor tensioning, Synchronicity

1. Introduction

In recent years, with the rapid growth of demand for mineral resources in social and economic development, the problem of energy shortage has become more and more prominent. Therefore, it has become an important strategic choice for China to search for mineral resources deep inside the earth[1] .

The hoists can be divided into two categories according to the different working methods: winding hoists and friction hoists. For friction hoists, as the mining depth increases, the load on the wire rope will be affected by the increased mass of the balanced tail rope. And when the depth exceeds 1200m, in order to ensure the safety of mining, people generally use winding hoists.

The winding hoists are divided into single-rope winding and multi-rope winding, and with the mining depth increase, the single-rope hoists require a higher power to drive motor and a larger diameter reel, which will bring more difficulties in manufacturing and installation, while the multi-rope winding hoists can avoid the relevant shortcomings. As for single-rope hoists, the tail rope is too heavy and the reel diameter is too large, while the steel rope is too thick and not easy to bend. Therefore, this study focuses on the study of multi-rope winding hoists for mining and hoisting in deep wells above 1200m.

Unlike conventional hoisting devices, multi-rope winding hoists have two or more rope zones on the main shaft. Studies on multi-rope winding hoists include: Wang Ji [2] used Hamilton principle to simulate and analyze the dynamic behavior of multi-rope hoists under unbalanced factors such as transport eccentricity and inconsistent reel radius, and proposed the design principle to restrict the inconformity of dynamic tension between cables. Wu[3] analyzed the dynamic characteristics of super deep multi point winding hoisting system under non-synchronous movements of the wire rope with the help of Lagrange equation .Due to the dynamic load of high speed and heavy load, the wire rope is deformed, thus making the tension difference among the synchronous ropes increase, which can be called *Deformation detuning of the wire rope*, and the *Deformation detuning of the wire rope* makes the synchronous transportation of the hoists more difficult. Under complex working conditions, the wire ropes may break due to uneven force, resulting in the risk of cage fall. In view of the fact that in multi-rope hoists, the problem of differential tension becomes more critical as the number of wound ropes increases[4],the design of a double-rope hoists can balance the problems of oversized single-rope reel diameters and poor multi-rope wire rope tension, so this study focuses on the double-rope hoists.

Studies about wire rope winding tension differences are:

Li[5]studied the difference in tension between the two ropes of a double-rope winding reel due to manufacturing

* Corresponding author: 202030412035@stu.ujn.edu.cn

deviations and differences in the depth of the winding groove, resulting in an unsynchronized lifting of the two ropes. Kaczmarczyk,S [6]-[7] established a vibration differential equation by means of Hamilton's principle to investigate the vibration phenomena of wire ropes in deep wells. Wang Ji[2] analyzed the effect of wire rope length error on the wire rope tension difference by establishing the system dynamics equations. Peng[8] investigated whether COF between wire ropes changes with load and sliding speed under lubricated and dry friction conditions. Yao[9] studied the step changes of hoisting velocity and acceleration caused by coil and layer crossover of rope winding on the LeBus reel.

The existing research on the tension difference of the double-rope winding hoists mainly revolve around the wire rope. The reel will produce a large amount of deformation after the force, so that the wire rope produces a large tension difference, thus affecting the synchronization of the wire rope hoist, but the results of this research are rare. The main findings of the current research on hoist reels are : JM Singly[10]-[11] derived a mathematical theory based on plates and shells and also derived stress distribution functions for welded coils with lengths greater than $\sqrt{4R\delta}$ and for welded coils with lengths less than $\sqrt{4R\delta}$. S.Otto[12] used FEA-Simulations to analyze the radial pressure of a multi-layer wire rope on the reel and the axial thrust on the rope stopper during heavy loads in a hoisting system. P.Dietz [13]-[14] studied the pressure acting on the reel during the winding process of multi-layer wire rope, considering the reduction of inter-layer tension and radial deformation of the wire rope caused by the winding of each layer, and deduced the general formula for calculating the pressure on the reel. Mangalekar S[15] carried out a FEA-Simulations under maximum load conditions on the reel and designed different reel models to derive the optimum reel structure. Wolny [16]studied the stress distribution in webs, barrel shells and other components in a reel and analyzed the hazardous areas on the barrel shell. Jin Minjie [17] analyzed the deformation of the rolls based on the theory of elastic foundation beams. The above research results have been analyzed mainly for single-rope winding hoists, and less analysis has been carried out on the forces on the reels of double-rope winding hoists.

Most of the above literatures for the roll deformation amount are obtained by the traditional theory calculation, and the calculation results are rough and have some error. With the continuous development of science and technology, more advanced theories and methods have been obtained. Benjamin Torres[18] proposed the use of a new FBG strain transducer fixed to the surface of the structure under test and showed that the measured strain differed from the actual strain by less than 2.5%. Chen[19] used Pro/E and ANSYS to build a solid model of the reel and solve for the deformation and stress distribution of the reel. Hu[20] integrated use of OptiStruct, HyperMesh software functional modules and used finite element static analysis to identify the shortcomings of the manhole on the reel and adjusted the structure through topology optimization design. Yu [21] analyzed the stress-strain of the reel based on ANSYS and carried out a topology optimization analysis of the reel. In this study, based on the above research, the reel model of double-rope winding hoists will be constructed by 3D software.And the deformation of the reel axially will be analyzed by FEA-Simulations, and the maximum deformation of the reel will be reduced and the synchronization and smoothness of double-rope winding hoists will be improved through the improvement of the internal structure of the reel.

2. Modeling of the double-cable winding hoister

Fig. 1 shows a typical 3D model of a double-rope winding hoist reel, and the main material of the reel is Q345A.It is mainly composed of brake disc, reel, rope exit hole, manhole, rope stopper, support column, ribs, two rope wrapping areas and two support wheels. Where d_1 is the diameter of the rope exit hole, d_2 is the diameter of the reel, d_3 is the diameter of the support column. While δ is the thickness of the reel wall, a and c respectively represent the length and width of the thicker ribs . And l_1 is the shortest distance between the two support wheels at the initial moment, while d represents the diameter of the wire rope. The parameters are shown in Table 1.

Table 1. Double rope winding hoist-2JKD-8×4.2 Sample machine parameters

Parameter Name	Value/Unit	Parameter Name	Value/Unit
Diameter of rope outlet hole d_1	200/mm	Diameter of support column d_3	500/mm
Thickness of cylinder shell δ	120/mm	Reinforcement width c	240/mm
Reel diameter d_2	8/m	Pivot wheel distance l_1	3400/mm
Reinforcement length a	350/mm	Wire rope diameter d	180/mm

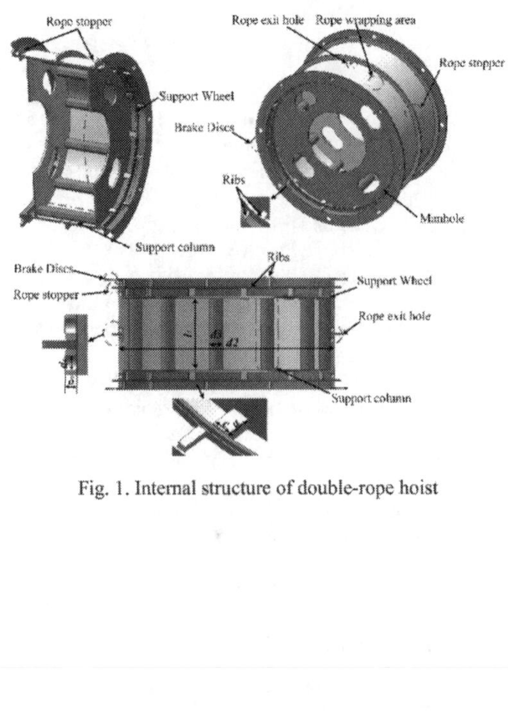

Fig. 1. Internal structure of double-rope hoist

https://doi.org/10.1051/matecconf/202338201014

3. Derivation of the double rope winder formula

The double-rope multi-layer hoists use the structure of a thick-shell bullet support, the structure of which is shown in Fig. 2. The thick-shell bullet support has two rope wrapping areas on the left and right, two rope exit holes in each rope-wrapping area and one support wheel.

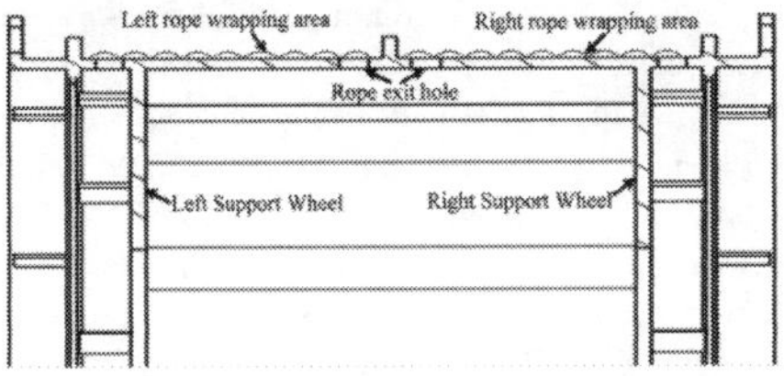

Fig. 2. Thick shell bullet support structure

The double-rope winding hoists are connected to the same cage for hoisting. When the wire rope is at rest before the hoisting starts, the total tension Q_0 of the two wire ropes is shown in Eq.(1),

$$Q_0=\left(S+S_0+pH\right)g \tag{1}$$

where S represents the rated load. S_0 represents lifting vessel mass. p represents unit mass of steel wire rope. H represents wire rope overhang height. g represents gravitational acceleration.

The total tension on the rope is F_n, where n is the number of wraps, when n=1, the rope is under total tension as shown in Eq.(2),

$$F_n=Q_0=\left(S+S_0+pH\right)g \tag{2}$$

After the wire rope winding a circle, its total tension[2]is affected by the lifting height, lifting speed, lifting acceleration, lifting process of longitudinal vibration[6]-[7] and other factors. At this time the total tension of the lifting system is shown in the Eq.(3),

$$F_n=Q_0-4(n-1)\pi Rpg+\frac{1}{2}(K-1)Sg+\left[\frac{Q_0}{g}-4(n-1)\pi Rp+2l_s p+M_1\right]a_{n-1} \tag{3}$$

where F_n represents total tension of the lifting system. R represents reel radius. K represents tank road lifting resistance coefficient. l_s represents length of dangling wire rope. M_1 represents shifting mass of the sky wheels. a_{n-1} represents lifting acceleration after n-1 turns of winding.

Steel wire rope winding process also needs to consider the single-layer and multi-layer winding tension reduction amount on the impact of the radial force of the reel. As single-layer winding, to start the research by rope ring one winding position , the reel radial variable is x_{11},while winding the nth circle.The ith circle corresponding to the wire rope tension reduction amount is $\triangle F_{in}$. It is shown in Fig.3 and Eq.(4),

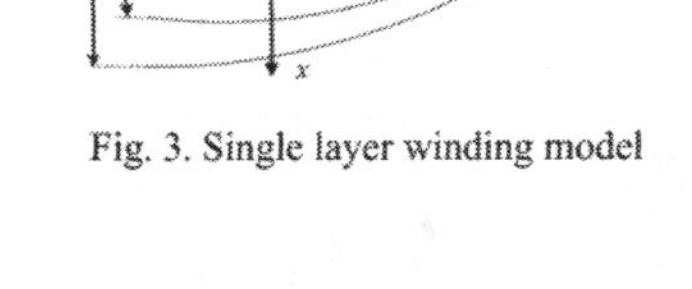

Fig. 3. Single layer winding model

$$\Delta F_{in}=\frac{x_{in}}{R-x_{i(i+1)}-x_{i(i+2)\cdots}\cdots-x_{i(n-2)}-x_{i(n-1)}}E_S A_S \quad (i\neq 1) \tag{4}$$

where $\triangle F_{in}$ represents reduction in tension on lap i at lap n. x_{in} represents radial deformation at the position of the nth winding turn i . E_S represents longitudinal modulus of elasticity of steel wire rope.

In multi-layer winding[13]-[14] process, the outer layer of wire rope makes the radial force of the reel increase while radial force reduces the radius of the reel, so the inner layer of wire rope appears circumferential relaxation.And the inner layer of wire rope tension is reduced, so the inner layer of radial pressure on the reel is also reduced. Set a total of m winding layer. It corresponds to the i layer of the actual tension size for F'_{im} .The tension size is shown in Eq.(5),

$$F'_{im}=F_i-F_{i\text{-}(i+1)}-F_{i\text{-}(i+2)}-\cdots\cdots-F_{i\text{-}(m-1)}-F_{i\text{-}m} \tag{5}$$

where F_i represents initial tension of the ith layer of wire rope. $F_{i\text{-}m}$ represents tension reduction of layer i caused by winding of mth layer. F'_{im} represents the actual tension of the ith layer.

Multiple layers of wire rope stacked with forces as shown in Fig. 4.

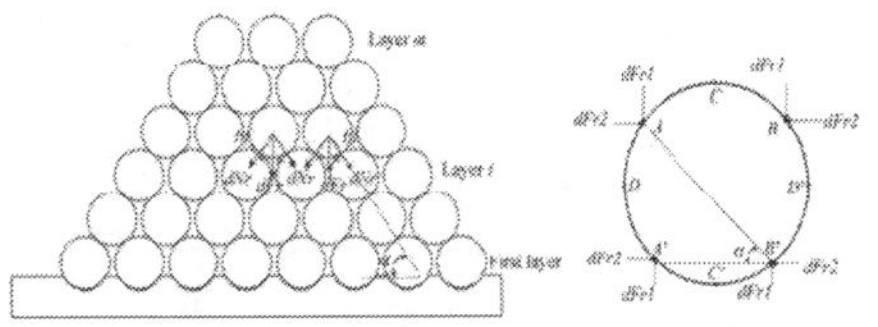

(a) Multi-layer winding force schematic

(b) Separate analysis of multi-layer winding

Fig. 4. Multiple layers of wire rope stacked under stress

The ith layer of wire rope is subjected to the mth layer of the force decomposed into the radial and axial components of the reel as shown in Eqs.(6) and (7),

$$dF_{r1}=\frac{dF_r}{2}=\frac{\left(F_m-F_{(i+1)\text{-}m}-F_{(i+2)\text{-}m}-\cdots\cdots-F_{(m-1)\text{-}m}\right)d\theta}{2} \tag{6}$$

$$dF_{r2}=dN_r\left(\cos\alpha-\mu_r\sin\alpha\right)=\frac{dF_r\left(\cos\alpha-\mu_r\sin\alpha\right)}{2\left(\sin\alpha+\mu_r\cos\alpha\right)} \tag{7}$$

where dF_r represents force at m^{th} level on the ith layer. dF_{r1} represents radial component force of layer i on the reel. dF_{r2} represents axial component force of layer i on the reel. μ_r represents friction coefficient between steel wire ropes. N_r represents the support force between the wire rope to generate friction.

The reduction of the chord AA' after the force is applied to the chord length AA' is shown in Eq.(8),

$$\Delta'_{i\text{-}m}=\frac{2\sin\alpha dF_{r1}}{E_r dl_i}=\frac{2\sin\alpha\left(F_m-F_{(i+1)\text{-}m}-F_{(i+2)\text{-}m}-\cdots\cdots-F_{(m-1)\text{-}m}\right)}{E_r[2R+(1+2(i-1)\sin\alpha)d]} \tag{8}$$

where E_r represents transverse modulus of elasticity of steel wire rope. α represents wire rope stacking angle. d represents the wire rope diameter. dl_i represents the arc length of the wire rope corresponding to the ith layer winding $d\theta$.

Similarly, the chord length AA' is subjected to a force of dF_{r2}, corresponding to the extension of the chord AA' under the action of F_{r2} as shown in Eq.(9),

$$\Delta_{i\cdot m}^{''} = \sin\alpha v_r \frac{2dF_{r2}}{E_r dl_i} = \frac{2v_r \sin\alpha\left(F_m - F_{(i+1)\cdot m} - F_{(i+2)\cdot m} - \cdots\cdots - F_{(m-1)\cdot m}\right)\cot(\alpha+\gamma)}{E_r[2R + (1+2(i-1)\sin\alpha)d]} \quad (9)$$

where v_r represents poisson's ratio of steel wire rope, and $\gamma = arctan(\mu_r)$.

The amount of radial change produced by the ith layer of wire rope is shown in Eq.(10),

$$\Delta_{im} = \Delta_{im}^{'} - \Delta_{im}^{''} \quad (10)$$

Set the mth layer winding caused by the radial deformation of the reel $\triangle R_m$, the ith layer wire rope corresponding to the total winding radius reduction is shown in Eq.(11),

$$R_{m\cdot m} = \Delta R_m + \Delta_{1m} + \Delta_{2m} + \ldots + \frac{1}{2}\Delta_{im} \quad (11)$$

where $\triangle R_m$ represents radial deformation of the reel caused by the mth layer winding. $R_{m\cdot m}$ represents reduction of the total winding radius corresponding to the ith layer of wire rope.

From Hooke's law, it is known that the mth layer of wire rope winding causes the reduction of the ith layer of wire rope tension as shown in Eq.(12),

$$F_{i\cdot m} = \frac{Z_{i\cdot m}}{2R + (1+2(i-1)\sin\alpha)d} = \frac{2\pi R_{m\cdot m}}{2R + (1+2(i-1)\sin\alpha)d} \quad (12)$$

where $Z_{i\cdot m}$ represents the ith layer corresponding to the circumferential length reduction, and $Z_{i\cdot m} = 2\Pi R_{m\cdot m}$.

The study of the reel can be based on the model of the thin-walled circular shell[10]-[11], as the spacing between the wire rope is much smaller than the diameter of the reel, the winding process of the wire rope can be regarded as a closed ring of uniform texture [22] acting on the reel. As the wire rope of tension F_n winds to the reel , the reel acts on the combined force of the wire rope as shown in Fig. 5(a). As the unit beam of the reel can be regarded as an elastic base beam [17] , so the force analysis of the reel can be based on this. Take a study of the unit beam.The unit beam along the radial direction produced $\triangle R$ deformation. In Fig. 5(b) in the reel boundary from the solid line to the dashed line, circumferentially it generated circumferential compressive stress σ_θ , and reel force deformation is shown in Fig. 5(b).

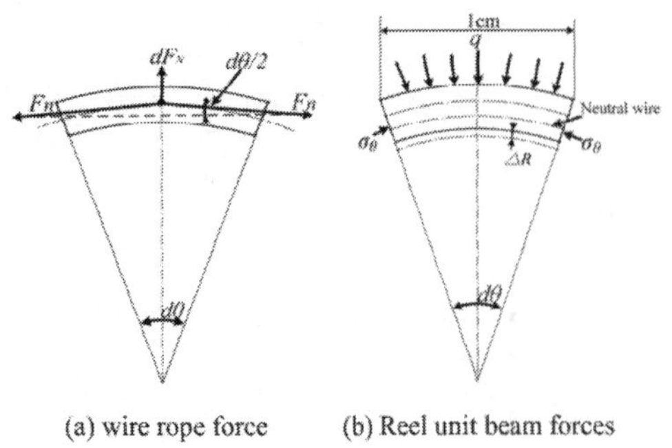

(a) wire rope force (b) Reel unit beam forces

Fig. 5. Force analysis of reel and wire rope

The support force on the wire rope is shown in Eq.(13),

$$dF_N = 2F_n \sin\left(\frac{d\theta}{2}\right) = 2F_n \frac{d\theta}{2} = F_n d\theta \quad (13)$$

where dF_N represents the wire rope is supported by the force of the reel.

The set of loads to which the reel is subjected is shown in Eq.(14),

$$q = \frac{dF'_N}{Rd\theta t_0} = \frac{F_n d\theta}{Rd\theta t_0} = \frac{F_n}{Rt_0} \quad (14)$$

where dF'_N represents pressure in the reel, $dF'_N = dF_N$. q represents load set. t_0 represents pitch of wire rope, $t_0 = d+e$, d represents diameter of steel wire rope, e represents gap between adjacent rope loops.

Because $\triangle R$ and the diameter of the reel compared to the radial displacement of the reel is very small, it can be considered that the distance to the center of the circle at any place on the cross-section is equal to the radius of the reel R . Circumferential strain ε_θ is shown in Eq.(15),

$$\varepsilon_\theta = \frac{2\pi(R-\Delta R) - 2\pi R}{2\pi R} = -\frac{\Delta R}{R} \quad (15)$$

The balance of circumferential compressive stress and load set is shown in Eqs.(16) and (17),

$$2\sigma_\theta \delta \sin\left(\frac{d\theta}{2}\right) = -qRd\theta \quad (16)$$

$$\sigma_\theta = E_d \varepsilon_\theta = -E_d \frac{\Delta R}{R} \quad (17)$$

where δ represents thickness of the reel. E_d represents roll modulus of elasticity.

Where $Rd\theta = 1$, the support force on the reel is F_N as shown in Eq.(18),

$$F_N = 2\delta\sigma_\theta \sin\frac{d\theta}{2} = -\frac{E_d \delta \Delta R}{R^2} = k_1 \Delta R \quad (18)$$

where F_N represents support force on the rollers.k_1 represents foundation reaction coefficient.

The support forces on the reel, the shear forces, and the load set are shown in Fig. 6.

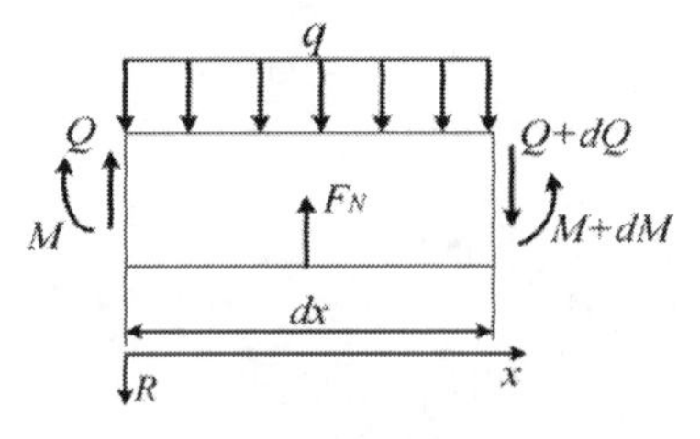

Fig. 6. Sketch of unit beam force analysis

The bending moment, shear force and circumferential force are shown in Eqs.(19) and (20),

$$\begin{cases} \sum Q = -Q + (q - k_1\Delta R)dx + Q + dQ = 0 \\ \sum M = (q - k_1\Delta R)dx\frac{dx}{2} + (Q+dQ)dx + M - (M+dM) = 0 \end{cases} \quad (19)$$

$$\begin{cases} M(X) = -D \cdot \frac{d^2\Delta R}{dx^2} \\ Q(X) = -D \cdot \frac{d^3\Delta R}{dx^3} \\ C_\theta = -\frac{E_d\delta\Delta R}{R} \end{cases} \quad (20)$$

where D represents flexural stiffness. $M(x)$ represents reel bending moment at any point along x. $Q(x)$ represents reel shear at any point along x. C_θ represents circumferential force.

4. Finite element analysis of reel deformation and forces

In this section, based on the above derivation, the finite element model of the reel will be constructed, and modal and static analyses will be carried out respectively to obtain the force and deformation of the reel, where the overall parameter values of the reel are shown in Table 2.

Table 2. Summary table of total parameters

Parameter Name	Value/Unit	Parameter Name	Value/Unit
Maximum static tension Q_0	1480/KN	Poisson's ratio of steel wire rope v_r	0.2
Lifting vessel mass S_0	80/t	Diameter of steel wire rope d	76/mm
Unit mass of steel wire rope p	21.4/kg/m	Pitch of wire rope t_0	78/mm
Wire rope overhang height H	1500/m	Roll modulus of elasticity E_d	2.1×10^5/Mpa
Gravitational acceleration g	9.8m/s^2	Friction coefficient of steel wire ropes μ_r	0.17
Lifting acceleration a_0	0.75/m/s^2	Width of rope winding area B	2.1/m
Length of dangling wire rope l_s	85/m	Cross-sectional area of steel wire rope A_s	2.21×10^3/mm^2
Number of winding layers n	3	Wire rope stacking angle α	59/°
Number of winding layers m	3	Reel radius R	4/m

Boundary constraints are applied to the two end faces of the support wheel and brake disc as fixed supports, and the two rope-wrapping areas of the reel are taken as cylindrical supports. Maximum static tension applied to the rope wrapping area .Under the above conditions, modal analysis is carried out as shown in Table 3.

Table 3. 1st to 6th order modal shapes

Order	Natural frequency (Hz)	Modal shape	Modal Feature
1	234.82		Large and unevenly distributed deformation in the middle
2	235.02		Large middle deformation and more uniform distribution of deformation
3	235.14		Intermediate deformation with a tendency to spread
4	235.36		Uneven distribution of intermediate deformation, with inconsistent peaks along the circumference
5	237.84		Intermediate deformation distribution with reduced deformation
6	238.05		Uneven distribution of total deformation along the circumference

In the lifting process the external load is constantly changing, but at the end of the brake disc and support wheel, the deformations are very small, so the end surfaces of the two parts will be used as boundary conditions. It can be imposed full restraint to the four end surfaces. In order to facilitate the study of the deformation and stress under the force of the reel, the maximum static tension of 1480kN of the wire rope at rest can be taken for the winding area of the reel. Maximum static tension will act on the winding areas as a circumferential force. The corresponding reel displacement cloud diagram and the corresponding equivalent force cloud diagram for the reel is shown in Fig. 7.

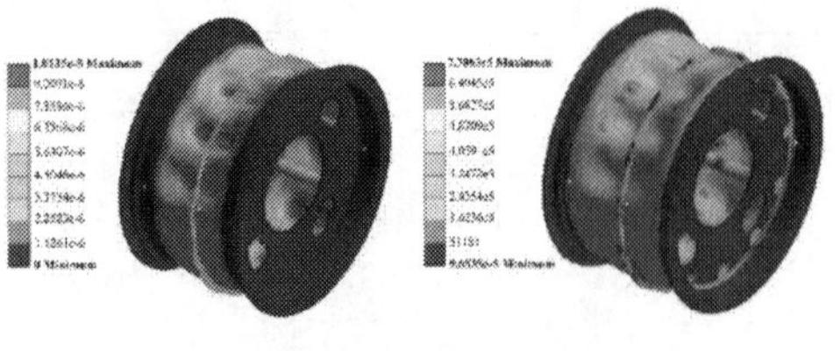

(a) Total deformation cloud diagram

(b) Equivalent Force cloud diagram

Fig. 7. Initial total deformation and equivalent force clouds

From Fig.7,we can get the conclusion that when the reel is in the work, the total deformation and equivalent force of the reel in the middle part are relatively large. Then the total deformation and equivalent force gradually decrease from the middle position to the two sides. With the maximum static tension of the wire rope, the maximum deformation of the reel is 1.0135×10^{-5}m and the maximum equivalent force is 7.3063×10^{5}Pa.When reel deformation is too large, it will lead to *Deformation detuning of the wire* rope, which will lead to the tension difference increase of the steel wire rope. This will eventually affect the synchronization of the hoist. As the area of greatest equivalent forces and deformations, the middle area plays the most significant impact on the wire rope in lifting process. So this study will focus on the analysis of the internal structure of the reel change for the impact on the middle of the reel. By changing the internal structure of the reel to reduce the amount of deformation and equivalent force in the middle of the reel.

In order to solve the above problems, this study optimizes the internal structure of the reel by increasing the diameter of the internal support column of the reel and obtains the total deformation and equivalent force clouds of the reel under the same force conditions as shown in Fig. 8

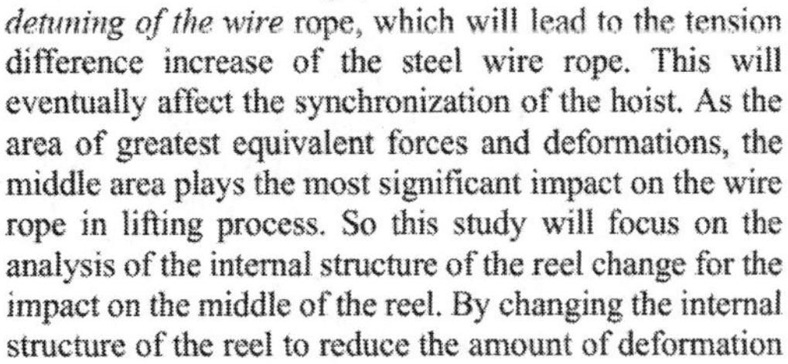

(a) Total deformation cloud diagram

(b) Equivalent Force cloud diagram

Fig. 8. Cloud diagram after changing internal structure

After improving the internal structure of the reel , it is found that the maximum deformation in the middle of the reel is 7.8803×10^{-6}m, which is about 22% lower than the original maximum deformation,and the maximum equivalent force becomes 5.7114×10^{5}Pa, which is about 21% lower than the original equivalent force value. Through numerical analysis, optimizing the diameter of the support column inside the reel can well reduce the reel equivalent force and deformation, then reduce the deformation of the reel and lower the tension difference of the wire rope and improve the synchronization and smoothness of the system in the working process.

5. Conclusion

For the application requirements of increasing mining depth, this study chooses the reel model of double-rope winding hoists .Under the maximum static tension of the wire rope, the study takes the reel static force and modal analysis by finite element analysis. It provides the largest equivalent force and deformation parts. And the internal structure of the reel is optimized and improved. The main conclusions are as follows .

1)This study focuses on the analysis of the impact of the reel structure on the wire rope tension difference. By using modal analysis, it gets the conclusion that the 1st to 6th order of the reel frequency is mainly concentrated in 236Hz or so. The maximum total deformation of the reel is mainly concentrated in the middle position, and with a small change in vibration frequency, the deformation of the middle part of the reel will occur in the circumferential movement. The deformation distribution is uneven and the deformation is large or small from time to time .

2) Through the static analysis of the reel, this study concludes that the maximum equivalent force and maximum deformation of the reel occurs in the middle position. After optimizing the diameter of the internal support column of the reel, the maximum total deformation of the reel is reduced from 1.0135×10^{-5}m to 7.8803×10^{-6}m, and the maximum equivalent force of the reel is reduced from 7.3063×10^{5}Pa to 5.7114×10^{5}Pa. The two improved parameters are 22% and 21% which are much lower than the original ones. The improvement of the internal structure of the reel has greatly reduced the deformation during the work, thus improving the synchronization and accuracy of the double rope hoisting and reducing the tension difference between the two ropes, providing a reference solution for the future development of double-rope winding hoists depth mining.

This study optimizes the internal structure of a double-rope winding hoists reel based on finite element analysis. The feasibility of the method is verified by using theoretical data and only provides an ideal state of improvement for the further developments of the hoist. The research method and mathematical model do not involve experimental verification, so in the future corresponding experiments can be designed for a more in-depth study.

References

1. Chang X, Peng Y, Zhu Z, et al. Tribological properties of winding hoisting rope between two layers with different sliding parameters. Adv Mech Eng 2016; 8: 1687814016679911.
2. Wang J, Pi Y, Hu Y, et al. Modeling and dynamic behavior analysis of a coupled multi-cable double drum winding hoister with flexible guides.

Mechanism and Machine Theory 2017; 108: 191–208.

3. Wu S, Gong X, He P. Dynamic analysis of multi point winding hoisting system under nonsynchronous movements of wire ropes in deep well. J Vibroeng 2021; 23: 512–527.
4. Antoniak J. Theoretical and run investigation of load distribution onto hoisting ropes in multi rope hoists. Sci.Issues Tech. Univ. Min.33 1968 (in Polish).
5. Li X, Zhu Z, Shen G, et al. Wire Tension Coordination Control of Electro-Hydraulic Servo Driven Double-Rope Winding Hoisting Systems Using a Hybrid Controller Combining the Flatness-Based Control and a Disturbance Observer. Symmetry-Basel 2021; 13: 716.
6. Kaczmarczyk S, Ostachowicz W. Transient vibration phenomena in deep mine hoisting cables. Part 1: Mathematical model. Journal of Sound and Vibration 2003;262(2): 219–244.
7. Kaczmarczyk S, Ostachowicz W. Transient vibration phenomena in deep mine hoisting cables. Part 2: Numerical simulation of the dynamic response. Journal of Sound and Vibration 2003;262(2) 245–289.
8. Peng Y, Chang X, Sun S, et al. The friction and wear properties of steel wire rope sliding against itself under impact load. Wear 2018; 400: 194–206.
9. Yao J, Deng X, Ma C, et al. Investigation of Dynamic Load in Superdeep Mine Hoisting Systems Induced by Drum Winding. Shock Vib 2021; 2021: 4756813.
10. Singh JM, Chattopadhyay A, Rao C. approach towards analysing stresses in afixed fixed ended long cylindrical winding drum. Mining Technology 1984;64:104-110.
11. Singh JM, Chattopadhyay A, Rao C. Stress analysis in a short cylindrical winding drum. Mining Technology 1984;66:400-403.
12. Otto S, Mupende I, Dietz P. Influence of the hoisting drum winding system on the end plate loads. In: Marjanovic D (ed) DESIGN 2002: Proceedings of the 7th International Design Conference, Vols 1 and 2. Zagreb: Univ Zagreb, Faculty Mechanical Engineering & Naval Architecture, pp. 1091–1096.
13. Dietz P. Design criteria for multilayer wound winch drums following lightweight design principles. In: Marjanovic D (ed) Design 2004: Proceedings of the 8th International Design Conference, Vols 1-3. Zagreb: Univ Zagreb, Faculty Mechanical Engineering & Naval Architecture, pp. 15–20.
14. Dietz P, Lohrengel A, Schwarzer T, et al. Problems related to the design of multi-layer drums for synthetic and hybrid ropes. In: Innovative ropes and rope applications: A celebration of 175 years of wire rope; Proceedings of the OIPEEC conference 2009 / 3rd International Stuttgart Ropedays.
15. Mangalekar S, Bankar V, Chaphale P. Design and Analysis of Central Drum in Mine Hoist.International Research Journal of Engineering and Technology 2016;03(06):1111-1114.
16. Stanisław WOLNY 和 Sławomir BADURA, 《Stress analysis in structural components of the Koepe pulley in hoisting installations》, Engineering Transactions 2012.
17. Minjie J, Hanbing W, Hongzhang C, et al. The optimum design of main shaft device of double reels for wrapped hoist. International Journal of Information and Systems Sciences 2005;1(3/4):398-405.
18. Torres B, Paya-Zaforteza I, Calderon PA, et al. Analysis of the strain transfer in a new FBG sensor for Structural Health Monitoring. Eng Struct 2011; 33: 539–548.
19. Chen S. The Optimum Design of A2JK hoist drum based on ANSYS. In: Wang JH, Qi JG (eds) Materials and Manufacturing, Pts 1 and 2. Stafa-Zurich: Trans Tech Publications Ltd, pp. 1028–1031.
20. Hu J, Li J-C, He X, et al. Large Mine Hoist Drum Topology Optimization Design. In: 2016 International Conference on Energy Development and Environmental Protection (edep 2016). Lancaster: Destech Publications, Inc, pp. 520–526.
21. Yu ZL, Li WM. CAE Optimization Design of Mine Hoist Spindle Device. AMR 2011; 299–300: 878–882.
22. Torrance B M. The design of winding drums. The South African Mechanical Engineer 1965,15:123-141.

Research Status and Optimization Methods of Zinc Ion Battery

Shurui Wang

College of Materials Science and Engineering, Shandong University, Jinan 250100, China

Abstract: Up against the energy shortage and aggravating environmental pollution, it is extremely urgent to develop renewable clean energy. With efficient energy storage and energy conversion, electrochemical energy storage is the key direction for the development of energy storage technology in the future. Besides, aqueous zinc ion battery has attracted researchers because of its low cost and high theoretical specific capacity. Cathode materials for aqueous zinc ion batteries are roughly divided into manganese-based compounds, vanadium-based compounds, Prussian blue analogues, etc, which usually use zinc metal as an anode. Electrolytes include solid hydrogel electrolytes and liquid ion electrolytes. However, some problems exist in cathode materials, such as elements dissolution and low discharge voltage, while anode materials have problems in zinc dendrite growth and side reactions, and water decomposition occurs in electrolytes. In recent years, researchers have devoted themselves to optimizing aqueous zinc ion batteries in different ways, so as to obtain their high performance. In this paper, the general situation of zinc ion battery is introduced at first, and then the research status is emphatically expounded from the perspectives of problems existing in cathode materials, anode materials, electrolyte, and their optimization methods, which provides references for developing high-performance aqueous zinc ion battery.

Keywords: Zinc Ion Battery; Optimization Method; Diffusion Kinetics; Side Reaction Inhibition.

1. Introduction

Recently, due to the continuous consumption of fossil fuels, energy shortage and the greenhouse effect have been aggravated, with the sustainable development of human society facing a severe test. On this basis, using renewable clean energy such as wind energy, solar energy, and tidal energy to gradually replace fossil energy has become the main energy revolution [1-4]. Therefore, efficient energy storage and energy conversion are imperative [5]. With high energy density, fast response, low cost, and convenient installation, electrochemical energy storage is the key direction for the development of energy storage technology in the future [4, 6]. In the rechargeable battery system, lithium-ion batteries have achieved great success because of their high energy density, good cycle stability, enduring cycle, and mature preparation technology. However, lithium resources are limited in nature, and high-cost flammable, and toxic organic electrolytes are unsafe, which easily leads to environmental problems, hindering the further development of lithium-ion batteries [7]. Compared with organic electrolytes with certain potential safety issues, aqueous electrolytes are safer and more reliable [8, 9]. More importantly, the ionic conductivity of aqueous electrolytes is two magnitude orders higher than that of organic electrolytes, which improves the power density and fast charging and discharging ability [5]. At present, aqueous rechargeable batteries based on different valence carriers have been developed, such as sodium ion batteries, potassium ion batteries, calcium ion batteries, zinc ion batteries, and so on [1]. Among many rechargeable aqueous batteries, zinc ion battery has attracted researchers. This is because there are abundant zinc resources in nature, and zinc cathode is low in cost and good in safety. More importantly, it has the advantages of low redox potential and high theoretical capacity (820 $mAh \cdot g^{-1}$) [10], which endows with the advantages of aqueous electrolyte and zinc cathode.

Cathode materials for aqueous zinc ion batteries can be roughly divided into manganese-based compounds, vanadium-based compounds, Prussian blue analogues, and other composite materials [11]. Zinc metal is usually used as the anode, and electrolytes include solid hydrogel electrolytes and liquid ion electrolytes. Because of the diverse potential electrolytes, zinc is advantageous in high redox potential, high safety, low toxicity, reversibility of zinc deposition/stripping, long cycle life, high volumetric energy density, etc., so aqueous zinc ion batteries are in the rapid development [12]. However, there are some problems in cathode materials, such as electrode degradation caused by manganese dissolution, low discharge voltage of vanadium-based compounds [13], zinc dendrite growth, water/oxygen-related side reactions [9, 14], and water decomposition in the electrolyte, etc.

The further application of aqueous zinc ion batteries still needs a lot of related research and improvement.

Based on this, this paper first introduces the general situation of aqueous zinc ion batteries, and then combines the latest research, with a focus on the current progress of cathode materials, anode materials, and electrolytes of aqueous zinc ion batteries. Finally, their future development is prospected, which provides references for preparing high-performance aqueous zinc ion batteries.

2. Cathode Material

Cathode materials can be roughly divided into manganese-based compounds, vanadium-based compounds, Prussian blue analogues, and other composite materials [11]. Manganese resources are abundant in the crust, and manganese-based oxides have tunnel and layered structures, which are beneficial to the diffusion kinetics of Zn^{2+} with high specific capacity. For example, MnO_2 has a theoretical capacity of 616 mAh • g^{-1} [15]. Vanadium-based compounds often have layered or channel structures, which is beneficial to the diffusion kinetics of Zn^{2+} as well as the intercalation and removal during charge and discharge. They, a kind of promising zinc ion storage materials, also have the advantages of high specific capacity (> 300 mAh • g^{-1}), rapid reaction kinetics, and multivalent state of vanadium [3]. As a transition metal, ferricyanate, with an open structure, the Prussian blue analogue is beneficial to the intercalation and removal of Zn^{2+}, realizing the cation storage with different valences [16]. In addition, organic electrode materials with low prices, less environmental pollution, and large theoretical capacity have great advantages in the storage of monovalent and divalent cations such as quinone compounds. Some synthetic nano-materials have also been proven to be used as cathode materials such as CNF/GNP composite aerogels, etc [17]. However, there are electrode degradation problems related to manganese dissolution in manganese-based compounds, which are caused by Jahn-Teller structural distortion, repeated structural phase transition, and irreversible phase formed by repeated intercalation and deprivation of Zn^{2+} in $ZnSO_4$ electrolyte. There are some problems in vanadium-based compounds, such as low discharge voltage, poor electronic conductivity, and electrode degradation [13]. The bound water and adsorbed water in Prussian blue analogues are unstable during the charging and discharging, with their low theoretical specific capacity [2]. The above defects seriously affect the performance of aqueous zinc ion batteries. In recent years, researchers have devoted themselves to preparing new cathode materials by doping other elements or defects, surface coating, or modification to improve the performance of aqueous zinc ion batteries [2].

2.1 Manganese-Based Materials

Manganese is the tenth most abundant element on the earth, and is a common transition metal element. With many different oxidation states, it can form multivalent manganese oxide phases with different atomic structures [7, 18]. As a cathode material for aqueous zinc ion batteries, manganese-based oxides have been widely studied in recent years. The common manganese-based oxides have various crystal types (α-, β-, γ-, δ-, λ-, R-type) of MnO_2, Mn_2O_3, and Mn_3O_4 [19]. The theoretical capacity of MnO_2 is 308 mAh/g and its voltage window is 0.8-1.8 V. The crystal structure can be divided into three types, including tunnel structure, spinel structure, and layered structure, which have the advantages of high natural abundance, low cost, low toxicity, high theoretical capacity, and environmental affinity [7, 13, 19]. $Mn2O_3$ is superior in energy density and simple in the synthesis process, while Mn_3O_4 has good cycling performance [20, 21]. However, when manganese-based oxides are used as cathode materials for aqueous zinc ion batteries, there are still some problems, such as structural deformation, inevitable manganese dissolution, and low inherent electronic conductivity [7, 22]. Therefore, researchers in recent years often use the doping of other elements, surface coating, composite materials building, defect engineering, and other different modification methods to optimize manganese-based cathode materials, so as to obtain high-performance aqueous zinc ion batteries [7]. Zhang et al. [21] prepared a Ni-doped Mn_2O_3 (NM) derived from NiMn layered double hydroxide (NiMn-LDH), which has a high specific capacity of 252 mAh/g, three times that of pure Mn_2O_3. When it was applied to the cathode of the Zn/NM20 battery, a high energy density of 327.5 Wh/kg was achieved under the condition of 0.14 kW/kg, and the capacity retention rate was 85.6% after 2500 cycles (Figure 1a). The results show that doping Ni^{2+} can promote electron rearrangement, improve the conductivity and reaction kinetics of nanomaterials, reduce the formation energy, stabilize the Mn-O bond, and improve the electrochemical performance of nanomaterials. Introducing cationic vacancies is effective to improve the low inherent conductivity of manganese-based oxide materials. These defect structures enrich the active sites of ion storage, reduce the electrostatic interaction between zinc ions and cathode materials, and effectively promote the diffusion kinetics of ions [19]. Bo et al. [19] optimized the traditional MnO cathode material by using the Al doping modified MnO. The aqueous zinc ion battery assembled by this material has a capacity retention rate of 89% after 500 cycles at 0.1 A/g and 1.0 A/g (Figure 1b). The results show that Al-doped modification introduces a large number of Mn vacancies, which effectively promotes the diffusion of zinc ions, increases the specific surface area and pore size of MnO, and makes the cathode wetter to the electrolyte. In addition, Al-MnO was transformed into R-MnO_2 during charge and discharge, which effectively inhibited the dissolution of manganese. Constructing nanostructures and hybridizing with conductive materials is another effective strategy to improve conductivity and adapt to structural changes during cycling, among which graphene has been widely used to hybridize with various electrode materials [23]. Wang et al. [23] prepared an independent mixed nanostructure composed of ultra-long manganese dioxide nanowires and graphene nanosheets, with a specific capacity of 317 mAh/g at 0.1 A/g (Figure 1c), which can provide a high energy density of 436 Wh/kg,

https://doi.org/10.1051/matecconf/202338201015

with 0.011% as the capacity attenuation rate of each cycle in 2000 cycles. The research shows that the unique 1D/2D nanostructure not only promotes strong electrochemical synergy, but also withstands severe mechanical deformation such as bending and folding. Cao et al. [24] prepared MnO_2/graphite nano-sheet hybrid by ball milling method. When it was used as a cathode of the aqueous zinc ion battery, the capacity retention rate was 80.8% after 1000 cycles at 0.1 A/g and 80 mAh/g at 1 A/g (Figure 1d), which was significantly better than pure MnO_2 cathode material. The reasons are as follows. ① The size of MnO_2 particles decreased significantly through the ball milling process, increasing specific surface area and exposing more active sites. ② The chemical bonding between MnO_2 and graphite nanosheets enhances the internal stability and interfacial adhesion of the material. ③ Graphite nanosheets have good mechanical properties and effectively restrain the volume change during cycling. Wang et al. [25] prepared a new type of γ-MnO_2-graphene composite material. When it was used in a $ZnMnO_2$ battery, the specific capacity was 301 mAh/g at 0.5 A/g and 95.8 mAh/g at 10 A/g, which was significantly better than the traditional MnO_2 cathode (Figure 1e). The results show that γ-MnO_2 with tunnel structure is beneficial to the intercalation/detachment process of zinc ions. Graphene has a large surface area, good electrical conductivity, and mechanical stability, which improves the poor electrical conductivity and poor cycle stability of manganese-based materials.

The construction of an electrode-electrolyte interface (EEI) is often used to improve the stability of the electrode, which also plays a good role in solving the manganese dissolution. Zuo et al. [26] synthesized a manganese sulfate nano-layer on MnO_2 cathode in situ by electrochemical treatment. After cyclic activation, the manganese sulfate nano-layer was transformed into δ-MnO_2 nano-layer (MON) interface phase with rich interlayer water, while MnO_2 coated with MON was used as cathode material for an aqueous zinc ion battery. After 1000 cycles at 10 A/g, the capacity retention rate was 85.9% (Figure 1f), which was significantly better than pure MnO_2 cathode. The results show that the interlayer phase of MOM is rich in interlayer water, which realizes the transition from zinc ion-dominated insertion to proton-dominated insertion, inhibits the dissolution of manganese, and improves the electrochemical performance. As for layered manganese-based materials, the introduction of metal ions (Ca^{2+}, Zn^{2+}, Na^{+}) and crystal water as interlayer columns can alleviate the poor cycle stability caused by structural deformation and manganese dissolution [27]. Sun et al. [27] prepared $Ca_{0.28}MnO_2 \cdot 0.5H_2O$ (CaMnO) electrode material by pre-inserting Ca^{2+} and water molecules into δ-MnO_2 as interlayer columns. In 1M electrolyte of $ZnSO_4$/0.1 M $MnSO_4$, CaMnO had no obvious capacity attenuation after 5000 cycles and the discharge capacity was 124 mAh/g at 3.5 A/g, showing excellent electrochemical performance. It is proved that the doping of water and calcium ions can essentially stabilize the structure of layered manganese-based materials.

In addition, phosphorylation may also be a means to solve the problem of manganese-based cathode materials. However, manganese phosphate has electrochemical hysteresis, which makes it difficult to find intercalation carriers for zinc ions with strong electrostatic interaction. In order to solve the above problems, Li et al. [28] prepared carbon-coated $Na_3Mn_2(P_2O_7)(PO_4)$ (NMPP@C) by liquid phase method and freeze drying technology, which showed a peak capacity of 150 mAh/g through constant current activation, with the final capacity stable at 90 mAh/g and 1.45 V as the discharge platform (Figure 1g). The aqueous zinc ion battery assembled by it shows excellent rate performance and good cycle performance. The results show that the structure of NMPP@C material is decomposed during constant current stimulation. Besides, the generated α-MnO_2 and $Zn_2P_2O \cdot 5H_2O$ are used as electrochemical reagents, α-MnO_2 as electron/proton host, and $Zn_2P_2O \cdot 5H_2O$ as zinc ion host.

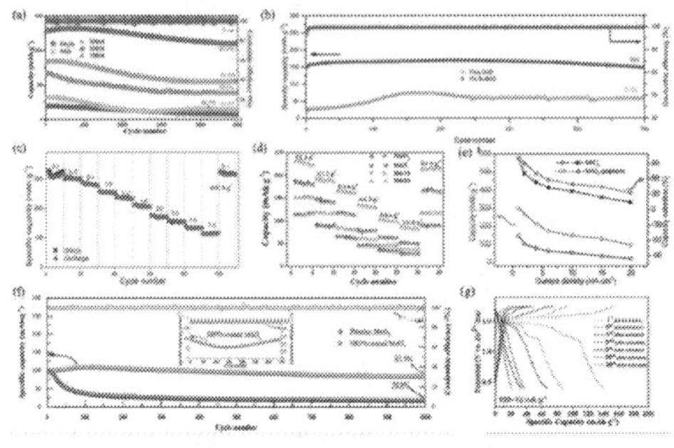

Figure. 1 (a) Cycle performance of Mn_2O_3, NM10, NM15, NM20, and NM30 [21]; (b) Long-term cycling performance of pure MnO and 5% Al-MnO at current density of 1 A/g [19]; (c) Rate performance of MnO_2/rGO (reduced graphene oxide) at various current densities [23]; (d) Rate performance of MnO_2, MG5, MG10, MG20 [24]; (e) Capacity and capacity retention of MnO_2 and MnO_2-graphene electrodes [25]; (f) Cycling performance of pure MnO_2 and MON-coated MnO_2 at current density of 10 A/g [26]; (g) Charge-discharge curves of NMPP@C at 0.1 C [28];

 https://doi.org/10.1051/matecconf/202338201015

Table 1 Performance of Different New Manganese-Based Electrode Materials in Batteries

Manganese-Based Electrode Material	Performance	References
Al-doped MnO	345 mAh/g at 0.1 A/g, 89% at 1.0 A/g after 500 cycles	[19]
Ni-doped Mn_2O_3	The high specific capacity is 252 mAh/g. When used in Zn/NM20 battery, the energy output of 327.5 Wh/kg is achieved at 0.14 kW/kg, and the capacity retention rate is 85.6% after 2500 cycles	[21]
Independent mixed nanostructure composed of manganese dioxide nanowires and graphene nanosheets	The specific capacity is 317 mAh/g at 0.1 A/g, which can withstand severe mechanical deformation of bending and folding. When used in zinc ion battery, the energy density is 436 Wh/kg, and the capacity decay rate of each cycle is 0.011% in 2000 cycles	[23]
MnO_2/graphite nanosheet hybrid	230 mAh/g at 0.1 A/g; 80 mAh/g at 1 A/g, the capacity retention rate is 80.8% after 1000 cycles	[24]
γ-MnO_2-graphene composites	301 mAh/g at 0.5 A/g, 95.8 mAh/g at 10 A/g	[25]
MnO_2 of δ-MnO_2 Nanolayer Coating with Rich Interlayer Water	The capacity retention rate was 85.9% after 1000 cycles at 10 A/g	[26]
$Ca_{0.28}MnO_2 \cdot 0.5H_2O$ (CaMnO)	In 1M electrolyte of $ZnSO_4$/0.1 M $MnSO_4$, there was no obvious capacity attenuation after 5000 cycles, and the discharge capacity was 124 mAh/g at 3.5 A/g	[27]
Carbon coated $Na_3Mn_2(P_2O_7)(PO_4)$ (NMPP@C)	The peak capacity is 150 mAh/g, the final capacity is 90 mAh/g, and the discharge platform is 1.45 V vs	[28]

2.2 Vanadium-Based Materials

Vanadium-based compounds have a variety of crystal structures, such as layered structure, tunnel structure, NASICON structure, etc., which are beneficial to realize the rapid transfer of multiple electrons, local electrical neutrality, and polarization alleviation caused by zinc ion intercalation. In addition, vanadium-based compounds have abundant oxidized states that can be converted to each other, which shows a more stable structure and a wider range of changes when used in aqueous zinc ion batteries [5]. However, the problems such as low average working voltage and easy structure collapse of vanadium-based compounds still need further study and solution [13]. Researchers often dope other metal ions, construct conductive coatings, use oxygen defect engineering, and different preparation methods to improve the performance of vanadium-based cathode materials [29]. Xu et al. [30] prepared a cathode material with excellent performance (V_2O_5@PEDOT/CC) by uniformly depositing V_2O_5@poly (3, 4-ethylene dioxythiophene) (PEDOT) hybrid nanosheet array on flexible carbon cloth (CC), which has a maximum capacity of 360 mAh/g at 0.1 A/g and a capacity retention rate of 97% after 600 cycles at 1 A/g (Figure 2a). The results show that the V_2O_5 nanosheet array enriches the active sites of zinc storage. PEDOT coating has good electrical conductivity and structural stability, which improves the diffusion kinetics of zinc ions and acts as a protective layer to inhibit structural collapse during the cycle. Su-Ho et al. [31] optimized the diffusion kinetics of zinc ions in V_2O_5 thin films by controlling the number of oxygen vacancies in the oxidation lattice. The prepared V_2O_5-x thin films were 57.2 mAh/g at 20C and 105.2 mAh/g at 5C (Figure 2b). The introduction of oxygen vacancies effectively reduced the strong electrostatic interaction between zinc ions and matrix oxide materials, improving the diffusion kinetics of zinc ions. Chen et al. [32] prepared hollow V_2O_5 nanosphere cathode by template-free solvothermal method, which was 327 mAh/g at 0.1 A/g, 146 mAh/g at 20 A/g (Figure 2c), and 147 mAh/g after 6000 cycles at 10 A/g, with a capacity retention rate of 69.7%. The results show that the hollow nanostructure can effectively reduce the structural stress during cycling, shorten the ion transfer path, and optimize the diffusion kinetics of zinc ions. Deng et al. [33] improved the reversibility of high-temperature argon-treated VO_2 (AVO) materials during deep circulation through an in-situ self-conversion electrochemical strategy. When the prepared AVO microsphere materials were used as cathodes of aqueous zinc ion batteries, they were 446 mAh/g at 0.1 A/g and 323 mAh/g at 10 A/g (Figure 2d), which could be stably circulated for more than 4000 times at 20 A/g, with an energy density of 308 Wh/kg. According to the research, hydrated V_2O_5 ($V_2O_5 \cdot nH_2O$) is formed due to the insertion of water molecules and electrochemical oxidation phase transition during the first charging process. As the main active substance in the subsequent cycle, this substance provides abundant active sites and can carry out highly reversible redox reactions.

By pre-intercalation of guest metal ions (Li^+, Na^+, Zn^{2+}, K^+), the interlayer spacing can be increased and the diffusion kinetics of zinc ions can be improved to reduce charge polarization effect and optimize the structural stability of vanadium-based electrode materials [34]. Zhang et al. [34] synthesized potassium vanadium oxide ($K_{0.23}V_2O_5$) with tunnel structure by hydrothermal method. When this material is used as cathode material, the capacity retention rate is 92.8% and the specific capacity is 103 mAh/g after 500 cycles under the condition of 2.0 A/g (Figure 2e). In addition, the ion diffusion rate is $1.88*10^{-9}$ -$2.6*10^{-8}$ cm^2/s. The research shows that the introduction of K^+ increases the interlayer spacing of vanadium-based electrode materials, which not only optimizes the diffusion kinetics of zinc ions, and improves the ion diffusion rate and conductivity, but also reduces the possibility of structural collapse.

More importantly, the researchers also found that high-performance cathode materials for zinc ion batteries can be obtained by combining pre-intercalation, constructing conductive coatings, and oxygen vacancy defect engineering [29]. Tan et al. [29] prepared a material that is poly (3, 4-ethylene dioxythiophene) (PEDOT) coated, manganese ion intercalated, and oxygen-enriched hydrated V_2O_5 (MnVOH@PEDOT) by one-step hydrothermal method. When the material was used as the cathode of aqueous zinc ion battery, it was 455.8 mAh/g at 0.2 A/g and 336.0 mAh/g at 8A/g (Figure 2f). After 2000 cycles, the capacity retention rate was 84.8% with good rate performance and cycle stability. The results show that PEDOT coating has high conductivity, which improves the conductivity of external electrons and effectively prevents vanadium oxide from dissolving. Manganese ion intercalation not only enlarges the interlayer spacing, improves the diffusion kinetics of zinc ion, but also plays the role of "pillar" and maintains the structural stability of the cathode. In addition, the oxygen vacancy introduced in V_2O_5 improves the conductivity of internal electrons.

Besides, in the research on the application of various metal vanadates (such as potassium vanadate, sodium vanadate, etc.) to aqueous zinc ion batteries, researchers found that compared with monovalent alkali metal cations (such as Na^+, K^+, etc.), polyvalent metal ions combined with oxygen atoms can produce different ionic bonds and layer sizes with better crystal structure [35]. Yang et al. [35] synthesized $Fe_2V_4O_{13}$ by simple solution chemical reaction. Through in-situ and non-in-situ characterization under different states of charge and discharge, it shows that the structure of the material is highly reversible, which is 380 mAh/g at 0.2 A/g (Figure 2g) and 83% capacity retention after 1000 cycles at 10 A/g (Figure 2h). The reversible phase transition from ferric vanadate to zinc vanadate and the intercalation/de-intercalation mechanism of zinc ion storage in tunnel structure were observed simultaneously, which provided a new direction for the design and research of zinc ion batteries. Under certain synthesis conditions, nanomaterials with unconventional phases can be prepared by adjusting the reaction kinetics, realizing the process from structural modulation of cathode materials to property improvement [36]. Qu et al. [36] prepared a hexagonal $Cs_{0.3}V_2O_5$ cathode material with a specific capacity of 543.8 mAh/g at 0.1 A/g and a capacity retention rate of 87.8% after 1000 cycles at 2 A/g (Figure 2i). The results show that compared with orthogonal V_2O_5, the hexagonal framework structure with Cs has strong stability and good electrical conductivity, and the open framework structure optimizes the diffusion kinetics of zinc ions.

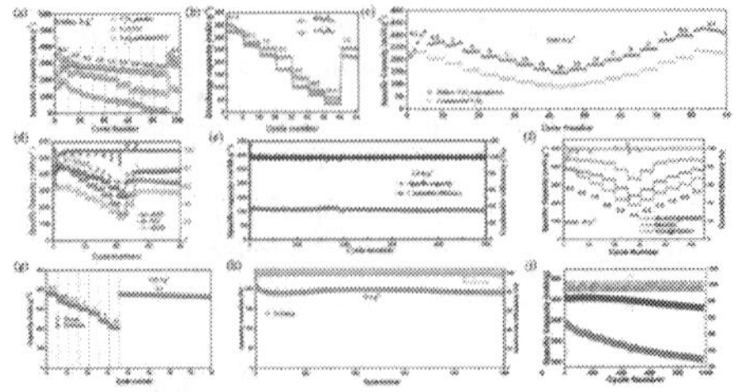

Figure. 2 (a) Rate performance of V_2O_5@PEDOT/CC, V_2O_5/CC, and V_2O_5 powders at different current densities [30]; (b) Rate performance of H-V_2O_{5-x} and L-V_2O_{5-x} at different current densities [31]; (c) Rate performance of hollow V_2O_5 nanospheres and commercial V_2O_5 [32]; (d) Rate performance of AVO (argon-treated VO2), VO (untreated VO2), and OVO (air-treated VO2) materials [33]; (e) Electrochemical properties of $K_{0.23}V_2O_5$ at 2.0 A/g [34]; (f) Rate performance of MnVOH@PEDOT, MnVOH, and VOH@PEDOT cathodes [29]; (g) Rate performance of $Fe_2V_4O_{13}$ cathode material at different current densities of 0.2-10.0 A/g [35]; (h) Cycling performance of $Fe_2V_4O_{13}$ cathode material at 10 A/g [35]; (i) Cycling performance of hexagonal $Cs_{0.3}V_2O_5$ cathode material at 2 A/g [36];

Table 2 Performance of Different New Vanadium-Based Electrode Materials in Batteries

Vanadium-Based Electrode Material	Performance	References
Material that is poly (3, 4-ethylene dioxythiophene) (PEDOT) coated, manganese ion intercalated, oxygen-enriched hydrated V_2O_5	455.8 mAh/g at 0.2 A/g, 336.0 mAh/g at 8A/g, and 84.8% capacity retention after 2000 cycles	[29]
V_2O_5@PEDOT/CC	360 mAh/g at 0.1 A/g, 232 mAh/g at 20 A/g, retention of 97% after 600 cycles at 1 A/g, and 89% after 1000 cycles at 5 A/g	[30]
V_2O_{5-x} thin film material	57.2 mAh/g at 20 C and 105.2 mAh/g at 5 C	[31]
Hollow V_2O_5 nanospheres	327 mAh/g at 0.1 A/g, 146 mAh/g at 20 A/g, 147 mAh/g after 6000 cycles at 10 A/g, and 69.7% retention	[32]
AVO microsphere material	446 mAh/g at 0.1 A/g, 323 mAh/g at 10 A/g, and more than 4000 cycles at 20 A/g, with an energy density of 308 Wh/kg	[33]
$K_{0.23}V_2O_5$	The capacity retention rate was 92.8% of 2.0 A/g, 103 mAh/g after 500 cycles	[34]
$Fe_2V_4O_{13}$	380 mAh/g at 0.2 A/g and 83% capacity retention rate after 1000 cycles at 10 A/g	[35]
Hexagonal $Cs_{0.3}V_2O_5$	543.8 mAh/g at 0.1 A/g, retention rate was 87.8% after 1000 cycles at 2 A/g	[36]

2.3 Prussian Blue Analogues

Hexaminometalate, commonly known as Prussian blue analogues (PBAs), is a material with an open frame and face-centered cubic lattice structure, which is high in solid diffusion coefficient (10^{-9}—10^{-8} cm^2/s) and ionic conductivity [16]. PBAs are considered a promising cathode material for aqueous zinc ion batteries because of their good rate performance, reversible insertion of various metal ions, good ionic conductivity, high safety, non-toxicity, and low cost [1, 16, 37]. Liu et al. [37] constructed a zinc ion secondary battery with a metal zinc anode, biological ionic liquid-water electrolyte, and nano-structured Prussian blue analogue cathode. At 10 mA/g, the discharge voltage platform of the battery was ~ 1.1 V, and the specific capacity was 120 mAh/g. The results show that the open framed PBAs cathode has large channels, which optimizes the reversible insertion/removal of zinc ions, and the zinc anode has good compatibility with biological ionic liquid-water electrolyte. However, PBAs have the problems of structural destruction and insufficient active sites during ion insertion/removal, which leads to low theoretical specific capacity and poor cycle stability [38]. On this basis, Zeng et al. [38] prepared a cathode material of PBA hollow sphere zinc ion battery that is cobalt-replaced and manganese-rich (CoMn-PBA HSs) by template ion exchange method. The specific capacity is 128.6 mAh/g at 0.05 A/g and 50.4 mAh/g at 2 A/g. After 1000 cycles, the capacity retention rate is 76.4% with high reversible capacity, good rate performance, and cycle stability. The results show that the manganese-rich hollow structure enriches the active sites, provides a large surface area, optimizes the diffusion kinetics of zinc ions, and adapts to the large volume change during repeated charge and discharge, inhibiting the structure collapse. Cobalt substitution further stabilizes the crystal structure and improves the cycle stability and cycle life.

2.4 Other Cathode Materials

Organic compounds and other new materials are also often used as cathodes for aqueous zinc ion batteries. Liu et al. [39] prepared a new type of metal-organic framework (MOF) cathode by solvothermal method. When the current density is 50 mA/g, the initial specific capacity is 90.7 mAh/g, which can be stably cycled more than 200 times at the current density of 1 A/g with high specific capacity and good cycle stability. In addition, the material can maintain high stability in the range of pH 0-12 for 30 days, which is beneficial to good recyclable zinc ion batteries. Cerium oxide has attracted researchers because of its unique crystal structure, bonds, properties, and changeable oxidation state. Zhang et al. [40] prepared a spindle-shaped cerium oxide (CeO2) particle by hydrothermal method and used it as cathode material of aqueous zinc ion batteries. The average specific discharge capacity of 500 cycles at the current density of 120 mA/g was 95.7 mAh/g, and the average discharge capacity of 100 cycles at the current density of 60 mA/g was 158.8 mAh/g.

3. Anode Material

Zinc is the most commonly used anode material in aqueous zinc ion batteries due to its abundant zinc resources, good chemical stability, low redox potential, and easy processing [22]. However, dendrite growth and water/oxygen side reactions are two key problems affecting the cycle performance of the zinc anode. The electrode position process of zinc tends to form irregular zinc dendrites. When zinc dendrites grow on a large scale, it is easy to penetrate the separator and cause a serious battery short circuit [9]. Secondly, the side reaction seriously harms the cycle performance of the aqueous zinc ion battery [14]. In order to solve the above problems, researchers have adopted surface modification, structural design of a new zinc anode, and electrolyte optimization to improve the performance of the zinc anode [41].

3.1 Surface Modification

Constructing an artificial protective layer on the zinc anode is a promising method to effectively inhibit dendrite growth of the zinc anode, which mainly includes two strategies including physical shielding and inducing uniform diffusion of ions. For example, nano-$CaCO_3$ coating and ZnF_2 coating can promote the redistribution of zinc ions and inhibit dendrite growth, and carbon nanotube-modified coating can construct a uniform electric field and induce uniform deposition of zinc [42, 43]. A direct method of physical shielding is to construct an artificial interface layer to inhibit the growth of zinc dendrites, which requires the artificial interface layer to have good mechanical properties. Another method is to induce uniform migration of zinc ions by ion tunneling confinement, chemisorption, and electrostatic interaction [43]. Lu et al. [42] constructed a layered double oxide (LDO) artificial interface protection coating to solve the above problems. The coulomb efficiency of the coating is 99.7% after 2800 cycles. Zn@LDO symmetric battery can stably cycle more than 6600 times under the condition of 5 mA/cm2 (Figure 3a) and more than 5000 times under the condition of 10 mA/cm2 (Figure 3b). After 1600 cycles at 1A/g, the capacity retention rate of Zn@LDO/α-MnO_2 was 86%. The results show that LDO immobilizes anions in $ZnSO_4$ electrolytes by electrostatic attraction, which reduces the concentration gradient of zinc ions and inhibits the growth of dendrites. In addition, zinc ions are adsorbed by interlayer anions, which inhibits the two-dimensional diffusion and aggregation on the anode surface, resulting in dendrite growth.

However, the side reaction caused by the direct contact between the electrolyte and anode is another problem that limits the electrochemical performance of zinc ion batteries, and unilateral inhibition of zinc dendrite growth and side reaction cannot completely solve the above problems [44]. This is because dendrite growth and side reactions promote each other, and dendrite formation increases the specific surface area of the anode, which intensifies the contact area between the anode and electrolyte, significantly accelerating the rate of hydrogen evolution reaction (HER). Besides, HER aggravates electrochemical corrosion. By-products of electrochemical corrosion will be unevenly deposited on the anode surface to promote dendrite growth [45]. Therefore, synergistic inhibition of dendrite growth and side reactions is very important. Liu et al. [44] constructed a hybrid surface modification coating on a zinc anode that is hydrophobic organic polyvinylidene fluoride (PVDF)-inorganic mesoporous silicon material (SBA15). When the coating is used in a zinc anode symmetric battery, it can stabilize the cycle for more than 1650h at 3 mA/cm^2; When the coating is applied to Zn||V_2O_5 battery (PVDF-SBA15@Zn||V_2O_5), the capacity retention rate is 82.14% after 1000 cycles, which is significantly better than 23.55% of traditional Zn||V_2O_5 battery (Figure 3c) and effectively inhibits zinc dendrite growth and side reactions. The results show that the uniformly distributed porous structure of SBA15 provides a uniform diffusion path for the diffusion of zinc ions and induces the uniform deposition and distribution of zinc ions on the zinc anode. Meanwhile, hydrophobic PVDF effectively inhibits the side reaction between free water molecules and zinc anode (Figure 3d).

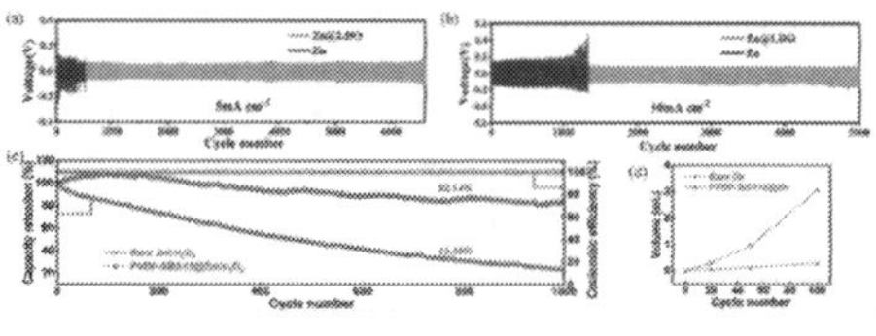

Figure. 3 (a) Long-term constant current cycle performance of Zn//Zn and Zn@LDO//Zn@LDO symmetric batteries at 5 mA/cm^2 [42]; (b) Long-term constant current cycle performance of Zn//Zn and Zn@LDO//Zn@LDO symmetric cells at 10 mA/cm^2 [42]; (c) Long-term cycle stability of Zn||V_2O_5 and PVDF-SBA15@Zn||V_2O_5 batteries at 1 A/g [44]; (d) Hydrogen evolution volumes of a bare zinc anode and PVDF-SBA15@Zn anode in different cycles [44];

3.2 Structure Design of New Zinc Anode

The structure design of a new zinc anode is another effective method to restrain the dendrite growth and side reaction of zinc. Constructing a three-dimensional zinc anode is promising to stabilize zinc anode. The three-dimensional structure increases reaction sites and the specific surface area of reaction sites while reducing the local electric field intensity during Zn deposition [46]. On this premise, it is a promising method to design a three-dimensional zinc anode with a gradient structure. The three-dimensional zinc anode with a gradient structure also improves the local charge transport dynamics and optimizes the Zn deposition process, effectively inducing the uniform deposition of Zn [47]. Cao et al. [47] reported an imprinted gradient zinc anode (polyvinylidene-Sn@Zn, PVDF-Sn@Zn), which is a double gradient structure that circulates stably for 1200h at 1 mA/cm^2 and 1 mAh/cm^2 (Figure 4a) and for 200h at 10 mA/cm^2 and 10 mAh/cm^2 (Figure 4b), effectively inhibiting zinc dendrite growth and side reactions (Figure 4c). The results show that the hydrophobic PVDF layer can effectively prevent the corrosion of Zn by electrolyte, and the hydrophilic conductive layer of Sn has a high redox potential, which inhibits the side reaction. Meanwhile, the gradient conductivity changes the electric field distribution at the bottom of the microchannel, the ion flux of zinc ions, and the local current density, thus realizing the controllable uniform deposition of zinc ions and preventing the short circuit caused by the uneven growth of dendrites at the top. The 3D printing technology is applied to the manufacture of a new three-dimensional structure of zinc anode, which improves the reversibility of zinc anode and zinc utilization rate [48]. Wu et al. [48] printed graphene array brackets (3DGs) with tubular structure (3DGT) and columnar structure (3DGP) current collectors in 3D through digital light processing technology. Under the condition of 2 mA/cm^2 (1 mAh/cm^2), the cycle time of 3DGT@Zn exceeds 1100h, the cycle time of 3DGP@Zn exceeds 900h, and the zinc utilization rate is 5%, which is significantly better than that of traditional zinc symmetric battery (Figure 4d and Figure 4e). More importantly, when 3DGs@Zn anode is used in a soft-pack battery with V_2O_5 cathode material, the capacity is 2.26 mAh/cm^2 under the conditions of N/P (1.74:1) and high zinc utilization rate (47.12%), which is significantly better than that of zinc foil//V_2O_5 soft-pack battery. The research shows that the tubular columnar structure of 3D printing is highly ordered, which can alleviate the significant volume change during reversible deposition/dissolution of zinc, induce uniform deposition of Zn, and effectively improve the utilization rate of zinc.

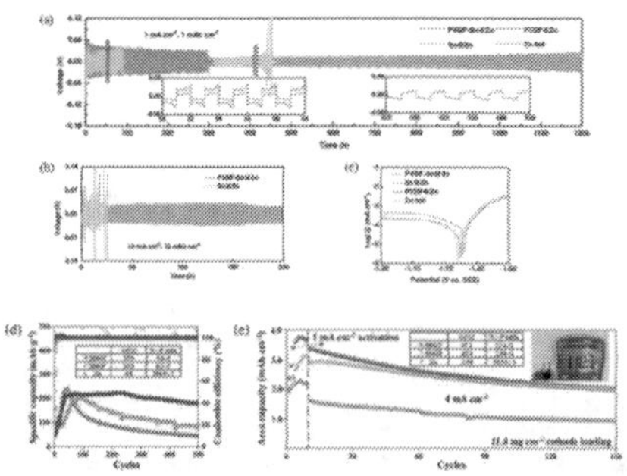

Figure. 4 (a) Working voltage image of zinc symmetric battery with a current density of 1 mA/cm^2 and capacity of 1 mAh/cm^2 [47]; (b) Working voltage image of zinc symmetric cells at 10 mA/cm^2 and 10 mAh/cm^2 [47]; (c) Corrosion curves of different electrodes [47]; (d) Cycle performance of 3DGT@Zn///V_2O_5, 3DGP @ Zn///V_2O_5, and Zn///V_2O_5 batteries at 4 A/g [48]; (e) Cycle performance of 3DGT @ Zn///V_2O_5, 3DGP @ Zn///V_2O_5, and Zn///V_2O_5 soft-pack batteries at 4 mA/cm^2 [48];

3.3 Electrolyte Optimization

Using electrolytic additives, such as Mxene, surfactants, small organic molecules, etc., is a simple method to effectively stabilize the solid electrolyte interface (SEI) film and inhibit the formation of zinc dendrites, which can achieve a longer cycle life of zinc anode [49]. Wu et al. [49] used 2-methylimidazole (Hmim) as an electrolyte additive to construct a strong inorganic/organic zinc-rich ($Zn_4SO_4(OH)_6$/Zn(Hmim)) solid electrolyte interface film

 https://doi.org/10.1051/matecconf/202338201015

in situ, thus realizing a highly stable dendrite-free zinc cathode. The coulomb efficiency of this dendrite-free zinc cathode reaches 97.32% (Figure 5a), which can be stably circulated for more than 2000h. When the electrolyte additive is used in Zn||V_2O_5 battery, it shows good cycle performance (Figure 5b). The results show that the zinc affinity of Hmim and its chelation with zinc ions together lead to the increase of initial nucleation sites and induce the homogeneous deposition of zinc. In addition, the change in the coordination environment reduces the ion diffusion coefficient and increases the nucleation overpotential, which makes the grain size smaller at the initial stage of zinc deposition. The organic/inorganic solid electrolyte interfacial film formed by in-situ construction has good chemical stability and thermal stability, which inhibits side reactions and dendrite growth. A variety of organic electrolyte additives, such as acetonitrile, dimethyl carbonate, silk peptide, saccharin, etc., have been developed as a method to inhibit the growth of zinc dendrites [50, 51] since organic molecule/polymer electrolyte additives can regulate the electrochemical behavior of the interface. Wang et al. [50] prepared a new colloidal electrolyte by dispersing oleic acid (OA) into a 2M $ZnSO_4$ solution. This colloidal electrolyte realizes highly reversible zinc deposition with a coulomb efficiency of 99.63% (Figure 5c). When the electrolyte is used in the $ZnMnO_2$ battery, it can be stably cycled more than 1000 times at 1 A/g (Figure 5d). The results show that oleic acid additive combines with zinc metal to form an oleic acid adsorption layer on the anode surface. This adsorption layer can homogenize zinc deposition, form a flat zinc metal anode, and separate the zinc metal negative electrode from the electrolyte, thus effectively inhibiting side reactions. Xu et al. [51] adopted a bifunctional zinc gluconate electrolyte as a new zinc salt for zinc ion batteries. Experiments show that only 0.3 M zinc gluconate (Figure 5e) is needed to integrate the cation source and anion interface, which obtains an efficient dendrite-free zinc anode interface without other electrolytes. According to the results, zinc gluconate can carry out a stable electrochemical process of zinc plating/stripping and provide zinc gluconate anions to build an artificially stable dendritic-free anion interface to realize dendritic-free zinc deposition.

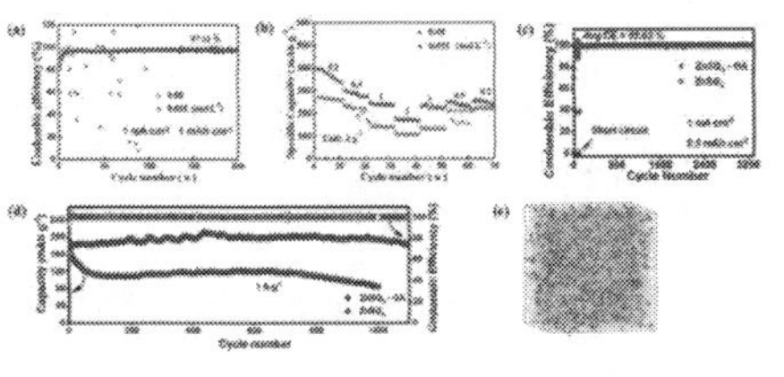

Figure. 5 (a) Coulomb efficiency of zinc plating/stripping at 1 mA/cm^2 (capacity 1 mAh/cm^2) [49]; (b) Cycle performance of Zn||V_2O_5 battery with and without Hmim [49]; (c) Coulomb efficiency of Zn||Cu battery in different electrolytes at 1 mA/cm^2 and 0.5 mAh/cm^2 [50]; (d) Cycle performance of Zn||MnO_2 battery at 1 A/g [50]; (e) MD-simulated three-dimensional images of zinc gluconate system at 0.3 M [51];

Table 3 Effect of Different New Zinc Anode on Zinc Ion Battery

Anode material	Performance	References
Zinc anode coated with layered double oxide (LDO)	Zn@LDO symmetric battery can stably cycle more than 6600 times at 5mA/cm^2 and more than 5000 times at 10 mA/cm^2. After 1600 cycles at 1A/g of Zn@LDO/α-MnO_2, the capacity retention rate was 86%	[42]
Zinc anode with hydrophobic organic polyvinylidene fluoride (PVDF)-inorganic mesoporous silicon material (SBA15) hybrid surface modification coating	PVDF-SBA@Zn||PVDF-SBA@Zn battery can cycle stably for more than 1650h at 3 mA/cm^2. The capacity retention rate of PVDF-SBA15@ Zn||V_2O_5 battery after 1000 cycles was 82.14%	[44]
Polyvinylidene-Sn@Zn	It was stable for 1200h at 1 mA/cm^2, 1 mAh/cm^2, and for 200h at 10 mA/cm^2, 10 mAh/cm^2	[47]
3DGs@Zn	At 2 mA/cm^2 (1 mAh/cm^2), the cycle time of 3DGT@Zn is over 1100h, and the cycle time of 3DGP@Zn is over 900h, and the utilization rate of zinc is 5%	[48]
Zn (2-methylimidazole electrolyte additive)	The coulomb efficiency reached 97.32%, and the cycle was stable for more than 2000h	[49]
Zn (new $ZnSO_4$-OA colloidal electrolyte)	The coulomb efficiency reaches 99.63%, and the cycle life exceeds 3340 times	[50]
Zn (zinc gluconate electrolyte)	After 125 cycles, the coulomb efficiency is 85%	[51]

4. Electrolyte

The electrolyte is another important factor that determines the performance of aqueous zinc ion batteries. Suitable electrolytes can keep good reversibility of zinc deposition/stripping and release a wide electrochemical window. $ZnSO_4$ and Zn (CF_3SO_3) $_2$ are the most common base electrolytes in aqueous zinc ion batteries [52]. $ZnSO_4$ electrolyte has a wide electrochemical window (0.2 ~ 2V wide voltage window) and good anode stability. Compared with $ZnSO_4$ electrolyte, Zn（CF_3SO_3）$_2$ has a wider electrochemical window (wide operating voltage ~ 2.0 V), better reversibility, and faster zinc deposition/dissolution kinetics [52, 53]. Common aqueous zinc ion electrolytes include solid hydrogel electrolytes and liquid ionic liquid. Among them, hydrogel electrolyte shows good flexibility, mechanical durability, temperature adaptability, and electrochemical stability [54]. However, electrolyte also faces severe challenges such as water decomposition reaction [55]. At present, researchers often add additives to electrolytes to improve their performance of electrolytes, which can also inhibit the dendrite growth of zinc anode and improve the reversibility and stability of the anode, preventing the dissolution of cathode materials [53, 56, 57]. Zhang et al. [58] used the $MgSO_4$ electrolyte as an additive in the $ZnSO_4$ electrolyte, and prepared $Mg_xV_2O_5$•nH_2O (MgVO) nanoribbon material as the cathode of the battery. The results show that when the ratio of electrolyte to additive concentration is 1: 1 (1.0 M $ZnSO_4$–1.0 M $MgSO_4$), the MgVO cathode has a specific capacity of 374 mAh/g at 100 mA/g (Figure 6a), a capacity retention rate of 90.3% after 200 cycles at 1 A/g (Figure 6b), and a reversible capacity of 175 mAh/g at 5 A/g. Zhao et al. [59] used a small amount of perfluorooctanoic acid (PFOA) as an additive, and the coulomb efficiency of the Zn||Cu battery using this additive was improved (Figure 6c). The zinc symmetric battery using this additive achieved reversible deposition for up to 2200h, and the Zn||$Na_{0.65}Mn_2O_4$ battery using this additive reached 153 mAh/g at 5 A/g (Figure 6d). The results show that the PFOA additive forms an adsorption layer at the interface between electrolyte and electrode, which redistributes the flux of zinc ions at the interface, inhibits the tendency of anions migrating to the negative electrode, and makes the zinc plating/stripping process of aqueous zinc ion battery more stable. The long perfluorocarbon chain in the PFOA additive has hydrophobic properties, which effectively inhibits the side reactions related to water and expands the working window of the electrolyte to 2.1 V.

In addition, in recent years, the research on flexible wearable devices puts forward new requirements for the development of non-traditional flexible batteries with light, thin, and integrated configuration [8]. Hydrogel electrolyte not only has mechanical properties such as flexibility and foldability of solid electrolyte, but also has diffusion characteristics and high conductivity of water-electrolyte [55], which is imperative for the development of flexible wearable aqueous zinc ion battery [60]. In addition, Li et al. [60] designed and synthesized poly (N, N'-dimethylacrylamide)/zinc trifluoromethane sulfonate (PDMAAm/Zn $(CF_3SO_3)_2$) ionic gel with high operating voltage window in ionic liquid solvent 1-butyl-3-methylimidazole trifluoromethane sulfonic acid. The ionic gel exhibited good mechanical strength and ionic conductivity with a drawing strain rate of 893.7%, a drawing strength of 151.0 kPa, and an ionic conductivity of 0.96 mS/cm (Figure 6e). After 5 rupture/healing cycles, the healing rate reached 87.2%, which showed excellent flexibility and cycle stability after healing under 808 nm near-infrared light irradiation. In addition, the battery has a high discharge voltage of 2.5 V and can stably cycle more than 1500 times at 3.0 A/g, which shows good electrochemical performance (Figure 6f).

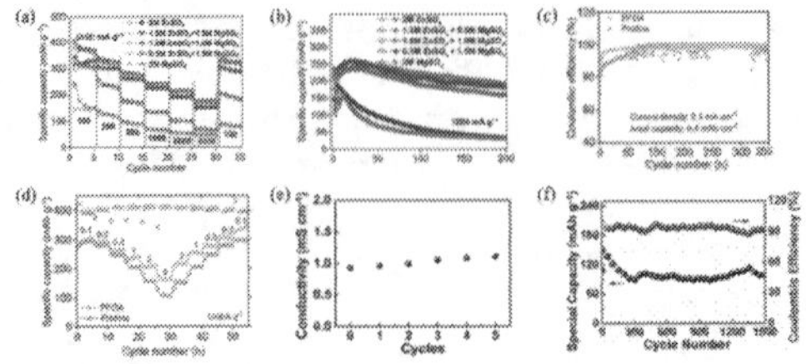

Figure. 6 (a) Rate performance of MgVO cathode in five electrolytes with different concentration ratios [58]; (b) Cycling performance of MgVO cathode in electrolytes with different concentration ratios at 1 A/g [58]; (c) Coulomb efficiency of Zn||Cu battery in different electrolytes [59]; (d) Rate performance of Zn||$Na_{0.65}Mn_2O_4$ battery in different electrolytes [59]; (e) Ionic conductivity of PDMAAm/Zn$(CF_3SO_3)_2$ ionic gel in multiple rupture/healing cycles [60]; (f) Long-term cycle performance and coulomb efficiency of PDMAAm/Zn$(CF_3SO_3)_2$ ion gel based zinc ion battery at 3.0 A/g [60];

5. Results and Prospects

In this paper, the research progress of aqueous zinc ion batteries is reviewed, which is embodied in cathode materials, anode materials, and electrolytes. As for cathode materials, we found that there are some problems in manganese-based cathode materials, such as structural deformation, manganese dissolution, and low inherent electronic conductivity. Adopting doping elements, surface coating, constructing high-conductive composites, defect engineering, and other methods can effectively inhibit manganese dissolution, alleviate volume change, and improve conductivity. The problems of vanadium-based cathode materials such as low average working voltage and easy collapse of structure can be solved by doping, surface coating, and defect engineering. In addition, it is found that Prussian blue analogues and other cathode materials still need further research and development. As for anode materials, in order to solve the dendrite growth and water/oxygen-related side reactions, we found that surface modification, structure optimization, and electrolyte optimization are more effective. As for the electrolyte, we found that the electrochemical performance of the electrolyte was effectively improved by adding additives. To sum up, this paper provides references for improving the research and development of cathode materials, anode materials, and electrolytes of aqueous zinc ion batteries, and in the meantime, looks forward to the future development of aqueous zinc ion batteries.

References

1. SELVAKUMARAN D, PAN A, LIANG S, et al. A review on recent developments and challenges of catherde materials for rechargeable aqueous Zn-ion patterns [J]. Journal of Materials Chemistry A, 2019, 7 (31): 18209-36.

2. SONG M, TAN H, CHAO D, et al. Recent Advances in Zn-Ion Batteries [J]. Advanced Functional Materials, 2018, 28 (41).
3. SHI Y, CHEN Y, SHI L, et al. An Overview and Future Perspectives of Rechargeable Zinc Batteries [J]. Small, 2020, 16 (23): e2000730.
4. ZHANG Q, LUAN J, TANG Y, et al. Interfacial Design of Dendrite-Free Zinc Anodes for Aquous Zinc-Ion Batteries [J]. Angew Chem Int Ed Engl, 2020, 59 (32): 13180-91.
5. LV T, PENG Y, ZHANG G, et al. How About Vanadium-Based Compounds as Cathode Materials for Aquous Zinc Ion Batteries? [J]. Adv Sci (Weinh), 2023: e2206907.
6. ZHAO Y, ZHU Y, ZHANG X. Challenges and perspectives for manganese-based oxides for advanced aqueous zinco-ion battles [J]. InfoMat, 2019, 2 (2): 237-60.
7. YADAV P, KUMARI N, RAI A K. A review on solutions to overcome the structural transformation of manganese dioxide-based cathodes for aqueous rechargeable zinc ion batteries [J]. Journal of Power Sources, 2023, 555.
8. HUANG S, ZHU J, TIAN J, et al. Recent Progress in the Electrolytes of Aquous Zinc-Ion Batteries [J]. Chemistry, 2019, 25 (64): 14480-94.
9. CAO Z, ZHUANG P, ZHANG X, et al. Strategies for Dendrite-Free Anode in Aquous Rechargeable Zinc Ion Batteries [J]. Advanced Energy Materials, 2020, 10 (30).
10. LI X, CHEN Z, YANG Y, et al. The phosphate cathodes for aqueous zinc-ion batteries [J]. Inorganic Chemistry Frontiers, 2022, 9 (16): 3986-98.
11. ZHANG Y, CHEN A, SUN J. Promise and challenge of vanadium-based cathodes for aqueous zinc-ion batteries [J]. Journal of Energy Chemistry, 2021, 54: 655-67.
12. MING J, GUO J, XIA C, et al. Zinc-ion battles: Materials, mechanisms, and applications [J]. Materials Science and Engineering: R: Reports, 2019, 135: 58-84.
13. MATHEW V, SAMBANDAM B, KIM S, et al. Manganese and Vanadium Oxide Cathodes for Aquous Rechargeable Zinc-Ion Batteries: A Focused View on Performance, Mechanism, and Developments [J]. ACS Energy Letters, 2020, 5 (7): 2376-400.
14. DU W, ANG E H, YANG Y, et al. Challenges in the material and structural design of zinc anode towards high-performance aqueous zinc-ion battles [J]. Energy & Environmental Science, 2020, 13 (10): 3330-60.
15. XIONG T, ZHANG Y, LEE W S V, et al. Defect Engineering in Manganese-Based Oxides for Aquous Rechargeable Zinc-Ion Batteries: A Review [J]. Advanced Energy Materials, 2020, 10 (34).
16. ZAMPARDI G, LA MANTIA F. Prussian blue analogues as aqueous Zn-ion battle elections: Current challenges and future perspectives [J]. Current Opinion in Electro Chemistry, 2020, 21: 84-92.
17. LU K. Construction of advanced energy storage systems through rational design of material combination and kinectice balance. Jinan: Dissertation For Doctoral Degree of Shandong University [D], 2018.
18. LIU Y, WU X. Strategies for constructing manganese-based oxide electrical materials for aqueous rechargeable zincon batters [J]. Chinese Chemical Letters, 2022, 33 (3): 1236-44.
19. HE B, HUANG J, JI P, et al. Al doped manganous oxide for high-performance aqueous Zn-ion batteries [J]. Journal of Power Sources, 2023, 554.
20. HAO J, MOU J, ZHANG J, et al. Electrically induced spinel-layered phase transition of Mn3O4 in high performance neutral aqueous rechargeable zinc battle [J]. Electro Chimica Acta, 2018, 259: 170-8.
21. ZHANG D, CAO J, ZHANG X, et al. Institution of Manganese Dissolution in Mn2O3Cathode with Controllable Ni2 + Incorporation for High-Performance Zinc Ion Battery [J]. Advanced Functional Materials, 2021, 31 (14).
22. TANG B, SHAN L, LIANG S, et al. Issues and opportunities facing aqueous zinc-ion batteries [J]. Energy & Environmental Science, 2019, 12 (11): 3288-304.
23. WANG J, WANG J G, LIU H, et al. A Highly Flexible and Lightweight MnO2/Graphene Membrane for Superior Zinc-Ion Batteries [J]. Advanced Functional Materials, 2020, 31 (7).
24. CAO J, ZHANG D, ZHANG X, et al. Mechanochemical reactions of MnO2 and graphic nanosheets as a durable zinc ion battle cathede [J]. Applied Surface Science, 2020, 534.
25. WANG C, ZENG Y, XIAO X, et al. γ-MnO2 nanorods/graphene composite as efficient cathode for advanced rechargeable aqueous zinc-ion battle [J]. Journal of Energy Chemistry, 2020, 43: 182-7.
26. ZUO Y, MENG T, TIAN H, et al. Enhanced H (+) Storage of a MnO (2) Cathode via a MnO (2) Nanolayer Interphase Transformed from Manganese Phosphate [J]. ACS Nano, 2023.
27. SUN T, NIAN Q, ZHENG S, et al. Layered Ca (0.28) MnO (2). 0.5 H (2) O as a High Performance Cathode for Aquous Zinc-Ion Battery [J]. Small, 2020, 16 (17): e2000597.
28. LI J, KUANG Q, WANG G, et al. Galvanostatic timed Na3Mn2 (P2O7) (PO4) as a high-voltage cathode material for aqueous zincion batters [J]. Electrochimica Acta, 2023, 441.
29. TAN S, SANG Z, YI Z, et al. Conductive coating, cation-intercalation, and oxygen vacancies co-modified vanadium oxides as high-rate and stable cathodes for aqueous zinco-ion batteries [J]. EcoMat, 2023.
30. XU D, WANG H, LI F, et al. Conformal Conducting Polymer Shells on V2O5 Nanoheet Arrays as a High-

Rate and Stable Zinc-Ion Battery Cathode [J]. Advanced Materials Interfaces, 2019, 6 (2).

31. CHO S-H, PARK J-S, KIM J H, et al. Oxygen-Related Defect Engineering of Amorous Vanadium Pentoxide Cathode for Achieving High-Performance Thin-Film Aqueous Zinc-Ion Batteries [J]. ACS Applied Energy Materials, 2023, 6 (5): 2719-27.
32. CHEN L, YANG Z, CUI F, et al. Enhanced rate and cycling performances of hollow V2O5 nanospheres for aqueous zinc ion battle cathede [J]. Applied Surface Science, 2020, 507.
33. DENG W, XU Z, LI G, et al. Self-Transformation Strategy Toward Vanadium Dioxide Cathode for Advanced Aqous Zinc Batteries [J]. Small, 2023: e2207754.
34. ZHANG W, TANG C, LAN B, et al. K0. 23V2O5 as a promotion catherde material for rechargeable aqueous zinc ion battles with excellent performance [J]. Journal of Alloys and Compounds, 2020, 819.
35. YANG W, YANG W, HUANG Y, et al. Reversible aqueous zinc-ion battle based on ferric vanadate cathode [J]. Chinese Chemical Letters, 2022, 33 (10): 4628-34.
36. QU G, GUO K, CHEN W, et al. Cs-Induced Phase Transformation of Vanadium Oxide for High-Performance Zinc-Ion Batteries [J]. Energy & Environmental Materials, 2023.
37. LIU Z, PULLETIKURTHI G, ENDRES F. A Prussian Blue/Zinc Secondary Battery with a Bio-Ionic Liquid-Water Mixture as Electrolyte [J]. ACS Appl Mater Interfaces, 2016, 8 (19): 12158-64.
38. ZENG Y, LU X F, ZHANG S L, et al. Construction of Co-Mn Prussian Blue Analog Hollow Spheres for Efficient Aqueous Zn-ion Batteries [J]. Angew Chem Int Ed Engl, 2021, 60 (41): 22189-94.
39. LIU Y, LI Z, HAN Y, et al. Highly Stable Metal-Organic Framework with Redox-Active Naphthalene Diimide Core as Cathode Material for Aqueous Zinc-Ion Batteries [J]. ChemSusChem, 2023: e202202305.
40. ZHANG J, JIA W, YANG H, et al. Cerium oxide as cathode material for aqueous zinc-ion battle [J]. Solid State Ionics, 2023, 391.
41. LI C, XIE X, LIANG S, et al. Issues and Future Perspective on Zinc Metal Anode for Rechargeable Aquous Zinc-ion Batteries [J]. Energy & Environmental Materials, 2020, 3 (2): 146-59.
42. LU X, ZHAO C, CHEN A, et al. Reducing Zn-ion concentration gradient by SO 42-immobilized interface coating for dendrite-free Zn anode [J]. Chemical Engineering Journal, 2023, 451.
43. YANG Q, LI Q, LIU Z, et al. Dendrites in Zn-Based Batteries [J]. Adv Mater, 2020, 32 (48): e2001854.
44. LIU M-C, TIAN C-Y, ZHANG D-T, et al. Design on modified-zinc anode with dendrite-and side actions-free by hydrophobic organic-inorganic hybrids for ultra-stable zinc ion batters [J]. Nano Energy, 2022, 103.
45. LIU Y, LIU Y, WU X. Toward Long-Life Aquous Zinc Ion Batteries by Constructing Stable Zinc Anodes [J]. Chem Rec, 2022, 22 (10): e202200088.
46. ZENG L, HE H, CHEN H, et al. 3D Printing Architecting Reservation-Integrated Anode for Dendrite-Free, Safe, and Durable Zn Batteries [J]. Advanced Energy Materials, 2022, 12 (12).
47. CAO Q, GAO Y, PU J, et al. Gradient design of imprinted anode for stable Zn-ion batters [J]. Nat Commun, 2023, 14 (1): 641.
48. WU B, GUO B, CHEN Y, et al. High Zinc Utilization Aquous Zinc Ion Batteries Enabled by 3D Printed Graphene Arrays [J]. Energy Storage Materials, 2023, 54: 75-84.
49. WU C, SUN C, REN K, et al. 2-methyl imidazole electrolyte additive enabling ultra-stable Zn anode [J]. Chemical Engineering Journal, 2023, 452.
50. WANG M, WU X, YANG D, et al. A colloidal aqueous electronic modulated by oleic acid for durable zinc metal anode [J]. Chemical Engineering Journal, 2023, 451.
51. XU X, SONG M, LI M, et al. A novel bifunctional zinc gluconate electrolyte for a stable Zn anode [J]. Chemical Engineering Journal, 2023, 454.
52. LIU H, ZHOU Q, XIA Q, et al. Interface challenges and optimization strategies for aqueous zinc-ion battles [J]. Journal of Energy Chemistry, 2023, 77: 642-59.
53. DU Y, LI Y, XU B B, et al. Electrolyte Salts and Additives Regulation Enables High Performance Aquous Zinc Ion Batteries: A Mini Review [J]. Small, 2022, 18 (43): e2104640.
54. MA R, XU Z, WANG X. Polymer Hydrogel Electrolytes for Flexible and Multifunctional Zincc-Ion Batteries and Capacitors [J]. Energy & Environmental Materials, 2023.
55. HAN M, LI T C, LI D S, et al. Electrolyte Modulation Strategies for High Performance Zinc Batteries [J]. Batteries & Supercaps, 2023.
56. LIU C, XIE X, LU B, et al. Electrolyte Strategies toward Better Zinc-Ion Batteries [J]. ACS Energy Letters, 2021, 6 (3): 1015-33.
57. GUO S, QIN L, ZHANG T, et al. Fundamentals and perspectives of electrolyte additives for aqueous zinc-ion batteries [J]. Energy Storage Materials, 2021, 34: 545-62.
58. ZHANG Y, LI H, HUANG S, et al. Rechargeable Aqueous Zinc-Ion Batteries in MgSO (4)/ZnSO (4) Hybrid Electrolytes [J]. Nanomicro Lett, 2020, 12 (1): 60.
59. ZHAO F, JING Z, GUO X, et al. Trace amounts of fluorinated surface additives enable high performance zinc-ion batteries [J]. Energy Storage Materials, 2022, 53: 638-45.
60. LI H, LIU Y, CHEN Z, et al. High voltage and healing flexible zinc ion battle based on ionogel electrolyte [J]. J Colloid Interface Sci, 2023, 639: 408-15.

Mechanism and control strategy of large-area concrete shrinkage cracks in the underground railway section

Meilin Yu*

Abstract: The cracks of the large-area concrete structure in the underground railway section directly affect the structural safety and service life. How to take effective measures to control concrete shrinkage cracks is a difficulty to be solved in the project.In view of the crack control demand of large-area concrete structure cracks in a vehicle section of Wuhan metro, based on the existing engineering cases of large-area concrete structure , shrinkage crack mechanism and control strategies, the corresponding measures are proposed to prevent crack production from two aspects of concrete raw materials and construction technology.The results show that through the regulation of concrete materials,maintenance,casting,post-pouring belt, temperature monitoring and other links, the reduction of shrinkage cracks of large-area concrete structure in the underground railway section is realized.These results can provide references for similar projects to lessen cracks, ensure structural safety and construction quality.

Key words: Underground railway section; large-area of concrete; shrinkage crack; control strategy; structural safety.

1. Introduction

In recent years, with the increasing improvement of Chinese infrastructure construction, urban underground rail transit has also developed rapidly.Because of the advantages of unaffected climate conditions, safety, speed and environmental protection, rail transit has become the main force of the transportation network in major cities.Statistically, by the end of 2022,55 cities have opened urban rail transit operation in Chinese mainland , reaching 10287.45 kilometers.China's urban railway construction will stay in the golden stage of development for a long time in the future.At present, more and more subway projects widely use large-area concrete structures.The characteristics of small thickness,huge plane size[1],large concrete pouring volume, cement hydration temperature rise make it easy to produce crack, affect the function of the structure, reduce the durability of structure, and greatly impact operation safety. So in the urban railway engineering construction,how to effectively control cracks gradually attracts more attention from the scientific research workers.

Based on the research results of control measures, mechanism and control technology in existing projects for large-area concrete shrinkage cracks, this paper takes a vehicle section of a subway project in Wuhan as the background, proposing preventive measures from two aspects of concrete material and construction technology, and discussing the technology for controlling large-area concrete cracks in subway section.

2. Application of large-area concrete cracks in the project

In the past few decades, with more and more projects of large concrete structures in China, the problem of large ground cracks has become more prominent.At present, Chinese engineers have had rich practical experience in large-area concrete structure projects, and put forward some effective measures to resist cracks, mainly from two aspects of construction materials and techniques.However, there is a lack of systematic theoretical research results on the control of such engineering cracks in China, and there is no clear criterion on the crack acceptance and repair methods of large-area concrete structures in the technical specifications and regulations of various industries in China.

The theoretical research of large-area concrete ground is not mature, but some effective practical methods have been obtained to solve the problem of concrete cracking.In recent years, concrete cracking has appeared in most large-area concrete structure cases in China. According to its crack situation, crack reasons and crack control measures, a simple summary (Table 1) is made, and several typical engineering cases are introduced in detail.

* Corresponding author: 864360253@qq.com

2.1 Case analyses of domestic large-area concrete structure

(1) Front Parking Building Project of Terminal 2 of Beijing Capital International Airport[2]

Terminal 2 of Beijing Capital International Airport is a hub for temporary airport parking.The parking building is rectangular, stretching 262.8m from north to south, 133.8m from east to west and covering an area of about 167,000 square meters.After inspections by the construction department, it is found that the beams, plates and column components of the parts in the main structure have different degrees of defects.There are many cracks on the surface of the beam members.The crack width is mostly between 0.05-0.7mm.Most beam bottoms has transverse cracks, irregular vertical cracks (shown in Figure 1 (a)); The destructive form of the plate members is mainly manifested as lots of cracks around the transverse main joint.The maximum width of the main cracks reached 0.5mm (shown in Figure 1 (b)); The main form of the column member cracks is vertical cracks and circular cracks along the column.The interwoven mesh cracks have a maximum width of 1.5mm(shown in Figure 1 (c)).After the quality detection and safety identification of the testing mechanism, the conclusion is that the temperature effect, shrinkage deformation and live loads on the floor lead to multiple cracks in the structure.

In terms of concrete structure design, the engineers give priority to different high-performance concrete, and control the temperature difference by mixing high-quality fine admixture and micro-expansion agent, strengthen the early construction and maintenance, setting up post-pouring belt and other technical measures, which has a remarkable effect of project crack control.

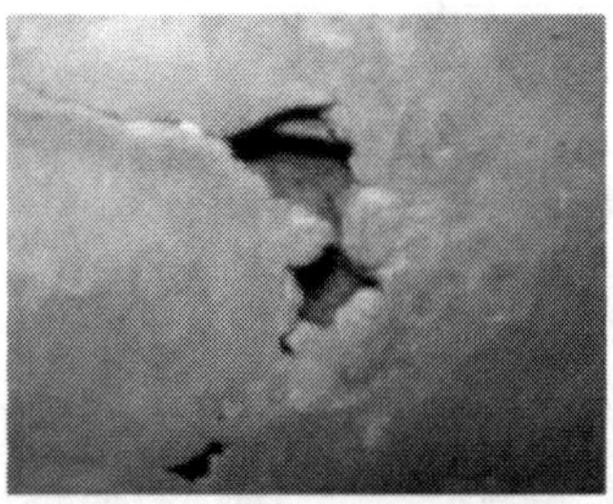

(a) Beam member cracks

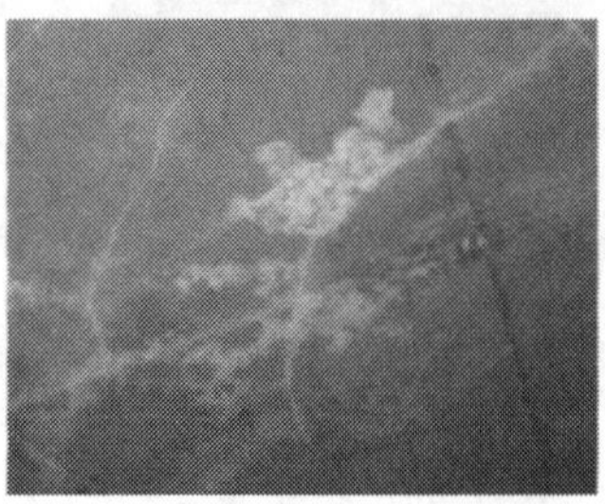

(b) plate member cracks

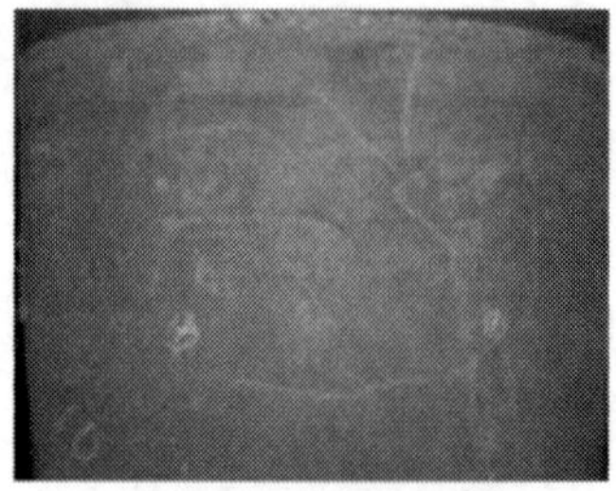

(c) column member cracks

Figure 1 Crack diagram of the front parking building of Terminal 2 of Beijing Capital International Airport

(2) Nanjing International Exhibition Center[3-4]

Nanjing International Exhibition Center integrates exhibition, conference, trade, information, entertainment and catering service functions as one, and is renowned for its modern facilities, complete set-ups and unique modeling.The total construction area of the project is 89,000 m^2.The building length is 292.8m, in width of 158.5m.The main structure of the project (Figure 2) is divided into two parts. The roof is a large arc steel arch with a span of 75m, and the second floor is a large exhibition hall (240m x 117m). Expansion joints are arranged in both longitudinal and transverse sections.

During the construction period, builders observe the G~J axis section of the 6 axis beam and the floor cracks, and see the cracks of prestressed and stressed concrete tension (Figure 3 and 4 respectively). Nearly 10 cracks appear in H-J axis before stressed tension in width between 0.1-0.35mm; 9 cracks in G-H axis in width between 0.07-0.25mm.Most of the cracks are distributed near the bull leg of the steel truss, perpendicular to the beam length direction and extending upward. The distribution is relatively even, with no concentrated cracks. The shape of the cracks is wide in the middle and narrow on both sides.These cracks are typical temperature cracks. Due to the elevated thermal temperature of concrete, the weak part of the strong electric pipe produces stress concentration, the stress exceeds the tensile strength of concrete, and then the structure produces cracks.

The construction department mainly strengthens the structural cracking resistance from the three aspects of concrete materials, concrete casting and maintenance measures.Concrete materials is low hydration thermal concrete with strength above C40.Moreover,it adds slow coagulation, water reduction and micro expansion into concrete. The pouring method is layer vibration, which shall be dense, the surface showing floating slurry and no longer subsidence; maintenance measures is taken for water storage.In the later process, after 4 months of observation, the temperature cracks were significantly reduced, and the construction scheme effect was obvious.

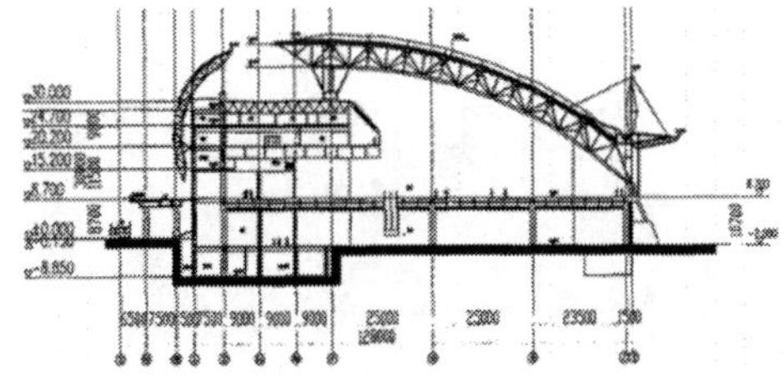

Figure 2 Section drawing of Nanjing International Exhibition Center

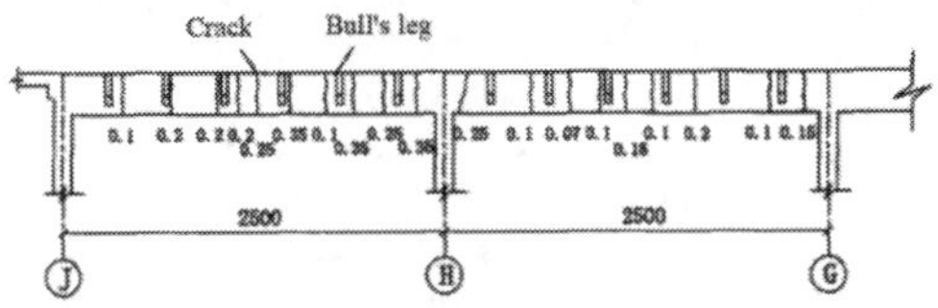

Figure 3 Pre-tension crack pattern of prestressed reinforced concrete

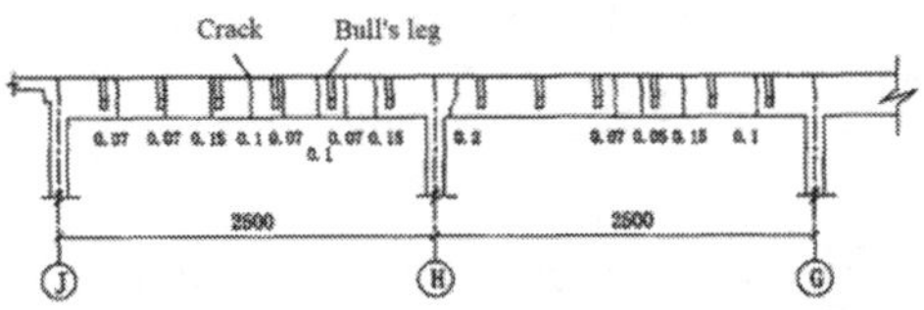

Fig. 4 Post-tensile cracks picture of prestressed reinforced concrete

2.2 Cracks and crack control measures in large-area structural projects

Table 1 Cracks and crack control measures in large-area structural projects

Project	Crack situation	Causes of cracks	Crack control measures
Qingdao Liuting International Airport Terminal Extension Project	Beam and plate cracks, the maximum crack width reaches 0.2mm	Temperature effect and shrinkage deformation , structural design	Set up the post-pouring belt and the prestressed band
Shenzhen Huawei Electric Power Plant	Small shrinkage cracks appear on the inside of the frame beam, the crack width is generally less than 0.1mm,, the crack distribution is relatively uniform, and the spacing is about 1500mm	Temperature effect, shrinkage deformation , construction technics, raw materials	Set up expansion reinforcement belt, expansion concrete, water storage
Johnson Controls Bohai Battery (Binzhou) plant	Beams, plates and columns all have cracks, and the maximum crack width reaches 0.3mm	Load over-limit, shrinkage deformation , temperature deformation and uneven settlement deformation	Control the matching ratio and concrete raw materials, control the pouring thickness of concrete, maintenance of heat preservation and moisturizing
Litong Square	The concrete surface runs through the cracks continuously.The middle section is wide, gradually thinning at both ends, and the crack width is 0.3-0.5 mm	Concrete material and construction technics	Use of polypropylene fibers anti-crack concrete, extend the mold removal time, maintenance of heat preservation and moisturizing
The underground garage of Shanghai Wanli community	Concrete bottom plate appears multiple longitudinal cracks, crack width is generally less than 0.3mm	Temperature deformation , shrinkage deformation , construction technics	Water storage, short and smooth transportation route, post-pouring belt and expansion belt, strengthen the temperature control of raw materials, and layered pouring and tamping
Xiawuqi Town Forest Park Artificial Lake	Cracks appear in the base plate,and the width is generally 0.3mm-0.5mm	Temperature deformation , shrinkage deformation , construction technics	Warehouse casting technology and cushion adopting segmented construction and ground seam technology
Shenzhen OCT Ecological Square	Cracks appear in the beams and columns.Cracks are perpendicular to the beam length, extending along the whole beam height.The crack width of the column edge section is 0.1mm-0.35mm,and the maximum is 0.25mm	Temperature deformation , shrinkage deformation , construction technics	Set post-pouring belt, prestressed tendons, expansion agent and expansion joints
Basement of Shanghai Binjiang Arc de Triomphe	Sints along the bottom plate distribution is uniform with no crack concentration.The average width of the crack is 0.25mm	Temperature deformation , shrinkage deformation , construction technics	Set post-pouring belt, expansion reinforcement belt, use low hydration thermal concrete and UEA—H high anti-cracking expansion agent

Combined with the above domestic large-area concrete structure projects,author can summarize the following main control measures:

(1)Most of the large-area concrete structures adopt prestressed reinforced concrete as the main measure to resist the temperature deformation and shrinkage deformation.Prestress technology can improve the crack resistance of the concrete in the later stage, but also well close the early concrete cracks. For the shrinkage cracks and temperature cracks in the early stage, the crack width of the cracks is reduced after the prestress tension to a certain extent.

(2)The choice of materials exerts significant influence on large concrete structure cracking. Using a new type of concrete can reduce the early hydration heat,enhance concrete compaction and decrease contraction value.Another choice is adding high-quality admixtures into the concrete,such as water reducer, expansion agent, etc.More applied expander agent in projects is UEA expansion agent which ensures that the concrete produces cubical expansion during the hardening process.When the structure is limited by the boundary conditions, a pretensile stress ,which can effectively offset the tensile stress inside the concrete structure to guarantee no cracking, occurs within the concrete.

(3)Post-pouring belt or expansion reinforcement belt is also a common anti-crack method, often as a compensation measure.The post-pouring belt can overcome the shrinkage caused by the temperature difference, and the gap is generally reserved in advance. After the internal hardening contraction of the concrete, the post-pouring belt part is done, which is conducive to protecting the integrity of the components.Combined with characteristics of the internal structure and the external environment, the distance between the rear pouring belt is usually controlled at 30-40m. If reasons of the building structure lead to no rear pouring belt, the expansion reinforcement belt is generally adopted.

(4)Construction technic is also a very important factor in the control of large-area concrete structure cracks.Generally, the production of cracks can be controlled from the aspects of concrete pouring methods, maintenance methods and construction methods. The common concrete construction technics to reduce cracks include layered pouring and layered tamping ,lowering the adverse impact of concrete temperature and shrinkage.

The crack control measures mentioned above play different roles in different engineering projects, but there is a lack of theoretical researches of various practical methods, and the protection mechanism isn't clear. Therefore, in the future, how to explain large-area concrete cracking and control problems in principle is the core research direction.

3. Research status on shrinkage crack mechanism of large-area concrete

Combined with domestic and foreign research results[5-8]and construction engineering cases, cracks can be divided into two kinds according to the causes of cracks: one is caused by the main deformation effect of the construction structure, such cracks account for about 80% -85% of the main cracks of all construction structures, such as temperature deformation, expansion deformation, uneven settlement of concrete; the other is caused by the load.The cracks only accounts for 1% -20% .Therefore, crack control should focus on structural deformation.This paper discusses the factors caused by large-area concrete structure cracks and mainly starts from shrinkage deformation, internal and external temperature difference change, construction technic, quality of raw materials and so on.

3.1 Contraction crack caused by concrete shrinkage deformation

The shrinkage deformation of concrete refers to the phenomenon of volume shrinkage due to the deformation when the concrete is not subject to external forces. Influencing factors include the water to ash ratio, the amount of gel material, the cement dosage, etc.The shrinkage of concrete can be divided into temperature contraction and non-temperature contraction[9]. Temperature contraction refers to the contraction caused by thermal expansion and cold contraction, and non-temperature contraction refers to the contraction deformation caused by non-temperature factors, including chemical contraction, dry contraction, plastic contraction, carbide shrinkage, etc. In engineering practice, the main consideration is temperature contraction and dry contraction[10]. While several other contraction types account for relatively small proportions in concrete shrinkage.The general shrinkage rate of concrete is 0.04-0.06%, while the stretch limit of concrete is 0.01-0.02%[11].Concrete is prone to crack when the tensile stress generated by the surrounding constraints exceeds the concrete limit tensile strength[12]. At present, scholars at home and abroad have studied many concrete shrinkage deformation prediction models, including ACI209[13]、CEB 90[14]、GL2000[15]、B3[16]、G-Z[17].The factors considered by each prediction model are different ,and hence the accuracy is also distinct.But all these do not determine the quality of a model.Factors considered by the various predictive models are listed in Table 2.As can be seen from Table 2, the component size, humidity, and concrete strength are the most considered factors in five models, while the other factors are considered by only a few models.Among them, the B3 model included the most comprehensive factors, and only the B3 model considered the component shape, but all models did not consider the effect of admixtures.After consulting the literature[18][19][20],it is learned that from the model prediction accuracy, the B3 results are the best.The theoretical calculation value results of ACI209, CEB 90 model are close to the test value results, GL2000 is secondary and G-Z is worse.B3 model is the semi-empirical and semi-theoretical formula, with the highest prediction accuracy compared to other models.

For the large-area concrete structure in the actual engineering cases, concrete shrinkage and dry shrinkage of road surface can cause plastic deformation, resulting in crack shrinkage problems in the structure.For the underground railway section,about 5h after the completion of concrete pouring, in the process of concrete contraction and condensation, the wind speed, high external temperature and the violent reaction of concrete surface cement to water, large water evaporation, result in concrete plastic shrinkage[21]。

https://doi.org/10.1051/matecconf/202338201016

Table 2 Impact factors considered in the concrete shrinkage deformation prediction model[20]

Influence factors	ACI209	CEB 90	GL2000	B3	G-Z
Bone material properties				√	
Cooperation ratio	√			√	
Plus agent					
Component size	√	√	√	√	√
Component shape				√	
Temperature		√		√	
Cement varieties		√	√	√	√
Humidity	√	√	√	√	√
Maintenance conditions				√	
Concrete strength	√	√	√	√	√

3.2 Contraction cracks caused by internal and external temperature difference

Temperature deformation is also one of the main reasons for the shrinkage cracks in the concrete in the underground railway section.Large-area concrete has the characteristics of large section size, small thickness and more cement dosage. In general, a large amount of hydration heat gathering inside the concrete structure is not easy to emit, leading to the continuous rising temperature of the concrete center.However, the concrete surface heat dissipation is fast, so the soil layer inside and outside concrete forms a large temperature gradient[22].At present, the domestic theoretical research mainly focuses on finite element analysis, and many scholars combine with engineering practice, compare theoretical data and engineering results of practical projects, and take effective measures to achieve the purpose of crack control.

Chen Yanzhuo[23]in the article applied ANSYS on the top of a city dig subway station to carry out temperature stress simulation.Through finite element simulation, temperature field analysis, stress field analysis,it shows that the temperature difference can lead to tensile stress in the plate when temperature is highest.Air heat exchange and large surface area lead to this phenomenon The core temperature is much lower than the temperature of the upper surface and is easy to prone to shrinkage cracks.

Yuan Kang[24] established a finite element model of a single room on the basis of the substrate temperature-contraction stress calculation formula through numerical simulation.It shows that the temperature-contraction stress is the main factor which causes the floor crack.When the temperature shrinkage stress is largest, the greater the temperature difference inside and outside the concrete, the greater the floor temperature contraction stress.So carrying out construction in the spring and autumn season is preferable,to achieve the purpose of temperature control.

Wu Wenwu[25]conducted analyses of Wuhan Yangluo Yangtze River Highway Bridge project.Combined the finite element analysis and the field test data, it suggests that when the temperature difference inside and outside of the concrete structure is not less than 25°C, a special concrete temperature control design of anchorages must be carried out.

Zuo Wenge[26]proposed that the temperature difference of concrete structure due to the environmental climate, structure, material hydration and other factors, uneven temperature difference is prone to make concrete crack in the early stage of concrete pouring.

Chen Liling[27]suggested that the reasons for concrete cracks in large area construction, among which the most common is that the temperature difference between inside and outside concrete causes cracks.In the concrete pouring,errors resulted by hydration time control will lead to a large amount of heat gathered inside the concrete component.The temperature difference inside and outside the concrete structure is too large,which results in tensile stress.If the tensile stress exceeds the crack resistance of the concrete, it is easy to cause different degrees of temperature cracks.

The above study shows that the studies of temperature control of the large-area concrete on temperature control have formed a series of mature results, mainly for test and demonstration from numerical simulation analysis, field monitoring and thermal calculation, but the results from the mechanism are less and not in-depth enough.

3.3 Contraction crack caused by construction technics

The construction technic is an important aspect of the research of large-area concrete cracks, which has a great impact on concrete shrinkage cracks, usually controlled and studied from several aspects of form work projects, maintenance measures, how to set post-pouring belt, and prestressing force technology.Aiming at the construction measures of large-scale areas of concrete, many scholars have carried out a lot of theoretical and experimental studies.

Lin Haishan[28]starting from the engineering examples of the subway station in Xiamen coastal area, the insufficient concrete pouring will lead to uncompacted concrete structure, weak parts are easy to crack and form larger cracks when there are loads, and excessive vibration will cause the separation of aggregates and cement slurry layers, and produce dry shrinkage phenomenon.

Wei Yulian[29]summarized previous results, and considering similar engineering construction experience for concrete structure crack control in Nanjing four bridge project.Through finite element analysis, combined with the structural characteristics of the anchor system, it shows that if the cooling pipe is not arranged in concrete,the cement hydration heat rise will eventually cause cracks.

Wei Min[30]studied causes of structural cracks in the large-area cast-in-place building cover. Combined with engineering cases, the use of templates with poor thermal conductivity and unreasonable disassembly time will lead to the increase of hydration heat and cracks in the construction stage.

Lin Min[31]summarized the typical engineering examples of guangdong Olympic sports center, Huawei scientific research center, Beijing east square.It shows that many large-area concrete projects exert prestressing force to resist the influence of shrinkage stress on cracks, when using prestressing force, post-pouring blocks provide the working surface with prestressing force to inhibit concrete from early shrinkage cracks.

Chen Jinrong[32]pointed out that road cloth or sacks are often used in projects to make water maintenance. In the high temperature environment, the internal water of the concrete is greatly evaporated.The temperature between inside and outside the concrete is large, which will cause the production of plastic cracks or temperature cracks.

Construction technics are the most important part of crack control.Not only because the construction technics can be controlled with greater flexibility, but also it has a decisive impact on the quality of its structure. Based on the actual engineering cases, most scholars analyze the impact of construction technics on concrete's characteristics.They put forward that insufficient vibration, backward maintenance measures, ignoring temperature control,wrong post-pouring belt location can lead to shrinkage cracks, so it is necessary to choose appropriate and effective construction measures,which can reduce the impact of cracks on engineering quality and life.

3.4 Quality of raw materials

In the large-area concrete structure, the type and dosage of cement directly affect the quality of the large-area concrete, the amount of the cement hydration heat, and the temperature stress[31]; Admixtures include water reducer, expansion agent, retarder, early strength agent and so on. The addition of concrete admixtures can significantly reduce the hydration heat of cement, prevent excessive temperature stress, and improve the early strength of concrete, and then improve the crack resistance of concrete[31]; Water cement ratio has a certain influence on the early self-contraction, hydration temperature rise and temperature deformation[33]. It can be seen that the type and dosage of cement, Water cement ratio, admixture variety and volume and so on,which have a great impact on the concrete cracks, so more and more scholars have also carried out related researches.

Tao Jianqiang[7]put forward that cement hydration heat is the main factor of temperature stress, so the variety and dosage of cement are key factors in the control of large-scale concrete cracks.Cement with high C3A 、 C3S produces more hydration heat,so it's better to choose low or medium heat cement; the more cement dosage, the more hydration heat of concrete.So that the temperature is higher inside the concrete.

Li Zhanyin[34]demonstrated that the water cement ratio has a direct effect on the hydration rate.When the water cement ratio (W/C) increases from 0.3 to 0.6,28d hydration heat significantly increases.The water cement ratio of the most unfavorable plastic shrinkage area of concrete is about 0.5. When W/C> 0.5, the crack area decreases with the water evaporation rate increasing.When W/C<0.5, the crack area increases with the water evaporation rate increasing.

Yang Changhui[35]pointed out that when the water cement ratio is high, the capillary pore diameter is large. The concrete mixture evaporates more water, the capillary develops at a small rate for a period of time and the heat liberation rate of hydration decreases;When the water cement ratio is low, the concrete integrity and cohesion are good, the plastic settlement is small, and the plastic shrinkage crack area is small.Therefore, plastic cracks caused by the rapid evaporation in the early stage should be avoided during construction.

Pan Lisha[36]focused on the study of the influence of four water reducers on the cement hydration.The mechanism of admixture action is to change the rate of hydration to affect the cement hydration heat, and the water reducer can significantly disperse the cement particles, improve the contact degrees of cement particles and water, and affect the cement hydration rate.

Wang Zilong[37]pointed out that the expanding agent can delay the condensation time of cement to compensate for the early shrinkage of concrete.

Domestic researchers studied the influence mechanism of concrete shrinkage cracks through construction material quality in many aspects, and the remarkable results can be summarized as two points: Additives are used to influence the heat of hydration of cement by changing the rate of hydration.The type and dosage of cement can directly affect the hydration heat.

The above study shows that the production of large-area concrete cracks is mainly caused by shrinkage deformation,difference changes of internal and external temperature, construction technics and raw material quality.A large number of scholars for large-area concrete cracks made a lot of theoretical analysis, field tests and engineering practice application.They made a lot of research results, but the overall theoretical level is not high and lack of comprehensive quantitative analysis and calculation.Most of the research results are based on projects and make qualitative mechanism analysis and put forward improvement measures,but most for bulk structure, less for large-area concrete.

4. Actual researches of large-area concrete shrinkage crack control in subway sections

Subway section is more difficult due to its complex construction technics, structural boundary conditions and complex geological conditions.The surface size of large-area concrete is large and the hydration heat produced after pouring is difficult to dissipate, so it's easy to produce greater temperature stress and deformation.Bottom plates, side walls, roof plates are constrained to varying degrees, concrete structure under loads and soil pressure.If the stress exceeds the structural tensile strength allowable value, it will harm the structural safety and building service life.Therefore, according to characteristics, it's necessary to control from the construction materials, construction methods, structural designs for different subway projects. How to conduct scientific analysis of concrete structure, how to take effective construction technics, how to choose better construction materials and how to reduce repair costs are important and difficult in projects.Solving these problems can bring huge economic and social benefits.Therefore, combined with numerical simulation, many researchers make in-depth analysis of the causes and distribution characteristics of cracks in subway sections and put forward effective measures to reduce shrinkage cracks from the perspective of construction designs.

Zhu Xianfa [38]relied on the Changzhou Urban Rail Transit Line 1 project.He made the mixing proportion design and the application of anti-cracking in the concrete structure of main parts of the subway station.It shows that the HME—V efficient cracking agent with the dual control function of temperature and expansion can effectively reduce the generation of cracks.

Ma Yongbo[39]put forward that the use of low thermal cement, fly ash of high quality , embedding the cooling water pipe in concrete, and controlling forming temperature is conducive to inhibiting the production of side wall cracks in the subway station.

Xie Xiaoli[8]found that section pouring concrete will cause temperature contraction and humidity contraction is not synchronized which lead to crack generation through the super long span structure concrete beam temperature and strain test.After controlling the raw material temperature of concrete, optimizing the concrete mixing proportion, reducing the concrete pouring temperature, layered continuous pouring technology,it shows that there is almost no crack in the subsequent projects.

Hu Dongkang and Hu Yixin[40-43]combined with a large number of practical cases from the basic technical requirements of the construction technics.They adopted new steel bar truss deck,sequence method, hoisting roof mesh and other technologies to reduce the generation of cracks and achieve the project goal.

Liao Dongming[44]took an example of Meijin station in Shenzhen metro line 9.From structural designs, temperature, construction materials, construction technics, he analyzed the causes of concrete crack formation.And from materials, mixing proportion, templates, concrete pouring technics, empty positions, regular inspection of foundation and completion acceptance,he put forward methods to limit concrete cracks.

Zhan Jianjun[45]took a station in jiangxi metro line 2 as the research object and built MIV—BP neural network models.For thin protective layers, large concrete blocks and unreasonable concrete mix proportion, after using steel mesh on the external wall, reducing concrete temperature,embedded cold water pipes in the side wall,adding admixture and other measures, the number of cracks in the station is significantly reduced.

Li Zhaoping[46]explored the cause and deformation law of the concrete cracks in the joint part of the subway station, and then concluded that the mortise and groove parts in the improved Tongue and groove type connector can jointly resist the bending moment, which is beneficial to reduce the generation of cracks.

Yu Yiming[47]made analyses of mixing proportion, cracking risk assessment, concrete pouring, vibration, mold removal and maintenance for concrete structure design of the side wall in a subway station.The test datas show that the use of cracking resistance of magnesium oxide with strong compensation performance can effectively reduce the number of cracks in the side wall.

Fang Kejun[48]took a subway station of Beijing as an example.According to the law of cracks, width and time,he summarized the reasons for cracks and put forward choosing anti-seepage concrete, reasonable use post-pouring belt and section pouring to control the production of concrete cracks.It shows that there is no concrete cracks and decreasing number of shrinkage cracks, which reaches the engineering goal.

Chen Yanzhuo[23]applied ANSYS to simulate the temperature stress of the roofs in a openly excavated subway station of a city.It shows that the difference of theoretical analysis and numerical simulation results is not obvious which means that preliminary design guidance of finite element software can be used to provide reference for crack control of similar projects.

Fan Jiangsu[49]made the force analysis and the shrinkage measurement of the roof structure for a openly excavated subway station .It proposes that 425 slag cement, post-pouring belt, layered pouring, layered tamping and other measures are taken to ensure the normal use of the roof.

To sum up, scholars at home and abroad combined engineering practice verification and numerical simulation of concrete cracks for the subway station.They found that the optimized design of mixing proportion, reducing temperature of concrete pouring , reasonable reinforcement, using expansion agent can effectively control large-areas concrete cracks.Due to its high durability and safety requirements of large-area concrete, its structural designs, construction control and concrete material selection are more complex, and the performance of different measures in different engineering backgrounds is also different.At present, the research on the large-area concrete shrinkage cracks in the subway section is less, and the effect of the optimized design still needs to be explored.Therefore, on the basis of previous researches, this paper focuses on large-area concrete crack control from the aspects of concrete material quality and construction technology based on a subway project in Wuhan,which has practical guiding significance for similar projects.

5. Project Overview

The subway station is a second-story island-style station, in which the large-area concrete beam runs north-south, the total length is 220m,the standard width is 17.6m,the roof cover thickness in the center of platform is about 1.5-2.9 m, and the depth of foundation pit bottoms in the station of standard section is about 21.8m.It is a large-area and cast-in-place concrete structure.The station adopts the method of pouring in segments, with a post-pouring belt at the interval of 50-70m, and the enclosure structure adopts a continuous wall that is 700mm in thickness.The main body of the station is constructed by the method of open excavation and shield method at both ends, and both ends are originating technique for the shield structure. The construction joints or deformation joints are set in the roof plate, side wall and bottom plate for waterproof treatment.The main structure of groove section is constructed from small mileages to large mileages. In the same groove section, the construction sequence of each component is bottom plate, side wall, middle plate, roof plate and middle column.

There was no crack in the early maintenance period. The age period of concrete is about 4 weeks. Some transverse cracks and vertical cracks appear on the roofs.There are a total of 18 cracks and the width of them is 0.3-0.7mm.They are mainly distributed in the four corners of the roof, extending around the station.Two floors are evenly distributed, roughly the same distribution.The platform side wall has 10 cracks, mainly including vertical cracks, through cracks.The crack width is 0.1-0.4mm and crack depth is between 142mm and 378mm.The shape is large in the middle and small at both ends, the tip mouth is at the top or bottom of the side wall.After the cracks in the structure are evaluated by relevant institutions, the conclusion is that the wall and plate cracks at the commuter station exceed the allowable range because interval of two concrete pouring is too long and the temperature difference during the temperature detection process exceeds 20°C.

The large-area reinforced concrete ground in the vehicle section is exposed to the external environment for a long time. The ground temperature change is greatly affected by the external environment.The external environment impact on the concrete floor temperature manifests the seasonal temperature difference.Floor concrete pouring is mainly from July to August.The temperature in Wuhan is high.The average highest temperature is 32.5°C and the average lowest temperature is 25.0°C.Concrete is affected by the external environment of frequent drying and wetting.Concrete condensation hardening causes a large temperature difference between the component surface and the internal concrete.Due to the area of the large-area concrete ground is large and the thickness is relatively small,the lower surface constraint effect is more obvious, which is more sensitive to contraction.It's easy to produce large deformation, and can lead to the production of shrinkage cracks.

(a) The interior of the subway

(b) Workers are doing reinforcement binding

Figure 5 Construction photos

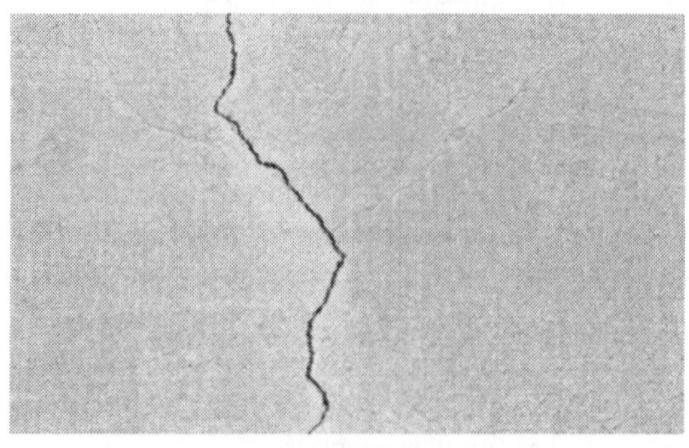

Figure 6. Vertical cracks

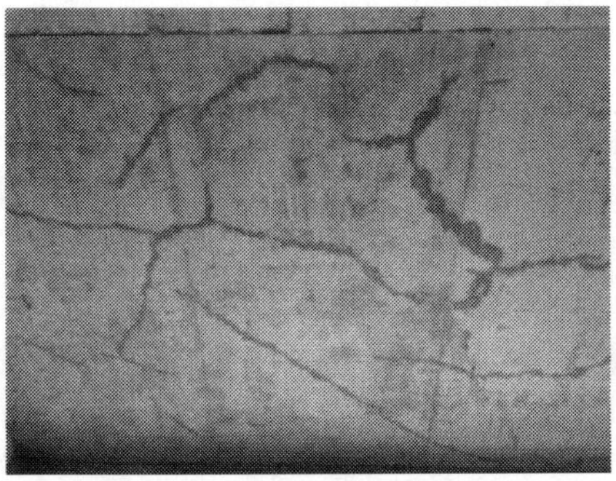

Figure 7 Through cracks

6. Prevention and control measures of large-area concrete shrinkage cracks in the subway section

6.1 Quality control of concrete materials

Before the construction operation, in view of the characteristics of the construction site, external environment climate change and other factors, in addition to strict sampling of raw materials for concrete mixes (cement, admixtures, sand, stone, etc.), it is also necessary to optimize the concrete mix ratio and choose the best mix ratio.Large-area concrete adopts the cement with low hydration heat and long condensation time, which can extend the time when the central point of the structure reaches the maximum temperature to reduce the hydration heat temperature in the center of the concrete structure[50]. At the same time, the cement admixture can be improved to enhance the concrete crack resistance ability.Mixed with selected quartzite aggregates, it is beneficial to reduce the probability of large-area cracks[51]. Under the condition that the strength of the concrete and the slump of the substrate can basically meet the design requirements, the content of admixture and aggregate increases and the amount of cement decreases.In addition,the water reducing agent, expansion agent, retarder can be added to large-area concrete according to the mix ratio, in order to improve the concrete performance.

6.2 Construction techniques

6.2.1 Measures of conservation

The concrete construction area of the subway section is large, so the construction and maintenance work need to be carried out at the same time.With concrete strength meeting designing strength, it's necessary to ensure that the concrete surface remains wet.The two-sided framework maintenance method is adopted to ensure good moisturizing and heat preservation performance. The appearance quality of the concrete after mold removal is very high, and there is no crack, which effectively solves the problem caused by large temperature difference.

6.2.2 Arrangement plan for post-pouring belt

The post-poured concrete strip mixed with a small amount of aluminum powder is adopted and use high-performance concrete.The pouring area shall be divided according to the location of the post-pouring belt in this project.The"slant layering method"is adopted to carry out concrete layering pouring. That is, The pouring sequence follows from high to low, with each layer of 300-500mm thickness of pouring, so as to ensure the continuous operation of concrete and pouring uniform.The vibration method adopts the "secondary vibration method", and the concrete insertion depth of the vibration rod is about 5cm, which eliminates the concrete surface gap and strengthens the concrete anti-crack performance.

6.2.3 Temperature monitoring

Temperature monitoring is a key link in the construction of large-area concrete. In addition to the temperature detection in the process of concrete pouring, it is also necessary to strengthen the detection of environmental temperature and internal and external temperature difference in the maintenance process.The temperature difference between inside and outside of the general concrete structure shall be controlled to not exceed 20°C.The project uses internal water pipe and pours cooling water to reduce the internal temperature,which effectively reduces the risk of concrete cracking.Using concrete tank truck with insulation layers to effectively reduce the heating of concrete during transportation.In order to reduce the mold temperature, pouring concrete is carried out as much as possible at night.

After taking the above series of anti-crack measures, there is no crack in the subsequent-poured concrete structure, and the ideal construction effect are achieved.The problem of shrinkage crack control of large-area concrete structure in underground railway section is successfully solved.

7. Conclusion

In this paper, on the basis of a large number of domestic and foreign references, the research results on the mechanism of shrinkage crack cracking in large-area concrete and its controlling methods at home and abroad are summarized; on the other hand, based on a vehicle section of the project of Wuhan city rail transit, the existing engineering experience and research results are flexibly applied to the actual project.Some conclusions are followed:

(1)In the analysis of the mechanism of shrinkage cracks in large-area concrete,combined theoretical and engineering researches and analyses,different scholars come to conclusions that the factors affecting large-area concrete cracks include internal and external temperature changes, shrinkage deformation, construction techniques, raw material quality, etc.Although Many research results have been achieved, there is a lack of systematic analyses and calculation.And the mechanistic analysis is also mostly for specific engineering cases, without systematic results.

(2)In large-area concrete crack controlling measures of underground railway sections , domestic and foreign scholars found that optimising the mix ratio, reducing the temperature of concrete pouring, reasonable reinforcement and other measures can effectively reduce the generation of shrinkage cracks.At present, there is few domestic researches related to the construction controlling measures for large-area concrete cracks in underground railway sections.So the optimal design still are to continue to explore.

(3)Based on the large-area concrete structure of the vehicle section of the metro project in Wuhan city rail transit, through the use of concrete material quality controlling, double-sided with formwork maintenance, sloping layered pouring, temperature monitoring and other measures, the crack treatment effect is obvious and can ensure normal operation of the metro.

The anti-cracking measures in this project can provide relevant guidance for the quality problems caused by the construction of large-area concrete structures in the future.

(4)The causes of large concrete cracks in underground railway sections are complex, and there are many factors to be considered.So there are many issues that have not been noted or studied more systematically during analyses. How to carry out real-time monitoring of cracks in key sections of underground railway sections, how to apply fibre concrete impermeability technology to underground railway sections, and how to use induced joints to control crack development are all important directions for future research.

References

1. Du Chuang, Wang Haozhong, Li Yanzhao, Zhang Chunxiao, Zhao Hailong.Current status of large-area concrete cracks [J].Building structure, 2018,48 (S1): 555-559.
2. Meng Ling.Analyses of ultra-long prestressed concrete frame structure [D].Huazhong University of Science and Technology, 2008.
3. Luo Bin.Study on prestressed large-area floor crack control [D].Jiangsu: Southeast University, 2000.
4. Li Jingen, Li Weibin, Yu Guojia, et al.Construction of extra-long and large pre-stressed building cover for Nanjing International Exhibition Centre [J].Architecture Technology, 2000,31 (12): 810-812.
5. Zhao Shunzeng, Li Changcheng, Jia Fujie, Wu Xunan.Study on self-healing of HCSA compensation shrinkage concrete [J].Concrete, 2018, {4} (09): 1-3 + 8.
6. Wang Tiemeng.Crack control of the engineering structure [M].Beijing:China Architecture & Building Press, 1997.
7. Tao Jianqiang, Li Huajian, Huang Jiamu, Huang Fali, Yi Zhonglai, Xie Yongjiang.Hydrated heat and crack control of massive concrete in railway engineering [J].Railway Engineering, 2018,58 (01): 146-149.
8. Xie Xiaoli, Zhu Huiying, Lu Linghuan, Yang Yang.Study on Crack Control of Ultra-Long Structural Concrete Beams at Metro Station [J].Concrete, 2017, {4} (03): 150-153.
9. Zhu Bofang, Wang Tongsheng, Ding Baoying, etc.The temperature stress and temperature control of hydraulic concrete structure [M]. Beijing: China WaterPower Press, 1976:23-25.
10. Luo Gang.Research and Application of Seamless Construction Technology of Ultra-Large Area Concrete Ground Structure [D].Chongqing University, 2004.
11. Wang Yongping.Concrete shrinkage test study [D].Tianjin: Tianjin University, 2010.
12. Xiao Shuhao.Contrtraction properties and early cracking analyses of concrete with double mixed fiber and expansion agent [D].Tianjin University, 2018.
13. ACI Committee 209(1992) . Prediction of creep, shrinkage and temperature effects in concrete structures . America: American Concrete Institute, 1992, 1-92
14. CEB International Concrete Council.1990 CEB-FIP mode specification (Concrete Structure).Structural Specification Room of China Institute of Architectural Sciences.Beijing: China Architecture & Building Press, 1991,57-70
15. N .J .Garder ,M .J .Lockman .Design Provisions for Drying Shrinkage and Creep of Normal -Strength Concrete .ACI Materials Journal ,2001,98(2):159-167
16. B.P .Bazant ,S .Baweja .Creep and Shrinkage Prediction Model for Analysis and Design of Concrete Structures-Model B3, RELEM Recommendation. Materials and Structures, 1995, 1(28):357—365
17. N.J.Garder, J.W.Zhao. Creep and Shrinkage Revisited. ACI Materials Journal, 1993,90(3):236—246
18. Ding Wensheng, Lv Zhitao, Meng Shaoping, et al.Analysis and comparison of prediction models of concrete shrinkage and creep [J].Bridge Construction, 2004 (6): 13-16.
19. Song Can, He Dazhi, Xu Wei.Selection of the high-performance concrete shrinkage prediction model [J].Concrete, 2011 (5): 40-42.
20. Tu Bing.Prediction model of concrete shrinkage and shrinkage effect of widening bridge structure [D].Hunan University, 2013.
21. Bai Yang.Cause analyses and countermeasures of concrete cracks in road and bridge construction[J].New Technology & New Products of China, 2020 (13): 93-94.
22. Sun Wenlai, Chen Lei.Cause analyses and preventive measures of large-area concrete ground cracks [J].Theoretical Research in Urban Construction, 2015,5 (12): 3353-3354.

23. Chen Yanzhuo, Jiang Zhiping, Geng Ruiqi.Temperature stress simulation of open-cut subway roof [J].China Science and Technology Information, 2008 (01): 37-38.
24. Yuan Kang.Analyses of shrinkage crack mechanism of cast-in-place plate of reinforced concrete [D].Shihezi University, 2007.
25. Wu Wenwu, Qu Ming, Jiang Yousheng, Deng Hai.Finite element analysis of temperature rise of mass concrete of Yangluo Yangtze River bridge [J].Journal of Wuhan University of Technology, 2006 (04): 95-97.
26. Zuo Gewen.Analysis of key points of supervision and control in mass concrete construction process [J].Housing and Real Estate, 2019 (22): 142.
27. Chen Liling.Crack treatment countermeasures in civil engineering construction [J].Scientific and Technological Innovation, 2019 (23): 99-100.
28. Lin Haishan.External wall cracks and leakage control of subway underground station in Littoral Area (Xiamen) [D].Xiamen University, 2019.
29. Wei Yulian.Research on the key technology of mass concrete construction and crack control of 4th Nanjing Yangtze Bridge[D].Southeastern University, 2018.
30. Wei Min.Structural crack mechanism and control technology of large-area cast-in-place building cover [D].Hohai University, 2004.
31. Lin Min.Research and practice of large-Area concrete crack control [D].Chongqing: Chongqing University, 2007.
32. Chen Jingrong.Analysis of causes of large-area concrete floor cracks and countermeasures[J]. Construction Materials & Decoration, 2016 (32): 1-2.
33. Xie Li, Wu Shengxing.Study on the effect of water-ash ratio on early shrinkage of concrete [J].Building Science, 2007,23 (1): 54-57.
34. Li Zhanyin, Dong Jihong.Effect of water-cement ratio on heat release model of cement hydration [J].Journal of Yancheng Institute of Technology(Natural Science Edition), 2011,24 (01): 71-73.
35. Yang Changhui, Wang Chuan, Wu Fang.Effect of water-ash ratio on plastic shrinkage cracks in concrete [J].Journal of Chongqing Jianzhu University, 2003 (02): 77-81.
36. Pan Lisha, Qiu Xueqing, Pang Yuxia, Zheng Dafeng, Yang Dongjie.Effects of water reducer on the hydration behavior of cement [J].Journal of the Chinese Ceramic Society, 2007 (10): 1369-1375.
37. Wang Zilong, Ding Jiantong, Cai Yuebo, et al.Effect of retarder on hydration exothermic heat of compensating shrinkage cementitious system [J].Bulletin of the Chinese Ceramic Society, 2017,36 (6): 2144-2149.
38. Zhu Xianfa, Ye Tiemin, Liu Deshun, Wang Ningning, Li Hua.Experimental Study on Concrete Crack Control of Urban Rail Transit Underground Station [J].Jiangsu Construction, 2017 (S1): 9-12.
39. Ma Yongbo.Generation mechanism and control method of cracks in the side wall of the subway station [J].Engineering and Technological Research, 2020,5 (02): 23-24.
40. Hu Dongkang, Zhao Haitao, He zenghui.Application of bin-jumping method in crack resistance of super-long mass concrete [J].China Concrete and Cement Products, 2018, {4} (07): 79-82.
41. Hu Yixin, Ye Chengshun, Bie Weiya, Wu Xingfa.Crack control technology of long-span steel-concrete composite gallery bridge floor slab [J].Construction technology, 2021,50 (10): 98-100 + 105.
42. D.P.Bentz. Blending different fineness cements to engineer the properties of cement-based materials[J]. Magazine of Concrete Research, 2010, 62(5):
43. Zhang JiaXuan,Lyu HaiMin,Shen ShuiLong,Hou DongWei.Investigation of Crack Control of Underground Concrete Structure with Expansive Additives[J].Journal of Materials in Civil Engineering, 2021,33(1):
44. Liao Dongming.Analyses and Preventive Measures for Cracking of Concrete Structure at Metro Station [J].Technology and Market, 2017,24 (08): 201-202.
45. Zhan Jianjun.Analyses and control factors of concrete cracks on side wall of subway station [J].Subgrade Engineering, 2018 (03): 113-116.
46. Li Zhaoping, Wang Chen, Su Huifeng, Shi Shuaifeng.Study on concrete cracks and joint deformation law of prefabricated subway station structure joint [J]. China Civil Engineering Journal, 2015,48 (S1): 409-413.
47. Yu Yiming, Li Liang, Xu Wenbing, Meng Hongfeng, Xu Qing.Mass concrete crack control technology of rail transit engineering side wall [J].China Harbour Engineering, 2019,39 (10): 17-20.
48. Fang Kejun, Kong Baode.Anti-cracking measures for platform slab concrete of metro station [J].Railway Investigation and Surveying , 2007 (01): 88-89.
49. Fan Jiangsu.Study on concrete roofs of subway station [J].West-China Exploration Engineering, 2011,23 (06): 204-205.
50. Zhou Liqiang.Analyses of hydration heat in bulk concrete [J].Architectural Engineering Technology and Besign, 2015 (8): 897-898.
51. Lu Miao.Research on Concrete Crack Control in Metro Civil Engineering Construction [J].Create Living, 2020 (5): 63-64.

 https://doi.org/10.1051/matecconf/202338201017

Research progress in nanotechnology for enhanced silicon-based and iron-based anodes for lithium-ion batteries

Yaqi Yao *

Shandong Normal University, Shandong, 250014, China

Abstract. Recent research has focused on making suitable anode materials for lithium-ion batteries and using nanotechnology to refine composite electrode materials with high reversible capacity and strong stability. However, much work remains to be done to develop these materials into commercially viable solutions. The anode materials must have high reversible capacity, a long lifetime, and the ability to accept and release lithium ions repeatedly. Nano-engineered composite anode materials based on silicon and iron have shown good promise but need improved electrochemical properties and longer effective lifetimes. This paper summarizes the performance characteristics of Si-based and Fe-based anodes and ways to improve their performance through nanoengineering.

Keywords. Lithium-ion batteries; Anode; Si; Fe; Nanostructures

1. Introduction

The discovery of rechargeable ("secondary") batteries marked a turning point in the history of artillery. In contrast to the primary battery, which produces an electric current and then the state cannot be reversed, the primary battery can no longer be used when one of the reactants is consumed. Primary batteries produce an electric current, and then the state cannot be reversed, and when one of the reactants is finished, the primary battery can no longer be used. A secondary battery is a rechargeable battery that can be reused and has reversibility. It mainly refers to the storm that can be used continuously by charging the active material after the battery has been discharged. Early batteries used water-based electrolytes such as dilute sulfuric acid as a solvent. Liquid batteries using water-based electrolytes not only occupy a large area and are difficult to carry but also can reach a maximum operating voltage of about 2 volts. In 1887, Englishman Wilhelm Hellesen invented the earliest dry-cell battery. Compared with the liquid battery, the electrolyte of the dry cell is in the form of a paste, which will not spill and is easy to carry, so it has gained wide application. As technology advances and electronic and mechanical applications diversify, consumer demand for batteries increases, demanding longer runtime, larger size, lighter weight, rechargeability, high safety, and low cost.

Lithium is one of the lightest elements in the periodic table, with a density of only 0.59 g/cm^{-3} and the lowest electrochemical potential (-3.04 V). These two advantages enable it to provide the highest voltage in the smallest volume, with a theoretical limit of up to 3860 mAh/g. Lithium batteries with direct lithium metal as the negative electrode are no longer used for various reasons and have been replaced by lithium-ion batteries, in which lithium can travel between two electrodes in the form of lithium ions. The widespread use of lithium-ion batteries in portable electronics, such as mobile phones, laptops, and even power tools, is evidence of their many benefits.[1]

Anode material now offers more capacity and flexibility because of recent breakthroughs in the nanostructure of electrode materials for lithium-ion batteries. These nanostructures can also enhance the performance of lithium-ion batteries by enabling them to utilize more dynamic materials, expand capacity, and augment operating voltage. On the other hand, nanomaterials can narrow the power density gap between electron transport and ion diffusion. In some instances, nanomaterials may be detrimentally affected by their diminutive size and expansive specific surface area. A larger specific area may lead to more noticeable side reactions with the electrolytes.[2]Nanostructured electrode materials thus present both chances and difficulties for augmenting energy storage.

The purpose of this review is to summarize the latest nanotechnology advances in anode materials for lithium-ion batteries, such as silicon-based electrodes and iron-based transition metal oxide electrodes, and to provide insight into how these different nano anode structures have enhanced lithium-ion battery performance in the past five years. The current and future anode materials are also presented with prospects and suggestions for strengthening lithium-ion batteries' energy density, multiplication capacity, and safety.

* Corresponding author: Yaoyaki1102@163.com

 https://doi.org/10.1051/matecconf/202338201017

2. Principle of lithium-ion battery

Lithium-ion batteries form dendrites that accumulate during charging and discharging, leading to uneven lithium deposition. The loops formed within the battery can thus be caused rapidly, releasing large amounts of heat to break down the electrolyte, potentially causing unforeseen disasters such as battery box explosions and fires. The basic principle of Li-ion battery is the removal/embedding of Li^+. Figure 1 shows the charging and discharging of a Li-ion battery, with $LiCoO_2$ and graphite serving as the anode and cathode materials and $LiPF_6$ serving as the electrolyte. The voltage used to charge the battery pushes Li^+ out of the storm. The electric field causes Li^+ to traverse the minuscule pores of the electrolyte and the diaphragm to the electrode's negative side. This electrode's graphite layer comprises numerous micro-pores in which the Li ions that reach the electrode are embedded.

When the battery is discharged, the Li^+ in the negative graphite layer will come off and move back to the positive electrode. Thus, the equation for the battery's response is:

Anode reaction： $C + xLi^+ + xe^- = CLi_x$

Cathode reaction： $LiCoO_2 = Li_{1-x}CoO_2 + xLi^+ + xe^-$

Total reaction： $LiCoO_2 + C = Li_{1-x}CoO_2 + CLi_x$

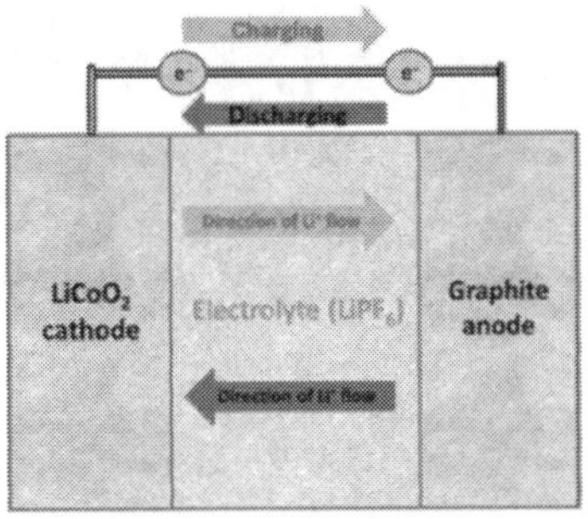

Figure 1. Graphite serves as the negative electrode of a lithium-ion battery, $LiCoO_2$ serves as the positive electrode, and LiPF6 serves as the electrolyte. [3]

Lithium-ion batteries also inevitably have lithium metal deposition during operation. The conventional liquid electrolytes used are not uniformly deposited at room temperature and high current densities and can form large numbers of lithium dendrites.[4] The generation of lithium dendrites is a fundamental factor affecting lithium-ion battery properties, and their growth leads to poor stabilization and low refundable capacity.

The components of the electrolyte are swiftly reduced as a result of the lithium metal's strongly negative redox potential, which results in the development of a solid electrolyte interface layer (SEI) on the anode's surface. By preventing direct contact between the negative electrode and the electrolyte, this SEI layer aids in the stability of the electrolyte and increases the lithium-ion battery's lifespan. SEI shields the electrode surface, thus obstructing the extraction and incorporation of Li^+ to a certain degree, resulting in a decrease in battery capacity.

Therefore, scientists have conducted extensive research to improve the performance of Li-ion batteries. Research on the anode of lithium-ion batteries has focused on enhancing the formation of lithium dendrites and the stability of SEI. Firstly the cathode material must meet three essential requirements: 1) high lithium storage capacity to ensure high weight/volume energy density; 2) low standard redox potential of the anode material to enable high cell voltage; 3) its electron/lithium ion conductivity must be outstanding in ensuring high rate performance. In recent years, there has been considerable research into silicon-based and transition metal compounds as anode materials.

3. Silicon-based electrodes

Because the theoretical capacity of conventional lithium metal cathode (3860mAh/g) reaches its limit, it cannot meet the market demand for higher output power of new generation lithium-ion batteries. Therefore, searching for alternative anodes with a high specific capacity, good cycling stability, and environmental friendliness has recently s been a hot issue in lithium-ion battery research. The remarkable specific power (4200 mAh/g), low reduction potential, abundance of natural reserves, and the ability to avoid forming dendrites have all made silicon (Si) a highly studied anode electrode material by scientists.

However, Si anode materials face two significant challenges; On the one hand, they usually have poor cycling stability, and repeated volume changes in charge and discharge can lead to the tearing of silicon particles and crushing of the solid electrolyte interface (SEI) layer is unstable and easily broken by dendrites; In contrast, silicon electrodes, with their lack of electrical conductivity, are not practical on their own, and are usually augmented with materials of higher electrical conductivity to augment the conductivity. Scientists have investigated many strategies to improve Si electrochemical performance within the nanodomain.

3.1 Controlling the size of Si to the nanometer level

Using nanostructures for metallic Si is a critical way to control the size of electrodes. After reducing the structure of Si to the nano level, it can be well adapted to the effects of volume changes during charging and discharging. Nanostructures can be divided into four aspects: 0D, 1D, 2D, and 3D. All four forms of nanostructures can improve the electrochemical performance, and the capacity is significantly enhanced compared to pure Si without any modification. Making Si in the 2D form will also improve the performance and stability of the negative electrode. Yang Ren prepared an anode material (Si-NSs@rGO) where ultrathin Si-NSs played an important role.[5] Due to their low cost and relatively small size, nano-sized particles can be used to enhance the physical properties of anode materials. Each of the four forms of nanostructures has its advantages and disadvantages. 0D nanoparticles have a large specific surface area, which can effectively promote Li+ deliquescence/embedding, and contact with the electrolyte leads to increased side reactions, usually with lower Coulomb efficiency and shorter battery life. The integrity and stability of the electrodes were maintained by the limited volume expansion of the 2D sheet nanostructures. However, overlap and stacking between nanosheets are easy to occur. While retaining the stability of the negative electrode structure, the porous 3D structure offers support areas to stop electrode disintegration. Nevertheless, it also causes increased side reactions with the electrolyte. Therefore, to maximize the performance of the anode material, the size and shape must be chosen.

3.2 Provides voids to accommodate Si volume changes

The Si nanoparticles will separate during charging and discharging due to the growth of the volume by squeezing each other until the particle size of Si drops to 150 nm, reaching the nanometer scale. Scientists employed nanoparticles with supportive qualities dispersed in the pores to strengthen the stability of the composite to address this issue. Hartmut Wiggers Synthesizing a Si/graphite/polyaniline composite (Figure 2) to enhance cycling stability, the conducting polymer polyaniline (PANI) was embedded with silicon and graphite particles.[6] Yao Zhang prepared Cu@MP-Si composites. The structure of MP-Si was effectively sustained by the Cu nanoparticles, dispersed in great abundance in the pores. Because of this, the cycling performance was greatly improved, with a capacity of 1559 mAh/g.[7] In testing the composite Sn@MP-Si anode electrode material, Chaoye Zhu found that nanoscale tin particles embedded in mesoporous silicon pores and close contact with silicon effectively supported the pore structure and built a conductive bridge network, improving performance.[8] Providing Si with a 3D porous nano spatial framework dramatically enhances performance and increases capacity.

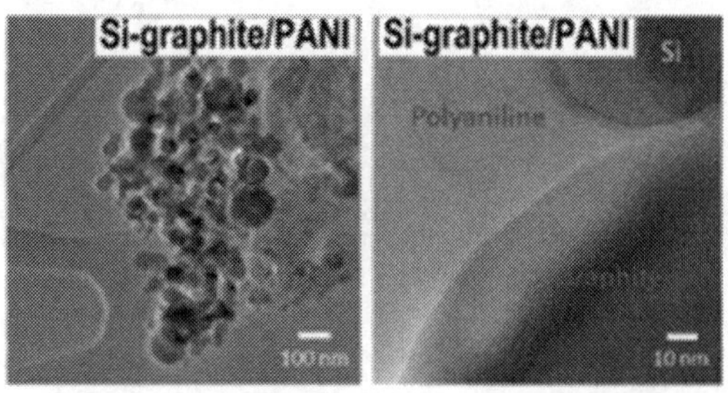

Figure 2 . The HR-TEM image of the Si/graphite hybrid shows that the silicon/graphite hybrid can be well embedded in a polyaniline conducting matrix. [6]

3.3 Combining nanoscale Si with a material with good electrical conductivity

A common way to enhance electrode performance is to blend nanoscale silicon with high electrical conductivity materials, typically metal or carbon. By making an alloy of Si and some metals (Ge, Cu, Zn, Sn, Mo, etc.), the conductivity of the alloy is controlled by the ratio of the two metals, so Nasr Bensalah et al. used this to make SiGe@Cu electrodes with specific charge capacity retention of 97.7%, 94.5%, 88.7% after 20, 30, 40 and 50 cycles, respectively and 81.2%.[9] Making SiGe alloys into thin films can better improve the chemical properties. Nasr Bensalah prepared MWCNT-Si/Ge nanocomposites as negative electrodes by controlling the thickness of Si and Ge film deposition by magnetron sputtering technique in the same year.[10] Kai Ma and Ning Lin discovered that when the ratio of Si to Ge was 4:1, the nanocomposite electrode created from Si and Ge particles in flocculent form had an initial capacity of 2040.3 mAh/g.[11] Another commonly used metal is Cu. Junying Zhang et al. introduced a Cu component with high electrical conductivity and synthesized Si@Cu composite anode material. [12] Thanks to the conductive copper network and nanowire structure, Juan Hong et al. improved the design of Si-Cu alloy, Si/Cu nanowires, with high discharge capacity and high capacity after several cycles.[13] There are also some doping of transition metals that can serve the purpose. Zizhong Chen doped Zn into the three-dimensional dendritic Si structure by chemical alloying method under mild conditions to make nanoscale porous materials of different proportions, which can provide enough space and a solid skeleton to relax the structural changes, which is conducive to relaxing the mechanical stress of Si, reducing its capacity degradation and ensuring sufficient electrode-electrolyte contact. [14]In addition, the doping of certain non-metals can also achieve the purpose of improving the electrode performance.

Carbon and silicon composites' electrical conductivity and SEI stability surpass that of pure silicon materials. Carbonaceous substances, including graphene, carbon nanofibers, and carbon nanotubes, are frequently employed; among them, reduced graphene oxide (rGO) stands out due to its superior physical and electrical conductivity. Optimization of electrode properties continues to evolve, and N atoms are now widely doped into carbon and silicon composites to optimize electrode properties. Studies have revealed that the electrochemical properties of N, S co-doped carbon materials are superior to those of N-doped carbon materials alone.[15]

3.4 Application of nanoscale surface coating technology

Nanoscale surface coatings that inhibit SEI layer growth and improve lithium-ion performance have been used in recent years to create replaceable coatings to replace carbonaceous coatings. Hongzheng Zhu et al. has developed a new surface coating material, aluminum nitride (AlO_xN_y), which achieves a 1297 mAh/g capacity after 140 cycles at a coating thickness of 2 nm (Figure 3A).[16] Received et al. uniformly covered the surface of the CMC-SiNS electrode with a CMC polymer layer (Figure 3B). The coating effectively inhibits SEI's growth, increases the electrode's service life, and exhibits excellent stability during charge/discharge cycles.[17] Na Li discovered that the encapsulated Al_2O_3 layer (Figure 3C) could stabilize the mesoporous Si nanosphere structure and SEI layer, leading to Si@Al_2O_3 nanospheres with higher Coulomb efficiency, cycling performance, and rate capability. This coating also effectively inhibits SEI growth, prolongs electrode service life, and displays remarkable stability during charge/discharge cycles.[18] As demonstrated in these studies, one new approach to improving the performance of lithium-ion batteries is to coat silicon materials with conductive materials sufficient to withstand specific tensile strains during silicon volume expansion, thereby improving the cycling stability and extending the life of silicon anodes.

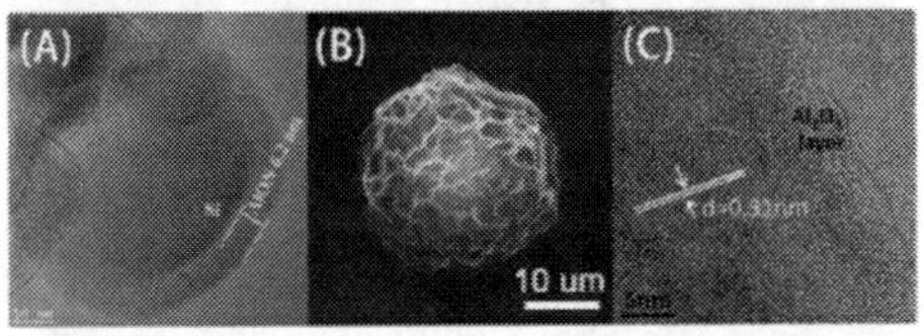

Figure 3 . (A) TEM image showing a uniform and conformal AlO_xN_y film coating on Si nanoparticles; [16](B) SiNS electrode uniformly coated with a CMC polymer layer; [17](C) HRTEM image of the Al_2O_3 layer encapsulated on mesoporous Si nanospheres.[18]

4. Iron-based electrodes

Transition metal oxides, composite materials, are frequently employed as anode materials for lithium-ion batteries. However, transition metal oxides are also unstable during long-time cycling. Designing transition metal oxide composites into nanostructures has been demonstrated to be a successful way of enhancing their performance, thus prompting the introduction of numerous strategies.[19]

Iron oxides are important transition metal oxides, among which Fe_3O_4 and Fe_2O_3 are commonly used as anode materials. However, Fe_3O_4 or Fe_2O_3 without modification has a huge volume change during charge/discharge, resulting in rapid capacity decay, loss of flexibility, and mechanical strength, which greatly hinders practical applications. Such problems have been investigated to achieve a long charge/discharge cycle process for iron oxides as the anode of Li-ion batteries.[20,21] One effective approach is constructing Fe_3O_4-based particles with specific morphology and structure; another strategy combines nanoscale Fe_3O_4 or Fe_2O_3 with some conductive materials.

4.1 Construction of Fe_3O_4-based or Fe_2O_3-based nanoparticles with different morphologies and structures

Jiaxin Li and Yunhui Li prepared solid, hollow, and porous Fe_3O_4 nanospheres on carbon nanotubes. They found that Fe_3O_4 with different nanostructures had different electrochemical properties, illustrating the importance of Fe_3O_4 morphology and structure.[22] Qichao Wu synthesized a novel cauliflower-shaped Fe_3O_4 nanoparticle and distributed it uniformly between a two-dimensional sandwich-like rGO and carbon layer to form a C/ Fe_3O_4/rGO composite, which, because of its unique structure, has a long lifetime. [23] Hollow Fe3O4 nanospheres prepared by Haonan Zhang provide a reversible capacity of 631 mA/g at 2 A/g due to their unique structure. Hui Lv et al. prepared an anemone-like Fe_3O_4 composed of nanorods, which was also experimentally confirmed to have good performance.[24] Yuexian Li, Jian Song, Yueyan Pan, Chengang Luo, et al. improved the previous complex preparation process and proposed a novel and facile one-pot reaction method to prepare porous carbon-coated Fe3O4 nano fusions (Fe_3O_4@PC), which, due to its unique spindle structure, reduced the disintegration of the electrode structure and performed well, exhibiting 896.3 and 592.9 mAh/g after 400 cycles at 200 and 1000 mA/g, respectively. [26]Not only that, but other morphological composites also have good properties.

Fe_2O_3-based materials can also be used to improve the electrochemical properties. Investigating the electrochemical properties of Fe_2O_3 nanoparticles and nanorods synthesized through hydrothermal synthesis, Jay Singh found that the particle morphology significantly affected the electrochemical properties of Fe_2O_3. The Fe_2O_3 nanoparticles exhibited remarkable electrochemical properties compared to the nanorod morphology, which may be attributed to their high structural stability and small agglomerates.[27] Various morphologies and structures of Fe_2O_3 nanoparticles have also been fabricated. Different morphological nanostructures augment the contact between the electrode material and electrolyte, cushion the mechanical strain caused by Li^+ detachment/embedding, or construct a framework structure to offer support.

4.2 Combining nanoscale Fe_3O_4 or Fe_2O_3 with some conductive material

Fe_3O_4/Fe_2O_3 is often doped with carbonaceous materials to improve electrode performance.[28] Meng Yua and Liuyang Sun et al. found more practical ways to obtain low-cost, functional nanomaterials for energy storage applications, using FeCl3·6H2O and NH4H2PO4 as raw materials and dopamine hydrochloride as a carbon source to synthesize carbon-coated Fe3O4 nanosheets (R-Fe3O4@C).[30] Pengxiang Zhao et al. has synthesized Fe3O4@reduced graphene oxide (Fe3O4@rGO) coupled structures in situ with good performance and a high specific capacity of 1296.8 mAh/g.[31] Rui Zhang and Shouchun Bao synthesized rod-shaped porous carbon skeleton-loaded magnetite nanoparticle composites (Fe3O4/RPC) exhibiting enhanced lithium-ion storage capacity [32] Ya-Jing Duan et al. assembled and synthesized hollow graphene spheres (HGSs) and combined them with Fe3O4 to make Fe3O4/HGSs composites that exhibited excellent electrochemical properties as anode materials.[33] Xiuling Shi et al. prepared a high-performance anode material made by rational encapsulation of multi-walled carbon nanotubes (MWNTs) in hollow Fe3O4 nanotubes followed by the application of a carbon coating, providing a high capacity of 758 mAh/g at the 500th cycle.[34]

New developments have also been made in recent years regarding doping Fe2O3-based materials. Bo Wang used Fe-MOFs as precursors to synthesize the Fe2O3@OMC material, which exhibited excellent capacity and stability due to the uniform distribution of Fe2O3 nanoparticles on the carbon substrate.[35] Lixia Ma prepared a prepared α-Fe2O3/graphite composite with better electrochemical performance by embedding Fe2O3 nanoparticles between expanded graphite layers. Due to the extended structure of graphite, the adverse effect of the volume change of α-Fe2O3 grains during charging and discharging can be effectively mitigated[36]. Graphene oxide (TGO) is also excellent carbon-based material; Xiaoru Meng et al. prepared thermal graphene oxide with increased specific surface area by simple heat treatment of graphene feedstock in an air-nitrogen atmosphere, which adsorbed a large number of Fe(OH)3 nanoparticles under the effect of mutual electrostatic attraction. The Fe2O3/TGO composite was then prepared by hydrothermal method. The TGO, with its high surface area and electrical conductivity, which reduces the impact of volume expansion, is equipped with large pores between them, thus allowing for the transport of lithium ions.[37]

N atoms have been widely explored to modify the band gap of carbonaceous materials or alter their surface chemistry by creating more active sites in the carbon skeleton. Thereby further improving their electrochemical properties, porous carbon mesh coating and in situ doped N synergistically can electrode enhance stability. Zhikang Liu et al. reported a new strategy for the simultaneous construction of in situ N-doped and porous carbon frame-coated Fe3O4 nano polyhedra with new Fe3O4@N-C composites with long lifetime.[38] N-doped carbon-encapsulated porous Fe3O4 nanocages (Fe3O4@C nanocages) were synthesized, also with good performance.

5. Conclusions and perspective

This paper details the performance advantages and disadvantages of silicon-based and iron-based anodes and ways to improve their performance through nanoengineering. It summarises some of the research advances made in recent years. Next, we summarise and outlook at these two materials, hoping that our work will help the field of anode materials for lithium-ion batteries. Silicon is considered the most promising alternative to graphite anodes because it is abundant, inexpensive, has better working potential than graphite, and has good environmental compatibility. Transition metal oxides also offer the advantages of high theoretical specific capacity, simple development processes, large reserves, and encouraging electrode materials. The biggest challenges facing both materials on their way to commercialization are addressing the severe volume expansion in charge and discharge and the irreversible capacity loss caused by unstable SEI films resulting from electrolyte decomposition. With the rise of nanoengineering, the problems of silicon anodes have been effectively alleviated.

Recent research into the utilization of silica-based and iron-based composite nanomaterials as anode materials for lithium-ion batteries has presented novel ideas for discovering superior alternative anodes. Each resulting fabric has shown the ability to outperform graphite. Still, improvements in performance using nanoengineering are far from a perfect solution, and a combination of approaches is needed to develop a material that balances good capacity with durability.

The development of lithium-ion batteries is currently facing a massive challenge in terms of what anode materials we will use and how to produce these materials commercially on a large scale and efficiently. Therefore, the lithium storage properties of these materials and the related electrochemical reaction mechanisms need to be investigated in depth, and an economically viable preparation method needs to be found.

References

1. Luo, Y., Tang, Y., Zheng, S., Yan, Y., Xue, H., & Pang, H. 2018. Dual anode materials for lithium- and sodium-ion batteries. Journal of Materials Chemistry A, 6 (10) : 4236-4259.
2. Wang, F., Wu, X., Li, C., Zhu, Y., Fu, L., Wu, Y., & Liu, X. 2016. Nanostructured positive electrode materials for post-lithium ion batteries. Energy & Environmental Science, 9(12): 3570-3611.
3. Balogun, M.-S., Qiu, W., Luo, Y., Meng, H., Mai, W., Onasanya, A., Olaniyi, T. K., & Tong, Y. 2016. A review of the development of full cell lithium-ion batteries: The impact of nanostructured anode materials. Nano Research, 9(10): 2823-2851.
4. Takeda, Y., Yamamoto, O., & Imanishi, N. 2016. Lithium Dendrite Formation on a Lithium Metal Anode from Liquid, Polymer and Solid Electrolytes. Electrochemistry, 84 (4) : 210-218.
5. Ren, Y., Xiang, L., Yin, X., Xiao, R., Zuo, P., Gao, Y., Yin, G., & Du, C. 2022. Ultrathin Si Nanosheets Dispersed in Graphene Matrix Enable Stable Interface and High Rate Capability of Anode for Lithium-ion Batteries. Advanced Functional Materials, 32 (16) .
6. Wiggers, H., Sehlleier, Y. H., Kunze, F., Xiao, L., Schnurre, S. M., & Schulz, C. 2020. Self-assembled nano-silicon/graphite hybrid embedded in a conductive polyaniline matrix for the performance enhancement of industrial applicable lithium-ion battery anodes. Solid State Ionics, 344.
7. Zhang, Y., Liu, Z., Zhu, C., Guo, X., Liu, W., & Qu, Y. 2021. Boosting anode performance of mesoporous Si by embedding copper nanoparticles. Journal of Alloys and Compounds, 850.
8. Zhu, C., Zhang, Y., Ma, Z., Zhu, Y., & Li, L. 2020. Mesoporous-Si embedded and anchored by hierarchical Sn nanoparticles as promising anode for lithium-ion batteries. Journal of Alloys and Compounds, 832.
9. Bensalah, N., Matalkeh, M., Mustafa, N. K., & Merabet, H. 2019. Binary Si–Ge Alloys as High-Capacity Anodes for Li-Ion Batteries. physica status solidi (a) , 217 (1) .

10. Bensalah, N., Kamand, F. Z., Mustafa, N., & Matalqeh, M. 2019. Silicon–Germanium bilayer sputtered onto a carbon nanotube sheet as anode material for lithium–ion batteries. Journal of Alloys and Compounds, 811.
11. Ma, K., & Lin, N. 2019. The controllable synthesis of Si/Ge composites with a synergistic effect for enhanced Li storage performance. Inorganic Chemistry Frontiers, 6 (7) : 1897-1903.
12. Zhang, J., Hou, Z.-L., Zhang, X., & Li, C. 2022. Si@Cu composite anode material prepared by magnetron sputtering for high-capacity lithium-ion batteries. International Journal of Hydrogen Energy, 47 (7) : 4766-4771.
13. Hong, J., Cheng, K., Xu, G., Stapelberg, M., Kuai, Y., Sun, P., Qu, S., Zhang, Z., Geng, Q., Wu, Z., Zhu, M., & Braun, P. V. 2021. Novel silicon/copper nanowires as high-performance anodes for lithium ion batteries. Journal of Alloys and Compounds, 875.
14. Chen, Z., Wang, X., Jian, T., Hou, J., Zhou, J., & Xu, C. 2020. One-step mild fabrication of branch-like multimodal porous Si/Zn composites as high performance anodes for Li-ion batteries. Solid State Ionics, 354
15. He, Q., Wu, Q., Wang, X., Fu, S., Huang, S., Tong, S., Cao, Y., Liu, Z., & Wu, M. 2021. An Anode Material for Lithium Storage: Si@N,S-Doped Carbon Synthesized via In Situ Self-Polymerization. ACS Applied Energy Materials, 4 (4) : 3555-3562.
16. Zhu, H., Shiraz, M. H. A., Liu, L., Zhang, Y., & Liu, J. 2022. Atomic layer deposited aluminum oxynitride coating for high-performance Si anode in lithium-ion batteries. Applied Surface Science, 578.
17. Park, S.-W., Ha, J. H., Park, J. M., Cho, B. W., & Choi, H.-J. 2021. Enhanced capacity retention based silicon nanosheets electrode by CMC coating for lithium-ion batteries. Electronic Materials Letters, 17 (3) : 268-276.
18. Li, N., Yi, Z., Lin, N., & Qian, Y. 2019. An Al (2) O (3) coating layer on mesoporous Si nanospheres for stable solid electrolyte interphase and high-rate capacity for lithium ion batteries. Nanoscale, 11 (36) : 16781-16787.
19. Ye, H., Zheng, G., Yang, X., Zhang, D., Zhang, Y., Yan, S., You, L., Hou, S., & Huang, Z. 2021. Application of different carbon-based transition metal oxide composite materials in lithium-ion batteries. Journal of Electroanalytical Chemistry, 898.
20. Wang, F., Wang, C., Chen, H., Zhang, W., Jiang, R., Yan, Z., Huang, Z., Zhou, H., & Kuang, Y. 2019. A composite of Fe3O4@C and multilevel porous carbon as high-rate and long-life anode materials for lithium ion batteries. Nanotechnology, 30(33).
21. Hong, M., Su, Y., Zhou, C., Yao, L., Hu, J., Yang, Z., Zhang, L., Zhou, Z., Hu, N., & Zhang, Y. 2019. Scalable synthesis of γ-Fe2O3/CNT composite as high-performance anode material for lithium-ion batteries. Journal of Alloys and Compounds, 770: 116-124.
22. Li, J., Li, Y., Chen, X., Kierzek, K., Shi, X., Chu, P. K., Tang, T., & Mijowska, E. 2019. Selective Synthesis of Magnetite Nanospheres with Controllable Morphologies on CNTs and Application to Lithium-Ion Batteries. Physica Status Solidi a-Applications and Materials Science, 216(11).
23. Wu, Q., Jiang, R., & Liu, H. 2020. Carbon layer encapsulated Fe3O4@Reduced graphene oxide lithium battery anodes with long cycle performance. Ceramics International, 46 (8) : 12732-12739.
24. Zhang, H. 2020. Facile synthesis of hollow Fe3O4 nanospheres and their application as drug delivery carrier and anode for lithium-ion batteries. International Journal of Electrochemical Science: 4789-4797.
25. Lv, H., Feng, X., Bi, H., & Song, X. 2022. Anemone-shaped Fe3O4 Micro-structures as the Anode Materials in High Electrochemical Performance Li-ion Batteries. International Journal of Electrochemical Science, 17(5).
26. i, Y., Song, J., Pan, Y., Luo, C., Chen, J., Sui, Z., & Tian, Q. 2023. Integration strategy for facile fabrication of porous carbon coated Fe3O4 nanospindles with enhanced lithium storage. Journal of Alloys and Compounds, 935.
27. Singh, J., Lee, S., Kim, S., Singh, S. P., Kim, J., & Rai, A. K. 2021. Improved lithium storage in Fe2O3 nanoparticles over nanorods morphology. Solid State Ionics, 362.
28. Li, C., Su, H., Zhang, K., Liu, Z., Wang, H., & Li, D. 2021. FeOOH derived urchin-like Fe2O3@C as superior anode for sodium ion storage. Journal of Alloys and Compounds, 858: 157714.
29. Zhao, X., Jia, Y., & Liu, Z.-H. 2019. GO-graphene ink-derived hierarchical 3D-graphene architecture supported Fe3O4 nanodots as high-performance electrodes for lithium/sodium storage and supercapacitors. Journal of Colloid and Interface Science, 536: 463-473.
30. Yu, M., Sun, L., & Ning, X. 2021. Controllable synthesis of carbon-coated Fe3O4 nanorings with high Li/Na storage performance. Journal of Alloys and Compounds, 878.
31. Zhao, P., Jiang, L., Li, P., Xiong, B., Zhou, N., Liu, C., Jia, J., Ma, G., & Zhang, M. 2023. Tailored engineering of Fe3O4 and reduced graphene oxide coupled architecture to realize the full potential as electrode materials for lithium-ion batteries. Journal of Colloid and Interface Science, 634: 737-746.
32. Zhang, R., Bao, S., Tan, Q., Li, B., Wang, C., Shan, L., Wang, C., & Xu, B. 2021. Facile synthesis of a rod-like porous carbon framework confined magnetite nanoparticle composite for superior lithium-ion storage. J Colloid Interface Sci, 600: 602-612.

33. Duan, Y.-J., Zhao, D.-L., Liu, X.-H., Yang, H.-X., Meng, W.-J., Zhao, M., Tian, X.-M., & Han, X.-Y. 2019. Novel design of Fe3O4/hollow graphene spheres composite for high performance lithium-ion battery anodes. Journal of Alloys and Compounds, 779: 466-473.
34. Shi, X., Yao, Q., Wu, H., Zhao, Y., & Guan, L. 2019. Rational design of multi-walled carbon nanotube@hollow Fe (3) O (4) @C coaxial nanotubes as long-cycle-life lithium ion battery anodes. Nanotechnology, 30 (46) : 465402.
35. Wang, B., Luan, S., Peng, Y., Zhou, J., Hou, L., & Gao, F. 2021. High electrochemical performance of Fe (2) O (3) @OMC for lithium-ions batteries. Nanotechnology, 32 (12) : 125403.
36. Ma, L., Wang, Z., Tian, S., Liu, X., Li, Z., Huang, J., Deng, X., & Huang, Y. 2020. The alpha-Fe (2) O (3) /graphite anode composites with enhanced electrochemical performance for lithium-ion batteries. Nanotechnology, 31 (43) : 435404.
37. Meng, X., Huang, J., Zhu, G., Xu, Y., Zhu, S., Li, Q., Chen, M., & Lin, M.-C. 2022. Fe2O3 nanoparticles anchored on thermally oxidized graphene for boosting lithium storage properties. Journal of Solid State Chemistry, 315.
38. Liu, Z., Liu, L., Zhao, Z., He, J., Wang, S., & Xiong, C. 2020. Structural engineering of porous N-doped carbon-coated Fe3O4 framework by controlling coordination for superior lithium-ion full-cell. Applied Surface Science, 526.
39. Wu, W., Wei, Y., Chen, H., Wei, K., Li, Z., He, J., Deng, L., Yao, L., & Yang, H. 2021. In-situ encapsulation of α-Fe2O3 nanoparticles into ZnFe2O4 micro-sized capsules as high-performance lithium-ion battery anodes. Journal of Materials Science & Technology, 75: 110-117.
40. Li, W., Yang, F., Rui, Y., & Tang, B. 2019. Strong covalent interaction Fe2O3/nitrogen-doped porous carbon fiber hybrids as free-standing anodes for lithium-ion batteries. Journal of Materials Science, 54 (8) : 6500-6514.

https://doi.org/10.1051/matecconf/202338201018

Research and Improvement of Electromechanical Transient Performance of Power System Based on SIMULINK

Duo Wang *

School of Electrical and Control Engineering, North University of China, Taiyuan, China.

Abstract: The research on electromechanical transient stability of power systems is of great significance. Analyzing the parameters of the simulation system and its improved system with typical short-circuit faults, this paper studies the transient stability of power systems and its improvement. Firstly, based on the SIMULINK of MATLAB, a double-loop simulation with the single-machine-infinite-bus system is built. The simulation experiments of the system under the circumstances of three-phase short circuit and single-phase grounding short circuit faults are then carried out to analyze the parameters such as limit cutting time and voltage. Besides, this paper explores how the controllable series capacitor compensation device and reducing mechanical power output by prime mover improve the transient stability of the system. Experiments show that the stability after fault is under conditions where the fault is removed before the critical clearing time, which can increase the threshold to gain more response time for relay protection devices. Thus, the value of researching and improving system transient stability is proved.

Keywords: Electromechanical Transient Stability; Short Circuit Fault; Critical Clearing Time; MATLAB

1. Basic Theory

1.1 Electromechanical Transient Stability of Power System

The research on electromechanical transient stability of power systems focuses on whether the system can return to the original state or transit to a new stable state after being disturbed in a stable operating mode. The operation of the system before and after disturbance is shown in Fig.1 [1].

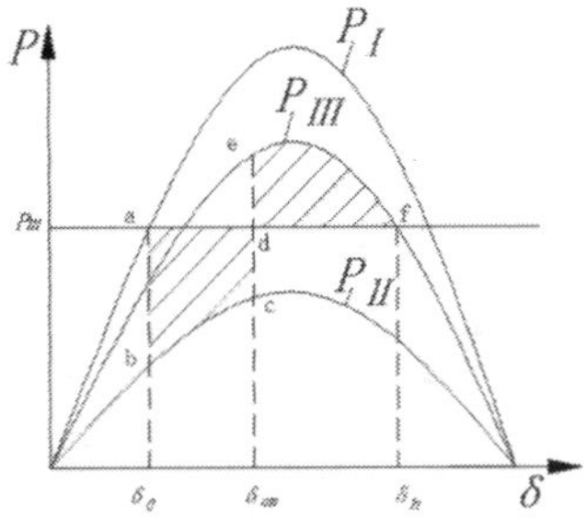

Fig.1 Power Characteristic Curve of Normal Operation, Failure, and Fault Removal of the Simple System

In Fig.1, P_m is the mechanical power output by the prime mover, P_I, P_{II}, and P_{III} are power characteristic curves of the generator during normal operation, after a short circuit fault, and after the timely removal of a short circuit. According to the equal area criterion [1], the system is in transient stability at the critical clearing angle δ_{cm}, and the corresponding fault clearing time is the critical clearing time t_{cm}.

1.2 Common Faults of Power System

Faults such as short and open circuits often occur during the power system operation, with the short circuit as the worst one. Thus, it is most representative to carry out experiments under this condition. In a three-phase system, short circuits are divided into as follows, including the three-phase short circuit $f^{(3)}$, two-phase short circuit $f^{(2)}$, two-phase earth fault $f^{(1,1)}$, and single-phase earth fault $f^{(1)}$. In this paper, the most typical $f^{(3)}$ and $f^{(1)}$ are used for research, as shown in Fig.2 and Fig.3.

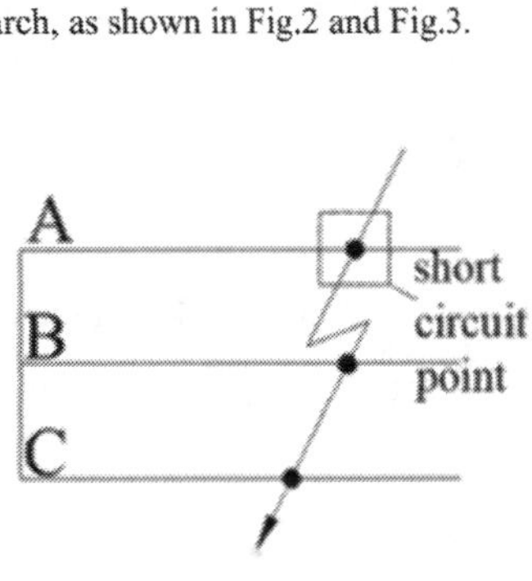

Fig.2 Three-Phase Short Circuit

* Corresponding author: 2015040406@st.nuc.edu.cn

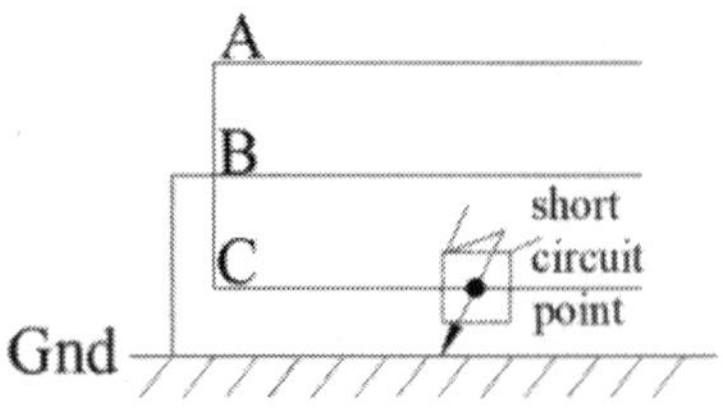

Fig.3 Phase Earth Fault

After the fault occurs, some output mechanical power of the prime mover will be stored in the rotor of the generator [1], which spares more time for the system to recover stability. Therefore, the system will remain stable if relay protection devices such as circuit breakers and automatic reclosure are used to remove the fault before t_{cm}, otherwise, it will lose stability.

1.3 Measures to Improve the Transient Stability of the System

It is of great engineering significance to improve transient stability. Generally, the following measures can be adopted.

Realize quick fault removal and adopt an automatic reclosing device.

Reduce the difference in generator power after disturbance, such as decreasing the mechanical power output by the prime mover.

Control dynamic elements that can be controlled in the system. For example, a controllable series capacitor compensation device is used in the line [2].

The generator's difference between mechanical power and electromagnetic power is the main cause of failing transient stability after a large disturbance, so it is usually considered to reduce such a difference first [1]. Meanwhile, the controllable series capacitor compensation device can reduce the total equivalent reactance of the system and improve the transfer capability. Thus, both methods can improve transient stability. This paper will focus on the effects of two methods on improving the transient stability of power systems.

2. Construction of Simulation Model

2.1 Overall Design Framework of Simulation Model

The simulation model is constructed by SIMULINK in MATLAB, consisting of the main module with a double-loop power system featuring a single machine infinite bus, the fault module for setting short circuits, and the observation module for parameter measurement. The overall design framework diagram is shown in Fig.4.

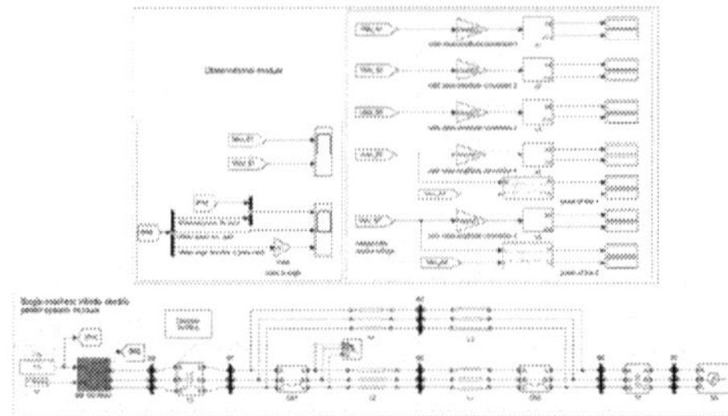

Fig.4 Overall Design Framework Diagram of the Simulation

2.2 Module of Single-Machine-Infinite-Bus Power System

The single-machine-infinite-bus power system constructed in this paper has two parallel lines, each of which is divided into two sections, with the fault point set in one of them. The two ends of the line are transformed by two D-Yg transformers to realize voltage transmission from *M2* to *G1*, as shown in Fig.5.

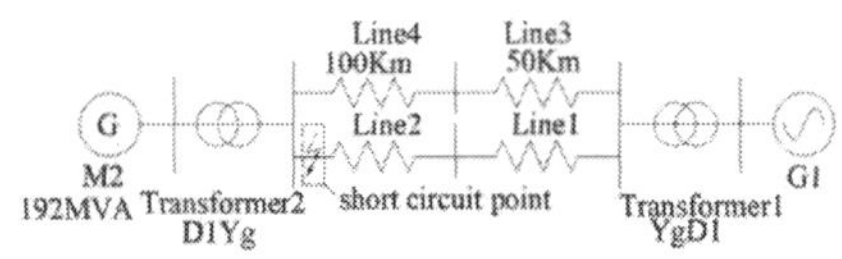

Fig.5 System Diagram

In Fig.5, the installed capacity of generator *M2* is 192MVA, the length of *Line4* and *Line2* is 100*Km*, and the length of *Line3* and *Line1* is 50*Km*. The short circuit point is on the side of *Line2* near *Transformer 2*, which is shown by a dashed box in Fig.5. The impedance parameters of the line are shown in Tab. 1.

Tab.1 Impedance Parameters of the Line

Impedance Parameters of Lines	*Line1*	*Line2*	*Line3*	*Line4*
Resistance (Ω)	0.1058	0.16928	0.1058	0.16928
Inductance (W)	2.3854e-3	2.2592e-3	2.3854e-3	2.2592e-3
Capacitance (F)	17.65e-9	15.24e-9	17.65e-9	15.24e-9

2.3 Fault Module

The short-circuit in the three-phase system is set by the component "*Three-Phase Fault*", and the short circuit type is set by checking the options in "*Fault between*". "*Phase A*", "*Phase B*", "*Phase C*" can be checked to set the three-phase short circuit. "*Phase A*" and "*Ground*" can be checked to set the Phase *A* earth fault. Set the total running time of the system to 10 seconds in "*Switching time*", and the failure occurs in the sixth second.

The setup and description of the fault are shown in Tab. 2.

https://doi.org/10.1051/matecconf/202338201018

Tab.2 Fault Description

Factor	Fault DC Line	Location of Fault	Type of Fault	Fault Clearing Mode
Setting Description	*Line2*	*Bus7*	*Phase A* Earth Fault, Three-Phase Short Circuit	Circuit Breaker *CB1*

2.4 Observation Module

The specific simulation settings of the observation module are shown in Tab. 3. The component "Gain" is used to convert the power angle of the generator from radian to angle, and the amplitude of each voltage to root-mean-square value. Besides, the component "Fourier" is used to convert the root-mean-square value to modulus and phase, with the component "Power" used to calculate the line power.

Tab.3 Observed Parameter

Object of Observation	Observation Parameter	Angle of Observation	Observation Instrument
Busx	*Vabc_Bx(x=1,2,4,5,7)*	Modulus value, phase angle	*Display*
Linex	*Px=Vabc_By*Iabc_By(x=1,2,y=5,7)*	Power	
M2	*d_theta, wm, Pe, Pm*	Power angle (δ), speed, power	*Scope*
G1	*Vabc_B1, Iabc_B1*	Voltage, current	

3. Fault Simulation

3.1 Overall Design of Simulation Experiment

The main design of the simulation experiment is divided into two sections. One is the study of electromechanical transient stability, which is mainly divided into stable operation experiments and experiments of fault and fault removal, constituting the content of Chapter 3. In addition, based on the experimental results in Chapter 3, two improved methods are proposed in Chapter 4, and the results of improving transient stability are analyzed. The flow chart of the simulation experiment is shown in Fig.6.

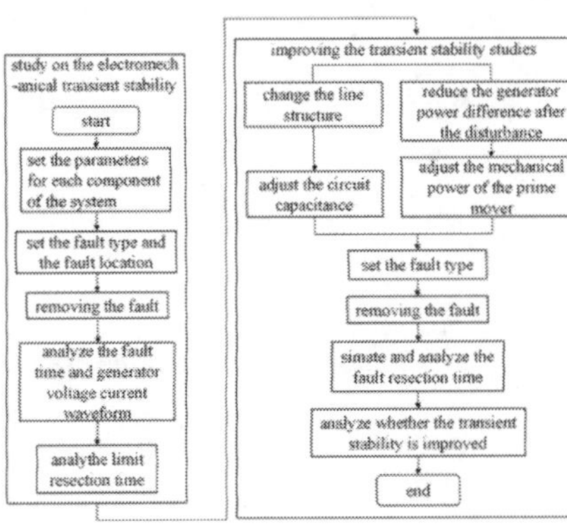

Fig.6 Simulation Flow Chart

3.2 Stable Operation

Stable operation of a power system is the state when the system works normally. If the system fails, the system will enter the transient process from stable operation [1]. When P_m is set to 1.632, a set of stable operation parameters of the system can be measured according to the simulation model, as shown in Tab. 4.

Tab.4 System Parameters at the Stable Operation

Object of Observation	*Vabc_B1*	*Vabc_B2*	*Vabc_B3*	*Vabc_B5*	*Vabc_B7*
Observation Value (pu)	0.075 ∠-7.3°	0.093 ∠14.8°	1.143 ∠33.3°	1.191 ∠36.8°	1.258 ∠43.0°
Object of Observation	*P2*	*Pe*	*P1*	*d_theta* (°)	*wm*
Observation Value (pu)	0.283+j0.519	1.637	0.319+j0.460	-24.8	0.999

The data shown in Tab. 4 are all standard values except the generator power angle, and the reference voltage and reference power are rated for the generator. Observing the waveform of the oscilloscope, it can be seen that the electromagnetic power of the generator is basically stable near 1.632, the mechanical power output by the prime mover. Besides, the rotating speed of the generator is kept near the rated value and the power angle of the generator tends to be a constant value with time. Therefore, this paper judges whether the fault is removed by whether the system can keep a stable state after the fault.

3.3 Fault Simulation of Three-Phase Short Circuit

An abnormal power supply will cause inconvenience and serious consequences to production and life, so it is vital to remove faults in time. When the fault occurs in the 6th second of the experiment, the parameters of *M2* and line are measured, so as to obtain the timely critical clearing time of the system. "Dichotomy" [3] is used to approximate the critical resection time. That is to say, if the system is just stable when the fault is removed at t_1, it is unstable when it is removed at t_2 with little difference between t_1 and t_2, then $t_{cm}=(t_2+t_1)/2$. We set the circuit breaker to clear the fault at 10/60 seconds, then accumulate the resection time for 5/60 seconds in turn repeatedly until it is observed that the system is no longer stable at 20/60 seconds. Then the critical resection time is calculated in [15/60, 16/60] by dichotomy approximation. The curves of generator parameters corresponding to t_1 and t_2 are shown in Fig.8 and Fig.9 respectively.

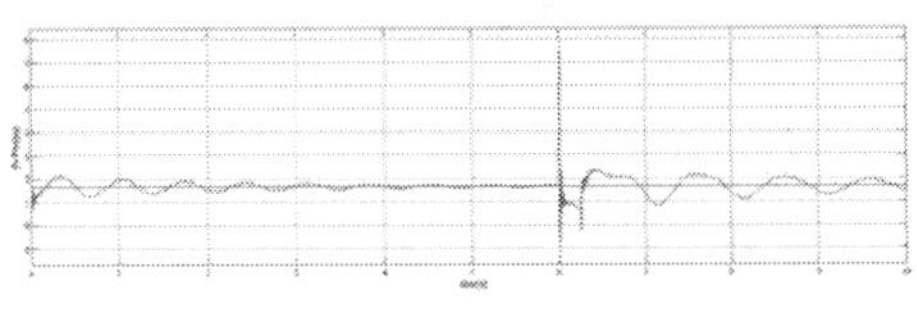

(a) Electromagnetic Power Pe

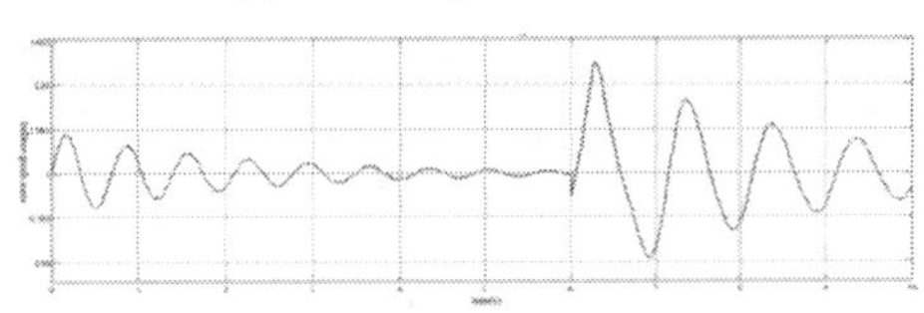

(b) Rotational Speed wm

 https://doi.org/10.1051/matecconf/202338201018

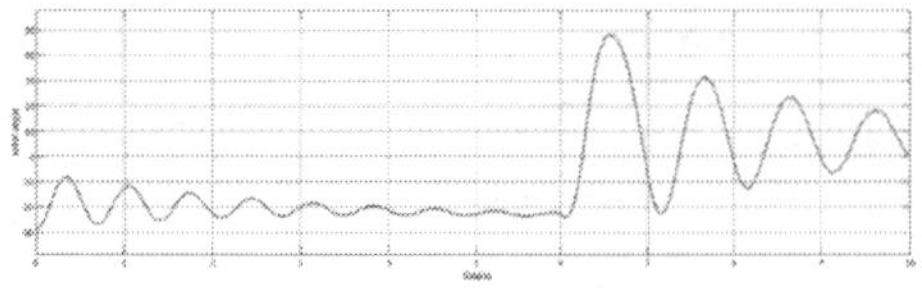

(c) Power Angle δ

Fig.8 Parameters of the Generator at 15 / 60 s

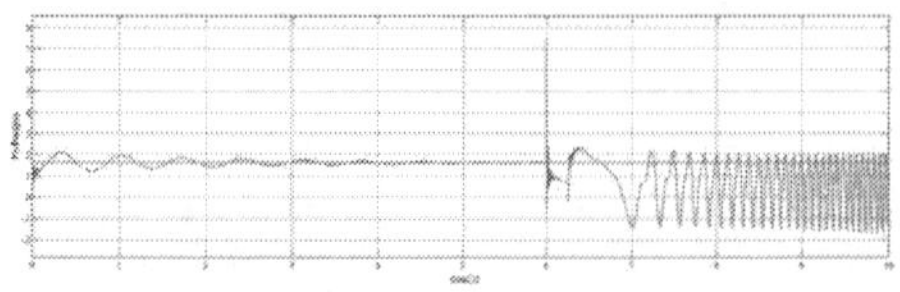

(a) Electromagnetic Power Pe

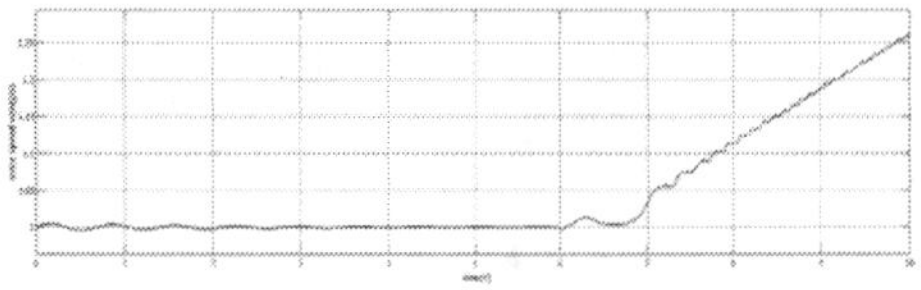

(b) Rotational Speed wm

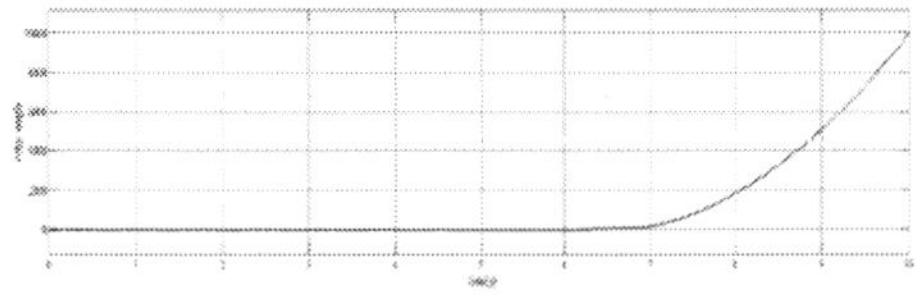

(c) Power Angle δ

Fig.9 Parameters of the Generator at 16 / 60 s

When the fault is removed at 15/60 seconds, the electromagnetic power of the generator is still stable at around 1.632, the speed is stable at about 1, and the power angle tends to be constant at 6°. In other words, the system can return to stability after self-adjustment. When the fault is removed at 16/60 seconds, the electromagnetic power of the generator is no longer stable at a fixed value. Besides, the changing speed and power angle of the generator tend to infinity with time, that is, the system is unstable at this time. Other parameters of the system at time t_1 and t_2 are shown in Tab. 5 respectively.

Tab.5 Parameters of the System at 15 / 60s and 16/60s

time (s)	*Vabc B1*	*Vabc B2*	*Vabc B4*	*Vabc B5*	*Vabc B7*	P_1	P_2
15 / 60	0.075 ∠-7.3°	0.080 ∠30.0°	1.071 ∠33.0°	0.001 ∠94.5°	1.113 ∠55.0°	0.005-j0.040	0.001-j0.001
16 / 60		0.035 ∠-86.0°	0.915 ∠22.7°	0.001 ∠88.6°	0.382 ∠-39.0°	Ami no 0	-0.001+j0.001

It can be seen from the above table that the voltage at the end of the line is always unchanged. After the fault is removed in 15/60 seconds, the voltage of Bus5 and the power of Line1 and Line2 decrease sharply compared with the steady state. It is caused by the circuit breaker clearing Line1 and Line2. However, when the fault is removed in 16/60 seconds, the parameters of the system change greatly and cannot be stable. The critical clearing time (t_{cm}) of 0.258 seconds can be approximately obtained by dichotomy and mean value method after 4 experiments with resection time of 10/60 s, 15/60 s, 30/60 s, and 16/60 s, with the calculation formula shown in formula (1). That is to say, after the three-phase short circuit occurs in the power system, the circuit breaker must remove the fault within no more than 0.258 s before the system returns to a stable state.

$$t_{cm} = \frac{t_1 + t_2}{2} = \frac{15/60 + 16/60}{2} = 0.258 \quad (1)$$

3.4 Simulation of Single-Phase Earth Fault

Single-phase earth fault is common in power systems. For example, a branch falling between phase A and the earth forms phase A earth fault. The circuit clearing time interval is [110/60, 115/60] by using the dichotomy method. The generator parameter curves corresponding to time t_1 and t_2 are shown in Fig.10 and Fig.11.

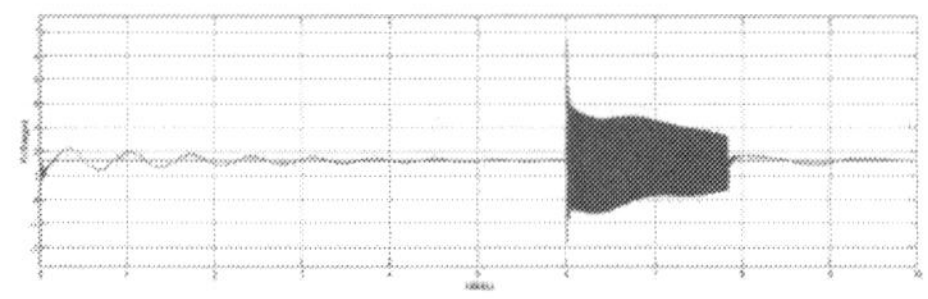

(a) Electromagnetic Power Pe

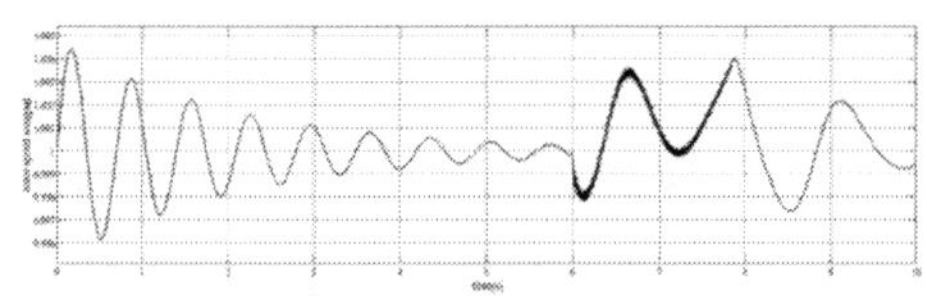

(b) Rotational Speed wm

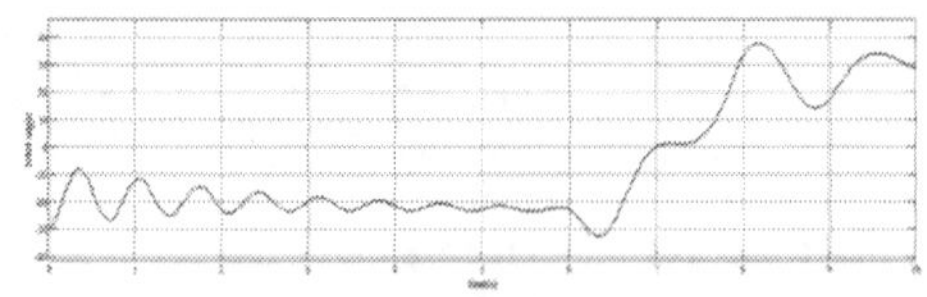

(c) Power Angle δ

Fig.10 Parameters of the Generator at 110 / 60 s

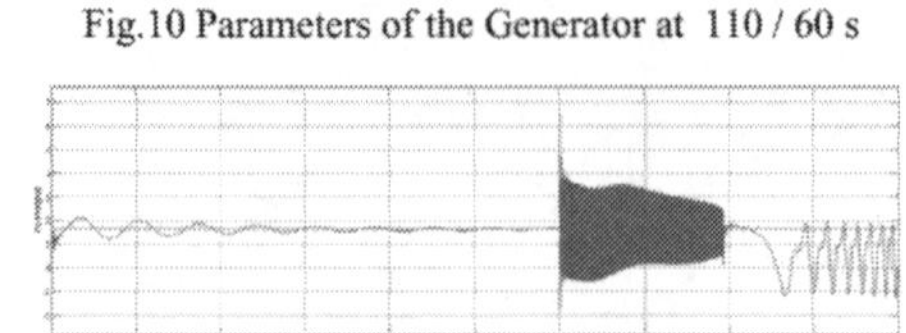

(a) Electromagnetic Power Pe

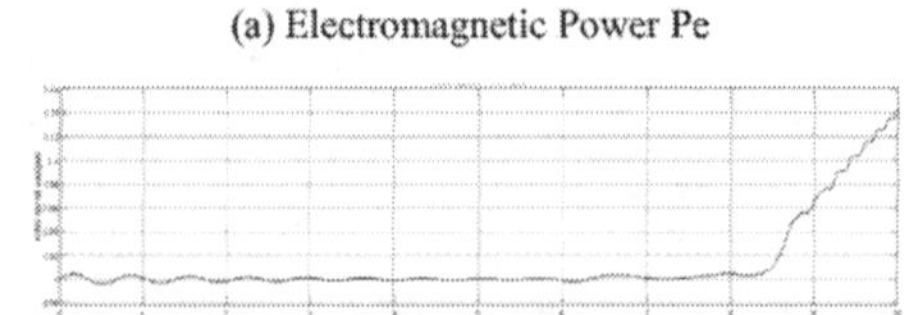

(b) Rotational Speed wm

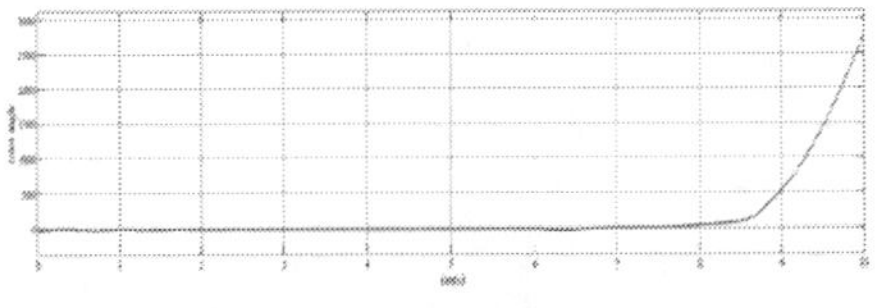

(c) Power Angle δ

Fig.11 Parameters of the Generator at 115 / 60 s

When the fault is removed at 110/60 seconds, the electromagnetic power of the generator is stable at around 1.632, the speed is stable at about 1, and the power angle tends to be constant at 20°, which shows that the system can remain stable after removing the fault. However, when the fault is removed at 115/60 seconds, a situation similar to that when the three-phase short circuit is removed at 16/60 seconds occurs, and the system is unstable. Other parameters at time t_1 and t_2 are shown in Tab. 6.

Tab.6 Parameters of the System at 15 / 60s and 16 / 60s(pu)

time (s)	*Vabc B1*	*Vabc B2*	*Vabc B4*	*Vabc B5*	*Vabc B7*	*P1*	*P2*
110 / 60	0.075 ∠-7.3°	0.068 ∠46.3°	1.024 ∠33.8°	0.001 ∠-25.6°	0.943 ∠65.6°	Amino 0	-0.001-j0.001
115 / 60		0.052 ∠-26.1°	1.020 ∠26.1°	0.001 ∠-47.7°	0.713 ∠17.0°	Amino 0	Amino 0

The analysis shows that the system parameters of a single-phase earth fault in critical conditions are similar to those of a three-phase fault. When the fault is removed at 110/60 s, the system is in a steady state, but when the fault is removed at 115/60 s, the system is unstable. When the resection time was 25/60 s, 50/60 s, 70/60 s, 90/60 s, 120/60 s, 100/60 s, 110/60 s, and 115/60 s, the critical clearing time was 1.875 seconds as shown in the formula (2).

$$t_{cm} = \frac{t_1 + t_2}{2} = \frac{110/60 + 115/60}{2} = 1.875 \quad (2)$$

By comparison, the critical clearing time of a single-phase earth fault is much longer than that of a three-phase short circuit, which shows that the fault bearing capacity of a three-phase short circuit is weaker and the requirements for relay protection devices are higher [4]. Therefore, it is necessary to further study how to improve the critical clearing time of three-phase short circuits.

4. Measures to Improve the Transient Stability of the System

The measures to improve the transient stability of the system are adding a controllable series capacitor compensation device and reducing the mechanical power output by the prime mover. The former is realized by increasing the capacitance of the line, that is, the capacitance of the four lines is slightly increased to 3e-8F. The latter is realized by modifying the mechanical power Pm of the prime mover, that is, adjusting the value of Pm from 1.632 to 1.55.
Then, simulations of the three-phase short circuit and single-phase earth fault are carried out for these methods, so as to obtain the corresponding critical clearing time. The recorded data are shown in Tab. 7.

Tab.7 Contrast Data (s)

Type / Fault	Original System	System After Series Capacitance	System After Pm Reduction
Phase A Earth Fault	1.875	3.083	2.417
Three-phase Short Circuit	0.258	0.375	0.417

It can be seen from the table that using the series capacitor compensation method, the critical clearing time is increased by 64.43% when a *Phase A* earth fault occurs, and by 45.35% when a three-phase short circuit occurs. In addition, by reducing the mechanical power of the prime mover, the critical clearing time is increased by 28.91% when a *Phase A* earth fault occurs, and by 61.63% when a three-phase short circuit occurs. Thus, these two methods are effective in improving the transient stability of the system. The specific choice should consider factors such as cost and convenience according to the actual situation [5].

5. Conclusion and Prospect

In this paper, the transient stability simulation model of the power system is constructed to simulate single-phase short circuits and three-phase short circuits. Besides, the changes in mechanical power, critical clearing angle, and other parameters of the system are studied. It is proven that the three-phase short circuit is the most serious fault, and the system will remain stable if the fault is removed before the critical clearing time corresponding to the critical clearing angle.
In addition, two methods improving the transient stability of the power system are simulated, which are reducing the mechanical power output of the prime mover and adding a controllable series capacitor compensation device.The correctness and effectiveness of these two methods are verified.
The conclusion is based on the ideal data, which can be used to further explore their application through the energy function method [6], artificial intelligence method [6], and so on in complex practical situations. With the universal application of artificial intelligence and big data analysis in the electrical field, the transient stability prediction method based on artificial intelligence came into being [6], which further promoted the data-driven artificial intelligence method and electric big data, deserving further in-depth study.

https://doi.org/10.1051/matecconf/202338201018

References

1. Fang, W. L., Li, J. H. & Wang, J. X. (2016). Transient analysis of electric power system. (fourth edition). Beijing: China Electric Power Press.
2. Sheng, N. J., Zhang, W. J., Peng, Z. Y., Chen, G., Cai, H. & Xu, Z. (2020). Research on the controllable series compensation device to improve the power flow problem of key transmission line. High Voltage Apparatus, 56(1), 148-154.
3. Xu, J. L. (2019). Research on dichotomy. New Century Intelligence, (77), 22-23.
4. Hu, K. Y. (2012). Research on relay protection device and requirements of power system. Gold Mountain, (01).
5. Chinnusamy, Emayavaramban, Amudha, Balachander, Ramkumar, M., Siva. (2019). Transient stability improvement in power system with smes and battery energy storage system. Mathematical & Computational Forestry & Natural Resource Sciences, 1, 132-143.
6. Feng, Z. K. (2021). Research on prediction method of power system transient stability based on deep learning. Beijing: Master Dissertation of Beijing Jiaotong University.

https://doi.org/10.1051/matecconf/202338201019

Research Progress of High Entropy Materials as Water Electrolysis Catalysts in Recent Years

Baizhen Wu*

School of Materials, Northwestern Polytechnical University, Xi'an 710129, China.

Abstract: Hydrogen energy has drawn considerable attention in recent years, among which the efficient production of hydrogen energy from water electrolysis has become a hot research topic in this field. The oxygen evolution reaction (OER) and the hydrogen evolution reaction (HER) in water electrolysis are the core of the catalytic efficiency. High entropy materials have excellent combinatorial properties due to their "cocktail effect", high entropy effect, slow diffusion effect, and lattice distortion effect. They have outstanding catalytic effect in the process of water electrolysis, and thus become a hot research topic in recent years. This review combines the two hot topics by introducing the up-to-date research progress of high entropy materials in catalyzing water electrolysis. First, evaluation indexes of electrocatalytic performance are explained. Second, this review demonstrates specific materials from the classification of high entropy alloys, high entropy oxides and nitrided high entropy materials, which includes the structure of the materials and their specific catalytic performance. Finally, it summarizes the research progress of high entropy materials in water electrolysis catalysis as well as gives an outlook on the future design direction of the materials.

Key words: high entropy materials; water electrolysis; electrochemical catalysts; hydrogen energy.

1. Introduction

The concept of high entropy materials was introduced in 1995[1], but it was not until after 2010 that high entropy materials entered a high-speed development phase when the exploration and applications about it became more and more intensive. High entropy materials are solid solutions composed of five or more principal elements, in which the elemental molar ratios are equal or approximately equal. They can be categorized as high entropy alloys[2], high entropy oxides[3], nitrided high entropy materials[4], etc. Due to the tunability of their principal elements, high entropy materials can use the properties of different main elements to adjust their properties, thus possessing high toughness[5], corrosion resistance[6], fatigue resistance[7], etc. This phenomenon is often referred to as the "cocktail effect"[8]. In addition, high entropy materials also have high entropy effect in thermodynamics[9], slow diffusion effect in kinetics[10], and lattice distortion effect in structure[11].

The efficient production of hydrogen energy[12] can mitigate the problems of depletion of traditional energy sources and environmental pollution, among which the method of water electrolysis has been attracting much attention. The half reaction of hydrogen evolution is the main source of hydrogen. How to improve the efficiency of hydrogen production, while focusing on the half reaction of hydrogen evolution, should also focus on the other side of the oxygen evolution half reaction[13]. In the past few years, research on high entropy materials in the direction of OER as well as HER catalysis[14] has been a hot topic, and these studies are very important for the goal of efficient preparation of hydrogen energy. Conventional non-noble metal electrode materials, such as Fe-Ni nanoparticles[15], Co-Cu alloys[16], and $TiO2$[17] have high overpotentials[18] required for electrolysis and other indicators such as Tafel slope[19] and catalytic stability[20] are not satisfactory as well. In contrast, noble metal materials, such as Ru and Ir[21,22], possess better catalytic properties but do not bring excessive economic effects due to their high cost. To solve the above problems, elements in high-entropy materials can be properly adjusted to achieve good catalytic performance using combinations of non-noble or noble metals, which possess lower overpotential, smaller Tafel slope, and more durable catalytic activity, resulting in more efficient and economical catalysis. It is worth noting that different surface morphologies of high entropy materials also affect their applications. For example, there is a significant difference in catalytic effect between the powder form[23] and the nanofiber form[24]. This paper compiles the research results of high entropy materials as catalysts for water electrolysis in recent years, analyzes and prospects how to achieve better electrocatalytic performance of high entropy materials, with the hope of bringing more inspiration for the research in this direction.

* Corresponding author: matt0820@mail.nwpu.edu.cn

2. Evaluation indexes

2.1 Overpotential

Overpotential is the potential difference of electrodes, also called superpotential. It is the difference between the non-equilibrium potential and the equilibrium potential of the electrode reaction. The electrochemical measurements mentioned in this review are performed on an electrochemical workstation using a typical three-electrode system with the sample electrode, Hg/HgO electrode or Ag/AgCl electrode and graphite rod as the working electrode, reference electrode and counter electrode, respectively. The reaction environment is a 1.0 M KOH solution at room temperature. During the test, linear scanning voltammetry (LSV) is used to apply a linear potential between the counter electrode and the working electrode. The commonly used potential scan rate is between 0.001 V s^{-1} and 0.1 V s^{-1}.

In order to disregard the potential difference between electrolytes and make the experimental processing easier, the result needs calibrating. The calibrated potential is labeled as "vs. RHE" in the abscissa. In this review, the potential at a current density of 10 mA cm^{-2} is taken as the overpotential of a material. The smaller the value of the overpotential, the better the catalytic performance of the material.

2.2 Tafel slope

The Tafel slope is obtained from the LSV curve. The ordinate of Tafel plot (Potential (V vs. RHE)) is the abscissa of LSV plot (Potential (V vs. RHE)), and the abscissa of Tafel plot (log(j/mA cm^{-2})) is the absolute value of the ordinate of LSV plot (Current density (mA cm^{-2})) transformed by taking the logarithm of the base of ten. A straight line is obtained after the transformation. The Tafel fitting equation can be obtained as follows:

$$\eta = a + b\log(j) \quad (2.1)$$

where η is the overpotential, j is the corresponding current density, a is constant and b is the Tafel slope in mV dec^{-1}. The Tafel slope indicates the change in overpotential requires for a 10-fold change in current. It is easy to see that the smaller the Tafel slope, the smaller the voltage rise for the same current change, and thus the better the catalytic performance of the material.

2.3 Stability

Catalytic stability is also an important indicator of the electrocatalytic performance of a material. There are generally chronopotential, chronocurrent, and cyclic voltammetry (CV) methods. In the chronopotential method , the U-t curve under long time catalytic operation is measured by keeping the current constant and observing the change of potential during the operation. If the change is small, the catalytic activity of the material is well maintained. The same is true for the chronocurrent method. Cyclic voltammetry method is to treat the material with successive CV scanning and measure the change of linear polarization curve before and after the treatment. If the discrepancy of the two curves is minor or no discrepancy, the catalytic activity of the material is considered to be well maintained and thus has high stability.

3. Research progress in high entropy materials

High entropy materials have excellent OER and HER catalytic properties. In this part, we will first introduce the preparation methods of different materials in recent years. Next, evaluation indexes such as the overpotential, Tafel slope, catalytic stability, and charge transfer resistance (R_{ct}) are selected to evaluate the catalytic properties of the materials. Finally, summary about their excellent structural characteristics is given.

3.1 High entropy alloys

In 2020, Tang et al.[25] prepared a porous CoCrFeNiMo high-entropy alloy by microwave sintering method. The research team used metal powder as the raw material and Mg powder as the space holder. The powder was pressed into discs and then went through microwave sintering to obtain CoCrFeNiMo. By analyzing the diffraction peaks of XRD patterns (as in Fig. 1a), it was found that the CoCrFeNiMo alloy has a face-centered cubic structure and most of the Mg has evaporated. Under transmission electron microscope (as in Fig. 1b), it was observed that a large number of stacking faults and twin crystal structures exist, which provide enormous active sites for electrocatalytic OER. Experimental groups were set up with Mg content as the variable, and x in CoCrFeNiMo-xMg were 0, 10, and 20, respectively. After LSV scanning (as in Figure 1c, d), the overpotentials of CoCrFeNiMo-20Mg and CoCrFeNiMo-0Mg at a current density of 10 mA cm^{-2} were 220 mV and 270 mV, and the Tafel slopes were 65.7 and 59.0 mV dec^{-1}, respectively. By comparing the polarization curves before and after CV (as in Fig. 1e), it was found that the discrepancy was minor, confirming the high stability of the electrocatalysts.

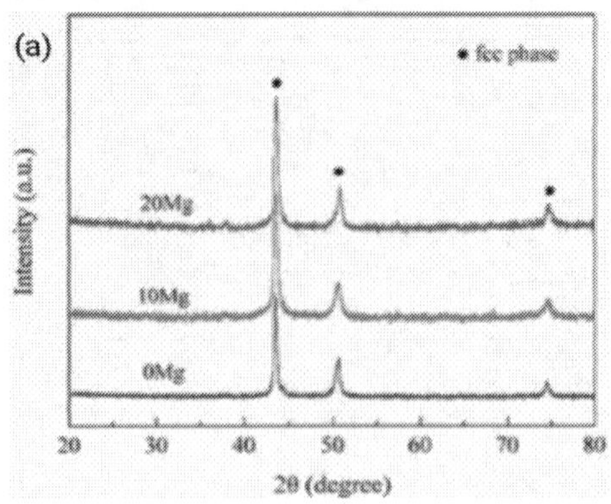

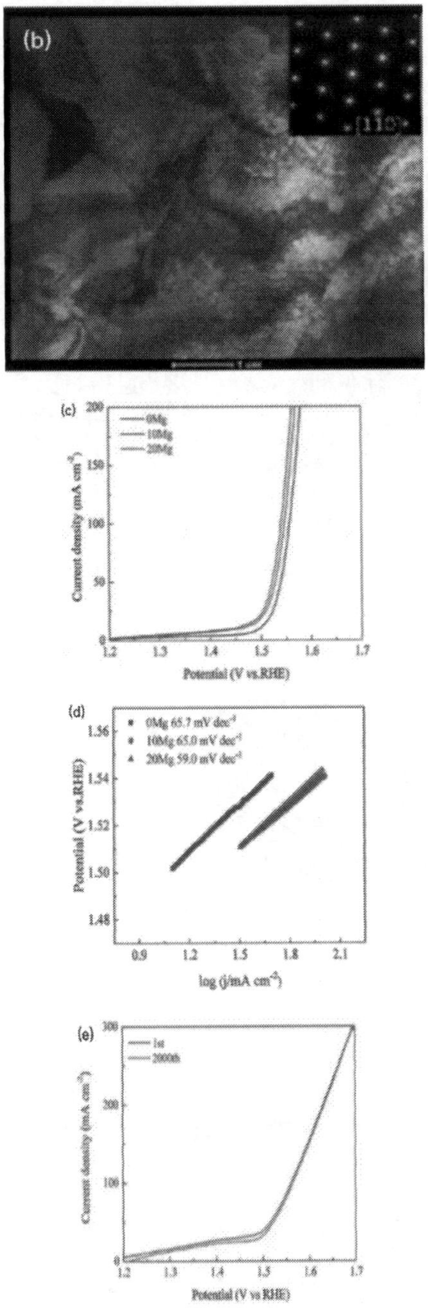

Fig. 1 (a) XRD patterns of CoCrFeNiMo-xMg (x = 0, 10, 20); (b) TEM image of CoCrFeNiMo; (c) LSV curves; (d) Tafel plots; (e) LSV curves before and after 2000 cycles.

Unlike Tang who directly used metal powder as raw materials, Jian et al.[26] used metal oxide powder to make precursors and made them into discs by spark plasma sintering (SPS). They obtained the high entropy alloy $Fe_{0.5}CoNiCuZn_x$ powder by electrolysis using the precursors as cathodes and Na_2CO_3 and K_2CO_3 as electrolytes. After thermodynamic analysis, the theoretical decomposition voltages of Na_2CO_3 and K_2CO_3 are higher than all oxides, allowing electrolysis of oxide mixtures with the electrolytes keeping stable. By adjusting the composition of the oxide precursors, they could change the Zn content in the HEA. After XRD analysis (as in Fig. 2a), it was found that with the increase of Zn content, the HEA grain size decreased, the crystal planar spacing increased, and thus improved OER catalytic performance. At a current density of 10 mA cm^{-2}, the overpotential of HEA($Zn_{0.8}$) was 340 mV, which was lower than that of HEA(Zn_0) (380 mV) and commercial IrO_2/Ta_2O_5-Ti (360 mV) (as in Fig. 2b). The Tafel slope decreased with increasing Zn to 84 mV dec^{-1} (HEA(Zn_0)), 58 mV dec^{-1} (HEA($Zn_{0.2}$)), 51 mV dec^{-1} (HEA($Zn_{0.5}$)) and 48 mV dec^{-1} (HEA($Zn_{0.8}$)), respectively (as in Fig. 2c).

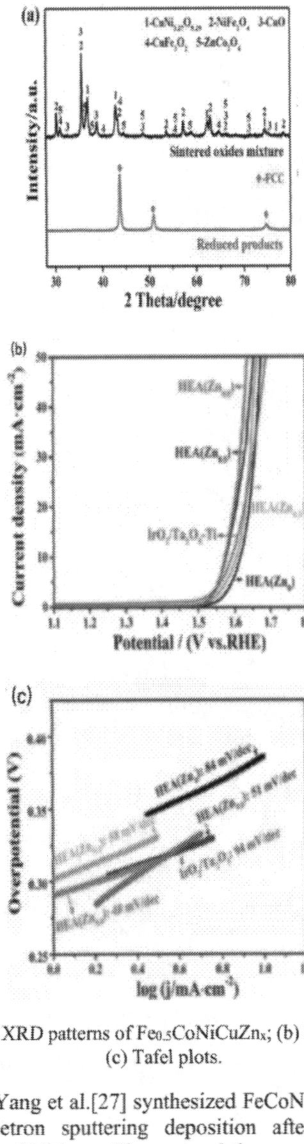

Fig. 2 (a) XRD patterns of $Fe_{0.5}CoNiCuZn_x$; (b) LSV curves; (c) Tafel plots.

In 2022, Yang et al.[27] synthesized FeCoNiCrMn films by magnetron sputtering deposition after ultrasonic treatment of Ni base. They named the samples as NF5, NF15, and NF30 according to the deposition time of 5 min,

15 min, and 30 min. Each sample was then self-reconstructed by CV scanning, and the treated samples were named as NF5CV, NF15CV and NF30CV. After electrocatalytic test (as in Fig. 3a), the overpotentials of NF5CV, NF15CV and NF30CV were 291mv, 282mv and 293mv respectively. The overpotential decreased with the increase of deposition time, probably because the film peeled off under the internal stress with the increase of deposition time. The Tafel slope of NF15CV was 64.3 mVdec^{-1} (as in Fig. 3b), which indicated a good electron transfer rate. After the stability test by chronopotential method for 16 h (as in Fig. 3c), the electronic structure of the NF15CV surface was changed and the overpotential increased only 0.15% at a current density of 10 mA cm^{-2}, showing an excellent electrochemical stability.

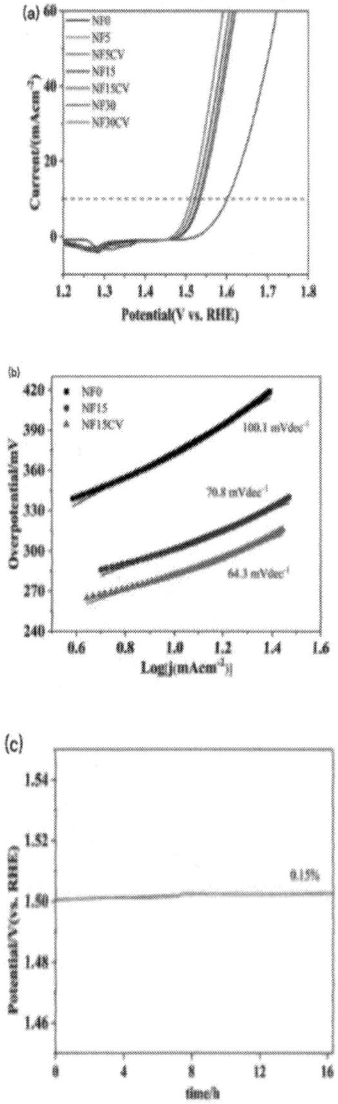

Fig. 3 (a) LSV curves for FeCoNiCrMn; (b) Tafel plots; (c) Chronopotentiometric stability study for OER of NF15CV at a current density of 10 mA cm^{-2}.

3.2 High entropy oxides

In 2021, Sun et al.[28] used the citric acid-nitrate method. They first compressed the nitrates of Cr, Mn, Fe, Co, Ni into discs and then made spinel structured high entropy oxide (S-HEO) $(Cr_{0.2}Mn_{0.2}Fe_{0.2}Co_{0.2}Ni_{0.2})_3O_4$ by heating at 1100°C. The atoms in S-HEO are able to flexibly occupy the point positions in the tetrahedral and octahedral structure (as in Fig. 4a), so it can be used as a conversion anode material for lithium-sodium ion batteries, where the reduction peak of the anode is elevated during the CV process. To measure its accurate electrocatalytic performance, the team performed self-reconstruction for S-HEO by CV scanning and found that element Fe showed the major catalytic performance during OER. S-HEO contains abundant cations, which can modulate the electronic structure and provide a large number of intermediates for OER. Thus its OER performance is much better than that of unitary or binary materials. After LSV scanning (as in Fig. 4b, c, d), S-HEO has a low overpotential of 322 mV at a current density of 10 mA cm^{-2} and its Tafel slope is 54.5 mV dec^{-1}. It still retains 89.6% of current density after 20 h operation at 1.5 V, showing excellent OER catalytic performance.

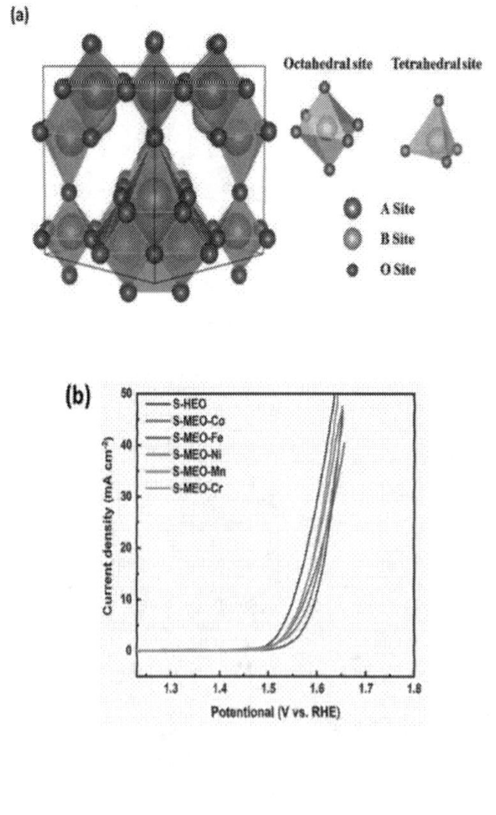

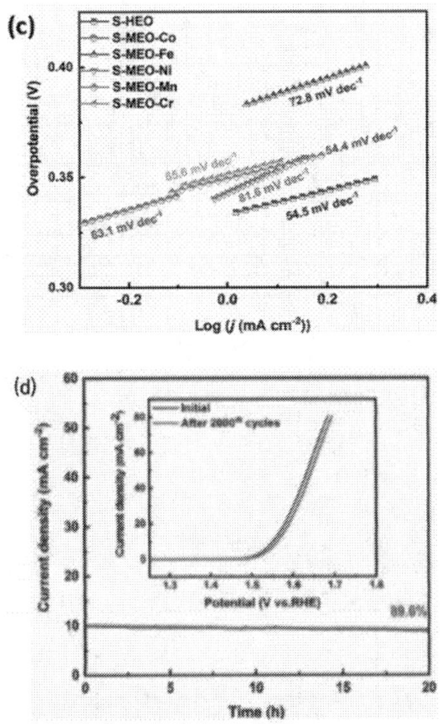

Fig. 4 (a) Tetrahedral and octahedral structures; (b) LSV curves of S-HEO; (c) Tafel plots; (d) i-t curves and polarization curves.

Compared with studying the OER catalytic performance of a specific material, Kim et al.[29] explored the influence of specific elements on the electrocatalytic performance of high-entropy oxide $(Mg_{0.2}Fe_{0.2}Co_{0.2}Ni_{0.2}Cu_{0.2})O$. The research team powdered the metal nitrate precursor and calcined it at 1000 °C. They also prepared medium entropy oxides (MEOs) by removing one of the elements for comparison. Through XRD analysis (as in Fig. 5a), it was found that both HEO and MEOs have a rock-salt structure. As the concentration of Cu^{2+} increased, the local environment of Cu^{2+} would evolve from octahedral to rhombohedral, resulting in local structural distortion. According to the XPS spectrum of O 1s (as in Fig. 5b), the proportion of trivalent cations (Me(III)) in MEO(w/o Cu) was larger than that of divalent cations (Me(II)), This phenomenon meant there were more defect sites with low oxygen coordination numbers and oxygen vacancies in MEO(w/o Cu), making it more conducive to OER. According to the LSV scanning and the Tafel plots (as in Fig. 5c, d), the overpotentials of MEO (w/o Cu) and HEO at a current density of 10mA cm^{-2} were 454mV and 1000mV, respectively, and the Tafel slopes were 60mV dec^{-1} and 182mV dec^{-1}. It might be that the Cu^{2+} hindered the transformation of Fe^{2+} and Co^{2+} from divalent to trivalent, thus causing its OER performance to decline.

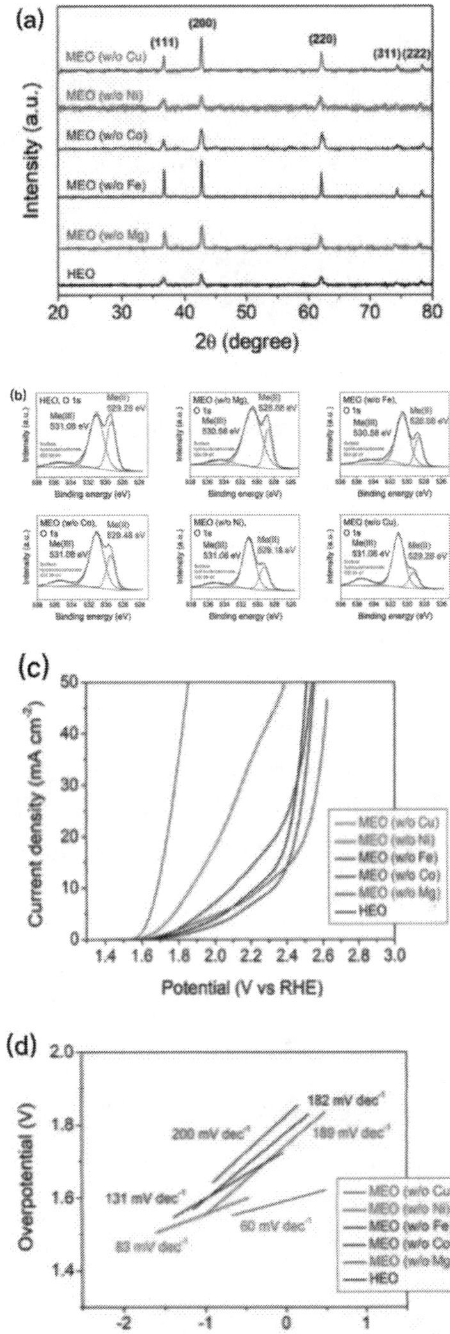

Fig. 5 (a) XRD patters of HEO and MEO; (b) XPS spectrum of O 1s; (c) LSV curves; (d) Tafel plots.

Similarly, Thi et al.[30] also explored the influence of different elements on the electrocatalytic performance of high-entropy oxides. The team dissolved lanthanum nitrate and transition metal sulfides in distilled water and prepared $La(CrMnFeCo_2Ni)O_3$ (labeled as L5M). They doubled the metal sulfide concentration to obtain the experimental groups L5M2Ni, L5M2Co, L5M2Fe, L5M2Mn and L5M2Cr. According to XRD analysis (as in Fig. 6a), all the samples showed an orthorhombic perovskite structure, that is, ABO_3 structure, in which A is a rare earth element or alkaline earth element, and B is composed of transition metal elements. After the electrocatalytic performance test (as in Fig. 6b, c), L5M2Co showed the best OER catalytic performance, with the overpotential of only 325mV at a current density of 10mA cm^{-2}, and the Tafel slope of only 51.2mV dec^{-1}. According to Fig. 6d, as the ratio of Co^{3+}/Co^{2+} increased, the overall OER performance of all samples became better. The octahedral structure of Co in perovskite can absorb electrons from water molecules. Fe is adjacent to Co in the periodic table, thus Co^{3+} can be well activated due to charge effects and spin effects. It can be seen from the Fig. 6d that L5M2Ni also showed good OER performance, but its Co^{3+}/Co^{2+} ratio is not high. This may be explained by the high amount of Mn^{3+} and Mn^{4+} in L5M2Ni. According to related research[31], samples consisting of $Mn^{3.5+}$ to $Mn^{3.7+}$ own excellent OER performance.

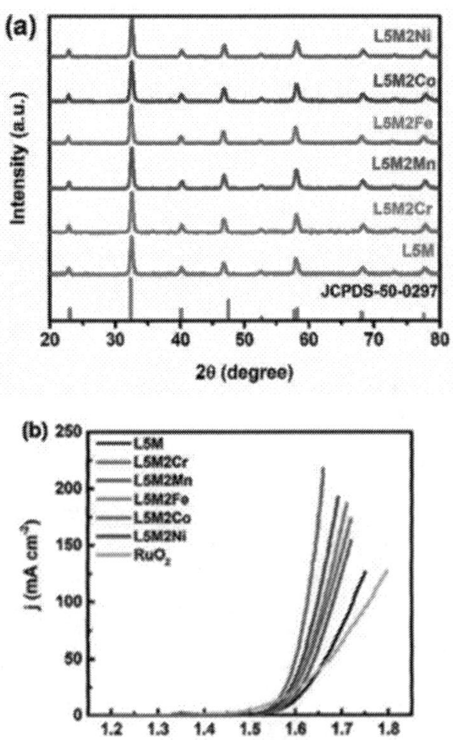

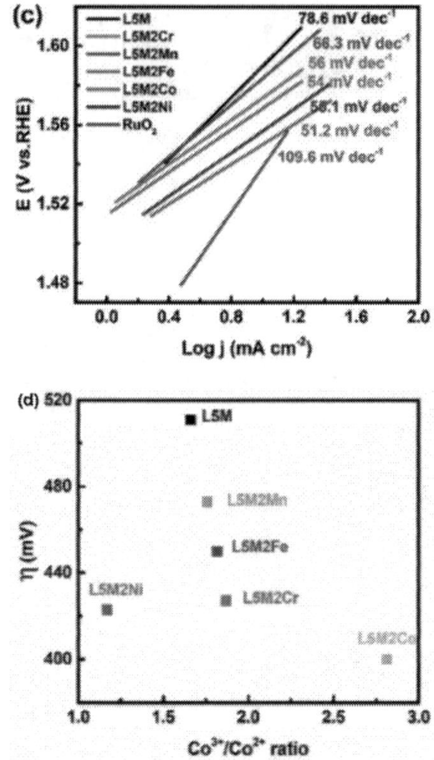

Fig. 6 (a) XRD patterns of L5M, L5M2Ni, L5M2Co, L5M2Fe, L5M2Mn and L5M2Cr; (b) LSV curves; (c) Tafel plots; (d) At a current density of 10mA cm^{-2}, the overpotential dependence versus Co^{3+}/Co^{2+} ratio.

3.3 Nitrided high entropy materials

In 2021, Liu et al.[32] prepared FeCoNiCuMnN/CC, FeCoNiN/CC, FeCoNiCuN/CC and FeCoNiMnN/CC by using metal nitrates as raw materials ad calcining them on carbon cloth in NH_3 atmosphere. Through scanning electron microscope observation (as in Fig. 7a, b), the morphology of nanowires and nanosheets in the sample was clearly visible. These nanostructures had rough surfaces and abundant curving sites, which could well improve the electrocatalytic performance. Through LSV scanning (as in Fig. 7c, d), the team found FeCoNiN/CC had the best electrocatalytic performance, with an overpotential of only 220mV at a current density of 10mA cm^{-2}, while FeCoNiMnN/CC had the smallest Tafel slope of 113mV dec^{-1}. The experiment also explored the effect of different annealing temperatures on the catalysis, and found that the HER catalytic performance was the best when the annealing temperature was 400 ℃ . The overpotential was only 182mV, the Tafel slope was 113mV dec^{-1}, and it had excellent durability.

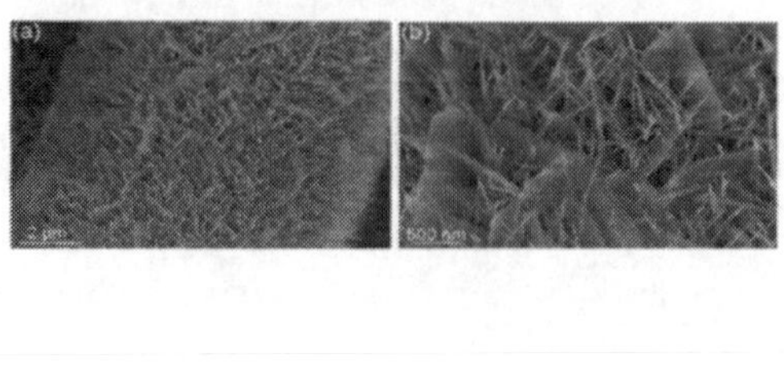

https://doi.org/10.1051/matecconf/202338201019

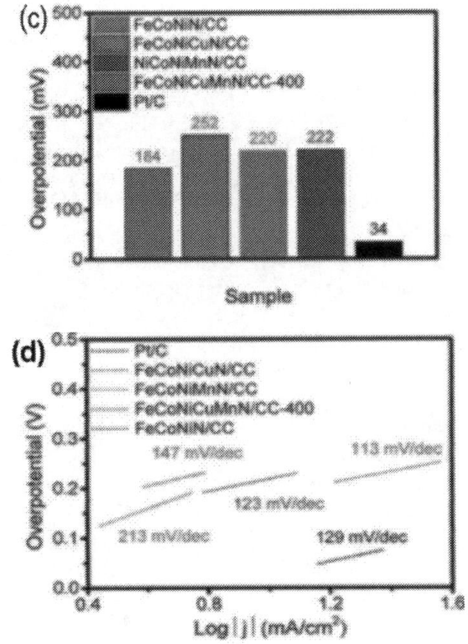

Fig. 7 (a and b) SEM images of FeCoNiCuMnN/CC; (c) Overpotentials at a current density of 10 mA cm^{-2}; (d) Tafel plots.

Compared with directly manufacturing high-entropy nitrides, Gao et al.[33] found another way to nitride high-entropy alloys to improve their performance. The team first prepared CoCrFeNiMn coating (HEA) on the surface of stainless steel by high-velocity oxygen fuel (HVOF) spraying method, and then performed plasma nitriding treatment on the sample to obtain N-HEA. After XRD analysis (as in Fig. 8a), it was found that there was no significant difference between the diffraction peaks of HEA and N-HEA, which confirmed that no other nitrides were formed during the experiment but only the alloy surface was remodeled. After nitriding treatment, more nitrogen atoms entered the interstices of the metal oxide, thereby improving the electrical conductivity and electron transfer rate. This treatment not only increased the surface active area, but also enhanced its intrinsic catalytic performance, which was beneficial to the OH^- absorption. After the OER catalytic performance test (as in Fig. 8b, c), the overpotential of N-HEA was 307mV at a current density of 10mA cm^{-2}, which was lower than that of HEA at 363mV. Its Tafel slope was 32mV dec^{-1}, which appeared significantly lower than that of HEA (61.4mV dec^{-1}). Through the Nyquist plots obtained by EIS analysis (as in Fig. 8d), the charge transfer resistance (R_{ct}) of N-HEA was only 3.98Ω, which was much lower than that of HEA at 36.27Ω, confirming the improvement of electron transfer rate after nitriding treatment.

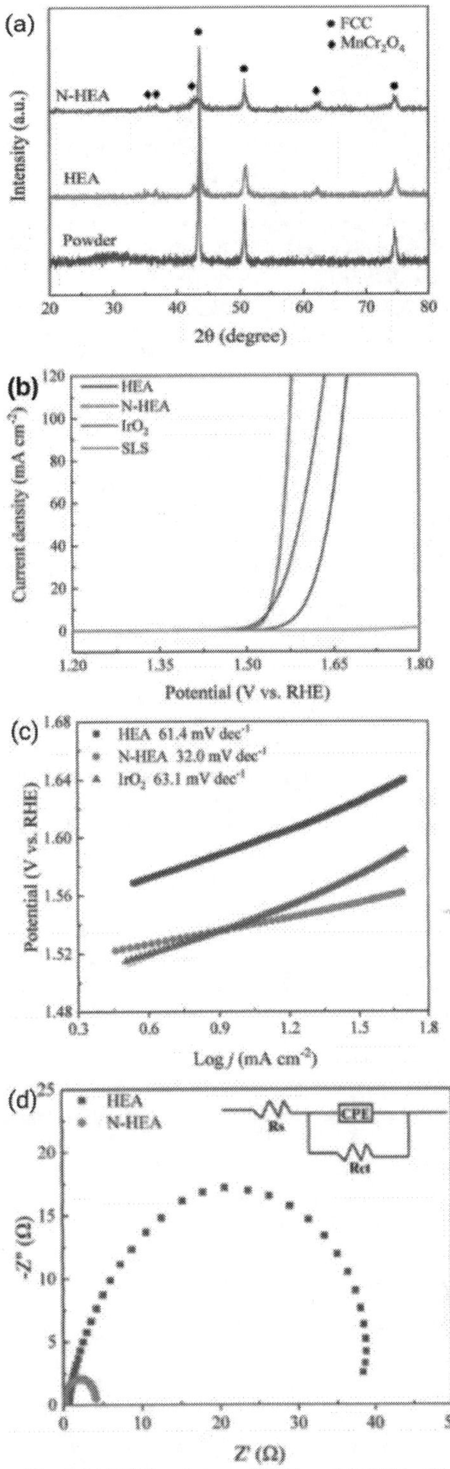

Fig. 8 (a) XRD patterns of HEA and N-HEA; (b) LSV curves; (c) Tafel plots; (d) Nyquist plots.

4. Conclusion

The efficient preparation of hydrogen energy can well alleviate the problems of traditional fossil energy depletion and environmental pollution. The development of high entropy materials provides more solutions for hydrogen production through water electrolysis. In this review, the preparation methods of high entropy alloys, high entropy oxides, nitrided high entropy materials and the catalytic performance of water electrolysis in recent years are summarized. Overpotential, Tafel slope, stability, charge transfer resistance (R_{ct}) and other indicators are chosen to evaluate the electrocatalytic performance of materials. Many of them show outstanding OER/HER catalytic performance, thus having very good application prospects as water electrolysis catalysts. Meanwhile, this paper also displays the structure, morphology, element composition and other characteristics of some materials with the hope of further exploring the underlying mechanism for their good OER/HER catalytic performance. These analysis can provide design ideas for future high entropy materials as catalysts for water electrolysis.

Although high entropy materials have developed rapidly in recent years, only a small part of this field has been excavated. Great challenges and opportunities still lie ahead. The future research of high entropy materials in this field can be considered from the following 3 aspects:

(a) Regulating element composition. Due to the "cocktail effect" of high entropy materials, different element combinations can well adjust the electronic structure, making the material exhibit strong "hydrophilicity". They can absorb more water molecules or hydroxide ions, thereby achieving better OER/ HER catalytic performance.

(b) Surface treatment. For a high entropy material with given elements, its electrocatalytic performance can still be improved through physical or chemical treatment, such as increasing defect sites in the material and nitriding the material.

(c) In-depth study of high entropy oxides. Judging from this review, the catalytic performance of high entropy oxides is generally better than that of high entropy alloys. This indicates that the future research on catalytic materials for water electrolysis should be advanced towards high entropy oxides in order to reduce their production costs and realize their large-scale applications.

References

1. Ming-Hung Tsaia [*] and Jien-Wei Yehb[*], High-Entropy Alloys: A Critical Review [J], Materials Research Letters, 2014, 2(3): 107-123.
2. Easo P. George [1,2*], Dierk Raabe[3] and Robert O. Ritchie [4,5], High-entropy alloys [J], Nature Reviews Materials, 2019, 4(8): 515-534.
3. Abhishek Sarkar [a,b,*], Ben Breitung [b,*] and Horst Hahn[a,b,c,d], High entropy oxides: The role of entropy, enthalpy and synergy [J], Scripta Materialia, 2020, 187: 43-48.
4. Olivia F. Dippo[1], Neda Mesgarzadeh[1], Tyler J. Harrington[1], Grant D. Schrader[2] and Kenneth S.Vecchio[1,2*], Bulk high-entropy nitrides and carbonitrides [J], Scientific Reports, 2020, 10.
5. L. Jiang [a], Y.P. Lu [b], M. Song [a], C. Lu [c,*], K. Sun [d], Z.Q. Cao [b], T.M. Wang [b], F. Gao [a,d], L.M. Wang [a,d,**], A promising $CoFeNi_2V_{0.5}Mo_{0.2}$ high entropy alloy with exceptional ductility [J], Scripta Materialia, 2019, 165: 128-133.
6. Ana Maria Zemanate [a], Alberto Moreira Jorge Júnior [b,c], Gisele Ferreira de Lima Andreani [a], Virginie Roche [c], Katia Regina Cardoso [a,*], Corrosion behavior of AlCoCrFeNix high entropy alloys [J], Electrochimica Acta, 2023, 441.
7. K. Liu[a], S.S. Nene[a], M. Frank[a], S. Sinha[a], R.S. Mishra[a,b,*], Extremely high fatigue resistance in an ultrafine grained high entropy alloy [J], Applied Materials Today, 2019, 15: 525-530.
8. Li Wang [a], Liuting Zhang [a,d,*], Xiong Lu [b], Fuying Wu [a,*], Xuan Sun [c], Hu Zhao [d], Qian Li [b,c,*], Surprising cocktail effect in high entropy alloys on catalyzing magnesium hydride for solid-state hydrogen storage [J], Chemical Engineering Journal, 2023, 465.
9. J. Ding [a], A. Inoue [a, b, c, d, *], Y. Han [a], F.L. Kong [b], S.L. Zhu [a], Z. Wang [e], E. Shalaan [c], F. Al-Marzouki [c], High entropy effect on structure and properties of (Fe,Co,Ni,Cr)-B amorphous alloys [J], Journal of Alloys and Compounds, 2017, 696: 345-352.
10. Witold Kucza [a, *], Juliusz Da̧browa [a, **], Grzegorz Cieslak [b], Katarzyna Berent [c], Tadeusz Kulik [b], Marek Danielewski [a], Studies of "sluggish diffusion" effect in Co-Cr-Fe-Mn-Ni, Co-Cr-Fe-Ni and Co-Fe-Mn-Ni high entropy alloys; determination of tracer diffusivities by combinatorial approach [J], Journal of Alloys and Compounds, 2018, 731: 920-928.
11. Chao-Chun Yen [a, b], Guan-Rong Huang [c], Yun-Cheng Tan [b], Han-Wen Yeh [b], Da-Ji Luo [a, d], Kang-Tien Hsieh [f], E-Wen Huang [b], Jien-Wei Yeh [a, e], Su-Jien Lin [a, e], Chun-Chieh Wang [d], Chin-Lung Kuo [f], Shou-Yi Chang [a, e, **] and Yu-Chieh Lo [b, e, *], Lattice distortion effect on elastic anisotropy of high entropy alloys [J], Journal of Alloys and Compounds, 2020, 818.
12. J.O. Abe [a,b,*], A.P.I. Popoola [a], E. Ajenifuja [a,b,c] and O.M. Popoola [b], Hydrogen energy, economy and storage: Review and recommendation [J], International Journal of Hydrogen Energy, 2019, 44(29): 15072-15086.
13. Xiaopeng Wang, Haoyin Zhong, Shibo Xi,[*] Wee Siang Vincent Lee,[*] and Junmin Xue[*], Understanding of Oxygen Redox in the Oxygen Evolution Reaction [J], Advanced Materials, 2022, 34(50).
14. Zhaoyu Wang [a], Junhua You [a,*], Yao Zhao [a], Ruyue Yao [a], Guangyi Liu [a], Jinlin Lu [b], Shiyong Zhao [c], Research progress on high entropy alloys and high entropy derivatives as OER catalysts [J], Journal of Environmental Chemical Engineering, 2023, 11(1).

15. Jahangeer Ahmed, Tansir Ahamad, and Saad M. AlShehri, Iron–Nickel Nanoparticles as Bifunctional Catalysts in Water Electrolysis [J], ChemElectroChem, 2017, 4(5): 1222-1226.
16. Hoyoung Kim, Hyanjoo Park, Seonhwa Oh and Soo-Kil Kim, Facile electrochemical preparation of nonprecious Co-Cu alloy catalysts for hydrogen production in proton exchange membrane water electrolysis [J], International Journal of Energy Research, 2020, 44(4): 2833-2844.
17. Petr Mazur, Jakub Polonsky, Martin Paidar, Karel Bouzek, Non-conductive TiO_2 as the anode catalyst support for PEM water electrolysis [J], International Journal of Hydrogen Energy, 2012, 37(17): 12081-12088.
18. Jordan E. Nutting, James B. Gerken, Alexios G. Stamoulis, David L. Bruns, and Shannon S. Stahl[*], "How Should I Think about Voltage? What Is Overpotential?": Establishing an Organic Chemistry Intuition for Electrochemistry [J], The Journal of Organic Chemistry, 2021, 86(22): 15875-15885.
19. Stephen Fletcher, Tafel slopes from first principles [J], Journal of Solid State Electrochemistry, 2009, 13(4): 537-549.
20. Feng Zeng [a,b] , Chalachew Mebrahtu [b] , Longfei Liao [b] , Anna Katharina Beine [c] , Regina Palkovits [b,c,*] , Stability and deactivation of OER electrocatalysts: A review [J], Journal of Energy Chemistry, 2022, 69: 301-329.
21. Tobias Reier,* Mehtap Oezaslan, and Peter Strasser, Electrocatalytic Oxygen Evolution Reaction (OER) on Ru, Ir, and Pt Catalysts: A Comparative Study of Nanoparticles and Bulk Materials [J], ACS Catalysis, 2012, 2(8): 1765-1772.
22. Yingnan Qin, Zuochao Wang, Wenhao Yu, Yingjun Sun, Dan Wang, Jianping Lai, Shaojun Guo, and Lei Wang, High Valence M-Incorporated PdCu Nanoparticles (M = Ir, Rh, Ru) for Water Electrolysis in Alkaline Solution [J], Nano Letters, 2021, 21(13): 5774-5781.
23. Paolo Veronesi[a,*], Elena Colombini[a] , Roberto Rosa[a] , Cristina Leonelli[a] , Marco Garuti[b] , Microwave processing of high entropy alloys: A powder metallurgy approach [J], Chemical Engineering and Processing: Process Intensification, 2017, 122: 397-403.
24. Wei Li[1,2] , Yanan Sun[1,2] , Li Ye[1] , Weijian Han[1] , Fenghua Chen[1] , Jianling Zhang[3] , Tong Zhao[1] , Preparation of high entropy nitride ceramic nanofibers from liquid precursor for CO_2 photocatalytic reduction [J], Journal of the American Ceramic Society, 2022, 105(6): 3729-3734.
25. J. Tang, J.L. Xu[*], Z.G. Ye[*], X.B. Li, J.M. Luo[*], Microwave sintered porous CoCrFeNiMo high entropy alloy as an efficient electrocatalyst for alkaline oxygen evolution reaction [J], Journal of Materials Science & Technology, 2021, 79: 171-177.
26. Peishu Yang [a] , Yongzhi Shi [a] , Tongjun Xia [a] , Zhenyu Jiang [a] , Xinyu Ren [a] , Lisha Liang [a] , Qingyi Shao [b,*] , Kaigui Zhu [a,*] , Novel self-supporting thin film electrodes of FeCoNiCrMn high entropy alloy for excellent oxygen evolution reaction [J], Journal of Alloys and Compounds, 2023, 938.
27. Jian Huang[a,b] , Peilin Wang[a,b] , Peng Li [a,b] , Huayi Yin[a,b] , Dihua Wang[a,b,*], Regulating electrolytic $Fe_{0.5}CoNiCuZn_x$ high entropy alloy electrodes for oxygen evolution reactions in alkaline solution [J], Journal of Materials Science & Technology, 2021, 93: 110-118.
28. Zheng Sun, Yongjie Zhao [*] , Chen Sun, Qing Ni, Chengzhi Wang, Haibo Jin, High entropy spinel-structure oxide for electrochemical application [J], Chemical Engineering Journal, 2022, 431(4).
29. Kyung-Hwan Kim, Yun-Hyuk Choi [*], Effect of constituent cations on the electrocatalytic oxygen evolution reaction in high-entropy oxide $(Mg_{0.2}Fe_{0.2}Co_{0.2}Ni_{0.2}Cu_{0.2})O$ [J], Journal of Electroanalytical Chemistry, 2022, 922.
30. Thi Xuyen Nguyen, Yi-Cheng Liao, Chia-Chun Lin, Yen-Hsun Su, and Jyh-Ming Ting[*], Advanced High Entropy Perovskite Oxide Electrocatalyst for Oxygen Evolution Reaction [J], Advanced Functional Materials, 2021, 31(27).
31. I. Zaharieva,[*a] D. Gonza´lez-Flores,[a] B. Asfari,[a] C. Pasquini,[a] M. R. Mohammadi,[a] K. Klingan,[a] I. Zizak,[b] S. Loos,[a] P. Chernev[a] and H. Dau[*a], Water oxidation catalysis – role of redox and structural dynamics in biological photosynthesis and inorganic manganese oxides [J], Energy & Environmental Science, 2016, 9(7): 2433-2443.
32. Chen Liu, Han Zhu, [*] Shuanglong Lu, Fang Duan and Mingliang Du [*], High entropy alloy nitrides with integrated nanowire/nanosheet architecture for efficient alkaline hydrogen evolution reactions [J], New Journal of Chemistry, 2021,45(47): 22255-22260.
33. G.J. Gao [a] , J.L. Xu [a,*] , J. Tang [a] , L.W. Zhang [b,**], Y.C. Ma [a] , J.M. Luo [a,***] , Plasma nitrided CoCrFeNiMn high entropy alloy coating as a self-supporting electrode for oxygen evolution reaction [J], Journal of Materials Research and Technology, 2023, 23: 5357-5367.

Highly dispersed Pt with CoFeP hollow nanocages for efficient oxygen evolution reaction

Jingyi Xie, Bin Dong

College of Chemistry and Chemical Engineering, State Key Laboratory of Heavy Oil Processing, China University of Petroleum (East China), Qingdao 266580, China.

Abstract. The strong binding energy of Pt with oxygen intermediates results in its low OER efficiency at the anode. Herein, Pt-doped CoFeP (Pt-CoFeP) material has been synthesized using CoFe-PBA as precursor through facile etching and phosphating processes. Interestingly, Pt-CoFeP with hollow cage framework exhibits large surface area after etching, which plays a domain-limiting role for the high dispersion of platinum atoms and facilitates the oxygen evolution reaction (OER). The target sample Pt-CoFeP only needs low overpotentials of 311.7 and 343.1 mV to deliver the current density of 10 and 50 mA cm-2, which is superior to commercial RuO2. The excellent activity can be attributed to large specific surface area in hollow framework and highly dispersed platinum atoms. And its great durability is evaluated through chronoamperometry measurement for 12 hours. This work provides a potential method to design electrocatalysts with complex structure loaded with highly dispersed Pt particles derived from Prussian blue analogues (PBAs) for water splitting.

Keywords: Prussian blue analogues, Phosphating, Platinum dispersion, oxygen evolution reaction.

1. Introduction

The practical application of renewable energy conversion is hindered by the sluggish kinetics of the oxygen evolution reaction (OER) [1]. Therefore, the design of low-cost and high activity OER catalysts is one of the key determinants to realize efficient energy conversion through water electrolysis [2]. Over the past decade, significant progress has been made in the exploration of novel OER electro-catalyst as replacements for expensive IrO_2 and RuO_2. For instance, lots of transition metal phosphides (TMPs) have attracted much attention from researchers, due to their good electrochemical stability. Chen et al. established that the utilization of nitrogen-doped graphene as a substrate for CoP nanosheet growth (CoP@NG) resulted in a lower overpotential of 354 mV at 10 mA cm^{-2} for OER, which is lower than that of bare CoP (395 mV) [3]. However, there is still potential for improvement in catalytic performance of CoP.

As a subclass of metal organic frameworks (MOFs), Prussian blue analogues (PBAs) possess bimetal centers, which offer a potential avenue for the formation of bimetallic phosphides. Moreover, the ordered structure of PBAs provides a possibility for designing complex and confined functional structures. For example, Xu et al. synthesized NiFeP using NiFe PBA as the precursor, which exhibited excellent oxygen evolution performance [4].

Inspired by the above considerations, we synthesized CoFeP with uniformly dispersed platinum atoms by using a CoFe PBA as the precursor and employing a combination of chemical etching and gas-phase phosphidation methods. The obtained CoFeP dopped with Pt (Pt-CoFeP) shows outstanding oxygen evolution activity, which may be enhanced by the hollow cage framework exposed numerous active site, as well as the optimized electron density between Co/Fe and Pt. Relative electrochemical experiments have been explored. Pt-CoFeP only needs a low overpotential of 311.7 mV to achieve the current density of 10 mA cm^{-2}, which is superior to commercial RuO_2.

2. Experimental Procedures

2.1 Materials Preparation

Chemicals: Potassium ferricyanide ($K_3[Fe(CN)_6]$, ≥ 99%), Cobalt (II) chloride hexahydrate ($CoCl_2 \cdot 6H_2O$, ≥ 99%), Trisodium citrate dihydrate ($C_6H_5Na_3O_7 \cdot 2H_2O$), Chloroplatinic acid hexahydrate ($H_2PtCl_6 \cdot 6H_2O$), L-ascorbic acid ($C_6H_8O_6$), Potassium hydroxide (KOH) were purchased from Sinopharm Chemical Reagent Co., Ltd..

Preparation of e-CoFe PBA: In a typic synthesis process, mixed solutions containing 60 mL of 1.2 mmol $K_3[Fe(CN)_6]$ in aqueous solution (Solution A) and 60 mL

 https://doi.org/10.1051/matecconf/202338201020

of 1.8 mmol $CoCl_2 \cdot 6H_2O$ and 2.7 mmol $Na_3C_6H_5O_7 \cdot 2H_2O$ in aqueous solution (Solution B) were stirred vigorously and aged at 25 °C for 24 h. The resulting CoFe PBA was collected via centrifugation, washed with deionized water, and dried at 60 °C. Then, e-CoFe PBA was obtained by etching in 1.0 M KOH solution.

Preparation of Pt-CoFeP: The e-CoFe PBA was mixed with 0.6 mL chloroplatinic acid and 1.2 mL L-ascorbic acid in 10 mL deionized water. After 4 h in an oven at 80 °C, Pt-CoFe PBA was obtained by centrifugation. After phosphating at 350 °C for 2 h, the final catalyst Pt-CoFeP was successfully synthesized.

2.2 Physical Characterization

The morphologies of all samples were characterized via SEM (FEI QuantaFEG 450, 20 kV). TEM was performed on Hitachi HT7700 system. Crystal structures were analyzed by XRD (Bruker D8) with a scanning rate of 5° /min. The valence states of elements were analyzed by XPS (ESCALAB 250Xi) which carried out on the spectrometer with an Cu Kα X-ray source.

2.3 Electrochemical measurements

Electrochemical measurements were performed on a Gamry Reference 3000 electrochemical workstation with a three-electrode system. The working, counter and reference electrode are the target catalysts, a platinum sheet and Hg/HgO electrode, repectively. The polarization curves were performed in 1 M KOH electrolyte with a scan rate of 5 mV s^{-1}. The double-layer capacitances (C_{dl}) were tested by cyclic voltammetry (CV), with scan rates ranging from 40 to 120 mV s^{-1}. Electrochemical impedance spectroscopy (EIS) measurements were conducted over a frequency range of 0.1 to 10^5 Hz with an amplitude of 5mV. Long-term stability was assessed through the chronopotentiometry test under 10 mA cm^{-2} for 12 h.

3. Results and Discussion

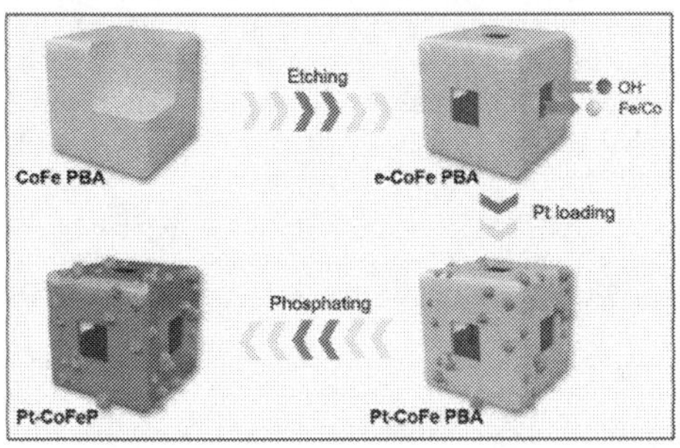

Figure 1: Schematic illustration of synthesis procedure for Pt-CoFeP hollow cage.

Hollow cage of Pt-CoFeP with open holes in each face-centered position are successfully fabricated through alkali etching, platinum (Pt) loading and phosphating processes (Figure 1). The potassium hydroxide is supposed to act as a cavity creator during the cube-to-cage conversion process via the site-selective chemical etching [5]. Furthermore, the unique morphology plays a positive role in the high dispersion of Pt atoms.

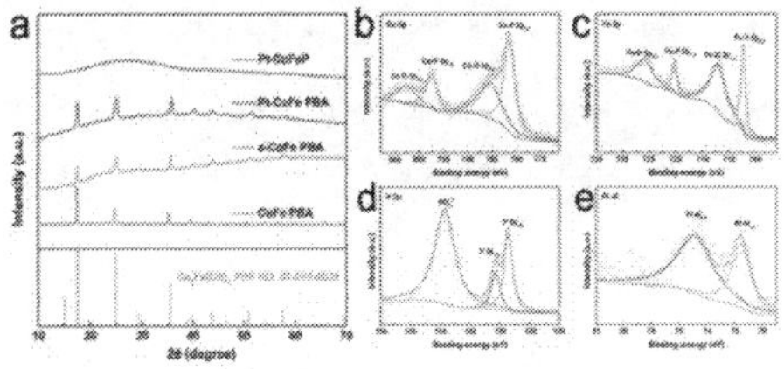

Figure 2: (a) XRD patterns of Pt-CoFeP and other contrast samples. XPS spectra of Pt-CoFeP cage: (b) Co 2p; (c) Fe 2p; (d) P 2p; (e) Pt 4f.

The crystalline phase structure of Pt-CoFeP and other relative samples are characterized by X-ray diffraction (XRD). As illustrated in Figure 2a, the diffraction peaks of CoFe PBA approximated at 17.513°, 24.865°, 35.452°, 43.788°, 51.009° and 57.554° belong to (200), (220), (400), (422), (440) and (620) crystal planes of $Co_2Fe(CN)_6$ (PDF No.01-075-0039), which is similar with that of e-CoFe PBA. Compared with the peaks of CoFe PBA and e-CoFe PBA, there exhibits a significant shift ,demonstrating that the introduction of Pt caused small lattice distortion [6]. However, no distinct dominating diffraction peaks can be observed, implying the amorphous charateristics of Pt-CoFeP [7].

To further explore the chemical structure of Pt-CoFeP, X-ray photoelectron spectroscopy (XPS) is carried out. As depicted in Figure 2b, two strong peaks at the binding energy of 781.6 and 797.5 eV are assigned to the Co $2p_{3/2}$ and Co $2p_{1/2}$ of Co-P, respectively. The peaks at 785.7 and 802.7 eV belong to the oxidized Co species [8]. Meanwhile, in the hign-resolution spectrum of Fe 2p (Figure 2c), peaks located at 707.3 and 720.2 eV are attributed to Fe $2p_{3/2}$ and Fe $2p_{1/2}$ of Fe-P, respectively. And the peaks at 712.3 and 725.7 eVcan be ascribed to the oxidized Fe species [9]. In addition, the peaks (Figure 2d) at 129.5 and 130.3 eV are assignable to P $2p_{3/2}$ and P $2p_{1/2}$, respectively. While the peak at 133.7 eV is attributed to the oxidized P species [10]. These results implying the formation of CoFeP. Furthermore, a pair of weak peaks at 71.6 (Pt $4f_{7/2}$) and 74.9(Pt $4f_{5/2}$) can be observed in Pt 4f, proving the existence of Pt. Notablely, the characteristic peak of Pt exhibits positive shift, which indicates the electron transfer between Pt and Co/Fe [11]. This phenomenon implys that the optimized electron structure of Pt doped CoFeP may be one of important factors to enhance catalytic activity.

Figure 3: SEM images. (a,b) CoFe PBA, (c,d) e-CoFe PBA, (e,f) Pt-CoFeP.

In order to reveal the structures of Pt-CoFeP and other contrast samples, scanning electron microscopy (SEM). As shown in Figure 3a and b, CoFe PBA are uniform dispersed nanocubes with the size of around 180-200 nm. While after alkali corrosion, the surface of e-CoFe PBA becomes rough, and significant holes can be observed on each face (Figure 3c, d). The large exposed surface area makes it possibel to provide more catalytic active sites, promoting the activity of traget catalyst. After loading Pt atoms and phosphating, Pt-CoFeP roufhly maintains the structure of nanocages.

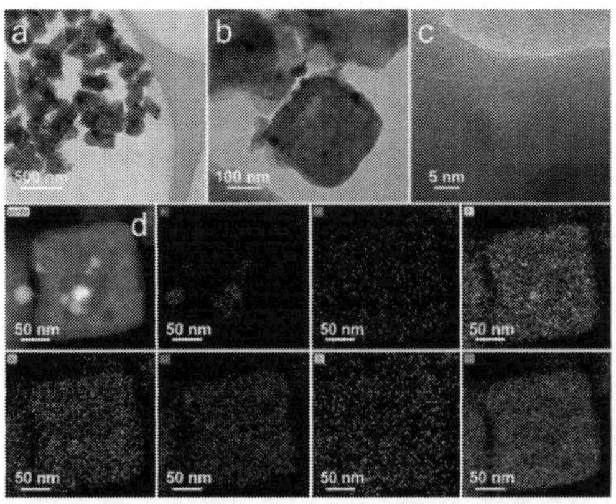

Figure 4: (a,b) TEM images; (c) HRTEM image; (d) HAADF-STEM image and corresponding elemental mapping of Pt-CoFeP.

The transmission electron microscope (TEM) and High-resolution TEM (HRTEM) are tested to further explore the compositions of Pt-CoFeP. As shown in Figure 4a and b, porous structure can be clearly observed. Notablely, the inside of Pt-CoFeP exhibits hollow framework, which can provide numerous catalytic active site for oxygen evolution process. The abundance of dispersed black particles observed in the image is indicative of the presence of platinum. However, no significant lattice fringle can be observed in the HRTEM image (Figure 4c), indicating a low degree of crystallinity in the Pt-CoFeP, which is consistent with the conclusion drawn from XRD analysis. The high-angle annular dark-field scanning transmission electron microscopic (HAADF-STEM) images and relative elemental mapping (Figure 4d) illustrates that Co, Fe, P, C, N and O are uniformly dispersed in Pt-CoFeP. On the other hand, Pt is mainly confined within the structure of nanocages, providing further evidence for its doping and confined growth.

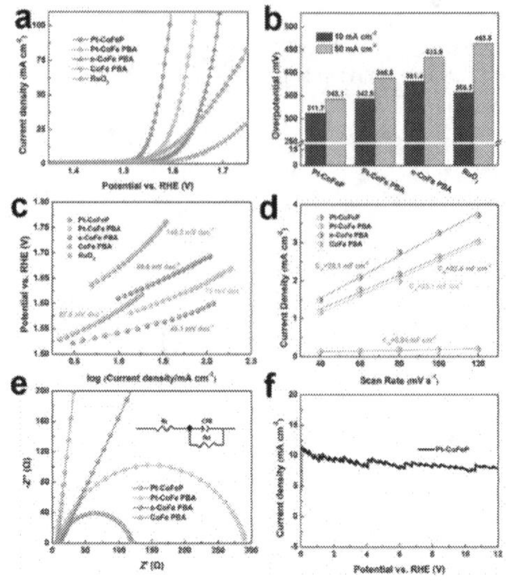

Figure 5: OER electrochemical properties of Pt-CoFeP and other contrast samples in 1 M KOH. (a) LSV curves; (b) overpotentials at 10 and 50 mA cm^{-2}; (c) corresponding Tafel plots; (d) double-layer capacitances (Cdl); (e) Nyquist plots; (f) Chronoamperometry curve for 12 h.

The electrocatalytic activities of Pt-CoFeP and other contrast samples toward OER under 1 M KOH electrolyte are evaluted by different electrochemical approaches. As demonstrated in Figure 5a, linear scan voltammetry (LSV) curve of Pt-CoFeP is tested with the scan rate of 5 mV s^{-1} on the reversible hydrogen electrode (RHE). The contrast samples, including Pt-CoFe PBA, e-CoFe PBA, CoFe PBA and commercial RuO_2 are also tested for comparison. The overpotentials (Figure 5b) of Pt-CoFeP (η_{10}=311.7 mV, η_{50}=343.1 mV) are much lower than that of other contrast samples, especially for commercial RuO_2. Besides, Pt-CoFeP possesses the lowest Tafel slope (48.1 mV dec^{-1}) in Figure 5c, indicating the greatly optimized reaction kinetics during OER process. The C_{dl} of Pt-CoFeP electrode (Figure 5d) is estimated to be 22.8 mF cm^{-2}, which indicates more exposed active site to enhance the water splitting process. In order to further explore the conductivity, charge-transfer resistance (R_{ct}) kinetics of Pt-CoFeP. R_{ct} of Pt-CoFeP (118.7 Ω) is much lower than that of other contrast samples. Thus, the optimized Pt-Co/Fe electronic structure and highly dispersed and confined growth of platinum, synergistically contributes to the improvement of the oxygen evolution reaction performance. In addition, the current density (10 mA cm^{-2}) of Pt-CoFeP electrode maintains nearly invariable after 12 h long-term durability measurement.

4. Conclusion

In summary, the hollow framework of Pt-CoFeP cage with open holes is successfully synthesized through alkali etching and phosphating processes. The target sample Pt-CoFeP, deriving from CoFe PBA shows excellent OER performance, which only requires 311.7 mV to reach the current density of 10 mA cm^{-2}. The enhanced activity may be due to the synergistic effect of optimized Pt-Co/Fe electronic structure and highly dispersed Pt atoms. This work may provide a reference for the design of complex structures for highly dispersed Pt atoms loading.

Acknowledgements

This work is financially supported by National Natural Science Foundation of China (52174283 and 52274308) and Innovation Fund Project for Graduate Student of China University of Petroleum (East China) (22CX04023A) and the Fundamental Research Funds for the Central Universities.

References

1. W. Li, C. Wang, X. Lu, Integrated transition metal and compounds with carbon nanomaterials for electrochemical water splitting, Journal of Materials Chemistry A, vol. 9 (2021), 3786-3827.
2. W. Zhang, K. Zhou, Ultrathin two-dimensional nanostructured materials for highly efficient water oxidation, Small, vol. 13 (2017), 1700806.
3. Y. Lu, W. Hou, D. Yang, et al. CoP nanosheets in-situ grown on N-doped graphene as an efficient and stable bifunctional electrocatalyst for hydrogen and oxygen evolution reactions, Electrochimica Acta, vol. 307 (2019), 543-552.
4. H. H. Zou, C. Z. Yuan, H. Y. Zou, et al. Bimetallic phosphide hollow nanocubes derived from a prussian-blue-analog used as highperformance catalysts for the oxygen evolution reaction, Catalysis Science & Technology, vol. 7 (2017), 1549.
5. J. Xie, X. Zhang, N. Yu, et al. Accelerating Fe sites saturation coverage through Bi-metal dynamic balances on double-layer hollow MOF nanocages for oxygen evolution, Materials Today Physics, vol. 27 (2022), 100778.
6. W. Chen, Y. Xie, X. Gao, et al. Simultaneous optimization of CoIr alloy nanoparticles and 2D graphitic-N doped carbon support in CoIr@CN by Ir doping for enhanced oxygen and hydrogen evolution reactions, Journal of Materials Chemistry A, vol. 10 (2022), 15543.
7. X. Zhang, Y. Wu, Y. Sun, et al. CoFeP hollow cube as advanced electrocatalyst for water oxidation, Inorganic Chemistry Frontiers, vol. 6 (2019), 604.
8. Y. Deng, Y. Cao, Y. Xia, et al. Self-templated synthesis of CoFeP@C cage-in-cage superlattices for enhanced electrocatalytic water splitting, Advanced Energy Materials, vol. 12 (2022), 2202394.
9. J. T. Ren, Y. S. Wang, L. Chen, et al. Binary FeNi phosphides dispersed on N,P-doped carbon nanosheets for highly efficient overall water splitting and rechargeable Zn-air batteries, Chemical Engineering Journal, vol. 389 (2020), 124408.
10. J. Xie, Z. Liu, J. Li, et al. Fe-doped CoP core–shell structure with open cages as efficient electrocatalyst for oxygen evolution, Journal of Energy Chemistry, vol. 48 (2020), 328-333.
11. Z. Jiang, J. Ren, Y. Li, et al. Low-cost high-performance hydrogen evolution electrocatalysts based on Pt-CoP polyhedra with low Pt loading in both alkaline and neutral media, Dalton Transactions, vol. 48 (2019), 8920.

Ultrahigh density defects of Mn3O4 boosting oxygen evolution reaction

Yanan Zhou, Yongming Chai and Bin Dong

College of Chemistry and Chemical engineering, China University of Petroleum (East China), Qingdao 266580, China

Abstract. Surface defects have a positive effect on the improvement of electrocatalytic performance due to the ability to introduce half-metallicity into selected semiconductors. Herein, we prepare a defect-rich Ni-doped Mn_3O_4 (Ni-Mn_3O_4) material by a simple hydrothermal and impregnation method. Ni-Mn_3O_4 exhibits a unique morphology consisting of ultrathin nanoarrays and nanoflowers embellished at the top, which endows the material with large specific surface area and reactive sites and contributes to enhanced oxygen evolution reaction (OER). The obtained Ni-Mn_3O_4 shows improved OER activity with the overpotential of 290 mV at the current density of 20 mA cm^{-2}. This remarkable OER performance can be attributed to the multi-dimensional interface and ultra-high density defects, leading to the formation of abundant active sites, thereby inducing half-metallicity properties and suitable intermediate adsorption energies. This work is expected to provide reference and guidance for the design and synthesis of high-performance electrocatalysts.

Keywords: Mn_3O_4, Ni doping, defects, OER.

1. Introduction

Climate change is a major global challenge facing humanity today. In order to actively respond to climate change, carbon neutrality has become an important issue worldwide. Hydrogen energy is a clean, efficient, renewable and ideal energy source, which can be efficiently converted into electricity and is regarded as one of the most promising energy sources. Hydrogen production by water electrolysis is an important approach to achieving industrial production of green hydrogen [1,2]. The sluggish kinetics of oxygen evolution reaction (OER), a crucial half-reaction of water electrolysis, hinders the improvement of reaction efficiency[3]. Therefore, the search for inexpensive, abundant and excellent OER catalysts has become a hot research topic in recent years. Compared with noble metal catalysts, 3d transition metal oxides with higher practical value have more electronic freedom to regulate the kinetics of OER reaction[4].

The Manganese-based oxides are appealing materials for OER owing to the wide range of oxidation states[5], such as MnO_2, Mn_3O_4, Mn_2O_3 and MnO, however, they have not been studied as intensively as FeCoNi. Among them, the spinel Mn_3O_4 exhibits favorable electrical conductivity due to the excellent electron transfer between oxidation states +2 and +3 [6]. The origin of the active sites of OER in Mn_3O_4 has been thoroughly investigated by Tiwdry and his teams. They confirmed by experiments and theoretical calculations that it is the Mn(iii) site present on the (112)-oriented surface of Mn_3O_4 that is highly active for OER[7]. Actually, the conductivity and the number of active sites cannot be enhanced enough for electrocatalysts, which can be realized by defect construction and atomic doping[8-11]. For example, Zhou et al. induced lattice distortion of Mn_3O_4 by Ni doping, which also led to the production of distorted octahedral Mn(iii) species and increased the number of active sites in the OER[8]. Nevertheless, the electrocatalytic OER activity of Mn_3O_4 is still far behind that of precious metals and transition metals such as Fe and Ni, and it is necessary to further improve the electrochemical performance.

Herein, we have successfully prepared spinel Mn_3O_4 on nickel foam (NF) by a novel in situ growth process and realized Ni doping on Mn_3O_4 (Ni-Mn_3O_4) via a sample dipping. The obtained Ni-Mn_3O_4 exhibits outstanding OER activity in alkaline media. The overpotential required at 20 mA cm^{-2} is 290 mV. The higher catalytic activity can be attributed to the unique multidimensional structure with two-dimensional nanoflowers dotting with one-dimensional nanosheets. Such morphology can expose abundant active Mn^{3+} sites and provide large active surface area. Besides, the ultrahigh density defects also contribute to the increase of active center and conductivity. This defect-engineering and atom doping strategy provide a feasible approach for the large-scale development of Mn-based material.

https://doi.org/10.1051/matecconf/202338201021

2. Experimental Section

2.1 Synthesis of Mn_3O_4

Mn_3O_4 was synthesized using a simple hydrothermal method. 2 mmol of $KMnO_4$ was dissolved in 40 ml of deionized water under mechanical stirring. The resulting dark purple solution was then transferred to a 50 ml Teflon-lined autoclave containing a clean piece of NF (1 cm × 4 cm) (0.11 g). The autoclave was then heated at 180 °C for 3 hours. After natural cooling to room temperature, the samples were washed several times with deionized water and ethanol and then dried in a vacuum drying oven.

2.2 Synthesis of Ni_x-Mn_3O_4

The above Mn_3O_4 was dipped into the 20 ml solution containing x M (x=0.05, 0.15 and 0.20) $Ni(NO_3)_2 \cdot 6H_2O$ for 2 h at 80 °C. The prepared Ni-Mn_3O_4 was washed with deionized water and dried for later use.

2.3 Characterization

X-ray diffraction (XRD) was performed on a Bruker D8 Advanced X-ray diffractometer Cu Kα radiation (λ=0.1542 nm). Scanning electron microscopy (SEM) images were obtained via the JEM 2011 F equipment. Transmission electron microscopy (TEM) images and related element mapping were detected via the FEI Tecni G20 device. X-ray photoelectron spectroscopy (XPS) was detected on Thermo Scientific Escalab 250 photo-electron spectrometer.

2.4 Electrochemical measurements

Electrochemical measurement was performed on Gamry 3000 workstation using a three-electrode system in 1.0 M KOH. The obtained samples, Pt wire and saturated calomel electrode are chosen as work electrode, counter electrode and reference electrode, respectively. All the potentials were converted to reversible hydrogen electrode (RHE) according to the Nernst equation: $E_{RHE} = E_{SCE} + 0.2415 + 0.0592pH$. Polarization curves were collected with a scan rate of 5 mV s^{-1}. The electrochemical impedance spectroscopy (EIS) was measured under an open circuit potential from 0.1 Hz to 100 Hz at the current voltage amplitude of 5 mV. Chronopotentiometric method was used to test the stability of electrocatalysts at 10 mA cm^{-2}.

3. Results and discussion

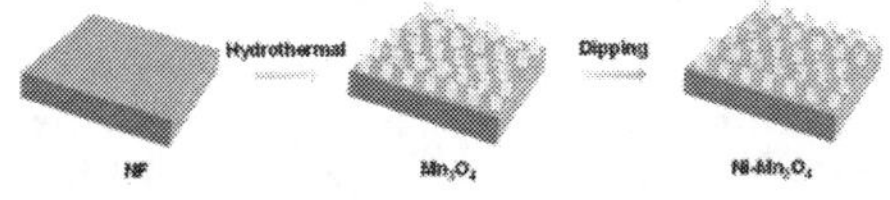

Figure 1: Schematic diagram of the preparation process of Ni-doped Mn_3O_4.

As shown in Figure 1, Ni-Mn_3O_4 was synthesized by a simple hydrothermal reaction and subsequent solution impregnation. Firstly, the color of NF changed from silver to black during the hydrothermal treatment in the $KMnO_4$ solution. Subsequently, the Ni-doped Mn_3O_4 was fabricated by immersing the obtained Mn_3O_4 in an aqueous solution containing Ni^{2+} ions.

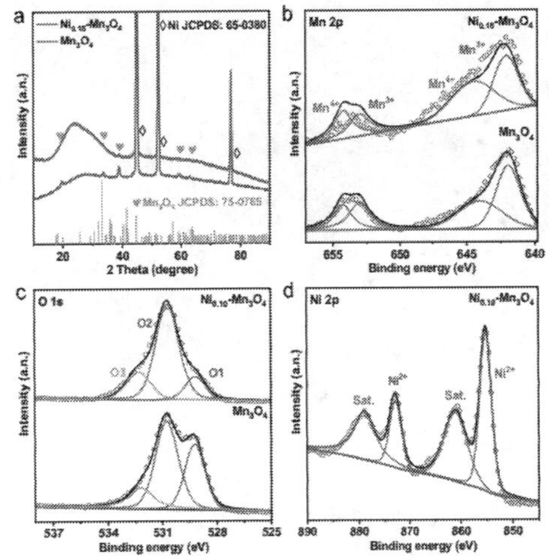

Figure 2: (a) XRD of Mn_3O_4 and Ni-Mn_3O_4. (b) Mn 2p, (c) O 1s and (d) Ni 2p of Ni-Mn_3O_4.

The crystal structures of Mn_3O_4 and Ni-Mn_3O_4 were investigated by XRD. As presented in Figure 2a, in addition to the three strong peaks at 44.9°, 52.5° and 76.9° belonging to metallic Ni (JCPDS: 65-0380), the characteristic peaks of the (020), (023), (041), (152) and (153) crystal planes of orthorhombic Mn_3O_4 (JCPDS: 75-0765) can be observed at 18.7°, 33.5°, 38.9°, 59.4°, 63.1°, respectively. After being treated in Ni^{2+} solution for 2 h, no Ni-related peaks can be discerned. Besides, these peaks shift to higher degree and become weaker, indicating that Ni atom has been successfully introduced and that the incorporation of Ni leads to the generation of more defects. Further, XPS was applied to identify the chemical properties and composition of the surface of Mn_3O_4 and Ni-Mn_3O_4 samples. In Mn 2p region (Figure 2b), the peaks located at 642.0 and 652.9 eV can be assigned to Mn^{3+}, and another set of peaks at 644.5 and 654.2 eV come from the Mn^{4+} species. It is noted that the atomic ratio Mn^{3+}:Mn^{4+} of Mn_3O_4 was about 1.15, which is higher than that of Ni-Mn_3O_4 (1.12). The increase in active Mn^{3+} content will inevitably and dramatically encourage the enhancement of OER activity. O 1s XPS is composed of three peaks located at 529. 2 (O1), 530.7 (O2) and 532.1 eV (O3), resulting from lattice oxygen, defective oxygen and absorbed oxygen on the surface, respectively. Furthermore, as shown in Figure 2d, the Ni 2p spectrum consists of four peaks, where the peaks at 854.9 and 872.7 eV are related to the Ni $2p_{3/2}$ and $2p_{1/2}$ of Ni^{2+} ions. Additionally, the satellite peaks situated at 860.8 and 878.9 eV also verify the presence of Ni^{2+}.

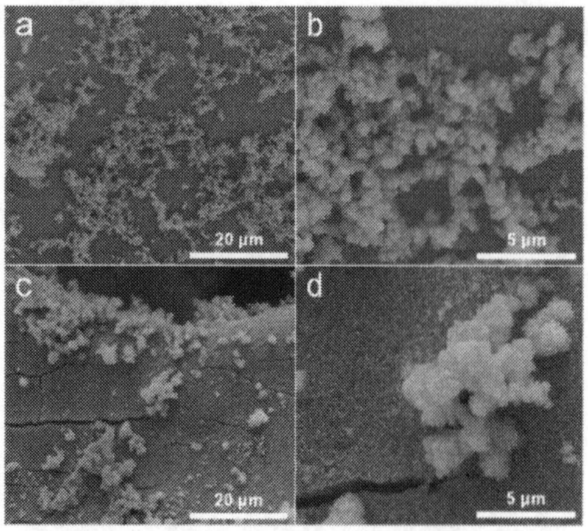

Figure 3: SEM images of (a,b) Mn_3O_4 and (c,d) Ni-Mn_3O_4.

To check the morphology of Mn_3O_4 and Ni-Mn_3O_4 samples, SEM images were detected and displayed in Figure 3. For Mn_3O_4, it can be clearly observed that the nanosheet arrays densely distribute over the entire NF, dotted with nanoflower microspheres with a diameter of ~450 nm. The coupling of one- and two-dimensional structures undoubtedly greatly increases the electrochemically active specific surface area, exposing an abundance of active sites. The structure of the nanosheets and nanoflowers after the impregnation treatment did not change significantly except that the surface became rough and the diameter of the nanoflowers expanded to ~1000 nm. This transition will further stimulate the reactivity by increasing the surface area.
In addition, Figure 4a gives the high-resolution TEM (HRTEM) images, proving the generation of ultrahigh density defects for Ni-Mn_3O_4. The existence of a large number of dislocations and lattice mismatches can modulate the catalyst energy band structure and adsorption properties of the material, which is expected to further drive the OER performance enhancement. The insertion in Figure 4a is the high-angle annular dark field (HAADF) image, confirming the lamellar structure with wavy folds. The corresponding element mappings are also collected, suggesting the homogenous dispersion of Ni, Mn and O in the ultrathin Ni-Mn_3O_4 nanosheets.

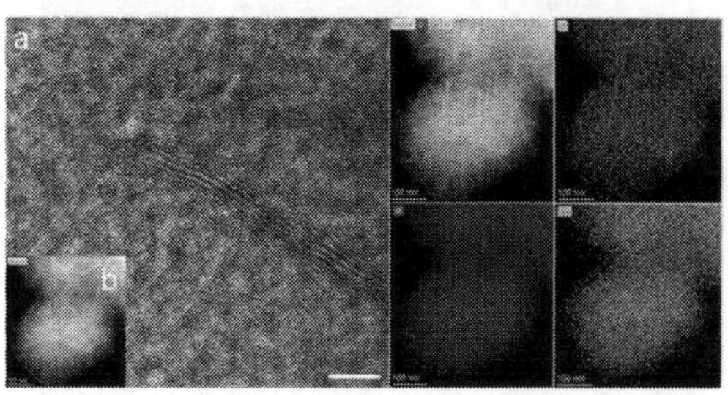

Figure 4: (a) HREM image and (b) HAADF image and corresponding element mapping of Ni-Mn_3O_4.

As discussed above, the ultrahigh density defects and large surface area would be propitious to water oxidation. Therefore, the electrochemical OER performance of synthesized samples was measured in alkaline media using a standard three-electrode system. As shown in Figure 5a, all the samples exhibit OER activity higher than commercial RuO_2. The Ni^{2+} content was first optimized and the maximum promotion effect was observed when the Ni^{2+} concentration in the solution was 1.0 M, conferring the highest OER activity to $Ni_{0.10}$-Mn_3O_4. At the current density of 20 and 50 mA cm^{-2}, the needed overpotential is 290 and 353 mV, which is a 52 mV (50 mA cm^{-2}) reduction from the Mn_3O_4 precursor.

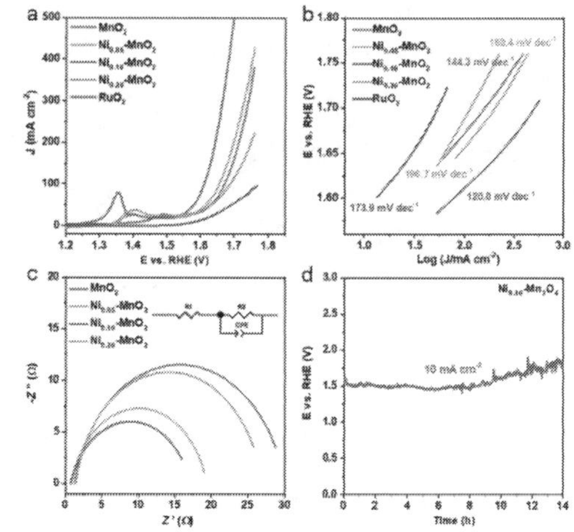

Figure 5: Electrocatalytic OER performances of the samples. (a) polarization curves, (b) Tafel plots, (c) EIS and (d) v-t curves at 10 mA cm-2.

Moreover, the Tafel slopes were also calculated by fitting the polarization curves to study the reaction kinetics (Figure 5b). The Tafel slope of $Ni_{0.10}$-Mn_3O_4 is 120.0 mV dec^{-1}, lower than Mn_3O_4 (144.3 mV dec^{-1}) and other contrast samples. Such result indicates Ni doping expedites the catalytic kinetics of Mn_3O_4. Moreover, in order to gain deeper insight into the excellent OER activity, EIS studies were carried out and the measured impedance data was fitted according to an equivalent electrical circuit (inset of Figure 5c). The R1, R2 and CPE represent electrolyte solution resistance (R_s), charge-transfer impedance (R_{ct}) and constant phase angle components associated with electric double layer capacitance. Encouragingly, $Ni_{0.10}$-Mn_3O_4 also possesses the smallest impedance with the R_{ct} of 16.4 Ω, lower than Mn_3O_4 (28.8 Ω) $Ni_{0.05}$-Mn_3O_4 (18.2 Ω) $Ni_{0.20}$-Mn_3O_4 (25.5 Ω), which implies that $Ni_{0.10}$-Mn_3O_4 owns a high electrical conductivity and fast transfer rate. Given that operational stability is a critical indicator for electrocatalysts, we further analyzed the long-term stability of the best-performing $Ni_{0.10}$-Mn_3O_4. The *v-t* curve shows that $Ni_{0.10}$-Mn_3O_4 can run stably for at least 14 h at the constant current density of 10 mA cm^{-2}, confirming the excellent durability. In a nutshell, the ultrahigh density defects and multiscale structure endow $Ni_{0.10}$-Mn_3O_4 with boosted catalytic performance for OER.

https://doi.org/10.1051/matecconf/202338201021

4. Conclusion

In summary, the Ni-doped Mn_3O_4 nanosheet arrays have been successfully prepared through an in situ growth method and impregnation method. The obtained $Ni_{0.10}$-Mn_3O_4 delivers the current density of 20 mA cm^{-2} at 290 mV. The superior OER activity would result from the following two factors. Firstly, the ultrahigh density defects and Ni doping reinforce electron transfer and surface adsorption. Secondly, the coupling of one-dimensional nanosheets and two-dimensional nanoflowers enlarges the active surface area and exposes more active catalytic sites. In addition, such a structure facilitates sufficient contact between the $Ni_{0.10}$-Mn_3O_4 catalyst and electrolyte. This work pinpoints the potential of defect engineering and atom doping in the development of earth-abundant elemental electrocatalysts for water oxidation.

Acknowledgements

This work is financially supported by the National Natural Science Foundation of China (52174283 and 52274308) and the Innovation Fund Project for Graduate Student of China University of Petroleum (East China) (No. 22CX04027A) and the Fundamental Research Funds for the Central Universities.

References

1. X. Liu, R.T. Guo, K. Ni, et al. Reconstruction-Determined Alkaline Water Electrolysis at Industrial Temperatures, Advanced Material, vol. 32 (2020), 2001136.
2. M.H. Ning, F.H. Zhang, L.B. Wu, et al. Boosting efficient alkaline fresh water and seawater electrolysis via electrochemical reconstruction, Energy & Environmental Science vol. 15 (2022), 3945-3957.
3. J. Wang, Y. Gao, H. Kong, et al. Non-precious-metal catalysts for alkaline water electrolysis: operando characterizations, theoretical calculations, and recent advances, Chemical Society Reviews, vol. 49 (2020), 9154-9196.
4. S.Q. Jiang, H.L. Suo, X.B. Zheng, et al. Lightest Metal Leads to Big Change: Lithium-Mediated Metal Oxides for Oxygen Evolution Reaction, Advanced Energy Material, vol. 12 (2022), 2201934.
5. P. Wang, S.Q. Zhang, Z.B. Wang, et al. Manganese-based oxide electrocatalysts for the oxygen evolution reaction: a review, Journal of Materials Chemistry A, vol. 11 (2023), 5476-5494.
6. W. Shi, W. Lee, J.M. Xue, et al. Recent Development of Mn-based Oxides as Zinc-Ion Battery Cathode, Chemsuschem vol. 14 (2021), 1634-1658.
7. C.C. Gowda, A. Mathur, A. Parui, et al. Understanding the electrocatalysis OER and ORR activity of ultrathin spinel Mn3O4, Journal of Industrial and Engineering Chemistry, vol. 113 (2022),153-160.
8. Q. Huang, X. Zhong, Q. Zhang, et al. Co3O4/Mn3O4 hybrid catalysts with heterointerfaces as bifunctional catalysts for Zn-air batteries, Journal of Energy Chemistry, vol. 68 (2022), 679-687.
9. H. Xue, A. Meng, T. Yang, et al. Controllable oxygen vacancies and morphology engineering: Ultra-high HER/OER activity under base–acid conditions and outstanding antibacterial properties, Journal of Energy Chemistry, vol. 71 (2022), 639-651.
10. K. Ham, S. Hong, S. Kang, et al. Extensive Active-Site Formation in Trirutile CoSb2O6 by Oxygen Vacancy for Oxygen Evolution Reaction in Anion Exchange Membrane Water Splitting, ACS Energy Letters, vol. 6 (2021), 364-370.
11. Y. Yan, J. Lin, T. Xu, et al. Atomic-Level Platinum Filling into Ni-Vacancies of Dual-Deficient NiO for Boosting Electrocatalytic Hydrogen Evolution, Advanced Energy Material, vol. 12 (2022), 2200434.

Sulphur doping of nickel ferrite nanosheet array for enhanced water oxidation

Haijun Liu, Bin Dong

College of Chemistry and Chemical Engineering, State Key Laboratory of Heavy Oil Processing, China University of Petroleum (East China), Qingdao 266580, China.

Abstract. Industrial water electrolysis requires highly-active and ampere-current-bearing oxygen evolution reaction (OER) catalysts, but achieving such a large operating current density at low overpotentials in available OER catalysts still remains a grand challenge. Herein, we report a facile and conventional strategy to grow vertically non-metallic modified nickel ferrite nanosheets on Fe foam (S-NiFe2O4/IF) as large-current-density OER catalysts. The vertically aligned nanosheet arrays can offer large electrochemical surface area, and the spacing between nanoarrays can allow for effective electrolyte access to the active sites. The introduction of S enhanced the adsorption and dissociation to water and optimized the adsorption of OER intermediates. As a result, S-NiFe2O4/IF exhibits superior OER activity with an overpotential of 287 and 326 mV to achieve 100 and 500 mA cm−2, respectively, and displays robust stability in alkaline media. In addition, the synthesis strategy developed here can be applied to other mixed transition metal oxides with similar self-supported earth-abundant nanoarrays for advanced electrocatalysis in energy-related reactions.

Keywords: Nickel ferrite, Sulphur doping, Nanosheet array, Water oxidation.

1. Introduction

Oxygen evolution reaction (OER) is the key to limit the efficiency of water electrolysis due to its slow kinetic process, and the development of efficient electrocatalysts is regarded as an effective way to overcome this bottleneck [1]. Currently, the research hotspots for high-efficiency electrolytic water catalysts focus on the noble metals and transition metal compounds. Precious metal OER catalysts, such as IrO_2, RuO_2, etc., are currently considered as the best materials for OER catalysts. However, they are expensive, scarce in resources, and not available for large-scale industrial applications [2]. Therefore, it is important to explore low-cost, high-activity non-precious metal catalysts to improve OER efficiency [3].

In recent years, transition metal oxides (TMOs), such as spinel, perovskite, and hydroxide have emerged as promising OER catalysts due to their rich intrinsic redox pairs, abundant reserves, and low-cost [4]. However, TMOs have inherent low electrical conductivity and limited active sites that still need to be addressed to further boost the catalytic activity. Various methods, such as fine inter-surface design [5], coupling with carbon materials [6], and doping with heteroatoms [7] have been used to improve the OER electrocatalytic performance of TMOs. Among these methods, heteroatoms doping is considered to be the most promising method to improve the electrocatalytic performance of TMOs. In addition, self-supported nanoarrays can offer large electrochemical surface area, and the spacing between nanoarrays can allow for effective electrolyte access to the active sites [8]. Moreover, in situ growth of the nanoarray on a conducting substrate allows for rapid charge transfer between the current collector and the active site while preventing delamination of the active material [9].

Inspired by these criteria, we herein highlight the synthesis of sulfur-incorporated nickel ferrite nanosheets array built grown on IF (S-$NiFe_2O_4$/IF) via hydrothermal reaction. As a free-standing electrode, S-$NiFe_2O_4$/IF exhibits remarkable catalytic activity and durability for water oxidation in alkaline medium. Moreover, the facile synthetic approach proposed in this study can be easily extended as a general approach for producing other mixed metal oxides for various applications, such as supercapacitors, metal-air batteries, and photocatalysis.

2. Experimental Procedures

2.1 Materials Preparation

Chemicals: Iron foam (IF, 1.0 mm in thickness) was purchased from Kunshan Lvchuang Electrical Co Ltd. Nickel chloride hexahydrate ($NiCl_2 \cdot 6H_2O$), thiourea (CH_4N_2S) and potassium hydroxide (KOH) were purchased from Sinopharm Chemical Reagent Co., Ltd.,

Shanghai, China. Deionized (DI) water was used in all experiments.

Preparation of S-$NiFe_2O_4$/IF: The solution was prepared by dissolving 0.1 mmol of $NiCl_2 \cdot 6H_2O$ into 60 mL of 0.4 M CH_4N_2S aqueous solution, then gentle stirring for 10 min to form a homogeneous solution. Then solution was transferred into a 100 mL Teflonlined stainless steel autoclave with a piece of IF (20 mm × 50 mm) and maintained at 160 °C for 2 h in an oven. After the autoclave cooled slowly to room temperature, the sample was taken out and washed with water and ethanol thoroughly, followed by vacuum drying.

Preparation of $NiFe_2O_4$/IF: For comparison, $NiFe_2O_4$/IF was also prepared by hydrothermal method. The $NiFe_2O_4$/IF was synthesized using a similar procedure to S-$NiFe_2O_4$/IF, except for the removal of CH_4N_2S during the hydrothermal treatment.

2.2 Physical Characterization

The crystalline structures of the samples were confirmed by powder X-ray diffraction (XRD) using a Philips X'Pert diffractometer with Cu Kα radiation (λ = 0.15418 nm). The structures and morphologies were collected on a scanning electron microscope (Hitachi S-4800) and a high resolution transmission electron microscope (FEI Tecnai G2, 200 kV). The corresponding energy-dispersive spectroscopy (EDS) mapping analyses were performed on representative areas. X-ray photoelectron spectroscopy (XPS) were carried out on an Escalab 250Xi system.

2.3 Electrochemical measurements

OER tests were conducted in a three-electrode configuration with an electrochemical workstation (Gamry Reference 1000 Instrument) at room temperature. The obtained sample, platinum column and Hg/HgO (1.0 M KOH) electrode served as the working electrode, counter electrode and reference electrode, respectively. All of the electrochemical measurements were conducted in 1.0 M KOH solution for OER. The recorded potentials were converted to the potential versus reversible hydrogen electrode (RHE) according to the Nernst equation: $E_{RHE} = E_{Hg/HgO} + 0.059 \times pH + 0.098$. Linear sweep voltammetry (LSV) measurements were performed with a scan rate of 5.0 $mV \cdot s^{-1}$. The Tafel plots ($\log j$ vs η) were obtained from the corresponding LSV curves, according to the Tafel equation ($\eta = b \log j + a$, where η is the overpotential, j is the current density and b is the Tafel slope). The electrochemical active surface area of catalyst is estimated from electrochemical double-layer capacitance (Cdl), which is obtained by measuring double-layer charging from the CVs at different scan rates in non-Faradaic potential range in 1.0 M KOH. The plots then fit to linear regression and the slope is just the C_{dl}. Electrochemical impedance spectroscopy (EIS) measured in the frequency from 10^5 Hz to 0.01 Hz. The durability of catalysts was obtained by conducting chronopotentiometry curve.

3. Results and Discussion

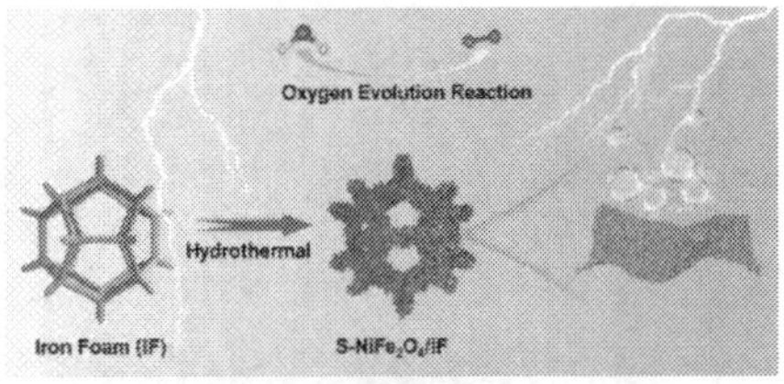

Figure 1: Schematic illustration of the preparation of S-$NiFe_2O_4$/IF.

Typically, self-supported S-$NiFe_2O_4$/IF nanosheet arrays on IF were prepared by a one-step synthesis, the hydrothermal reaction, as shown in Figure 1. IF was used as the substrate because it provides a large surface area and a high conductivity. The morphological, structural, and compositional features of the as-prepared S-$NiFe_2O_4$/IF were characterized. The structural information of the as-fabricated S-$NiFe_2O_4$/IF was studied by X-ray diffraction (XRD), as shown in the Figure 2a. The XRD pattern shows sharp diffraction peaks at 30.3°, 35.7°, 43.4°, 57.4°, and 63.0°, which can be ascribed to (220), (311), (400), (511) and (440) reflections from $NiFe_2O_4$ (JCPDS card no. 00–054-0964), respectively [10]. In addition, the diffraction peak located at 44.7° is attributed to the (110) plane of Fe (03-065-4899) from IF. Due to the low sulfur source utilization of the sulfur-containing material and the low hydrothermal reaction temperature, no additional peaks were observed [11]. Figure 2b displays the scanning electron microscopy (SEM) images of the IF. It can be seen that the surface of IF is smooth. SEM image of S-$NiFe_2O_4$/IF (Figure 2c, and Figure 3) exhibit full coverage of bare IF with the S-$NiFe_2O_4$ nanosheet composed with uniform separated nanoparticles. This type of catalytic arrangement is conducive to fast electron transfer kinetics, mainly due to its high surface exposure and sufficient interfacial contact. In order to examine the specific microstructure of S-$NiFe_2O_4$, transmission electron microscopy (TEM) observations were carried out. TEM image of S-$NiFe_2O_4$/IF product (Figure 2d) reveals that the S-$NiFe_2O_4$ NPs with size of ≈ 50 nm are uniformly loaded on the IF substrate without obvious aggregation. This construction of nanosheets composed of nanoparticles can increase the exposed active sites and thus improve the catalytic activity of OER. The high-resolution TEM (HRTEM) image (Figure 2e) exhibits the typical lattice fringes with distance of 0.147 nm and 0.251 nm, corresponded to the $NiFe_2O_4$ (440) and (311) crystal plane, respectively. The SAED pattern exhibits several light circles attributed to (440) and (311) planes of $NiFe_2O_4$ phase, demonstrating the well-crystalline structure (Figure 2f). The energy dispersive spectrum (EDS) mapping (Figure 2g) indicates the homogeneous distribution of Ni, Fe, O and S elements within the 2D nanosheets. In addition, the uniform distribution of S elements in the entire section demonstrates the successful doping of heteroatoms into $NiFe_2O_4$/IF.

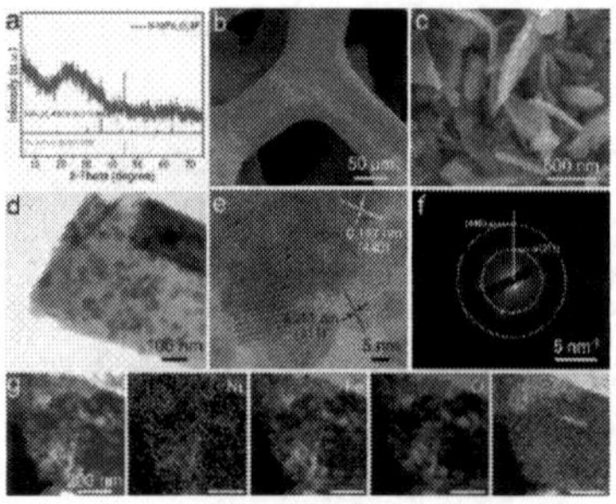

Figure 2: (a) XRD pattern of S-$NiFe_2O_4$/IF. (b) SEM image of IF. (c) SEM image of S-$NiFe_2O_4$/IF. (d) TEM image of S-$NiFe_2O_4$/IF. (e) HRTEM image of S-$NiFe_2O_4$/IF. (f) SAED image of S-$NiFe_2O_4$/IF. (g) HAADF-STEM-EDS mapping of S-$NiFe_2O_4$/IF.

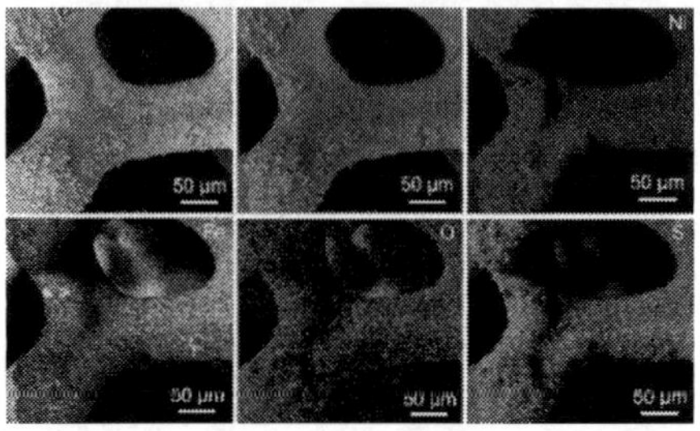

Figure 3: The SEM mapping of S-$NiFe_2O_4$/IF.

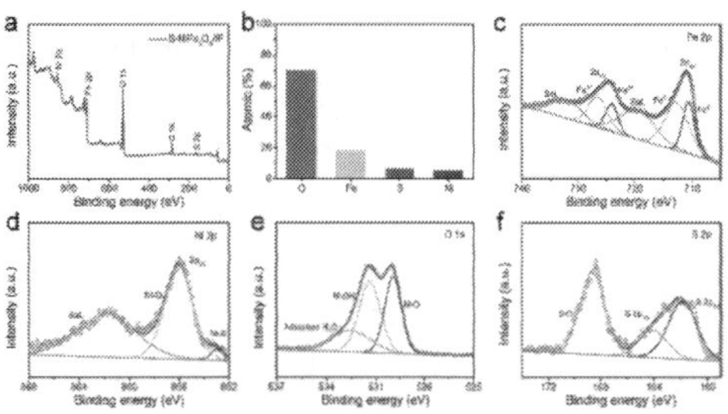

Figure 4: (a) XPS survey of S-$NiFe_2O_4$/IF. (b) The content of O, Fe, S and Ni elements. XPS spectra of S-$NiFe_2O_4$/IF in (c) Fe 2p region, (d) Ni 2p region, (e) O 1s region and (f) S 2p region.

To determine the surface chemical state and electronic structure, an X-ray photoelectron spectroscopy (XPS) survey was recorded. The resulting high–resolution spectra from the Fe 2p, Ni 2p, O 1s and S 2p core level regions are collected in Figure 4. According to the XPS spectra of S-$NiFe_2O_4$/IF, Ni, Fe, S, O elements are proved in S-$NiFe_2O_4$/IF (Figure 4a), agreeing with the EDX-mapping results. The content of O, Fe, S and Ni in the as-prepared S-$NiFe_2O_4$/IF determined by XPS (Figure 4b) were 70.3 %, 18.3 %, 6.4 %, and 5.0 %, respectively. In the Fe 2p high-resolution XPS spectrum of S-$NiFe_2O_4$/IF (Figure 4c), two spin-orbit doublets at binding energies at ca. 711.0 and 724.6 eV can be attributed to the Fe $2p_{3/2}$ and Fe $2p_{1/2}$, respectively. One doublet of which at 710.6 and 724.0 eV was belonged to Fe^{2+} and another one located at 712.6 and 726.5 eV was belonged to Fe^{3+}. The other peaks at 719.5 and 732.0 eV were attributed to satellite peak for S-$NiFe_2O_4$/IF. Furthermore, the profile of the Ni $2p_{3/2}$ spectrum can be deconvoluted into three peaks, where the bonding energies at 852.9, 856.0 and 861.8 eV attributed to the characteristic of Ni-S, Ni-O species and shake-up satellites (denoted Sat.), respectively. As shown in Figure 4e, the O 1s spectrum is further fitted with four doublets, including M-O at 529.9 eV, M-OH at 531.4 eV, and adsorbed H_2O at 532.5 eV, respectively. A clear chemisorbed water signal appears, showing the excellent adsorption capacity of as-prepared catalyst for oxygen-containing intermediates. Figure 4f shows S 2p high-resolution spectra of S-$NiFe_2O_4$. The peak at the binding energy of 164.0 eV and 161.9 eV in S-$NiFe_2O_4$ is owing to the S $2p_{1/2}$ and S $2p_{3/2}$, respectively. The other peaks of S 2p at the higher binding energy (~168.6 eV) are recognized as the oxidized S species (S–O). All the above results demonstrate the successful introduction of S atoms into $NiFe_2O_4$ crystal on the surface of the IF.

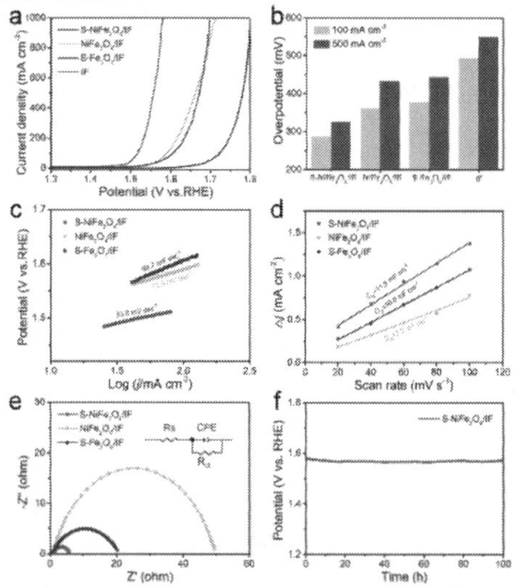

Figure 5: OER electrochemical properties of S-$NiFe_2O_4$/IF and other contrast samples in 1.0 M KOH. (a) Polarization curves; (b) overpotentials under 100 and 500 mA cm^{-2}; (c) Tafel plots; (d) the double layer capacitances (C_{dl}) for estimating electrochemical specific area (ECSA); (e) Nyquist plots; (f) the Chronoamperometry curve of S-$NiFe_2O_4$/IF for 100 h.

The electrocatalytic performance of S-$NiFe_2O_4$/IF catalyst for OER was evaluated on a conventional three-electrode system under alkaline pH conditions of 1.0 M KOH. The S-$NiFe_2O_4$/IF electrode requires the lowest overpotential ($\approx$ 287 mV) to reach 100 mA cm^{-2} compared to the other electrodes (Figure 5a,b). OER is related to electrochemical kinetics:The smaller the slope of the Tafel plot, the faster the reaction rate. The Tafel slopes of S-$NiFe_2O_4$/IF, $NiFe_2O_4$/IF and S-Fe_3O_4 were 53.8 mV dec^{-1}, 73.0 mV dec^{-1}, and 99.7 mV dec^{-1}, respectively, indicating the fastest OER rates of S-$NiFe_2O_4$/IF electrode (Figure 5c). We further evaluated the ECSA of as-prepared catalysts to understand the intrinsic surface-area activity, which is proportional to the C_{dl}. S-$NiFe_2O_4$/IF possess a C_{dl} value of 11.9 mF cm^{-2}, which is larger than that for $NiFe_2O_4$/IF (7.2 mF cm^{-2}) and S-Fe_3O_4 (10.0 mF cm^{-2}) (Figure 5d). This shows that the sulfidation process can increase the ECSA value, and sulfidation process is beneficial to further increase the number of exposed active sites. EIS at a constant overpotential of 0.55 V (Hg/HgO) was further applied to probe the electrode kinetics under the OER process. As shown in Figure 4e, the Nyquist plots in the high-frequency range present standard semicircles, whose diameters are determined by the charge-transfer resistance (R_{ct}). The R_{ct} of S-$NiFe_2O_4$/IF is about 5.7 Ω, which is much smaller than those of $NiFe_2O_4$/IF (≈ 49.2 Ω) and S-Fe_3O_4 (≈ 20.0 Ω) (Figure 5e), suggesting a rapid charge transfer rate in the S-$NiFe_2O_4$/IF catalyst. OER electrochemical stability is another critical factor in evaluating catalysts' performance [12]. Fig. 3e shows the long-term OER process stability test was carried out for 100 h at a fixed current density of 100 mA cm^{-2} (Figure 5f), and there was no evident voltage increase. In conclusion, the S-$NiFe_2O_4$/IF catalyst has excellent OER stability in 1.0 M KOH.

4. Conclusion

In conclusion, a facile and reliable strategy was proposed and successfully demonstrated to synthesize S-$NiFe_2O_4$/IF via hydrothermal method. Benefitting from the rationally designed 3D hierarchical structure and optimized composition with good electrical conductivity, the as-synthesized S-$NiFe_2O_4$/IF has demonstrated a low overpotential of 287 mV at 100 mA cm^{-2}, a small Tafel slope of 53.8 mV dec^{-1}, and long-term stability over 200 h in alkaline OER process. The cost-effective Fe-based catalyst materials and low-cost facile preparation for mass production enable its potential application in practical water splitting.

Acknowledgements

This work is financially supported by National Natural Science Foundation of China (52174283 and 52274308).

References

1. Y. Sun, J. Wu, Y. Xie, et al. Dynamics of both active phase and catalysis pathway for spinel water-oxidation catalysts, Advanced Functional Materials, vol. 32 (2022), 2207116.
2. C. Kou, J. Zhou, H. Wang, et al. Boron pretreatment promotes phosphorization of FeNi catalysts for oxygen evolution, Applied Catalysis B: Environmental, vol. 330 (2023), 122598.
3. Y. Yan, K. Huang, J. Lin, et al. Charge redistribution in FeOOH nanoarray by ecological oxygen-reduction deposition for boosting electrocatalytic water oxidation, Applied Catalysis B: Environmental vol. 330 (2023), 122595.
4. J. Liu, M. Zheng, J. Li, et al. Lithiation-induced defect engineering to promote oxygen evolution reaction, Advanced Functional Materials, vol. 33 (2023), 2209753.
5. C. Wang, H. Lu, Z. Mao, et al. Bimetal schottky heterojunction boosting energy-saving hydrogen production from alkaline water via urea electrocatalysis, Advanced Functional Materials, vol. 30 (2020), 2000556.
6. L. Yan, Y. Xu, P. Chen, et al. A freestanding 3D heterostructure film stitched by MOF-derived carbon nanotube microsphere superstructure and reduced graphene oxide sheets: a superior multifunctional electrode for overall water splitting and Zn–air batteries, Advanced Material, vol. (2020), 2003313.
7. L. Yu, L. Wu, B. McElhenny, et al. Ultrafast room-temperature synthesis of porous S-doped Ni/Fe (oxy)hydroxide electrodes for oxygen evolution catalysis in seawater splitting, Energy & Environmental Science, vol. 13 (2020), 3439.
8. J. Liu, D. Zhu, Y. Zheng, et al. Self-supported earth-abundant nanoarrays as efficient and robust electrocatalysts for energy-related reactions, ACS Catalysis, vol. 8 (2018), 6707–6732
9. T. Ma, S. Dai, S. Qiao, Self-supported electrocatalysts for advanced energy conversion processes, Materials Today, 19 (2016), 265-273.
10. Z. L. Cao, Z. Li, K. Su, et al. Rational design of hollow oxygen deficiency-enriched NiFe2O4@ N/rGO as bifunctional electrocatalysts for overall water splitting, Journal of Energy Chemistry, vol. 54 (2021), 595–603.
11. H. Liu, R. Luan, L. Li, et al. Sulphur-dopant induced breaking of the scaling relation on low-valence Ni sites in nickel ferrite nanocones for water oxidation with industrial-level current density, Chemical Engineering Journal, vol. 461 (2023), 141714.
12. J. Chi, X. Zeng, X. Shang, Embedding RhPx in N, P co-doped carbon nanoshells through synergetic phosphorization and pyrolysis for efficient hydrogen evolution, Advanced Functional Materials, vol. 29 (2019), 1901790.

A coupled radial-torsional lumped parameter model of rubber bushing for multibody dynamic simulations

Chang Cao, Xingxing Feng*, Min Liu

Wuhan Second Ship Design and Research Institute, Wuhan, Hubei 430205, China

Abstract. Many constitutive models' finite element (FE) models and lumped parameter models included have been presented for rubber bushings. In general, the FE models cannot be directly used in mechanical multibody systems with bushings, and the lumped parameter models can not consider the coupled modes of deformation of bushings. For this reason, a two-dimensional lumped parameter model for rubber bushings is presented, which considers the coupled radial-torsional deformation and can be easily implemented into mechanical multibody systems. FE method is used to simulate the rubber bushing subjected to a coupled radial-torsional deformation. A procedure to determine the model parameters is presented. The proposed coupled bushing model and a traditional one-dimensional bushing model are applied into a simple multibody system and compared. Simulation results show the coupled radial-torsional behavior of the bushing.

Keywords: Rubber bushing; constitutive model; coupled deformation; multibody dynamics; parameter identification

1. Introduction

Rubber bushings are widely used in vehicle or machinery systems. Such filler-reinforced rubber bushing presents highly nonlinear mechanical properties such as frequency dependency [1], amplitude dependency (Payne effect [2] or Fletcher-Gent effect [3]), Mullins' effect [4], pre-strain dependency [5] [6] and temperature dependency [7] [8], etc. Amount of constitutive models have been proposed for rubber bushings, including lumped parameter models and finite element (FE) models. Most lumped parameter models and FE models are based on the overlay concept. The basic idea of the overlay concept is that the rubber material behavior can be split into a number of parallel fractions, each one with conventional mechanical properties (hyper-elastic, elastoplastic and viscoelastic), so its complex behavior is achieved by suitable combination of simple constitutive models [9].

Lumped parameter models can globally represent the behavior of the rubber bushings using a relatively small number of parameters. In lumped parameter models, frequency dependency is usually produced by chains of springs and dashpots. Austrell [10] used several Maxwell elements put in parallel to represent frequency dependency. Berg [11] used Zener model to represent the elastic and viscous forces in a nonlinear rubber spring model. Scheiblegger et al. [12] used the so-called dual Kelvin-Voigt to include the frequency dependency in the models of bushings and cab mounts. The basic layout of the model goes back to the approach of Pfeffer and Höfer [13].

As the models include more springs and dashpots, the complexity and number of parameters of models are highly increased. As an alternative for obtaining a good representation of the frequency dependency of rubber-like material, fractional derivative models require much less parameters. The fractional derivative models were widely applied in rubber isolators [14] [15] and railway elastic joint [16], etc.

In lumped parameter models, amplitude dependency is usually produced by friction elements. Coulomb friction elements are firstly used [1][17][18]. A special mention should be given to the Berg's friction element [11] which involves only two parameters, and has been combined with different viscoelastic to reproduce the dynamic behavior of rubber components [14][15][16]. Other friction models include Bouc-Wen model [19][20][21][22], rate-dependent triboelastic model [23][24] and friction models suggested by Scheiblegger [12], etc.

For FE models, the nonlinear elasticity of rubber bushings are usually modelled by hyper-elastic material using strain energy functions, where Mooney-Rivlin [25][26], Yeoh [27][28] , Ogden [29] , Arruda-Boyce [30] forms are the popular ones. The viscoelasticity or frequency dependency is usually modelled by generalized Maxwell models [31][32][33][34] or fractional derivative viscoelastic models [35][36][37][38]. There are two main trends to model the amplitude dependency of rubber bushing in FE models. The first trend is to use model of Kraus [39][40], which is based on microstructural description and reorganization rates of the carbon-black

* Corresponding author: 1310231571@qq.com

https://doi.org/10.1051/matecconf/202338201023

network. The second and more popular trend is the use of friction models [35][41][42][43].

In real working conditions, the rubber bushings experience complex coupled modes of deformation, thus to investigate the coupled modes of deformation is of significant importance. Lee et al. [44] solved the boundary problem for coupled axial-torsional deformation mode for bushings, and developed an explicit relation in which the axial force and torsional moment were each expressed explicitly in terms of axial displacement and rotation. The major disadvantage of this approach is that the force-displacement relation is defined implicitly as the solution to a boundary value problem.

FE models are naturally used to consider the coupled modes of deformation. Kadlowec et al. attempted to determine this relation directly from experiments, in which bushing was subjected to a coupled radial-torsional deformation mode [45] or coupled axial-torsional deformation mode [46]. Based on the experimental studies of Kadlowec et al. [45][46], some FE models were presented to predict the behaviors of rubber bushings subjected to coupled deformations [47][48][49][50]; in each model the hysteresis or amplitude dependency was not included.

In general, the lumped parameter models cannot consider the coupled modes of deformation of the rubber bushings. However, the coupled bushing model for radial and torsional mode proposed by Ok et al. [51] should be mentioned. In the model, the Bouc-Wen model is used for the radial nonlinear characteristics and Kelvin-Voigt model for the torsional characteristics. A kinematic relation is suggested to consider the coupling effects of the radial bushing force and torsional moment.

In summary, the FE models can consider the coupled deformations of rubber bushings, but cannot be easily applied into mechanical systems. Most lumped parameter models can be easily applied into mechanical system models, but cannot consider the coupled deformation modes. Therefore, the major goal of this study is to develop a two-dimensional lumped parameter model which not only can be easily implemented into mechanical systems, but also can describe the coupled deformations of bushings.

In Section 2, the proposed two-dimensional bushing model is introduced. In Section 3, FE rubber bushing model is built and used to simulate the rubber bushing responses under coupled deformations as well as uniaxial deformations. In Section 4, the parameter identification procedure is explored. In Section 5, the proposed model is applied into a multibody system. In Section 6, main conclusion throughout the paper are drawn.

2. Coupled lumped parameter bushing model

2.1 Structure of the coupled model

The bushing considered here is mold-bonded and composed of a hollow elastomer cylinder contained between inner and outer steel sleeves. During use, the bushing sleeves undergo displacements and rotations relative to one another. A novel 2-dimenional lumped parameter model is proposed here which can include radial and torsional deformation for the cylindrical rubber bushings.

The assumptions of the model are summarized as follows. As shown in Fig.1, the outer steel sleeve and inner steel sleeve are considered as two rigid bodies. M points are distributed symmetrically on the inner sleeve and outer sleeve respectively, termed as inner points and outer points. For each pair of points, a spring-damper element is used to connect them. The structure of spring-damper element is introduced in Section 2.2.

As shown in Fig.2, let X_0Y_0 be the coordinate system of the inner sleeve whose origin is located at the center of the inner sleeve. As the inner sleeve is assumed to be fixed, X_0Y_0 can be considered as the global coordinate system. A body reference X_1Y_1 is established at the center of the outer sleeve, whose location and orientation with respect to the global coordinate system are defined by a set of coordinates called reference coordinates and denoted as $\mathbf{q}$. The vector $\mathbf{q}$ is written in a partitioned form as

$$\mathbf{q} = [\Delta^T, \psi]^T \tag{1}$$

where $\Delta = [x, y]^T$ is a set of Cartesian coordinates the define the location of the origin of the body reference and ψ is a set of rotational coordinate that describe the orientation of the body reference.

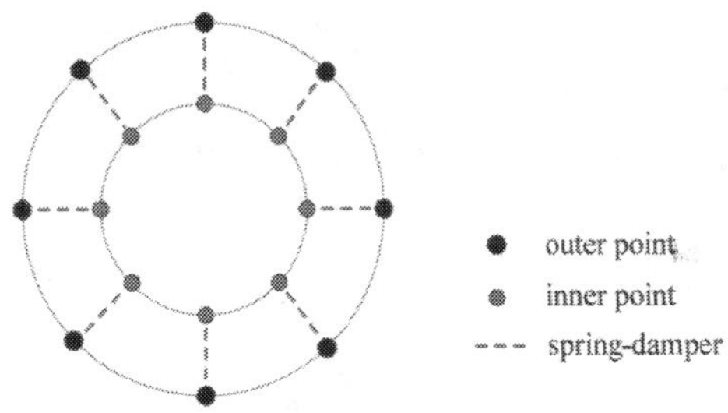

Fig.1 Structure of the coupled model

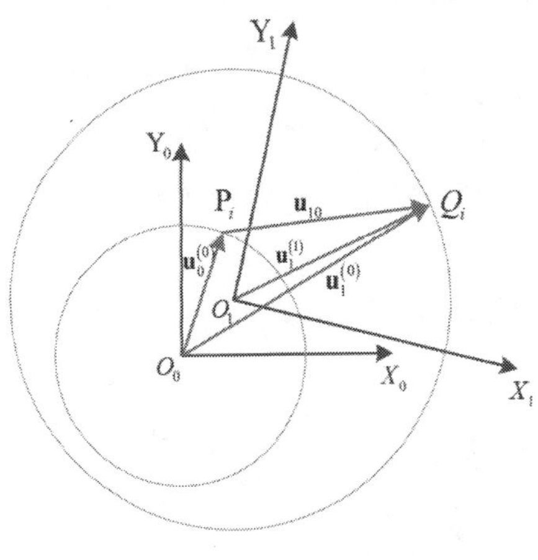

Fig.2 Relative motion between outer and inner sleeves

Assume that the radius of the outer sleeve is R; the radius of the inner sleeve is r. An arbitrary inner point P_i on the inner sleeve can be written in the X_0Y_0 coordinate system as

$$\mathbf{u}_0^{(0)} = [r\cos\frac{2\pi}{M}(i-1), r\sin\frac{2\pi}{M}(i-1)]^T \tag{2}$$

As mentioned above, M is number of points distributed in each layer; point P_i represents the ith point on the discretized inner sleeve. Correspondingly, an arbitrary outer point Q_i on the outer sleeve is written in the coordinate system X_1Y_1 as

$$\mathbf{u}_1^{(1)} = [R\cos\frac{2\pi}{M}(i-1), R\sin\frac{2\pi}{M}(i-1)]^T \tag{3}$$

In order to calculate the deformation of the spring-damper between points P_i and Q_i, $\mathbf{u}_1^{(1)}$ need to be transformed to the coordinate system X_0Y_0, yield

$$\mathbf{u}_1^{(0)} = \boldsymbol{\Delta} + \mathbf{A}^{(01)}\mathbf{u}_1^{(1)} \tag{4}$$

where $\mathbf{u}_1^{(0)}$ is the position vector of point Q_i in the coordinate system X0Y0; $\mathbf{A}^{(01)}$ denotes the transformation matrix from the reference body coordinate system to the global coordinate system. The transformation matrix $\mathbf{A}^{(01)}$ is expressed by

$$\mathbf{A}^{(01)} = \begin{bmatrix} \cos\psi & -\sin\psi \\ \sin\psi & \cos\psi \end{bmatrix} \tag{5}$$

After that $\mathbf{u}_1^{(0)}$ is obtained, the vector between points P_i and Q_i is expressed in the global coordinate system X0Y0

$$\mathbf{u}_{10} = \mathbf{u}_1^{(0)} - \mathbf{u}_0^{(0)} \tag{6}$$

As shown in Fig.3, assume that the distance between points P_i and Q_i is $|\mathbf{u}_{10}(t)|$ at present time step and $|\mathbf{u}_{10}(t-\Delta t)|$ at the last time step. The relative velocity between the points P_i and Q_i at present time is approximated by

$$|\dot{\mathbf{u}}_{10}(t)| = \frac{\mathbf{u}_{10}(t) - \mathbf{u}_{10}(t-\Delta t)}{\Delta t} \tag{7}$$

where t denotes the time and Δt denotes the small time step.

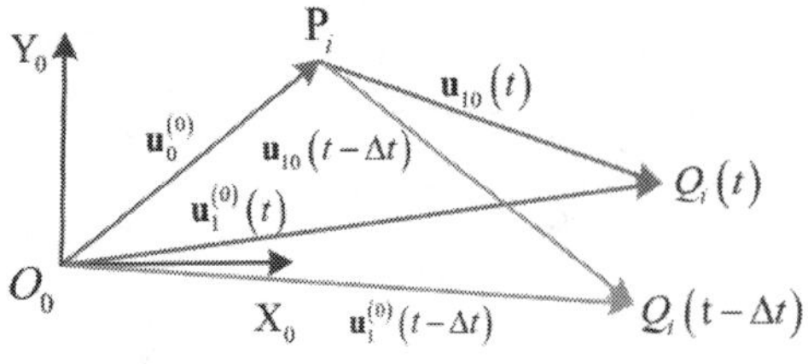

Fig.3 Relative velocity between outer and inner sleeve

The force of the spring-damper element depends on the constitutive relationship. It is noted that the spring-damper element can be formulated by a variety of known constitutive models, such as Kelvin-Voigt model, Maxwell model, Zener model, etc. All the spring-damper elements are assumed to be identical with the same model parameters. Without losing any generality, the force of the spring-damper element at present time t is denoted as $\mathbf{F}(|\mathbf{u}_{10}(t)|, |\dot{\mathbf{u}}_{10}(t)|, t)$.

After that the spring-damper force between points P_i and Q_i is obtained, all the spring-damper forces should be summed to get the total force which acts on the center of mass of the inner sleeve. First of all, the force $\mathbf{F}(|\mathbf{u}_{10}(t)|, |\dot{\mathbf{u}}_{10}(t)|, t)$ is decomposed in the global coordinate system in the form of

$$\mathbf{F}(|\mathbf{u}_{10}(t)|, |\dot{\mathbf{u}}_{10}(t)|, t) = F_x\boldsymbol{i} + F_y\boldsymbol{j} \tag{8}$$

where $\boldsymbol{i}$ and $\boldsymbol{j}$ are the unit vectors along the X_0 and Y_0 axes, respectively. The force $\mathbf{F}(|\mathbf{u}_{10}(t)|, |\dot{\mathbf{u}}_{10}(t)|, t)$ which acts on point P_i then is moved to the center of mass of the inner sleeve, point O_0. This movement introduces a moment vector of couple, yields

$$\mathbf{M} = \mathbf{u}_0^{(0)} \times \mathbf{F}(|\mathbf{u}_{10}(t)|, |\dot{\mathbf{u}}_{10}(t)|, t) = M_z\boldsymbol{k} \tag{9}$$

where $\boldsymbol{k} = \boldsymbol{i} \times \boldsymbol{j}$.

The total force which acts on the center of mass is

$$\mathbf{F}_{total} = \sum_{i=1}^{M}\mathbf{F} = \boldsymbol{i}\sum_{i=1}^{M}F_X + \boldsymbol{j}\sum_{i=1}^{M}F_Y \tag{10}$$

The total moment of couple which acts on the inner sleeve is

$$\mathbf{M}_{total} = \sum_{i=1}^{M}\mathbf{M} = \boldsymbol{k}\sum_{i=1}^{M}M_z \tag{11}$$

It is noted that Eq. (10) and (11) are used to describe the coupling effects between the radial and torsional deformations of the bushing. Eq.(10) and (11) represents the coupling part of the proposed bushing model. A modification is then made to achieve more stable and predictable behavior of the bushing model, as follows

$$F_X^* = F_X + k_x x + c_x\dot{x} \tag{12}$$

$$F_Y^* = F_Y + k_y y + c_y\dot{y} \tag{13}$$

$$M_Z^* = M_Z + k_z\psi + c_z\dot{\psi} \tag{14}$$

Eq. (12) to (14) represent the final formulation of the bushing model, where F_X^*, F_Y^*, M_Z^* represent the three force components of the bushing model, $k_x, k_y, k_z, c_x, c_y, c_z$ represent the parameters to be estimated. It can be seen that the force and moment of couple which act on the inner sleeve can be calculated through the general displacement vector $\mathbf{q} = [\Delta^T, \psi]^T$ which represents the deformations of the bushing. An important issue is to give the exact formulation of $\mathbf{F}(|\mathbf{u}_{10}(t)|, |\dot{\mathbf{u}}_{10}(t)|, t)$, the constitutive equation of the spring-damper element.

2.2 Spring-damper element

The spring-damper element can be expressed by any available one-dimensional lumped parameter rubber model. In this study, the spring-damper element is shown in Fig.4. The model is composed of two components:

$$F = F_e + F_v \tag{15}$$

where F_e represents the elastic force; F_v represents the viscous force; k_1 represents the main spring which can be linear or nonlinear; k_2 , c_1 and c_2 are linear spring and dashpot elements. The static friction force for the amplitude dependency is not considered in the spring-element as well as the whole bushing model.

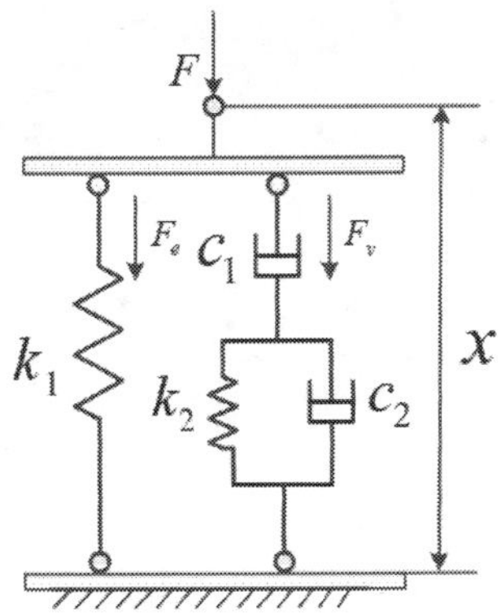

Fig.4. Spring-damper element

The elastic force F_e at the main spring k_1 is simply expressed by

$$F_e = k_1 x \tag{16}$$

The viscous force F_v at the dashpot is

$$F_v = c_1(\dot{x} - \dot{x}_1) = k_2 x_1 + c_2 \dot{x}_1 \tag{17}$$

where x is the overall displacement and x_1 is the displacement at the second spring k_2 ; x_1 is estimated by

$$\dot{x}_1 = \frac{c_1 \dot{x} - k_2 x_1}{c_1 + c_2} \tag{18}$$

The computational algorithm for the coupled bushing model is summarized in Fig.5.

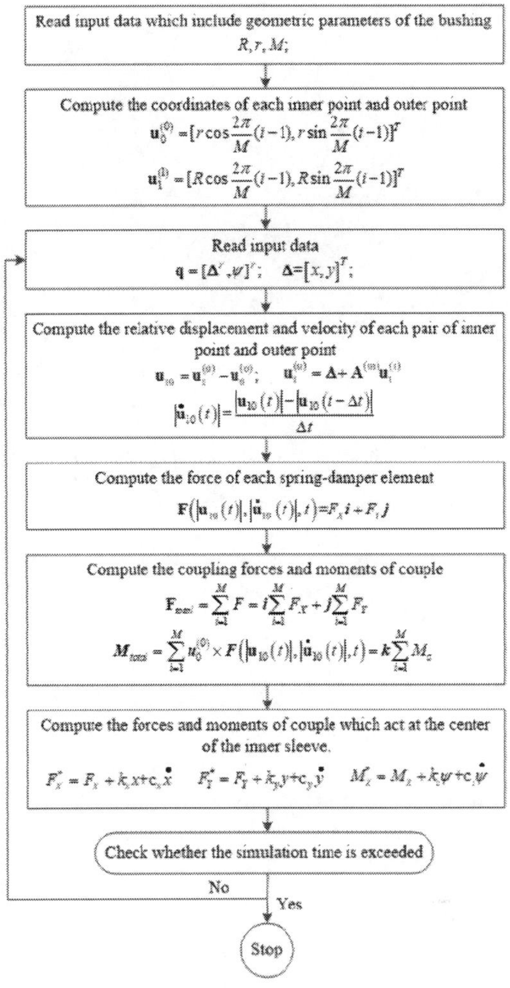

Fig.5 Computational algorithm for the 2-dimensional rubber bushing model

3. Finite element modelling and simulation

Limited to the lack of multi-axis rubber testing, FE method is used for modelling and simulation of rubber bushing subjected to coupled mode of deformation. FE methods are previously widely used to model rubber bushings subjected to coupled deformation modes and show reasonable accuracy [45-50].

3.1 FE rubber bushing model

The viscoelastic bushing model is adopted from Ref [52]. The bushing has an inner radius of r=13mm and outer radius of R=26mm. It is assumed that the bushing is long enough for plane strain deformation to occur. That is to say, only radial and torsional deformations are considered. The mesh is shown in Fig.6. The bushing is modelled with first-order reduced-integration elements.

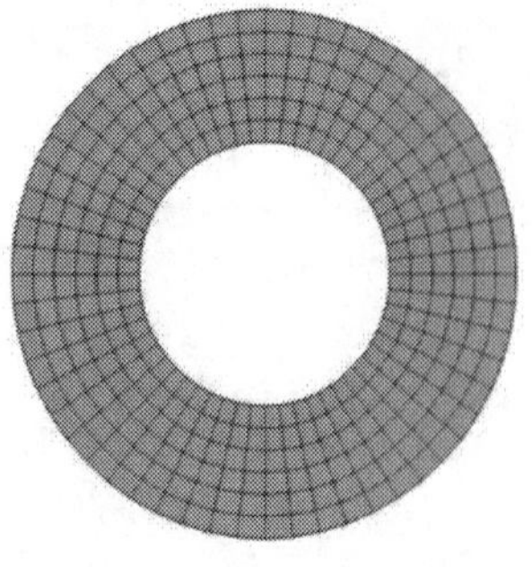

Fig.6 Finite element model of viscoelastic bushing

Similar with the proposed bushing model, the outer sleeve and inner sleeve of the FE model are modelled as rigid bodies. The outer sleeve is modelled by fixing displacement components at all the outside nodes. The inner sleeve is modelled by connecting all the inner nodes to a node located in the center of the model.

The material model is based on the overlay concept. The instantaneous behavior is defined by a Mooney-Rivlin model with the constants $C_{10} = 27.56MPa$, $C_{01} = 6.89MPa$ and $D_1 = 0.0029MPa^{-1}$. The viscous behavior is defined by a time-dependent shear modulus $G(\tau)$ and a time-dependent bulk $K(\tau)$, each of which is expanded in a Prony series

$$G(\tau) = G_0\left(g_\infty + \sum_{i=1}^{N_G} g_i e^{-\tau/\tau_i^G}\right)$$
$$K(\tau) = K_0\left(k_\infty + \sum_{i=1}^{N_K} k_i e^{-\tau/\tau_i^K}\right) \quad (19)$$

where G_0 and K_0 are instantaneous moduli; g_i and k_i are relative moduli. Note that $g_\infty + \sum_{i=1}^{N_G} g_i = k_\infty + \sum_{i=1}^{N_K} k_i = 1$. In general, it is assumed that $\tau_i = \tau_i^K = \tau_i^G$. The parameters are listed in Table 1.

Table 1.Material parameters for the FE model

i	k_i	g_i	τ_i (sec)
1	0.2	0.5	0.1
2	0.1	0.2	0.2

3.2 FE Simulation

The simulation tasks include one-axis simulation and two-axis simulation. Harmonic displacement inputs and step displacement inputs are imposed on each axis of the bushing. The simulation results are used for the parameter estimation of the proposed coupled lumped parameter model.

3.2.1 One-axis simulation with step input excitation

The rubber bushing shows relaxation behavior under step input excitation. The relaxation simulations are performed with different displacements and rise times, as shown in Table 2.

Table 2.Step inputs with different displacements and rise times

	Radial	**Torsional**
Displacement	2mm	10deg
Rise time	0.1s, 0.2s, 0.5s	0.1s, 0.2s, 0.5s

Fig.7 shows the radial forces with respect to the rise times under the displacement 2mm. The maximum bushing force increases with the decreasing rise time. Fig.8 shows the torsional torques with respect to the rise time under the angle 10deg. The maximum bushing torque increases with the decreasing rise time.

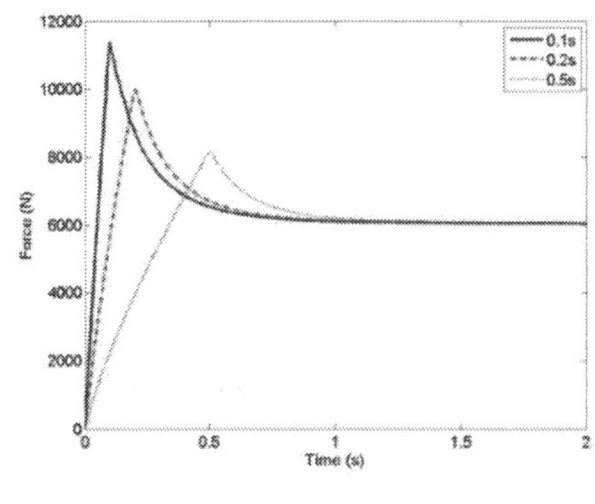

Fig.7 Radial forces with respect to the rise times

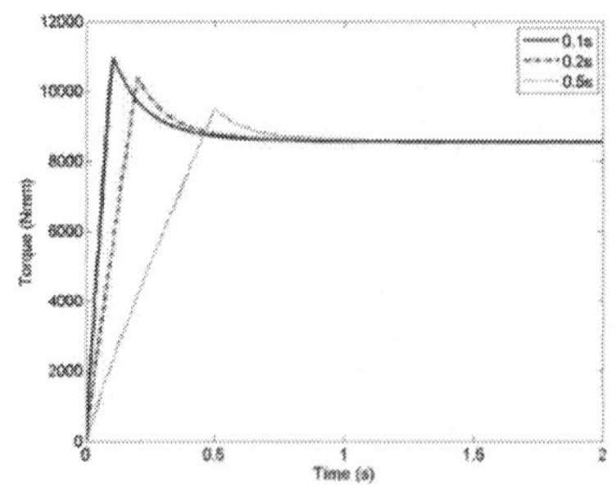

Fig.8 Torsional torques with respect to the rise times

3.2.2 One-axis simulation with harmonic excitation

The harmonic dynamic simulations are performed with different frequency, as shown in Table 3. It is noted that a static prestrain is applied before harmonic excitation. The prestrain is 2mm for the radial direction and 10 degrees for the torsional direction.

Table 3.Harmonic inputs with different frequencies

	Radial	Torsional
Prestrain	2mm	10deg
Amplitude	0.5mm	5deg
Frequency	1Hz, 5Hz, 10Hz	1Hz,5Hz,10Hz

Fig.9 shows the radial forces with different excitation frequencies. The force magnitude increases with the increasing frequency. Fig.10 shows the torsional torques with different excitation frequencies. Similarly, the torque magnitude increases with the increasing frequency.

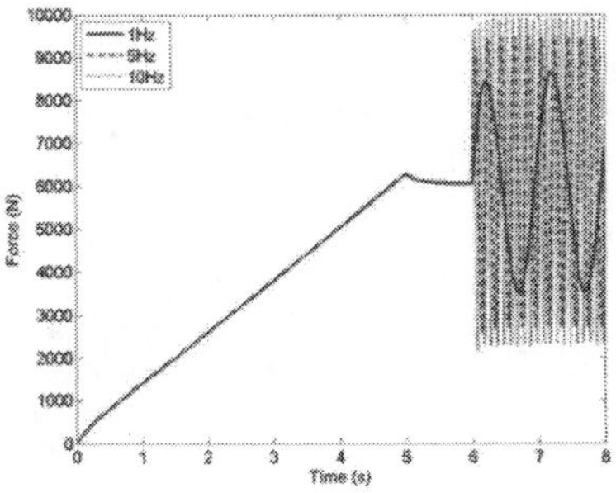

(a)

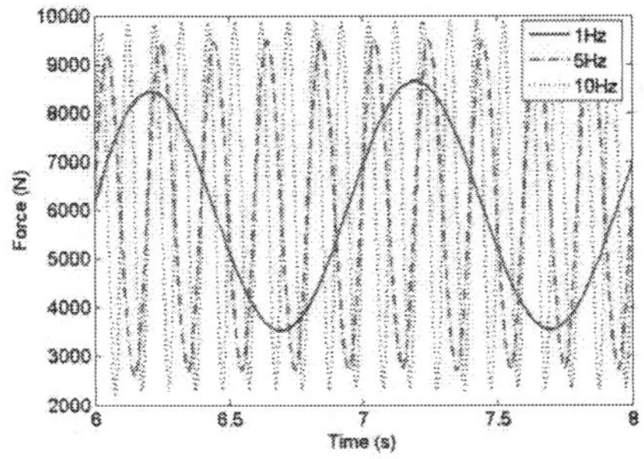

(b)

Fig.9 Radial forces under harmonic excitations: (a) global view; (b) local view

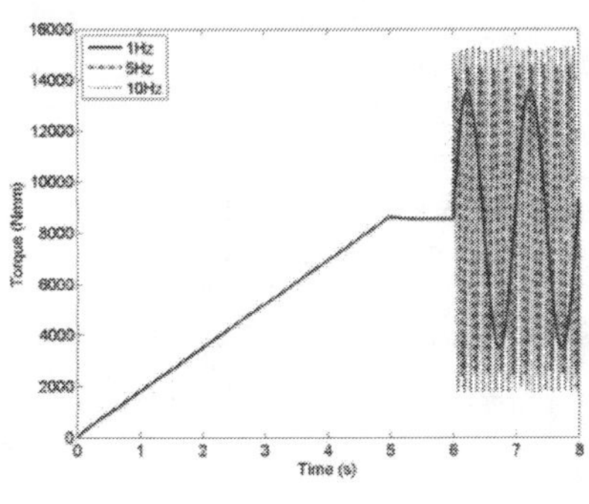

(a)

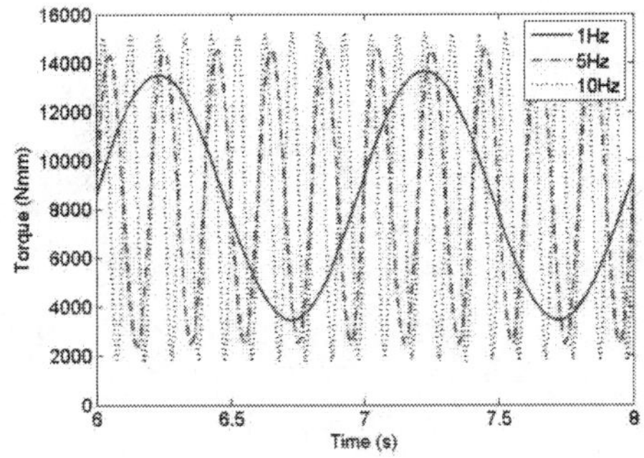

(b)

Fig.10 Torsional torques under harmonic excitations: (a) global view; (b) local view

3.2.3 Two-axis simulation with step input excitation

The specifications of coupled simulations with step inputs are shown in Table 4. In both the radial direction and torsional direction, step inputs are applied, simultaneously.

Table 4.Specifications of coupled simulation with step input

	Radial	Torsional
Displacement	1mm, 2mm, 3mm	10deg, 20deg, 30deg
Rise time	0.1s	0.1s

Fig.11 shows the bushing radial forces with a 2mm radial input and different torsional inputs (0deg, 10deg, 20deg and 30deg). It is found that the radial force is not sensitive to torsional angle input for this rubber bushing. Fig.12 shows the bushing torques with a 10deg torsional input and different radial inputs (0mm, 1mm, 2mm and 3mm). It is found that the torsional torque is sensitive to radial displacement input.

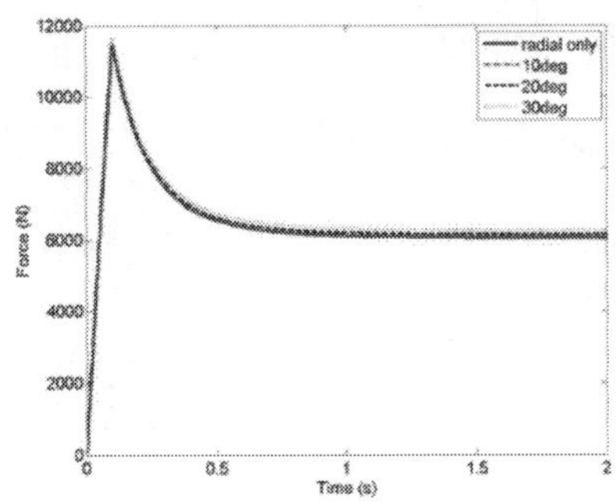

Fig.11 Radial forces with different step torsional angle inputs

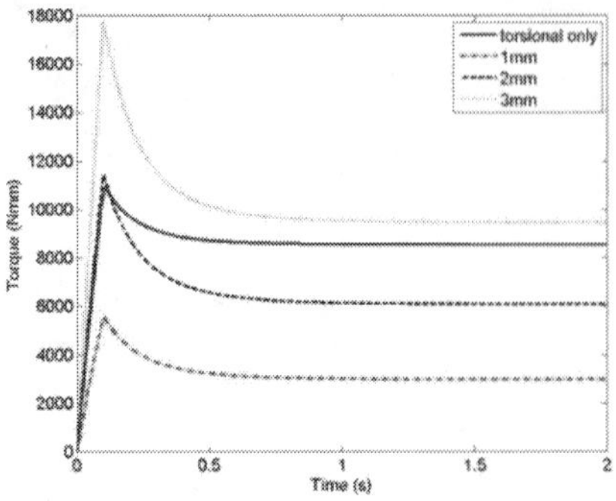

Fig.12 Torsional torques with different step radial displacement inputs

3.2.4 Two-axis simulation with harmonic excitation

The specifications of coupled simulation with harmonic excitation are shown in Table 5. The harmonic inputs are applied in both radial and torsional directions, simultaneously. It is noted that the pre-strains are applied in the radial direction and torsional direction before the harmonic excitations are applied.

Table 5.Specifications of coupled simulation with harmonic excitation

	Radial	**Torsional**
Prestrain	2mm	10deg
Amplitude	0.5mm	5deg
Frequency	5Hz, 10Hz	5Hz,10Hz

Fig.13 shows the radial forces with a radial harmonic input (5Hz) and different torsional harmonic inputs (5Hz, 10Hz). It is observed that the radial force is not sensitive to the torsional angle for this rubber bushing. Fig.14 shows the torsional torques with a torsional harmonic input (5Hz) and different radial harmonic inputs (5Hz, 10Hz). It is observed that the torsional torque is sensitive to the radial displacement.

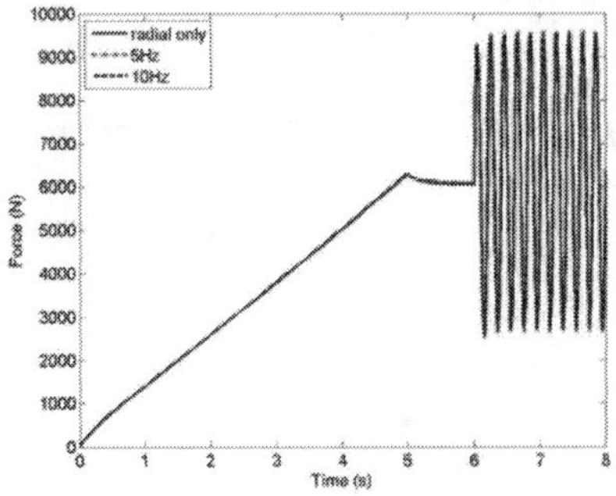

(a)

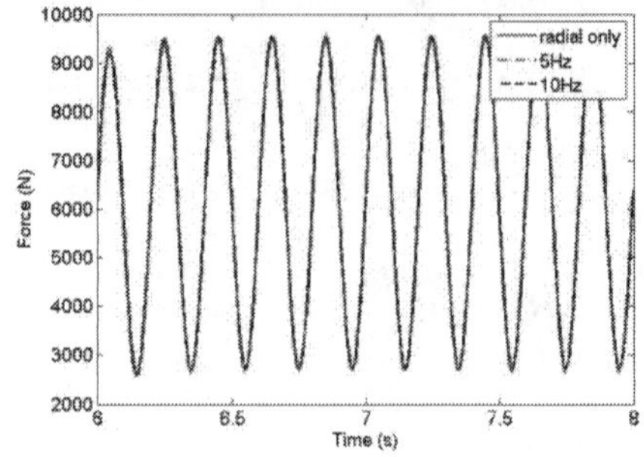

(b)

Fig.13 Radial forces with different harmonic torsional angle inputs: (a) global view; (b) local view

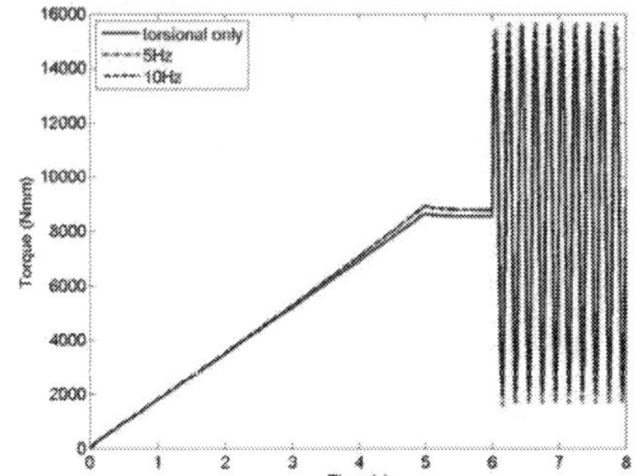

(a)

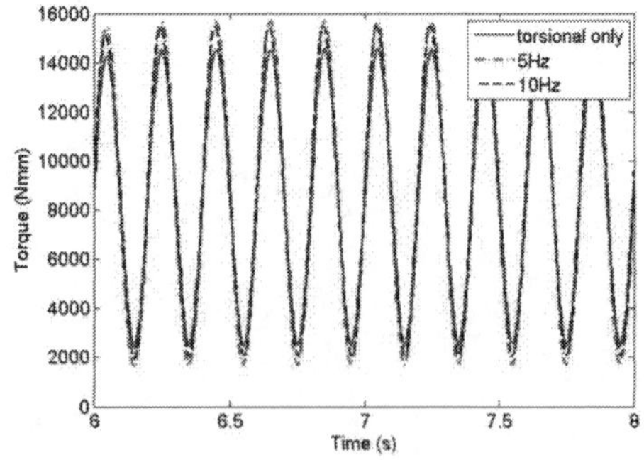

(b)

Fig.14 Torsional torques with different radial harmonic displacement inputs: (a) global view; (b) local view

For the bushing in consideration, the torsional torque is obviously affected by radial deformation. The radial force is also slightly affected by the torsional deformation. It again proves the coupling effects of the rubber bushing deformations. To develop the coupled bushing model is of significant importance.

4. Parameter identification

For the proposed bushing model, the parameter identification is a rather complex issue because it should include the radial, torsional and radial-torsional deformation modes of the bushing. The proposed bushing model is made up of the inner sleeve, outer sleeve and M spring-damper elements. The value of M is assumed to be 16. All the spring-damper elements are assumed to be identical. Thus the number of the parameters of the bushing model is relatively small. The parameters to be estimated include k_x,c_x,k_y,c_y, k_1, k_2, c_1 and c_2. Due to the symmetry of the rubber bushing, $k_x = k_y$, $c_x = c_y$, the final parameters to be estimated include k_x , c_x , k_1 , k_2 , c_1 and c_2 . A parameter identification procedure is explored in the following sections.

4.1 Identification from decoupled radial and torsional inputs

Radial inputs include step inputs and harmonic inputs. The parameter identification in response to radial step input is carried out by using the FE simulation with a displacement of 2mm and a rise time of 0.1s. The maximum bushing force and relaxation phenomenon should be included in the following objective functions:

$$E_1 = \left|F_{FE} - F_{\mathrm{Pro}}\right|_{t=T^*} \tag{20}$$

$$E_2 = \left|F_{FE} - F_{\mathrm{Pro}}\right|_{t=1.5s} \tag{21}$$

where F_{FE} denotes the FE model and F_{Pro} denotes the proposed model. Eq. (20) represents the maximum bushing force difference between the FE model and proposed model. Eq. (21) represents the bushing force difference at 1.5s.

The parameter identification to radial harmonic input is carried out by using the FE simulation with frequency of 1Hz and amplitude of 0.5mm. The corresponding objective function is as follows:

$$E_3 = \frac{1}{n}\sum_{i=1}^{n}\left|F_{FE} - F_{\mathrm{Pro}}\right| \tag{22}$$

where n is the number of data points.

Torsional inputs include step inputs and harmonic inputs. The parameter identification in response to torsional step input is carried out by using the FE simulation with a displacement of 10 degrees and a rise time of 0.1s. The maximum bushing torsional torque and relaxation phenomenon are included in the following objective functions:

$$E_4 = \left|M_{FE} - M_{\mathrm{Pro}}\right|_{t=T^*} \tag{23}$$

$$E_5 = \left|M_{FE} - M_{\mathrm{Pro}}\right|_{t=1.5s} \tag{24}$$

where M_{FE} denotes the FE model and M_{Pro} denotes the proposed model. Eq. (23) represents the maximum torsional torque difference between the FE model and proposed model. Eq. (24) represents the bushing torsional torque difference at 1.5s.

The parameter identification to torsional harmonic input is carried out by using the FE simulation with frequency of 1Hz and amplitude of 5 degrees. The corresponding objective function is as follows:

$$E_6 = \frac{1}{m}\sum_{i=1}^{m}\left|M_{FE} - M_{\mathrm{Pro}}\right| \tag{25}$$

where m is the number of data points. In order to improve the accuracy of the parameter identification results, the objective function for the combined radial and torsional inputs is used, as follows

$$e_1 = \sum_{i=1}^{6}\omega_i E_i \tag{26}$$

where $\omega_i, i = 1,2,\cdots 6$ are weighting factors such that $\sum_{i=1}^{6}\omega_i = 1$. It is noted that the radial and torsional inputs are applied separately in Eq. (26). The optimization problem is finished in the environment of MATLAB software. The gradient-based method, sequential quadratic programming is used to seek the minimum. The identification results based on uniaxial inputs are shown in Table 6.

Table 6 Parameter identification results from uniaxial radial and torsional inputs

Parameter	Value	Unit	Parameter	Value	Unit
k_x	977.2	N/mm	k_1	276.0	N/mm
c_x	5.3	N/(mm/s)	c_1	72.4	N/(mm/s)
k_z	13294.3	N·mm/rad	k_2	100.3	N/mm
c_z	1938.6	N·mm/(rad/s)	c_2	2.3	N/(mm/s)

Fig.15 to 18 shows the comparisons of FE model and proposed model under radial or torsional inputs. The simulation results of FE model and the proposed model are in good agreement. It means that the proposed model shows relative accurate behaviors under various excitations.

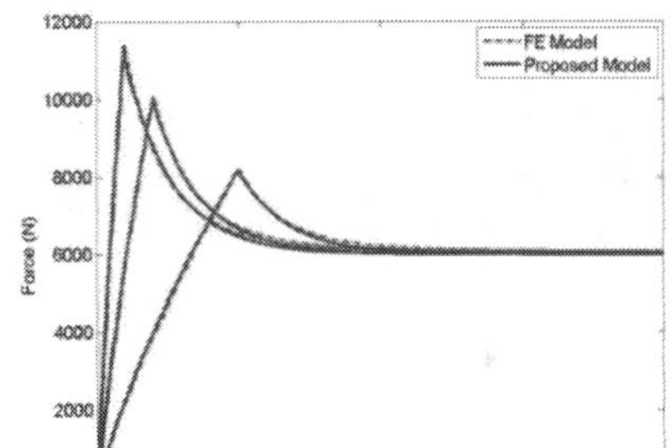

Fig.15 Radial force of FE model and proposed model (step input)

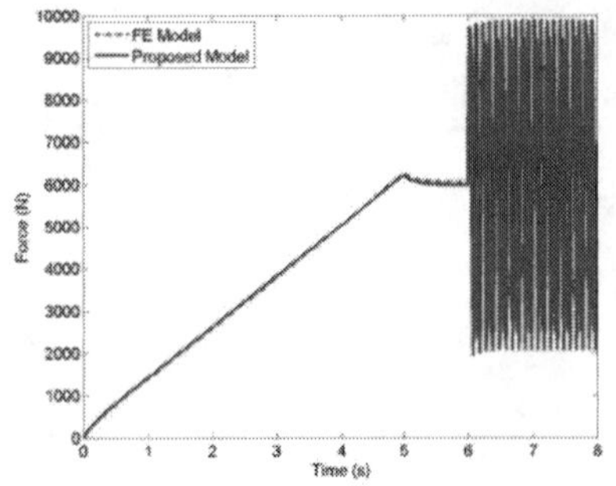

(a)

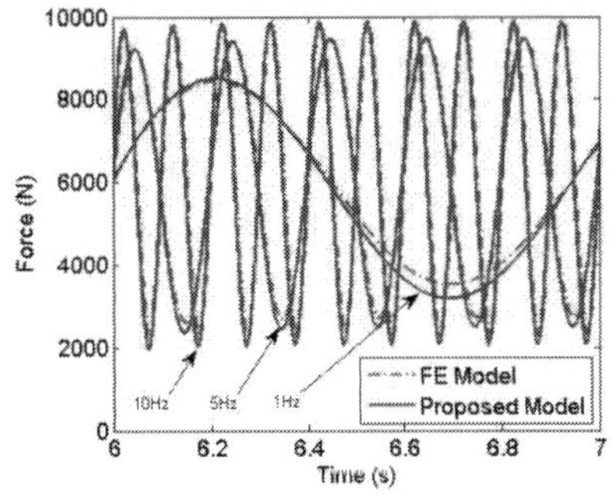

(b)

Fig.16 Radial force of FE model and proposed model (harmonic input): (a) global view; (b) local view

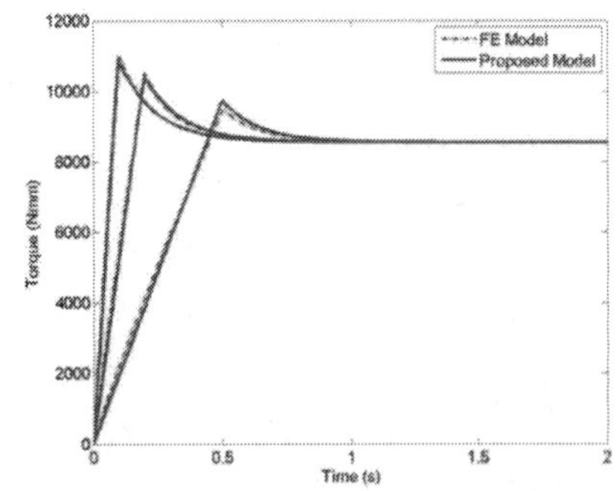

Fig.17 Torsional torque of FE model and proposed model (step input)

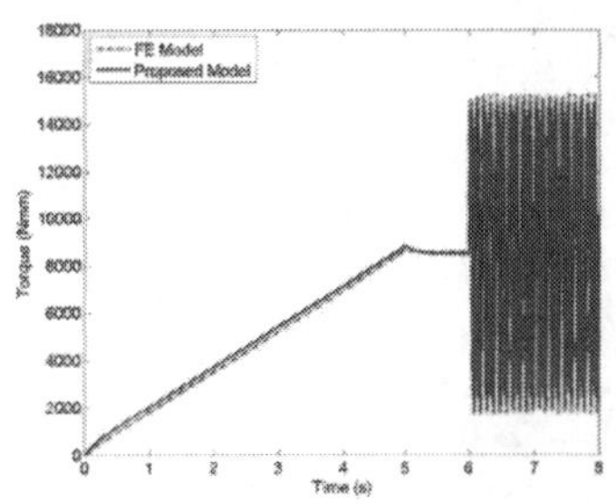

(a)

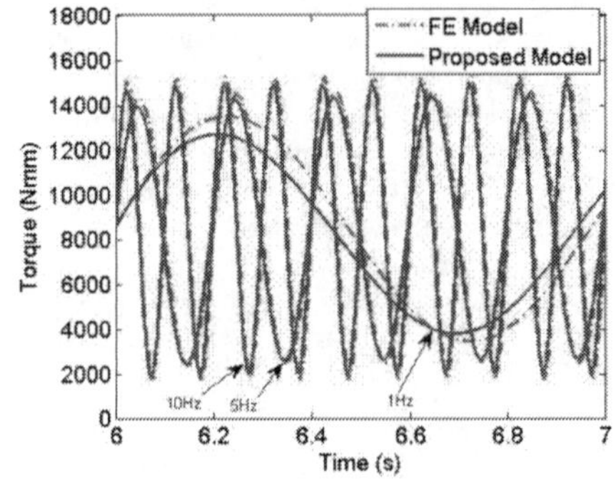

(b)

Fig.18 Torsional torque of FE model and proposed model (harmonic input): (a) global view; (b) local view

4.2 Identification from coupled radial and torsional inputs

The coupled radial and torsional inputs are not used in the parameter identification process in Section 4.1. In this section, the parameters obtained in Table 6 are slightly modified by optimizing the following objective function:

$$e_2 = \sum_{i=1}^{m} \left| M_{FE} - M_{\mathrm{Pro}} \right| \tag{27}$$

where M_{FE} and M_{pro} the bushing torque with a 10deg torsional input and a 1mm radial input. The initial values shown in Table 6 are used for this optimization problem, and each parameter is allowed only a 10% variation of its initial value. The final optimization results, i.e. the model parameters are determined and shown in Table 7. Figure 19 to 21 shows the bushing responses under different kinds of coupled excitations. Some errors are observed, but these errors can be accepted.

Table 7 Parameter identification results from coupled radial and torsional inputs

Parameter	Value	Unit	Parameter	Value	Unit
k_x	1021.2	N/mm	k_1	257.6	N/mm
c_x	8.4	N/(mm/s)	c_1	68.8	N/(mm/s)
k_z	12554.4	N·mm/rad	k_2	107.1	N/mm
c_z	1755.6	N·mm/(rad/s)	c_2	2.1	N/(mm/s)

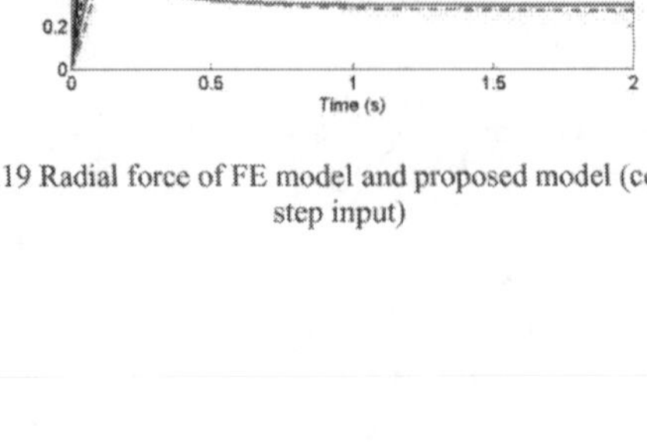

Fig.19 Radial force of FE model and proposed model (coupled step input)

https://doi.org/10.1051/matecconf/202338201023

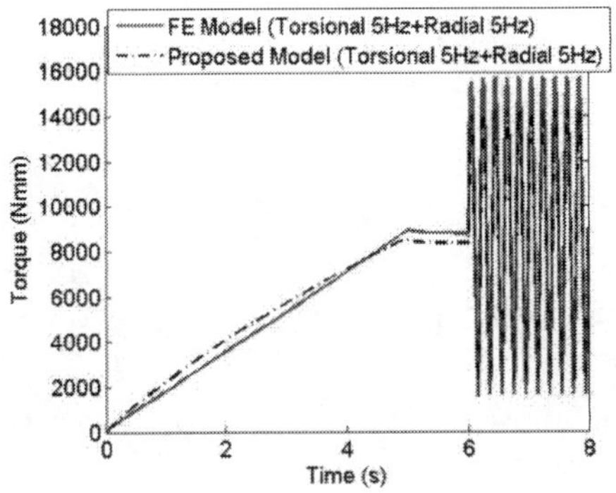

Fig.20 Torsional torque of FE model and proposed model (coupled harmonic input: Torsional 5Hz; Radial 5Hz)

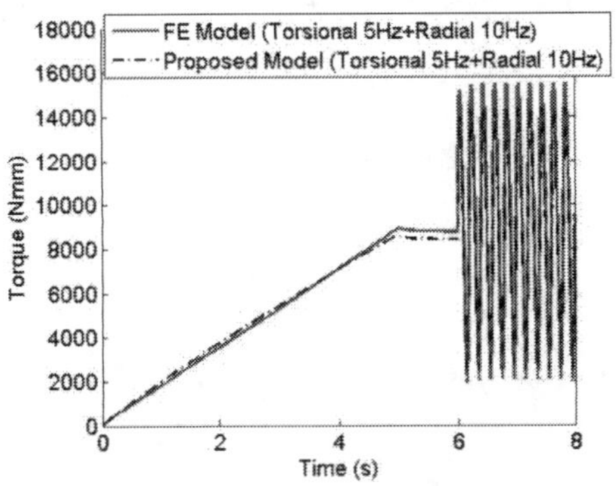

Fig.21 Torsional torque of FE model and proposed model (coupled harmonic input: Torsional 5Hz; Radial 10Hz)

Using the obtained parameters shown in Table 7, the rubber bushing responses under uniaxial excitations should be examined again. Fig.22 to 25 shows the rubber bushing behaviors under uniaxial step or harmonic inputs. The proposed model shows relative good agreement with the FE model. In comparison with Fig.15 to 18, the errors between the proposed model and FE model become larger, because coupling effects are taken into account. Overall, the proposed model with the parameters in Table 7 shows relative accurate and stable behaviors under various kinds of deformations.

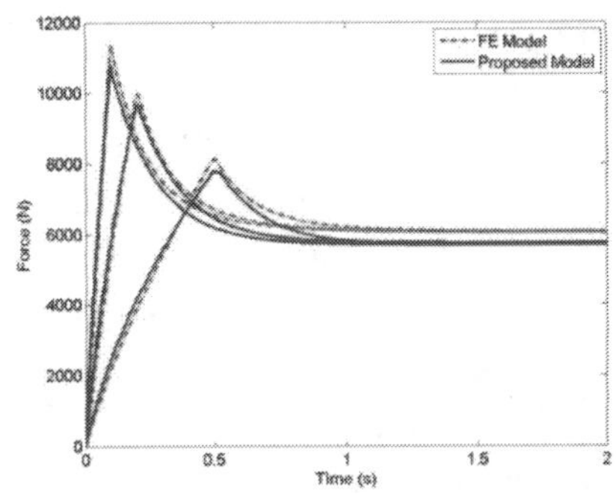

Fig.22 Radial force of FE model and proposed model (step input)

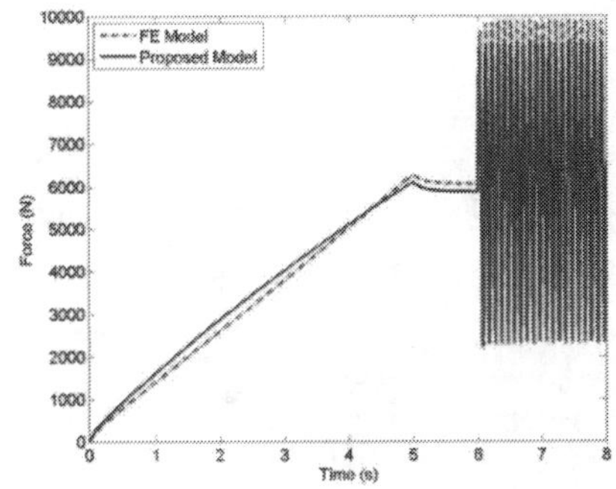

Fig.23 Radial force of FE model and proposed model (harmonic input)

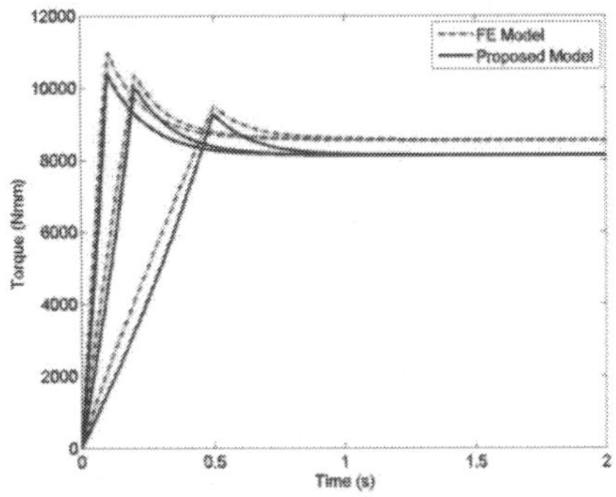

Fig.24 Torsional torque of FE model and proposed model (step input)

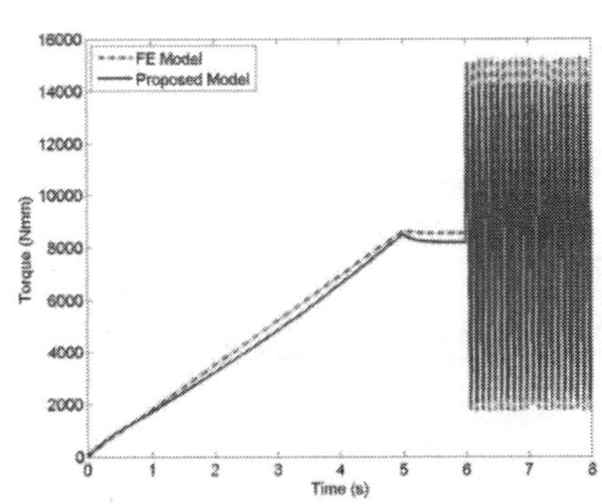

Fig.25 Torsional torque of FE model and proposed model (harmonic input)

5. Multibody dynamic simulations with rubber bushing

Both bearings and rubber bushings are widely used for the joints in multibody systems. The use of rubber bushing joints leads to a decrease of the number of kinematic constraints. One-dimensional models are widely presented and applied. A simple multibody model is used to illustrate the difference between the proposed two-

 https://doi.org/10.1051/matecconf/202338201023

dimensional bushing model and one-dimensional bushing models.

The schematic of a double pendulum is shown in Fig.26. Both rods are symmetric and homogeneous, and the centers of masses are at the midpoints. The parameters associated with the model are $l_1 = l_2 = 500mm$ and $m_1 = m_2 = 100kg$. The initial configuration of the model is $\theta_1 = \theta_2 = \pi / 4$. The rubber bushing joint is modelled by the proposed bushing model and one-dimensional bushing models, respectively. The one-dimensional bushing models include a radial model and a torsional model, whose constitutive structure is show in Fig.4. It is the same with the spring-damper element used in the proposed model. The parameters of one-dimensional models are estimated based on the corresponding FE simulations in Section 3.2.1 and 3.2.2. The parameters of the radial and torsional models are shown in Table 8.

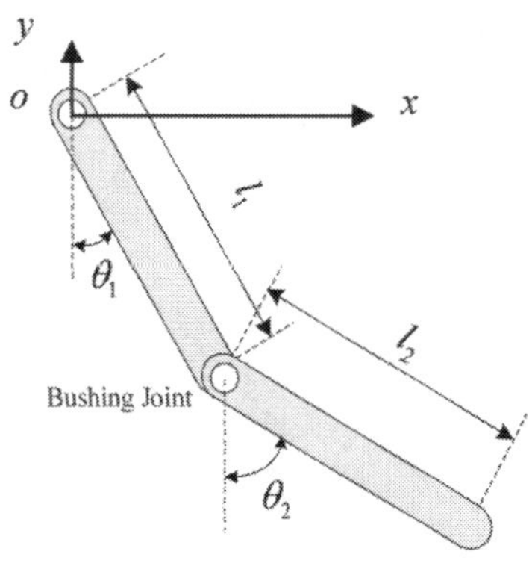

Fig.26 Multibody model with bushing joint

Table 8 Parameters of the one-dimensional rubber bushing models

	k_1	c_1	k_2	c_2
Radial model	2991.9 (N/mm)	568.7 (Ns/mm)	3623.0(N/mm)	16.5 (Ns/mm)
Torsional model	43335(N/rad)	23094(Ns/rad)	15236(N/rad)	567(Ns/rad)

The multibody model is established in software ADAMS. The user subroutine is programmed in FORTRAN to compute the forces associated with the proposed model as well as the one-dimensional models when a given displacement is applied to the rubber bushing. Using a GFOSUB user subroutine allows to implement the time integration algorithm directly in the subroutine. Thus the proposed model programmed in user subroutine is easily applied into other complex planar multibody systems.

Fig.26 to 28 shows the comparison of the proposed two-dimensional model and traditional one-dimensional model. For the rubber bushing considered in this paper, the radial force is not very sensitive to the torsional deformation, so the radial forces of the two models in Fig.26 and 27 do not show significant differences. However, the torsional torques shows obvious differences under coupled radial-torsional deformations. This simple application shows the importance of considering coupling effects of the rubber bushing deformations.

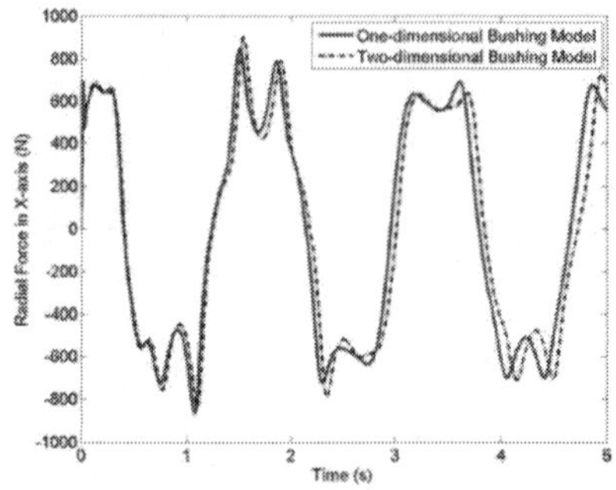

Fig.26 Radial force in the X-axis of two rubber bushing models

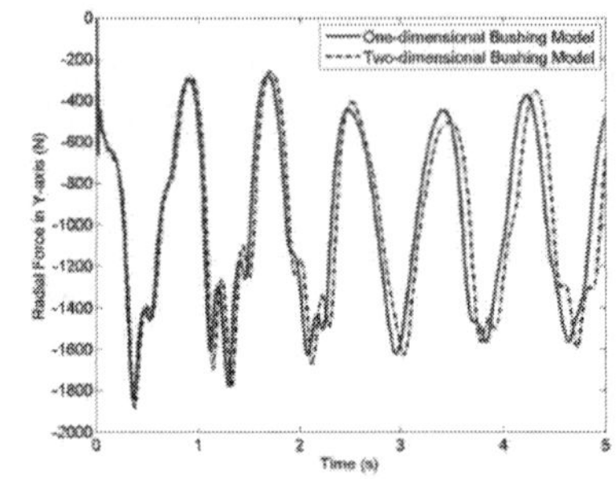

Fig.27 Radial force in the Y-axis of two rubber bushing models

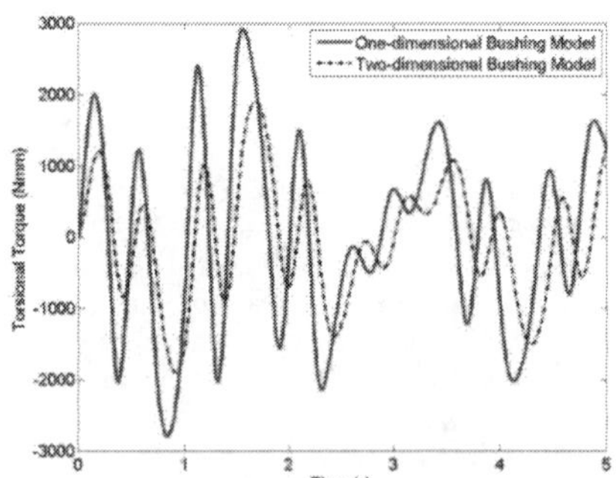

Fig.28 Torsional torque of two rubber bushing models

6. Conclusions

This work is motivated by the facts: the FE rubber models cannot be directly used in multibody systems, and the lumped parameter rubber bushing models generally cannot consider the coupled modes of deformation of bushings. This paper presents a two-dimensional lumped parameter model for rubber bushings and a parameter estimation procedure based on FE simulations. FE simulations illustrate the coupling effects between the radial and torsional deformations. The proposed model is applied into a simple multibody system, and the coupling effects of radial and torsional deformations of the rubber bushing are observed.

FE method is used to simulate the rubber bushing subjected to coupled deformations as well as uniaxial deformations. Ongoing work is addressed to carry out the multiaxial tests for rubber bushings.

Acknowledgements

This research was supported by the National Natural Science Foundation of China (11991034). This financial support is gratefully acknowledged.

References

1. Höfer P., Lion A., Modelling of frequency- and amplitude-dependent material properties of filler-reinforced rubber. Journal of the Mechanics and Physics of Solids, 57:500-520, 2009.
2. Payne A.R., Whittaker R.E., Low strain dynamic properties of filled rubbers. Rubber Chemistry and Technology, 44:440-478, 1971.
3. Fletcher W.P., Gent A.N., Non-linearity in the dynamic properties of vulcanized rubber compounds. Rubber Chemistry and Technology, 27(1):209-222, 1954.
4. Mullins L., Softening of rubber by deformation. Rubber Chemistry and Technology, 42:339-362, 1969.
5. Thorin A., Azoug A., Constantinescu A., Influence of prestrain on mechanical properties of highly-filled elastomers: measurement and modeling. Polymer Testing, 31: 978-986, 2012.
6. Wollscheid D., Lion A., Predeformation- and frequency-dependent material behavior of filler-reinforced rubber: experiments, constitutive modelling and parameter identification. International Journal of Solid and Structures, 50: 1217-1225, 2013.
7. Lion A., On the large deformation behavior of reinforced rubber at different temperatures. Journal of the Mechanics and Physics of Solids, 11/12:1805-1834, 1997.
8. Delattre A., Lejeunes S., Lacroix F., Méo S., On the dynamical behavior of filled rubbers at different temperatures: experimental characterization and constitutive modeling. International Journal of Solids and Structures, 90:178-193, 2016.
9. Gracia L.A., Liarte E., Pelegay J.L., Calvo B., Finite element simulation of the hysteretic behavior of an industrial rubber. Application to design of rubber components. Finite Elements in Analysis and Design, 46: 357-368, 2010.
10. Austrell P.E., Modelling of elasticity and damping for filled elastomers. Ph.D. Thesis, Lund University, Sweden, 1997.
11. Berg M., A non-linear rubber spring model for rail vehicle dynamics analysis. Vehicle System Dynamics, 30:197-212, 1998.
12. Scheiblegger C., Roy N., Silva P.O., et al., Non-linear modelling of bushings and cab mounts for calculation of durability loads. SAE Technical Paper 2014-01-0880, 2014.
13. Pfeffer P.E., Hofer K., Simple non-linear model for elastomer and hydro-mountings to optimise overall vehicle simulation. ATZ Worldwide, 2002-5, 2002.
14. Sjőberg M., Kari L., Non-linear behavior of a rubber isolator system using fractional derivatives. Vehicle System Dynamics, 37(3):217–236, 2002.
15. Sjőberg M., Kari L., Nonlinear isolator dynamics at finite deformations: An effective hyperelastic, fractional derivative, generalized friction model. Nonlinear Dynamics, 33:32-336, 2003.
16. Alonso A., Gil-Negrete N., Nieto J., et al., Development of a rubber component model suitable for being implemented in railway dynamic simulation programs. Journal of Sound and Vibration, 332:3032-3048, 2013.
17. Gil-Negrete N., On the modelling and dynamic stiffness prediction of rubber isolators. Ph.D Thesis, University of Navarra, Spain, 2004.
18. Ramrakhyani D.S., Lesieutre G.A., Smith E.C., Modelling of elastomeric materials using nonlinear fractional derivative and continuously yielding friction elements. International Journal of Solids and Structures, 41: 3929-3948, 2004.
19. Bouc R., Modèle mathéatique d'hystérésis, Acustica, 24:16-25, 1971.
20. Wen Y.K., Method of random vibration of hysteretic systems. Journal of the Engineering Mechanics Division, 102(2):249-26, 1976.
21. Spencer B.F., Dyke S.J., Sain M.K., Carlson J.D., Phenomenological model for magnetorheological dampers. Journal of Engineering Mechanics, 123(3):230-238, 1997.
22. Ok J.K., Yoo W.S., Sohn J.H., New nonlinear bushing model for general excitations using Bouc-Wen hysteretic model. International Journal of Automotive Technology, 9:183-190, 2008.
23. Coveney V.A., Johnson D.E., Turner D.M., Rate-dependent modeling of a highly filled vulcanizate. Rubber Chemistry Technology, 73:565-577, 2000.
24. Puel G., Bourgeteau B., Aubry D., Parameter identification of nonlinear time-dependent rubber bushings models towards their integration in multibody simulations of a vehicle chassis. Mechanical Systems and Signal Processing, 36:354-369, 2013.
25. Mooney M., A theory of large elastic deformation. Journal of Applied Physics, 11:582-592, 1940.
26. Rivlin R.S., Saunders D.W., Large elastic deformations of isotropic materials VII. Experiments on the deformation of rubber. Philosophical Transactions of the Royal Society Biological Sciences, A243:251-288, 1950.

27. Yeoh O.H., Some forms of the strain energy functions for rubber. Rubber Chemistry and Technology, 66:754-771, 1993.
28. Yeoh O.H., On the Ogden strain-energy function, Rubber Chemistry and Technology 70:175–182, 1997.
29. Ogden R.W., Non-linear elastic deformations. Ellis Horwood, Chichester, 1984.
30. Arruda E.M., Boyce M.C., A three-dimensional constitutive model for the large stretch behavior of rubber elastic materials. Journal of the Mechanics and Physics of Solids, 41:389–412, 1993.
31. Vandenbroucke A., Laurent H., Hocine N., Rio G., A hyperelasto visco hysteresis model for an elastomeric behavior: experimental and numerical investigations. Computational Materials Science, 48: 495-503, 2010.
32. Hasanpour K., Ziaei-Rad S., Finite element simulation of polymer behavior using a three-dimensional, finite deformation constitutive model. Computers and Structures, 86:1643-1655, 2008.
33. Lin R.C., Schomburg U., A finite elastic–viscoelastic–elastoplastic material law with damage: theoretical and numerical aspects. Computer Methods in Applied Mechanics and Engineering, 192(13):1591-1627, 2003.
34. Gil-Negrete N., Viñolas J., Kari L., A simplified methodology to predict the dynamic stiffness of carbon-black filled rubber isolators using a finite element code. Journal of Sound and Vibration, 296: 757-776, 2006.
35. Gil-Negrete N., Viñolas J., Kari L., A nonlinear rubber material model combining fractional order viscoelasticity and amplitude dependent effects. ASME Journal of Applied Mechanics, 76: 011009-1, 2009.
36. Enelund M., Mahler L., Runesson K., Josefson B., Formulation and integration of the standard linear viscoelastic solid with fractional order rate laws. International Journal of Solids and Structures, 36:2417-2442, 1999.
37. Adolfsson K., Enelund M., Fractional derivative viscoelasticity at large deformation. Nonlinear Dynamics, 33:301-321, 2003.
38. Schmidt A., Gaul L., Finite element formulation of viscoelastic constitutive equations using fractional time derivatives. Nonlinear Dynamics, 29:37-55, 2002.
39. Kraus G., Mechanical losses in carbon black filled rubbers. Journal of applied Polymer Science, 39:75-92, 1984.
40. Ulmer J.D., Strain dependence of dynamic mechanical properties of carbon black filled rubber compounds. Rubber Chemistry and Technology, 69:15-47, 1995.
41. Kaliske M., Rothert H., Constitutive approach to rate-independent properties of filled elastomers. International Journal of Solids and Structures, 35: 2057-2071, 1998.
42. Rabkin M., Brüger T., Hinsch P., Material model and experimental testing of rubber components under cyclic deformation. Proceedings of the Third Conference on Constitutive Models for Rubbers, 319-324, 2003.
43. Austrell E.,Olsson A.K., Jonsson M., A method to analyze the nonlinear dynamic behavior of carbon-black filled rubber components using standard FE codes. Proceedings of the Second Conference on Constitutive Models for Rubbers, 231-235, 2001.
44. Lee S.B., Wineman A., A model for non-linear viscoelastic coupled mode response of an elastomeric bushing. International Journal of Non-Linear Mechanics, 35:177-199, 2000.
45. Kadlowec J., Wineman A., Hulbert G., Elastomer bushing response: experiments and finite element modeling. Acta Mechanica, 163:25-38, 2003.
46. Kadlowec J., Gerrard D., Pearlman H., Coupled axial-torsional behavior of cylindrical elastomer bushings. Polymer Testing, 28:139-144, 2009.
47. Naghdabadi R., Baghani M., Arghavani J., A viscoelastic constitutive model for compressible polymers based on logarithmic strain and its finite element implementation. Finite Elements in Analysis and Design, 62:18-27, 2012.
48. Khajehsaeid H., Arghavani J., Naghdabadi R., Sohrabpour S., A visco- hyperelastic constitutive model for rubber-like materials: A rate-dependent relaxation time scheme. International Journal of Engineering Science, 79:44-58, 2014.
49. Hasanpour K., Ziaei-Rad S., Mahzoon M., A large deformation framework for compressible viscoelastic materials: Constitutive equations and finite element implementation. International Journal of Plasticity, 25: 1154-1176, 2009.
50. Khajehsaeid H., Baghani M., Naghdabadi R., Finite strain numerical analysis of elastomeric bushings under multi-axial loading: a compressible visco-hyperelastic approach. International Journal of Mechanics and Materials in Design, 9:385-399, 2013.
51. Ok J.K., Sohn J.H., Yoo W.S., Development of nonlinear coupled mode bushing model based on the Bouc-Wen hysteretic model. Proceedings of the ASME 2007 International Design Engineering Technical Conferences & Computers and Information in Engineering Conference, DETC2007-34721, Las Vegas, Nevada, USA, 2007.
52. Abaqus 6.13 Example Problems Guide, 2013.

Time Response and Natural Frequency Analysis for Structures with Interval Parameters

Xingxing Feng, Chang Cao, Min Liu*

Wuhan Second Ship Design and Research Institute, Wuhan, Hubei 430205, China

Abstract. This paper deals with the evaluations of lower and upper bounds of dynamic responses and natural frequencies for structures with uncertain interval parameters. The structural physical parameters, geometric dimensions and initial conditions can be considered as interval parameters. The modified Chebyshev interval method (MCIM) is presented to handle the uncertainties. In comparison with the Chebyshev interval method (CIM), MICM provides with tighter bounds of the time responses and natural frequencies without a significant increase in computation time. Monte Carlo method is also used to provide with exact bounds. MICM, CIM and Monte Carlo method are applied for structures with interval parameters, and the efficiency and accuracy of CIM and MCIM are verified.

Keywords: Interval uncertainty; structure; time response; natural frequency; Chebyshev interval method; finite element analysis

1. Introduction

Probabilistic uncertainty and imprecision in structural parameters and in environmental conditions and loads are challenging phenomena in engineering analyses [1]. For instance, variations of inertial, stiffness and damping properties due to uncertainties may significantly affect the dynamics characteristics of structures. Therefore, it is of significant importance to estimate the effects of geometrical and material uncertainties on structural dynamics. The most common approach to structures with uncertainties is to model the structural parameters as random or probabilistic variables. Under these circumstances, the probabilistic approaches demand the probability density functions of the structural parameters. The common probabilistic approaches include generalized polynomial chaos method [2,3] and Monte Carlo, etc. Unfortunately, the knowledge of probability density function is not always available. As alternative tools, the non-probabilistic approaches, such as fuzzy finite element analysis [4,5], interval model [6] and convex model [7,8], have been extensively used for handling uncertainties arising in engineering problems. Moens and Hannss [9] presented a general overview of the state-of-art in interval and fuzzy finite element analysis.

The interval model is widely used when only the bounds of uncertain parameters are required. Moore [10] and Alefeld [11] done the pioneering work in the field of interval analysis. When the uncertain parameters are modelled as interval parameters, the mass, stiffness and damping matrices of the structure are also interval matrices; the time responses and eigenvalues (natural frequencies) are solved by so-called interval analysis methods. Both time response problem and eigenvalue problem have attracted much research attention in the last decades. The major challenge or objective for interval analysis is the evaluation of the narrowest intervals enclosing all possible solutions.

The evaluation of natural frequencies or eigenvalues is crucial in vibration analysis of structures in the presence of uncertainty. Qiu et al. [12] introduced the eigenvalue inclusion principle to determine the lower and upper bounds on the eigenvalues due to interval parameters. Chen et al. [13] applied matrix perturbation theory for evaluating the bounds of eigenvalue of structures. Wang et al. [14] presented a modified interval perturbation finite element method for structural design, in which the interval matrix and vector were expanded by the Taylor series. Gao [15] presented the interval factor method for the natural frequency and mode shape analysis of truss structures with interval parameters. Li et al. [16] divided the structural eigenvalue problem with interval parameters into a series of QB (quadratic programming with box constraints) problems by the use of information of the second-order partial derivatives of eigenvalues. Sofi et al. [17] presented an efficient procedure to seek the bounds of the eigenvalues, where interval uncertainties were handled following the improved interval analysis via extra unitary interval [18,19]. The common drawback of the aforementioned approaches is the calculation of the derivatives of the mass and stiffness matrices w.r.t. interval variables. It is a rather difficult and time

* Corresponding author: 549931866@qq.com

consuming task to calculate the derivatives, especially for complex structures with multi uncertain variables.

The uncertainty analysis of time response of structures has received relatively less attention in comparison with evaluation of eigenvalues. The common interval analysis methods include mathematical programming methods [20–23], perturbation methods [24–26] and polynomial series expansion methods [6,27], etc. Qiu and Wang [24,25] presented first-order perturbation methods to estimate the range of the dynamic response of structures. Xia et al. [26] presented a perturbation method based on the vertex solution theorem for the first-order derivation of the dynamic response from its central value and avoided interval extension/overestimation problems. The main disadvantage of perturbation methods is the requirement to calculate the first derivatives of the system responses w.r.t. uncertain variables. To overcome this disadvantage, Wu et al. [6] presented the Chebyshev interval method for multibody dynamics to achieve shaper and tighter bounds for meaningful solutions. The Chebyshev interval method has been widely used for multibody systems [6,28,29] and vehicle dynamic system [30], etc. In comparison with most conventional interval models, Chebyshev interval method can reduce the interval overestimation and it doesn't require derivatives of system responses w.r.t. uncertain variables.

In this paper, the Chebyshev interval method (CIM) is modified to determine the upper and lower bounds of time responses and natural frequencies of structures in the presence of uncertainties. The modified Chebyshev interval method (MCIM) achieves more accurate interval bounds and the computation time is not significantly increased. The Monte Carlo method is used to provide with the exact interval bounds although it is time consuming. In Section 2, primary knowledge about interval analysis is reviewed. In Section 3, time response and eigenvalue problems for structures with interval parameters are formulated. In Section 4, the CIM is briefly reviewed and MCIM is presented. In Section 5, numerical examples are used to show the efficiency of CIM and MCIM. In Section 6, conclusions throughout the paper are drawn.

2. Preliminaries

An interval parameter represents a range of uncertainty. For each interval parameter, there are two numbers that represent the lower and upper bounds of the parameter. A real interval $[x]$ in real set R is defined as

$$[x]=[\underline{x},\overline{x}]=\{\underline{x}<x<\overline{x}\,|\,x\in R\} \quad (1)$$

where $\underline{x}$ is the lower bound and $\overline{x}$ the upper bound. In interval analysis, each mathematical operator has a new definition. Given two intervals $[a]=[\underline{a},\overline{a}]$ and $[b]=[\underline{b},\overline{b}]$, the four arithmetic operations are defined as follows

$$[a]*[b]=\{x*y\,|\,x\in[\underline{a},\overline{a}],y\in[\underline{b},\overline{b}]\},\ *\in\{+,-,\times,\div\} \quad (2)$$

Eq. (2) can be rewritten in a detailed manner as

$$\begin{aligned}&[a]+[b]=[\underline{a}+\underline{b},\overline{a}+\overline{b}]\\&[a]-[b]=[\underline{a}-\overline{b},\overline{a}-\underline{b}]\\&[a]\times[b]=[\min(\underline{a}\cdot\underline{b},\underline{a}\cdot\overline{b},\overline{a}\cdot\underline{b},\overline{a}\cdot\overline{b}),\max(\underline{a}\cdot\underline{b},\underline{a}\cdot\overline{b},\overline{a}\cdot\underline{b},\overline{a}\cdot\overline{b})]\\&[a]\div[b]=[\min(\underline{a}/\underline{b},\underline{a}/\overline{b},\overline{a}/\underline{b},\overline{a}/\overline{b}),\max(\underline{a}/\underline{b},\underline{a}/\overline{b},\overline{a}/\underline{b},\overline{a}/\overline{b})]\end{aligned} \quad (3)$$

Note that the division operation is not defined if $0\in[b]$ in Eq. (3).

Several interval variables can form an interval vector. For instance, vector $[\mathbf{x}]$ includes k interval variables: $[x_1],[x_2],\cdots,[x_k]$and each element has an interval bound, i.e. $[x_i]=[\underline{x_i},\overline{x_i}]$,$i=1,2,\cdots,k$.

A function containing one or more interval variables is called an interval function. To provide the exact bounds of interval function is the major objective in interval analysis. However, interval function has some overestimation caused by the intrinsic wrapping effect [31], meaning that the bounds of interval function are overestimated. Therefore, an interval inclusion function is more easily to be obtained. The interval inclusion function $[f]([\mathbf{x}])$ is an inclusion of interval function $f([\mathbf{x}])$ if

$$\forall[\mathbf{x}]\in IR^n, f([\mathbf{x}])\subset[f]([\mathbf{x}]) \quad (4)$$

It is not necessary to distinguish the interval function and interval inclusion function, because interval models or methods generally overestimate the bounds of the interval function. The interval methods used in this paper, CIM and MCIM have small overestimations.

3. Problem formulation

The equation of motion for a linear uncertain structure with n degrees of freedom is formulated as

$$\begin{cases}\mathbf{M}([\mathbf{x}])\ddot{\mathbf{u}}([\mathbf{x}],t)+\mathbf{C}([\mathbf{x}])\dot{\mathbf{u}}([\mathbf{x}],t)+\mathbf{K}([\mathbf{x}])\mathbf{u}([\mathbf{x}],t)=\mathbf{F}([\mathbf{x}],t)\\ \mathbf{u}([\mathbf{x}],0)=\mathbf{u}_0([\mathbf{x}]),\dot{\mathbf{u}}([\mathbf{x}],0)=\dot{\mathbf{u}}_0([\mathbf{x}])\end{cases} \quad (5)$$

where $\mathbf{M}([\mathbf{x}]),\mathbf{C}([\mathbf{x}]),\mathbf{K}([\mathbf{x}])$ are the mass, damping and stiffness matrices, respectively; $\mathbf{F}([\mathbf{x}],t)$ is the external load vector; $\mathbf{u}_0([\mathbf{x}])$ and $\dot{\mathbf{u}}_0([\mathbf{x}])$ are the initial conditions; $[\mathbf{x}]$ is the vector of k interval parameters $[x_1],[x_2],\cdots,[x_k]$ and each element has an interval bound, i.e. $[x_i]=[\underline{x_i},\overline{x_i}]$, $i=1,2,\cdots,k$. It is noted that the analytical formulation of Eq. (5) is always unknown, and only numerical solutions can be obtained by fixing the interval variables at given values. The time response of the structure is solved using Newmark algorithm [32].

The vibration analysis of un-damped linear structure with interval parameters leads to the generalized interval eigenvalue problem:

$$\mathbf{K}([\mathbf{x}])\boldsymbol{\phi}_j([\mathbf{x}])=\lambda_j([\mathbf{x}])\mathbf{M}([\mathbf{x}])\boldsymbol{\phi}_j([\mathbf{x}]),j=1,2,\cdots,n \quad (6)$$

where $\lambda_j([\mathbf{x}]) = \omega_j^2([\mathbf{x}])$ is the *j-th* squared interval natural frequency and $\boldsymbol{\phi}_j([\mathbf{x}])$ is the corresponding eigenvector.

The symbol $[\mathbf{x}] \in [\underline{\mathbf{x}}, \overline{\mathbf{x}}]$ represents a vector with k independent interval variables, and it may involve the uncertain factors resulting from geometry size, material properties and initial conditions. As a result, the solution of Eq. (5), i.e. $\mathbf{u}([\mathbf{x}], t)$ as well as solution of Eq. (6), i.e. $\lambda_j([\mathbf{x}])$ is considered as a function in terms of the interval variables. The major goal is to get the numerical solutions of $\mathbf{u}([\mathbf{x}], t)$ and $\lambda_j([\mathbf{x}])$ using interval analysis methods.

4. Interval analysis methods

This section introduces the basic idea of CIM briefly. For details, the readers can refer to [6]. Then MCIM is presented and introduced in details.

4.1 Chebyshev interval method (CIM)

A function defined in $C[a,b]$ can be approximated by the truncated Chebyshev series of degree n

$$f(x) \approx \frac{1}{2} f_0 + \sum_{i=1}^{n} f_i C_i(x) \quad (7)$$

The Chebyshev polynomials $C_i(x) = \cos(i\theta)$, $x = \cos\theta, \theta \in [0, \pi]$ are defined over $x \in [-1,1]$. The Chebyshev polynomial on $[a,b]$ is also defined by Eq. (7) through a simple transformation $(2x-(a+b))/(b-a) = \cos\theta$ .The constant coefficients f_i are calculated by

$$f_i = \frac{2}{\pi}\int_{-1}^{1} \rho(x) f(x) C_i(x) dx = \frac{2}{\pi}\int_{0}^{\pi} f(\cos\theta)\cos i\theta d\theta, i = 0,1,2,\ldots,n \quad (8)$$

where $\rho(x) = 1/\sqrt{1-x^2}$ is the weighting function. The Chebyshev interval function is defined in terms of interval variable $[x] = [\underline{x}, \overline{x}]$, as follows

$$f_{C_n}([x]) = \frac{1}{2} f_0 + \sum_{i=1}^{n} f_i C_i([x]) = \frac{1}{2} f_0 + \sum_{i=1}^{n} f_i [-1,1] = \frac{1}{2} f_0 + [-1,1]\sum_{i=1}^{n} |f_i| \quad (9)$$

For multidimensional problems, the original function $f(\mathbf{x}) = f(x_1, x_2, \cdots, x_k)$ is approximated by multi-dimensional Chebyshev polynomials of degree n, such that

$$f(\mathbf{x}) = f(x_1, x_2, \cdots, x_k) = \sum_{i_1=0}^{n}\sum_{i_2=0}^{n}\cdots\sum_{i_k=0}^{n} f_{i_1,i_2,\cdots,i_k} C_{i_1,i_2,\cdots,i_k}(x_1, x_2, \cdots, x_k) \quad (10)$$

The k-dimensional Chebyshev polynomial of $x_i = \cos\theta_i \in [-1,1]$ $i = 1,2,3,\ldots,k$ is defined by

$$C_{i_1,i_2,\ldots,i_k}(x_1, x_2, \ldots, x_k) = \cos(i_1\theta_1)\ldots\cos(i_k\theta_k) \quad (11)$$

Similarly, the multi-dimensional coefficients is calculated by

$$f_{i_1,i_2,\ldots,i_k} = \left(\frac{2}{\pi}\right)^k \times \int_0^{\pi}\ldots\int_0^{\pi} f(\cos\theta_1,\ldots,\cos\theta_k)\cos i_1\theta_1\ldots\cos i_k\theta_k d\theta_1\ldots d\theta_k \quad (12)$$

The k-dimensional Chebyshev interval function is defined as follows

$$f_n([x]) = \frac{1}{2^k} f_{0,0,\cdots,0} + \sum_{i_1=0}^{n}\sum_{i_2=0}^{n}\cdots\sum_{i_k=0}^{n} f_{i_1,i_2,\cdots,i_k} C_{i_1,i_2,\cdots,i_k}([x_1],[x_2],\cdots,[x_k])$$

$$= \frac{1}{2^k} f_{0,0,\cdots,0} + [-1,1]\sum_{i_1=0}^{n}\sum_{i_2=0}^{n}\cdots\sum_{i_k=0}^{n}\left|f_{i_1,i_2,\cdots,i_k}\right|,$$

$$i_1 + i_2 + \cdots + i_k \neq 0 \quad (13)$$

In Eq. (13), the property of Chebyshev polynomials $C_{i_1,i_2,\cdots,i_k}([x_1],[x_2],\cdots,[x_k]) \in [-1,1]$ is used so that $C_{i_1,i_2,\cdots,i_k}([x_1],[x_2],\cdots,[x_k]) = [-1,1]$. It is noted that the vector $[\mathbf{x}]$ includes k interval variables and each element has an interval bound. Using Eq. (13), the solutions of Eq. (5) and Eq. (6), i.e. $\mathbf{u}([\mathbf{x}], t)$ and $\lambda_j([\mathbf{x}])$ can be obtained.

4.2 Modified Chebyshev interval method (MCIM)

The process of constructing Chebyshev approximation polynomial in MCIM is the same as that in CIM. In MCIM, the least squares method is used to calculate the coefficients, this process would be the traditional RSM. Transform Eq. (10) to the following form

$$f(\mathbf{x}) = \sum_{i_1=0}^{n}\sum_{i_2=0}^{n}\cdots\sum_{i_k=0}^{n} f_{i_1,i_2,\cdots,i_k} C_{i_1,i_2,\cdots,i_k}(\mathbf{x}) = \boldsymbol{\beta}^T\boldsymbol{\alpha} =: \tilde{f}(\mathbf{x}) \quad (14)$$

where the coefficient vector

$$\boldsymbol{\beta} = [\beta_1, \cdots, \beta_s]^T = [f_{0,\cdots,0}, \cdots, f_{n,\cdots,n}]^T$$

and the polynomial basis vector

$\boldsymbol{\alpha} = [\alpha_1, \cdots, \alpha_s]^T = [C_{0,\cdots,0}(\mathbf{x}), \cdots, C_{n,\cdots,n}(\mathbf{x})]^T$, and $s = (n+1)^k$

The coefficient vector is calculated through the least squares method, i.e.

$$\boldsymbol{\beta} = \left(\mathbf{C}^T\mathbf{C}\right)^{-1}\mathbf{C}^T\mathbf{Y}, \text{ where } \mathbf{Y} = [f(\mathbf{x}_1), f(\mathbf{x}_2), \cdots, f(\mathbf{x}_s)]^T \quad (15)$$

where $\mathbf{Y}$ is the vector consisting of function values evaluated at the sampling points, and the matrix $C \in R^{s\times s}$ is composed by the values of polynomial basis vector at the sampling points

$$\mathrm{C} = \begin{bmatrix} \alpha_1(\mathbf{x}_1) & \cdots & \alpha_s(\mathbf{x}_1) \\ \vdots & \ddots & \vdots \\ \alpha_1(\mathbf{x}_s) & \cdots & \alpha_s(\mathbf{x}_s) \end{bmatrix} \quad (16)$$

where $\mathbf{x}_s$ denotes the sampling points, which are the tensor product of the sampling points in each dimension, such that

$$\mathbf{X} = \mathbf{X}_1 \otimes \cdots \otimes \mathbf{X}_k \quad (17)$$

where $\mathbf{X}_i$ denotes the sampling points of the *i-th* dimension (i=1,2,...,k), i.e. the zeros points of Chebyshev polynomial of order m. So the total number of sampling points is $s = m^k$. For numerical stability, m should be equal to or larger than n+1.

In CIM, the bounds of $f_n([\mathbf{x}])$ are calculated using the property of Chebyshev polynomials $C_{i_1,i_2,\cdots,i_k}([x_1],[x_2],\cdots,[x_k]) \in [-1,1]$, which simplifies the calculations but introduces errors because CIM assumes that $C_{i_1,i_2,\cdots,i_k}([x_1],[x_2],\cdots,[x_k]) = [-1,1]$. This assumption is reasonable and effective in most cases. In MCIM, the bounds of $f_n([\mathbf{x}])$ are calculated directly by Monte Carlo, yields

$$f_n([\mathbf{x}]) = [\min \tilde{f}(\mathbf{x}), \max \tilde{f}(\mathbf{x})] \quad (18)$$

In general, the original function such as Eq. (5) and (6) is extremely complicated; using Monte Carlo is time consuming. However, $\tilde{f}(\mathbf{x})$ is a polynomial model which is much less complicated than the original function, so using Monte Carlo is not time consuming.

4.3 Comparison of CIM and MCIM

This section offers qualitative comparison of CIM and MCIM.

(1) Both CIM and MCIM are based on Chebyshev polynomial approximation (Eq.10 and Eq.14). It is noted that both CIM and MCIM do not require the derivatives of the system responses w.r.t interval variables, which is an advantage over perturbation methods and Taylor series based methods.

(2) Both CIM and MCIM need to calculate the coefficients of the polynomials. The difference is that CIM uses multiple integral (Eq.12) to calculate the coefficients while MCIM uses least squares (Eq.15). The authors have found that the coefficients obtained by multiple integral and least squares method are nearly the same. For the problems with large number of interval variables, the least squares method is easier to be implemented.

(3) After getting the Chebyshev polynomial approximation formulation, interval function is obtained using the properties of Chebyshev polynomials in CIM (Eq.13), while using Monte Carlo method in MCIM (**Eq.18**). Thus CIM introduces overestimation and MCIM does not. It can be seen in the numerical examples that MCIM is more accurate than CIM.

5. Numerical examples

5.1 12-bar truss structure

The first example concerns a plane truss structure which is modelled with 8 nodes and 12 elements (Fig.1). The nominal structure is characterized by the following geometrical and mechanical properties: cross-sectional areas of the bars $A_e = 0.0001m^2$; Young's moduli of the bars $E_e = 2.1e11Pa$; nominal lengths of the bars specified in Fig.1 where $L_e = 1m$; material density ρ=7830kg / m^3. Each truss element has two DOFs at each node so that the structure possesses 12 DOFs.

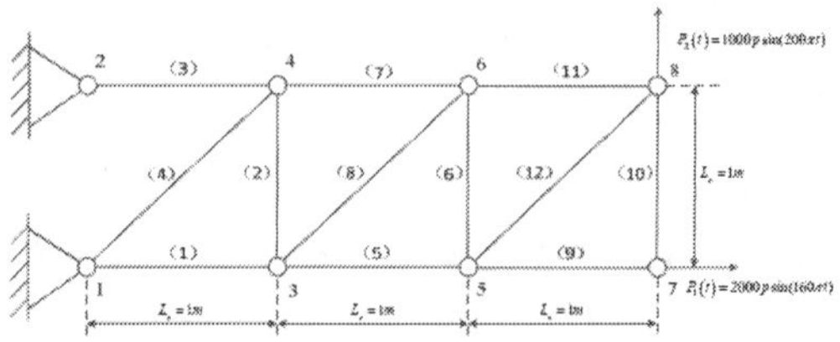

Fig.1 A schematic of truss structure

The 12-bar truss structure is subjected to nominal harmonic sinusoidal excitations $P_1(t) = 2000p\sin(160\pi t)$ and $P_2(t) = 1000p\sin(200\pi t)$, with initial conditions $\mathbf{u}_0([\mathbf{x}]) = \mathbf{0}$ and $\dot{\mathbf{u}}_0([\mathbf{x}]) = \mathbf{0}$. Rayleigh damping is assumed so that the damping matrix is given by $\mathbf{C} = \alpha\mathbf{M} + \beta\mathbf{K}$, where α=20s^{-1} and β=0.00001s. The Young's modulus, mass density and amplitude of the harmonic sinusoidal excitation are considered as interval parameters: $[E] = [1.995e11, 2.205e11]Pa$, $[\rho]$=[7438.5,8221.5]kg/m^3, $[p] = [0.98, 1.02]$.

The truss structure is solved for a period of 0.1s with a step size of 0.0001s using CIM, MCIM and Monte Carlo, respectively. Fourth-order Chebyshev polynomial is used for each interval variable, so 5 sampling points (zero points of fifth-order Chebyshev polynomial) is used for each interval variable. Since there are 3 interval variables, the total number of sampling points is 5^3=125 for CIM and MCIM. For the Monte Carlo method, 1000 samplings are used.

The time responses of the truss structure are shown in Fig.2 to 8. Fig.2 to 5 shows the horizontal and vertical displacements of nodes 7 and 8 under uncertain dynamic loads. Fig.6 to 8 shows the stresses at the truss elements. It is observed from Fig.2 to 8 that each time response has upper and lower bounds due to the interval parameters. Monte Carlo obtains the exact bounds without overestimations. CIM obtains slightly wider response intervals than Monte Carlo due to the inherent wrapping effects as mentioned in Section 2. In comparison with CIM, MCIM obtains fairly exact response intervals, meaning that MCIM controls the wrapping effects better.

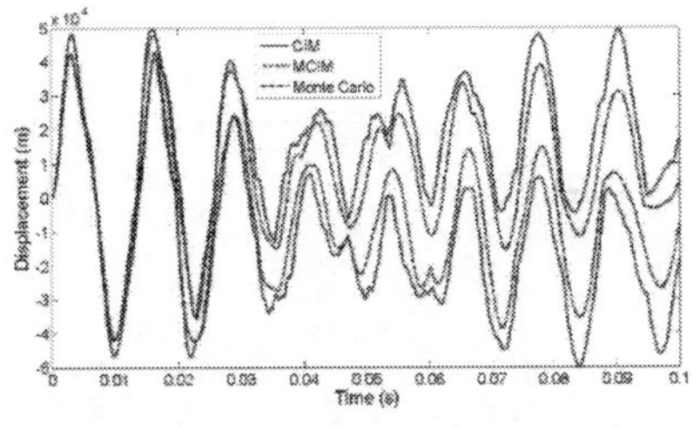

Fig.2 Horizontal displacement of node 7

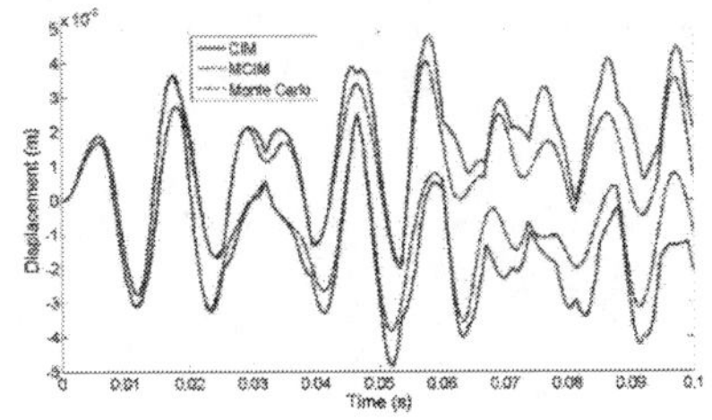

Fig.3 Vertical displacement of node 7

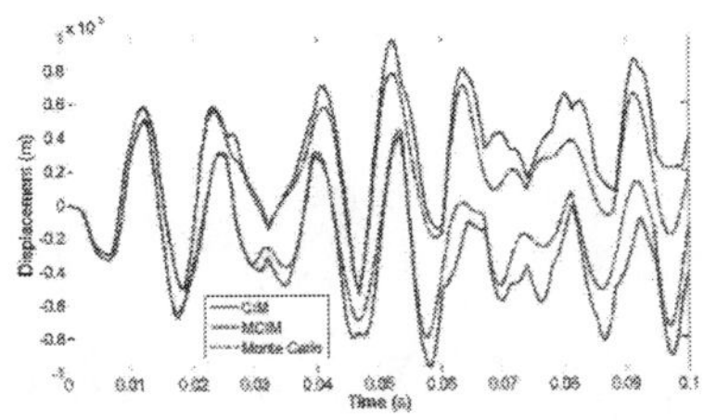

Fig.4 Horizontal displacement of node 8

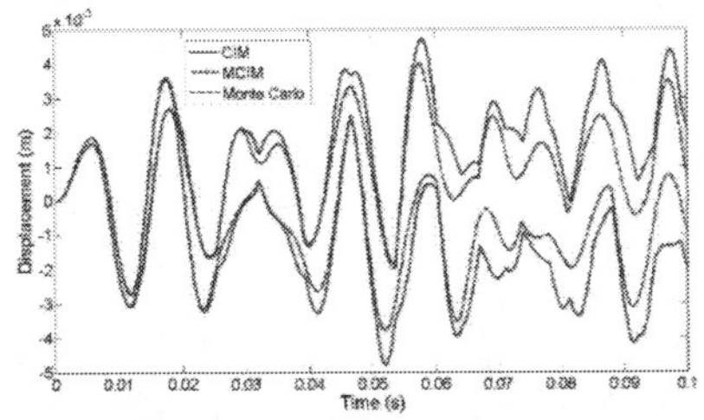

Fig.5 Vertical displacement of node 8

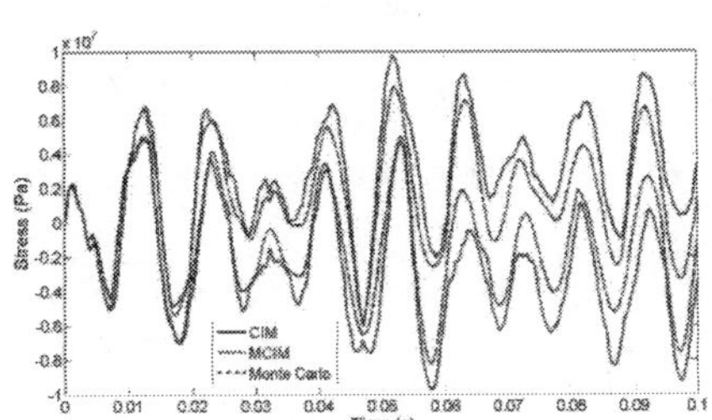

Fig.6 Stress of element 10

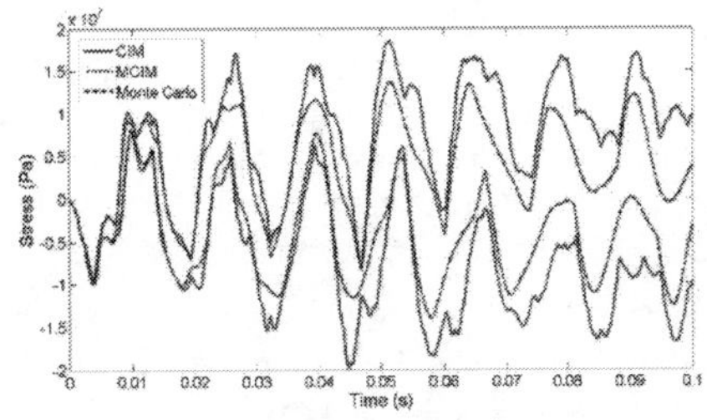

Fig.7 Stress of element 11

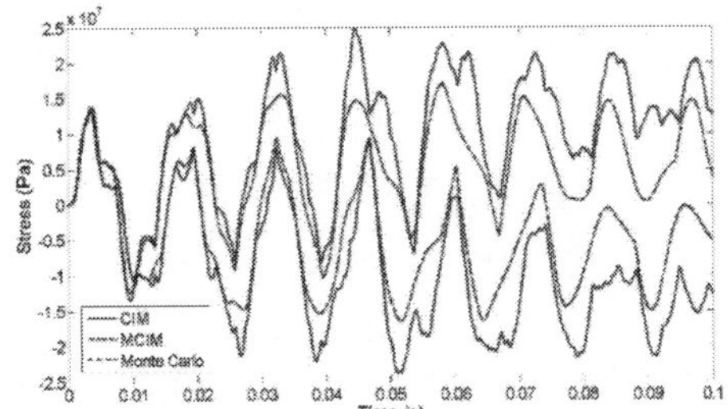

Fig.8 Stress of element 12

Table 1 shows the lower and upper bounds of the first 5 natural frequencies (eigenvalues). Since the eigenvalues are calculated through algebraic operations, the three methods obtain nearly the same natural frequency bounds. Particularly, MCIM obtains the exact results as the Monte Carlo does. Moreover, the width of natural frequency bound increases with the increasing modal order. It should be noted that although 3 interval parameters are considered in this example, only the Young's modulus and material density affect the bounds of eigenvalues.

Table 2 shows the total calculation times for time responses and eigenvalues. It is observed that Monte Carlo is the most time consuming and CIM is the least time consuming. MCIM costs about 5s longer than CIM for calculating the bounds of Chebyshev polynomials, but costs much less time than Monte Carlo.

Table 1 Natural frequencies of the truss structure (Unit: rad/s)

	CIM		**MCIM**		**Monte Carlo**	
j	$\underline{\omega}_j$	$\overline{\omega}_j$	$\underline{\omega}_j$	$\overline{\omega}_j$	$\underline{\omega}_j$	$\overline{\omega}_j$
1	421.6	467.2	422.6	467.1	422.6	467.1
2	1445.4	1601.7	1448.7	1601.2	1448.7	1601.2
3	1791.0	1984.7	1795.1	1984.1	1795.1	1984.1
4	2889.6	3202.1	2896.3	3201.2	2896.3	3201.2
5	3684.8	4083.4	3693.3	4082.1	3693.3	4082.1

Table 2 Calculation times for the truss structure

	CIM	**MCIM**	**Monte Carlo**
Calculation Time (s)	3.1575	8.0184	25.2412

https://doi.org/10.1051/matecconf/202338201024

5.2 Cantilever beam

The second example concerns a cantilever Euler-Bernoulli beam uniformly discretized into 10 elements, as shown in Fig.9. The beam is characterized by the following geometrical and mechanical nominal properties: total length $L = 2m$; Young's modulus $E = 3.0e10Pa$; rectangular cross-section area with width $b = 0.1m$ and height $h = 0.15m$; material density ρ=$2500kg / m^3$. Since each element has two DOFs at each node, the discretized beam possess 20 DOFs.

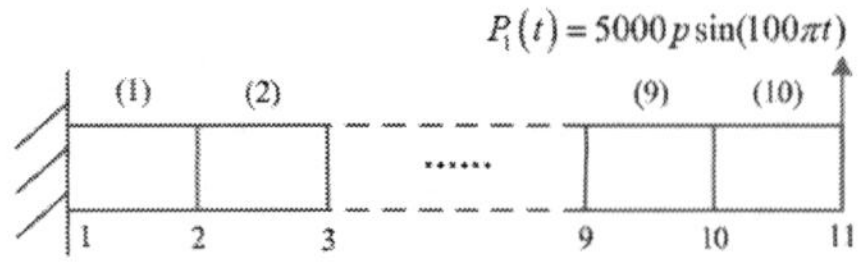

Fig.9 A schematic of cantilever beam

The beam is subjected to harmonic sinusoidal excitations $P_1(t) = 5000p\sin(100\pi t)$ with initial conditions $\mathbf{u}_0([\mathbf{x}]) = \mathbf{0}$ and $\dot{\mathbf{u}}_0([\mathbf{x}]) = \mathbf{0}$. Rayleigh damping is assumed so that the damping matrix is given by $\mathbf{C} = \alpha\mathbf{M} + \beta\mathbf{K}$, where α=$20s^{-1}$ and β=$0.00001s$. The Young's modulus, mass density and amplitude of the harmonic sinusoidal excitation are considered as interval parameters: $[E] = [2.85e10, 3.15e10]Pa$, $[\rho] = [2375, 2625]\text{kg/m}^3$, $[p] = [0.98, 1.02]$.

The finite element beam is solved for a period of 0.1s with a step size of 0.0001s using CIM, MCIM and Monte Carlo, respectively. Fourth-order Chebyshev polynomial is used for each interval variable, so the total number of sampling points is 5^3=125 for CIM and MCIM. For the Monte Carlo method, 1000 samplings are used.

Fig.10 to 13 shows the time responses of the truss structure. Fig.10 shows the dynamic deflection of the cantilever beam (vertical displacement of node 11) under uncertain dynamic loads. Fig.11 shows the shear force of element 1 under uncertain dynamic loads. Fig.12 and 13 show the bending moments of element 1 at the left end and the right end.

It is observed that each time response has upper and lower bounds due to the interval parameters. In theory, Monte Carlo provides with the exact bounds without overestimations. CIM obtains slightly wider response intervals than Monte Carlo due to the wrapping effects as mentioned in Section 2. MCIM obtains the nearly the same exact results with Monte Carlo. However, the overestimations of CIM are very small and can be accepted.

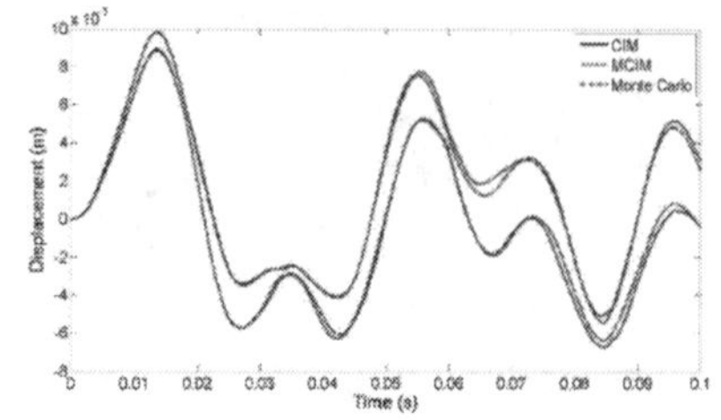

Fig.10 Vertical displacement of node 11

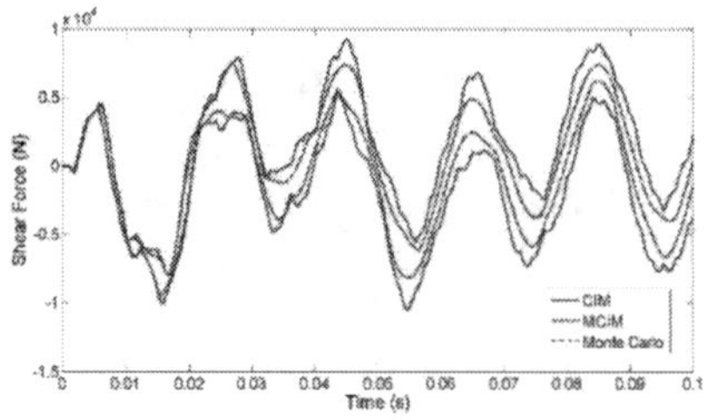

Fig.11 Shear force of element 1

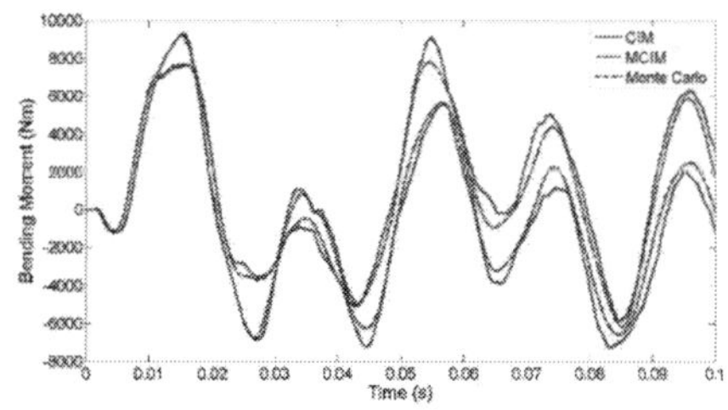

Fig.12 Bending moment of element 1 at the left end

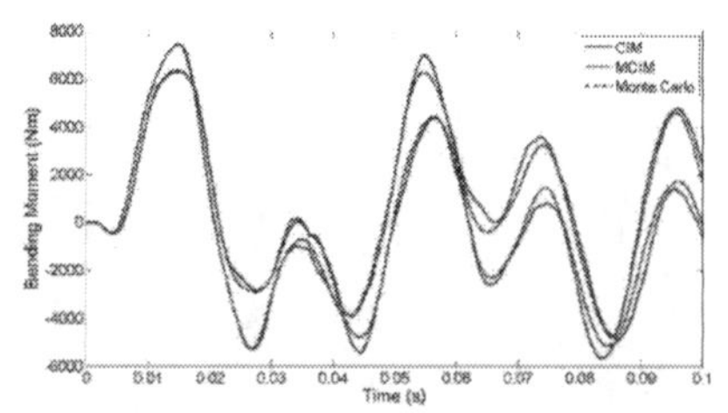

Fig.13 Bending moment of element 1 at the right end

Table 3 shows the lower and upper bounds of the first 5 natural frequencies for the beam. It is observed that the three methods obtain nearly the same natural frequency bounds. The results of MCIM and Monte Carlo are the same. From first- to fifth- order modal, the widths of natural frequencies are about 10, 80, 230, 450 and 740, respectively. So the width of the natural frequency bound increases with the increasing modal order.

Table 4 shows the total calculation time for time responses and natural frequencies. It is observed that Monte Carlo is most time-consuming and CIM is most computational cheapest. MCIM costs about 6s longer than CIM for calculating the bounds of Chebyshev polynomials, but costs much less time than Monte Carlo.

Table 3 Natural frequencies of the cantilever beam (Unit: rad/s)

	CIM		MCIM		Monte Carlo	
j	$\underline{\omega}_j$	$\overline{\omega}_j$	$\underline{\omega}_j$	$\overline{\omega}_j$	$\underline{\omega}_j$	$\overline{\omega}_j$
1	125.1	138.7	125.4	138.6	125.4	138.6
2	784.2	869.0	786.0	868.7	786.0	868.7
3	2196.2	2433.7	2201.3	2433.0	2201.3	2433.0
4	4306.7	4772.5	4316.6	4771.0	4316.6	4771.0
5	7130.4	7901.6	7146.9	7889.2	7146.9	7889.2

Table 4 Calculation times for the cantilever beam

	CIM	MCIM	Monte Carlo
Calculation Time (s)	3.3435	9.5789	23.4620

6. Conclusions

An interval analysis method termed as modified Chebyshev interval method (MCIM) is presented to estimate the time responses and eigenvalues of structures with interval parameters. MCIM is developed based on the Chebyshev interval method (CIM). MCIM, CIM and Monte Carlo method are applied for solving two structures in presence of interval parameters. These three methods are compared in terms of overestimation and computation time. In theory, Monte Carlo method provides with the exact bounds without any overestimation, but costs a long computation time. CIM costs much less time than Monte Carlo method, and introduces a slight of overestimations. MCIM can also provide with the exact results, but costs a little longer computation time than CIM. Even so, the computation time of MCIM is much shorter than that of Monte Carlo method. The overall conclusion is that both MCIM and CIM are effective and accurate interval analysis methods. It should be noted that both MCIM and CIM do not demand the derivatives of the system outputs w.r.t interval variables. So MCIM and CIM are suitable for complex systems where the calculation of derivatives is difficult. MCIM and CIM can also be used for black-box problems where the mathematical models of the systems are unknown.

Acknowledgements

This research was supported by the National Natural Science Foundation of China (11991034). This financial support is gratefully acknowledged.

References

1. M. Beer, S. Ferson, V. Kreinovich, Imprecise probabilities in engineering analyses, Mechanical Systems and Signal Processing. 37 (2013) 4–29. doi:10.1016/j.ymssp.2013.01.024.
2. A. Sandu, C. Sandu, M. Ahmadian, Modeling multibody systems with uncertainties. Part I: Theoretical and computational aspects, Multibody System Dynamics. 15 (2006) 369–391. doi:10.1007/s11044-006-9007-5.
3. C. Sandu, A. Sandu, M. Ahmadian, Modeling multibody systems with uncertainties. Part II: Numerical applications, Multibody System Dynamics. 15 (2006) 241–262. doi:10.1007/s11044-006-9008-4.
4. A.S. Balu, B.N. Rao, High dimensional model representation based formulations for fuzzy finite element analysis of structures, Finite Elements in Analysis and Design. 50 (2012) 217–230. doi:10.1016/j.finel.2011.09.012.
5. D. Behera, S. Chakraverty, Fuzzy finite element analysis of imprecisely defined structures with fuzzy nodal force, Engineering Applications of Artificial Intelligence. 26 (2013) 2458–2466. doi:10.1016/j.engappai.2013.07.021.
6. J. Wu, Z. Luo, Y. Zhang, N. Zhang, L. Chen, Interval uncertain method for multibody mechanical systems using Chebyshev inclusion functions, International Journal for Numerical Methods in Engineering. 95 (2013) 608–630. doi:10.1002/nme.4525.
7. C. Jiang, R.G. Bi, G.Y. Lu, X. Han, Structural reliability analysis using non-probabilistic convex model, Computer Methods in Applied Mechanics and Engineering. 254 (2013) 83–98. doi:10.1016/j.cma.2012.10.020.
8. C. Jiang, X. Han, G.Y. Lu, J. Liu, Z. Zhang, Y.C. Bai, Correlation analysis of non-probabilistic convex model and corresponding structural reliability technique, Computer Methods in Applied Mechanics and Engineering. 200 (2011) 2528–2546. doi:10.1016/j.cma.2011.04.007.
9. D. Moens, M. Hanss, Non-probabilistic finite element analysis for parametric uncertainty treatment in applied mechanics: Recent advances, Finite Elements in Analysis and Design. 47 (2011) 4–16. doi:10.1016/j.finel.2010.07.010.
10. R.E. Moore, Methods and applications of interval analysis, Prentice Hall, London, 1979.
11. G. Alefeld, Introductions to interval computations, Academic Press, New York, 1983.
12. Z. Qiu, X. Wang, M.I. Friswell, Eigenvalue bounds of structures with uncertain-but-bounded parameters, Journal of Sound and Vibration. 282 (2005) 297–312. doi:10.1016/j.jsv.2004.02.051.
13. S.H. Chen, H.D. Lian, X.W. Yang, Interval eigenvalue analysis for structures with interval parameters, Finite Elements in Analysis and Design. 39 (2003) 419–431. doi:10.1016/S0168-874X(02)00082-3.
14. C. Wang, Z. Qiu, X. Wang, D. Wu, Interval finite element analysis and reliability-based optimization of coupled structural-acoustic system with uncertain parameters, Finite Elements in Analysis and Design. 91 (2014) 108–114. doi:10.1016/j.finel.2014.07.014.
15. W. Gao, Interval natural frequency and mode shape analysis for truss structures with interval parameters, Finite Elements in Analysis and Design. 42 (2006) 471–477. doi:10.1016/j.finel.2005.09.003.

16. Q. Li, Z. Qiu, X. Zhang, Eigenvalue analysis of structures with interval parameters using the second-order Taylor series expansion and the DCA for QB, Applied Mathematical Modelling. 49 (2017) 680–690. doi:10.1016/j.apm.2017.02.041.
17. A. Sofi, G. Muscolino, I. Elishakoff, Natural frequencies of structures with interval parameters, Journal of Sound and Vibration. 347 (2015) 79–95. doi:10.1016/j.jsv.2015.02.037.
18. A. Sofi, E. Romeo, A novel Interval Finite Element Method based on the improved interval analysis, Computer Methods in Applied Mechanics and Engineering. 311 (2016) 671–697. doi:10.1016/j.cma.2016.09.009.
19. G. Muscolino, A. Sofi, Bounds for the stationary stochastic response of truss structures with uncertain-but-bounded parameters, Mechanical Systems and Signal Processing. 37 (2013) 163–181. doi:10.1016/j.ymssp.2012.06.016.
20. S. Tangaramvong, F. Tin-Loi, C. Yang, W. Gao, Interval analysis of nonlinear frames with uncertain connection properties, International Journal of Non-Linear Mechanics. 86 (2016) 83–95. doi:10.1016/j.ijnonlinmec.2016.07.006.
21. Z. Qiu, Z. Ni, An inequality model for solving interval dynamic response of structures with uncertain-but-bounded parameters, Applied Mathematical Modelling. 34 (2010) 2166–2177. doi:10.1016/j.apm.2009.10.028.
22. J. Feng, D. Wu, W. Gao, G. Li, Uncertainty analysis for structures with hybrid random and interval parameters using mathematical programming approach, Applied Mathematical Modelling. 48 (2017) 208–232. doi:10.1016/j.apm.2017.03.066.
23. C. Jiang, Z.G. Zhang, Q.F. Zhang, X. Han, H.C. Xie, J. Liu, A new nonlinear interval programming method for uncertain problems with dependent interval variables, European Journal of Operational Research. 238 (2014) 245–253. doi: 10.1016/j.ejor.2014.03.029.
24. Z. Qiu, X. Wang, Comparison of dynamic response of structures with uncertain-but-bounded parameters using non-probabilistic interval analysis method and probabilistic approach, International Journal of Solids and Structures. 40 (2003) 5423–5439. doi:10.1016/S0020-7683(03)00282-8.
25. Z. Qiu, X. Wang, Parameter perturbation method for dynamic responses of structures with uncertain-but-bounded parameters based on interval analysis, International Journal of Solids and Structures.42 (2005)4958–4970.doi:10.1016/j.ijsolstr.2005.02.023.
26. Y. Xia, Z. Qiu, M.I. Friswell, The time response of structures with bounded parameters and interval initial conditions, Journal of Sound and Vibration. 329 (2010) 353–365. doi:10.1016/j.jsv.2009.09.019.
27. S. Yin, D. Yu, H. Yin, B. Xia, Interval and random analysis for structure–acoustic systems with large uncertain-but-bounded parameters, Computer Methods in Applied Mechanics and Engineering. 305 (2016) 910–935. doi:10.1016/j.cma.2016.03.034.
28. J. Wu, Y. Zhang, L. Chen, Z. Luo, A Chebyshev interval method for nonlinear dynamic systems under uncertainty, Applied Mathematical Modelling. 37 (2013) 4578–4591. doi:10.1016/j.apm.2012.09.073.
29. Z. Wang, Q. Tian, H. Hu, Dynamics of spatial rigid–flexible multibody systems with uncertain interval parameters, Nonlinear Dynamics. 84 (2016) 527–548. doi:10.1007/s11071-015-2504-4.
30. [30] J. Wu, Z. Luo, N. Zhang, Y. Zhang, A new uncertain analysis method and its application in vehicle dynamics, Mechanical Systems and Signal Processing. 50–51 (2015) 659–675. doi:10.1016/j.ymssp.2014.05.036.
31. L. Jaulin, Applied interval analysis: with examples in parameter and sate estimation, robust control and robotics, Springer, New York, 2001.
32. N.M. Newmark, A method of computation for structural dynamics, Journal of the Engineering Mechanics Division. 85 (1959) 67–94.

Research on Distance Measurement Method of field driving Test system

Yao Yao*, Han Chao, Wu Yang, Dai Jinzhou, Xu Peng.

Beijing Institute of Metrology, Beijing 100029, China.

Abstract: The accuracy of driving test systems is very important. Aiming at the content and characteristics of the subject II examination items, we studied the distance measurement items and methods of the field driving test system, and test the distance examination items in the field driving test system. The results show that the measurement method studied in this paper can realize the accurate judgment of the field driving test system, providing technical support for the accuracy detection of the driving test system, and ensure the fairness and justice of the driving test.

Keywords: Driving test, field driving test system, distance measurement method, laser tracker.

1. Introduction

Driver test is an important source of road traffic safety. In the situation that the number of motor vehicles and the number of drivers grow rapidly, how to improve the driving quality of drivers through the test, so as to ensure the safety and smooth road traffic is of great significance. The evaluation of driving test has experienced the change from artificial evaluation to the combination of automatic computer evaluation and artificial evaluation in China[1-3]. Driving test systems is an intelligent evaluation system for driving test. It includes field driving test system and road driving test system, which corresponds to the subject 2 and subject 3 of GA 1026-1027 [4]. With the development of information, sensor and network technology, the field driving test system has realized the electronic evaluation. It's important to measure the driving test system before going use it, and it can be used only after the test is qualified. It also needs periodic or non-periodic measurement during the application, preventing the impact of inaccuracy. Therefore, as a specific measuring equipment, the field driving test system is faced with measurement verification, measurement value transfer and other detection. In order to ensure the accuracy and reliability of the test system and the objectivity and justice of the examination results, it is necessary to conduct regular detection and traceability of the key parameters which affecting the test results.

The field driving test system will conduct the examination and evaluation according to the test content and evaluation requirements of Subject 2 in GA1026-1027. The field driving test system should be meet the requirements of different types of vehicles, which are Class I: small cars, small automatic transmission cars, low speed trucks and small automatic passenger cars for the disabled, Class II: large buses, tractors, city buses, medium buses, large trucks, and Class III: three-wheeled automobile, ordinary three-wheeled motorcycle, ordinary two-wheeled motorcycle, mopeds. The test items include reversing into the garage, fixed point parking and starting on the ramp, right Angle turn, side parking, curve driving, etc. The related evaluation items of measurement characteristics include vehicle body out of the line, vehicle rear slip distance, vehicle body touching the warehouse side line, project completion time, stopover time, vehicle speed and so on. Although the test items of different types of vehicles are not identical, the content of the test is basically the same. According to the content of the evaluation items, it can be classified into distance, time and speed. Distance items investigate the distance between the vehicle and the sideline, and judge whether the vehicle is pressed or out of line.

2. Measurement

The field driving test system evaluates the distance items according to the real-time position of the car and the sidelines. The position of the car is reflected by the antenna above the car roof and the delineation point of the car frame (as shown in Figure 1). The sidelines of the field are also recorded in the driving system. Therefore, the field test system can get the real-time position of the car and the sidelines. However, the antenna misplaced before the use of field driving test system, and the deviation from its original position during application, the right distance between the car and the sidelines will be changed, resulting in a certain deviation and then affect the accuracy of the test evaluation. Therefore, it is necessary

* Corresponding author: yaoy@bjjl.cn

to develop a method to judge the distance evaluation items of field test system.

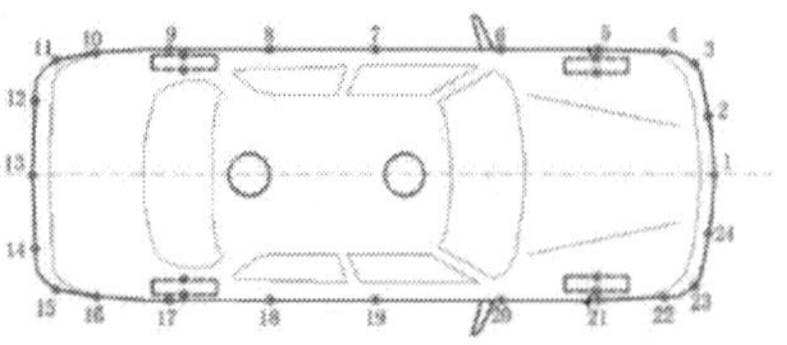

Fig.1 The delineation point of the car frame

The distance measurement items of the field test system can use laser tracker (as shown in Fig.2) for real-time detection. The laser tracker consists of host machine and target ball, which the host is equivalent to the coordinate dot and the target ball can feedback its own coordinate position[5-7]. The principle of laser tracker is to dynamically track the spatial position data of the target through two Angle encoders and one distance sensor[8]. The geometric measurement model of the laser tracker is a spherical coordinate system, including a laser interferometric length measurement unit, and a pitch Angle and level Angle measurement unit based on two high-precision encoders[9]. The measured parameters cannot be used directly and need to be converted to spatial Cartesian coordinates. The original parameters mainly include the distance l from the distant point of the coordinate system to the center of the target ball, the horizontal Angle α, and the pitch Angle β.

Fig. 2 The photo of laser tracker

The above spherical coordinate system is converted to Cartesian coordinate system through the space coordinate transformation relation, as shown in Eq. (1).

$$\begin{cases} x = f_x(l,\alpha,\beta) = l \cdot sin\alpha \cdot cos\beta \\ y = f_y(l,\alpha,\beta) = l \cdot sin\alpha \cdot sin\beta \\ z = f_z(l,\alpha,\beta) = l \cdot cos\alpha \end{cases} \quad (1)$$

The target ball is manually installed on the magnetic target seat, and the included Angle between it and the end is random. Therefore, the influence of target ball installation on the measurement accuracy error is random. The measurement uncertainty of laser tracker used in this paper is ±15μm+6μm/m. The systematic errors measured by laser tracker have obvious regularity and direction.

Let dl represents the systematic error in length measurement, dα and dβ represent the error in horizontal and pitch direction of Angle measurement, respectively. Then the spatial coordinate change relationship caused by sensor signal can be obtained according to the spatial coordinate measurement formula. Eq.(1) is developed by total differential at the ideal point (l0, α0, β0) in space, Eq.(2) can be obtained as shown below:

$$T(l_0,\alpha_0,\beta_0) = \begin{bmatrix} sin\alpha_0 cos\beta_0 & l_0 \cos\alpha_0 \cos\beta_0 & -l_0 \sin\alpha_0 \sin\beta_0 \\ sin\alpha_0 sin\beta_0 & l_0 \cos\alpha_0 \sin\beta_0 & l_0 \sin\alpha_0 \cos\beta_0 \\ \cos\alpha_0 & -l_0 \cos\alpha_0 & 0 \end{bmatrix} \quad (2)$$

In order to describe the measurement error of laser tracker more intuitively, the values in the above expansion formula are compared with the theoretical measurement values, as shown in Eq.(3).

$$\begin{bmatrix} dx \\ dy \\ dz \end{bmatrix} = \begin{bmatrix} f_x - f_{x0} \\ f_y - f_{y0} \\ f_z - f_{z0} \end{bmatrix} = T(l_0,\alpha_0,\beta_0) \cdot \begin{pmatrix} dl \\ d\alpha \\ d\beta \end{pmatrix} \quad (3)$$

T(l0, α0, β0) represents the coefficient matrix of the error matrix, which reflects that the measurement error in the Cartesian coordinate system is the result of the comprehensive action of the three measurement sensors. The coordinate transformation formula of any coordinate point in the space can be expressed by Eq. (4), which the $e = [e_x e_y e_z]^T$ represents the random measurement error and represents the Cartesian coordinates of the instruction point space. [10].

$$\begin{bmatrix} x \\ y \\ z \end{bmatrix} = \begin{bmatrix} x_0 \\ y_0 \\ z_0 \end{bmatrix} + T(l_0,\alpha_0,\beta_0) \begin{pmatrix} dR \\ d\alpha \\ d\beta \end{pmatrix} + \begin{bmatrix} e_x \\ e_y \\ e_z \end{bmatrix} \quad (4)$$

During detection, the laser tracker host is placed on the road of the field, and the target ball can be fixed at any position of the vehicle according to the demand, such as fixed on the edge of the vehicle profile that exits first. The vehicle profile can be obtained by software processing after the target ball is placed at multiple points of the vehicle. When gets the moment that the field test system decides "unqualified" or "score deduction", the distance from the vehicle to the sideline can be calculated by the software, and the test results of distance measurement items can be obtained, and the accuracy of distance measurement items can be judged accordingly.

Next, we introduce the specific measurement methods. Preparation before detection: Use laser tracker to determine the contour coordinates of the field and vehicle (wheel), and determine the boundary coordinates according to the requirements of the boundary lines in Table 1, Table 2 and Table 3 of GA/T1028.3-2017 [11]. According to the situation of the test items, the measurement can be classified into three types as follows:

1. The distance from body (wheel) to sideline

1.1 Select the contour of the vehicle (wheel) which is closest to the sideline, place the target ball on the contour line, and ensure that the target ball exits first from the edge line.

1.2 After confirming that the communication connection between the target ball and the laser tracker is normal, the driving vehicle simulates the process of vehicle (wheel) exits or on the sideline. The inspector captures the moment when the field test system prompts "score

deduction" or "unqualified" for the judgment result, and then records the time.
1.3 The same time of the laser tracker and step 1.2 is collected by the time synchronization software, and match the scanning coordinate points of target ball at this time. The distance between the vertical point on the ground and the sideline is obtained by the calculation of software, that is, the distance between the body (wheel) and the side line is obtained.
2. The distance from the front bumper of a car or the front axle of a motorcycle to the post line
2.1 Place the target ball on the contour line of the front bumper of the car or the front axle of the motorcycle to ensure that the target ball exceeds the post line first.
2.2 After confirming that the communication connection between the target ball and the laser tracker is normal, the driving vehicle simulates that the front bumper or the front axle of the motorcycle is not set on the post line after the vehicle stops, and the inspector captures the moment when the field test system judgment results indicate "score deducted" or "unqualified", and records the time.
2.3 Collect the same time of the laser tracker and step 2.2 through the time synchronization software, and match the scanning coordinate points of target ball at this time. It can obtain the distance of this point relative to the sideline by the calculation of software, that is, the distance between the front bumper of the car or the front axle of the motorcycle and the post line is obtained.
3. Rear slip distance of vehicle
3.1 Place the target ball at any point on the vehicle outline.
3.2 After confirming that the communication connection between the target ball and the laser tracker is normal, the surveyor records the scanning coordinate points of the target ball of the laser tracker when the vehicle stops at the fixed point of the ramp
3.3 After the vehicle stops, and then the vehicle starts, simulating the vehicle backsliding according to the content of the detection evaluation item. The inspector record the time when captures the decision of the test system. Then match the scanning coordinate point of the laser tracker.
3.4 The difference between the distance of the target ball in step 3.2 and step 3.3 is the vehicle backsliding distance, which can be obtained through software processing.

3. Results

Based on the above method, we measured the test items in different situations, and part of the results are displayed in Table 1. The results show that the method descried above can measure the real-time distance of the content of test item, and give guidance for the judge of the test items. Thus, the accuracy of the test system can be reflected. The method can be also applied to other aspects, which relates to the distance detection.

Tab.1 The results of measured test items

NO.	examination item	test item	situation	measured value	judgement
1	reverse stall parking	Distance from vehicle to sideline	Body out of the right depot position line	30.71 mm	○
			Body out of the right depot position line	36.87 mm	○
			Body out of the back depot position line	6.32 mm	○
2	Ramp designated stop and start	Rear slip distance of vehicle	The rear slip distance of the vehicle is 10 cm when starting	11.671 cm	○
3		Rear slip distance of vehicle	The rear slip distance of the vehicle is over 30 cm when starting	32.760 cm	○
4		Distance from wheel to sideline	The wheel crested the road	10.86 mm	○
5		Distance from body to sideline	Body distance from road edge exceeds 30 cm	31.137 cm	○
6		Distance from body to sideline	The vehicle body is 50 cm away from the road edge	52.001cm	○
Note: Fill "○" in the judgment column for qualified, "╳" for unqualified.					

In order to investigate the accuracy of measurement results, repeatability test was carried out. The first test content in Table 1 was selected and repeated for 10 times. The results are shown in Tab.2.

Tab.2 The results of 10 measurements

Measurement	1	2	3	4	5	6	7	8	9	10
Ruselts	30.90	31.05	30.86	30.79	30.75	31.00	30.81	30.70	30.75	30.87

Repeatability is calculated according to Bessel's formula, which shown as Eq.(5) :

$$S_A = \sqrt{\frac{1}{n-1}\sum_{i=1}^{n}(x_i - \bar{x})^2} \quad (5)$$

SA is repeatability, xi is the result of the ith measurement, x is the arithmetic mean of the 10 measurements, and n is the number of measurements. The calculated value of SA is 0.11 mm, which represents a good repeatability.

4. Conclusion

In order to evaluate the accuracy of distance test items in the field driving test system, we proposed a detection method that can judge its accuracy. Through the synchronous use of laser tracker and timing software, the examiner can obtain the real-time distance between the vehicle and the sideline in the field, so as to judge whether the vehicle meets the test requirements. The measurement accuracy of the test system can be obtained by comparing the measured real-time distance and evaluation conclusion with the distance evaluation result of the test system. The evaluation method proposed in this study is helpful to judge the accuracy of distance test items in the field driving test system, and also applicable to other aspects of real-time distance test.

References

1. Pan Han-zhong, Zhang Jun, Fan Zhi-xiang. The development and prospect of technical specifications for driving test in China[J]. Chinese Journal of Ergonomics, 2018, 24(3):1-6.
2. Hua Sha, Sun Wei, Yong Cheng-ming, Zhang Hao. Analysis of key indicators on motor vehicle driver examination system[J]. Intelligent Transportation, 2018, 51(2):76-80.
3. Zhao Xiao-lin. Intelligent driver training and examination services system[J]. Geomatics World, 2014, 1(21):91-96.
4. Ministry of Public Security of the People's Republic of China. GA 1026-2017 Driving test content and methods[S]. Beijing: Ministry of Public Security, PRC.
5. Li Chang-Kai, Liu Yu-jian. Measurement for moveable bridge shafts with laser tracker[J]. Value Engineering, 2022,36:105-107.
6. Wang Zhi-chen, Chen Tao, Cao Yu-Yan, Zhou Chao. Design and geometrical tolerance detection of center section of large-aperture telescope[J]. Optics and Precision Engineering, 2022, 23(30):3039-3044.
7. Yin Rui-duo, Zu Hong-fei, Chen Zhang-wei, Guo Xiao-wei, Yan Ping. Research on error influencing factors of coordinated measurement of Multi-Base station laser tracker[J]. 2022, 11(66):12-15.
8. J A Santos-Hernandez, M Valtierra-Rodriguez, J P Amezquita-Sanchez, et al. Hilbert filter based FPGA architecture for power quality monitoring[J]. Measurement, 2019, 147: 106819.
9. W Zhang, M Kumar, J Liu. Multi-parameter online measurement IoT system based on BP neural network algorithm[J]. Neural Computing and Applications, 2018, 31(12): 8147-8155.
10. Yin Ruiduo, Zu Hongfei, Chen Zhangwei, Guo Xiaowei, Yan Ping. Research on error influencing factors of coordinated measurement of Multi-Base station laser tracker[J].Metrology Science and Technology, 2022, 66(11):12-15.
11. Ministry of Public Security of the People's Republic of China. GA/T 1028.3-2017 General technical specifications for driving test systems-Part 3: Field driving test system[S]. Beijing: Ministry of Public Security, PRC.

* Corresponding author: yaoy@bjjl.cn

Intelligent mechanical manufacturing technology based on intelligent manufacturing technology

Kang Zheng

Hubei Engineering Institute, Huangshi, China

Abstract. With the development of intelligent manufacturing technology, the application of intelligent machinery manufacturing technology has been widely promoted. Intelligent machinery manufacturing technology is a technology that integrates advanced manufacturing technology and intelligent control technology. It can realize the intelligent control of the entire manufacturing process, improve the manufacturing efficiency and product quality, and reduce the manufacturing cost. Based on intelligent manufacturing technology, intelligent machinery manufacturing technology has been developed rapidly and has been widely applied in various fields such as automotive, aviation and so on. It can achieve the goals of automation, intelligence, high efficiency, and personalization, and promote the development of various industries towards intelligent manufacturing. However, there are still challenges in the development and application of intelligent machinery manufacturing technology, such as the high cost of equipment, the difficulty in controlling the manufacturing process, and the lack of skilled workers. Future research should focus on solving these challenges and promoting the further development and application of intelligent machinery manufacturing technology.

Keywords. Intelligent manufacturing technology; Intelligent mechanical manufacturing process, Automation, Optimization, Efficiency

1. Introduction

1.1 The background and significance of intelligent manufacturing technology

Intelligent manufacturing technology is a manufacturing model that integrates information technology, advanced manufacturing technology, automation technology, material science and technology, and other technologies, aiming to improve production efficiency, optimize resource utilization, and enhance product competitiveness. Intelligent manufacturing technology can not only achieve production automation, digitization, and flexibility but also improve production quality and efficiency through data analysis and optimization.

The background of intelligent manufacturing technology can be traced back to the development of digital manufacturing technology. Digital manufacturing technology has been widely used in the manufacturing industry since the 1980s. Digital manufacturing technology improves production efficiency and product quality through digital design, digital processing, and digital management. However, digital manufacturing technology still has some shortcomings, such as inflexible production lines and high production costs. Therefore, intelligent manufacturing technology emerged[1].

The significance of intelligent manufacturing technology is to improve the competitiveness of the manufacturing industry. With the continuous advancement of science and technology and the constant changes in the market, traditional manufacturing is facing increasing pressure. The introduction of intelligent manufacturing technology can enable traditional manufacturing to achieve digitization and automation, improve production efficiency and quality, and reduce production costs and resource consumption. At the same time, intelligent manufacturing technology can also achieve production flexibility and personalization, adapt to market changes, and enhance enterprise competitiveness.

In summary, the background and significance of intelligent manufacturing technology are evident. The introduction of intelligent manufacturing technology can enable manufacturing to achieve digitization, automation, flexibility, and personalization, improve production efficiency, quality, and competitiveness, and bring more social benefits.

1.2 The concept and development process of intelligent mechanical manufacturing technology.

Intelligent mechanical manufacturing technology refers to the upgraded and improved traditional mechanical manufacturing technology using intelligent technology to achieve digitalization, intelligence, and automation in production. Intelligent mechanical manufacturing technology can improve production efficiency, reduce production costs, and improve product quality and machining accuracy.

The development of intelligent mechanical manufacturing technology can be traced back to the 1980s when the development of numerical control and computer technology led to the emergence of intelligent mechanical manufacturing technology. With the continuous development of information technology, sensing technology, control technology, and artificial intelligence technology, intelligent mechanical manufacturing technology has been continuously upgraded and improved. Nowadays, intelligent mechanical manufacturing technology has become an essential part of intelligent manufacturing technology.

The development of intelligent mechanical manufacturing technology is of great significance to the future of the manufacturing industry. Intelligent mechanical manufacturing technology can effectively improve the automation level, digital level, and intelligence level of the manufacturing industry, enhancing the competitiveness of enterprises. Intelligent mechanical manufacturing technology can also promote the development of the manufacturing industry towards high-end, intelligent, green, and sustainable development, promoting the sustainable development of the manufacturing industry.

2. The key technologies of intelligent machinery manufacturing technology.

Intelligent machinery manufacturing technology involves various technical fields, and the key technologies include but are not limited to the following.

Artificial intelligence technology: including machine learning, deep learning, natural language processing, etc., can be used in production planning, process control, product design, etc. Machine vision technology: including image recognition, 3D imaging, etc., can be used in parts inspection, quality control, robot control, etc. Sensor technology: including temperature sensors, pressure sensors, vibration sensors, etc., can be used to monitor parameter changes during production and achieve intelligent monitoring and control. Cloud computing technology: can be used to achieve big data analysis, collaborative design, intelligent manufacturing, etc. Human-computer interaction technology: including speech recognition, gesture recognition, etc., can achieve human-computer interaction and improve production efficiency and quality. 3D printing technology: can achieve rapid prototyping, small batch production, customized production, etc. Network security technology: can ensure information security and network security during the production process[2].

3. The applications of intelligent machinery manufacturing technology.

3.1 Automobile production

The intelligent mechanical manufacturing technology can be applied to the intelligent upgrading of the entire automobile production line, including automation, intelligent logistics, intelligent quality control, and other aspects. This can improve the production efficiency and quality of automobiles and reduce production costs. Intelligent mechanical manufacturing technology can also be applied to the production and processing of automobile parts, such as engine components and body parts, to improve processing accuracy and production efficiency and achieve personalized customization of parts. The automation and intelligence of automobile body painting can also be realized through the application of intelligent mechanical manufacturing technology. For example, using robots for spraying operations can improve the efficiency and quality of painting while reducing painting costs. The intelligent mechanical manufacturing technology can be applied to the production and manufacturing of automobile intelligent driving systems, such as the manufacturing of sensors and control modules, to improve the reliability and safety of the driving system. In general, the application of intelligent mechanical manufacturing technology in automobile manufacturing can improve the production efficiency and quality of automobiles, reduce production costs, and achieve personalized customization of automobiles. With the continuous innovation and development of intelligent mechanical manufacturing technology, its application in automobile manufacturing will become increasingly widespread[3].

3.2 Aircraft production

The intelligent mechanical manufacturing process can achieve intelligent upgrading of the entire aircraft production line, including automation, intelligent logistics, and intelligent quality control on the production line. This can improve the efficiency and quality of aircraft production, and reduce production costs. The intelligent mechanical manufacturing process can be applied to the production and processing of aircraft components, such as engine components, body parts, etc., which can improve processing accuracy and production efficiency, and achieve personalized customization of parts. The intelligent mechanical manufacturing process can be applied to the automation and intelligence of aircraft body painting, for example, using robots for spraying operations, which can improve the efficiency and quality of painting and reduce the cost of painting. The intelligent mechanical manufacturing process can be applied to the production and manufacturing of aircraft intelligent driving systems, such as the manufacturing of sensors and control modules, which can improve the reliability and safety of the driving system. Overall, the application of intelligent mechanical manufacturing processes in aircraft manufacturing can improve the efficiency and quality of aircraft production, reduce production costs, and achieve personalized customization of aircraft[4]. With the continuous innovation and development of intelligent mechanical manufacturing technology, its application in aircraft manufacturing will become more and more widespread.

3.3 Other industries

Intelligent mechanical manufacturing technology can not only be applied to the automotive and aviation industries, but can also achieve intelligent manufacturing in other industries. In the electronics industry, intelligent mechanical manufacturing technology can be applied to the manufacturing of printed circuit boards, for example, using precision processing equipment to achieve high-precision printing. In the medical industry, intelligent mechanical manufacturing technology can be applied to the manufacturing of medical equipment, such as surgical robots and artificial hearts, which can improve surgical accuracy and safety. In the food industry, intelligent mechanical manufacturing technology can be applied to the automation and intelligence of food production lines, for example, using robots for food processing and packaging, which can improve production efficiency and quality, and reduce production costs. In the construction industry, intelligent mechanical manufacturing technology can be applied to the production and processing of building materials, such as the manufacturing of prefabricated concrete components, which can improve construction efficiency and quality.

Overall, the application of intelligent mechanical manufacturing technology in various industries can achieve automation, intelligence, high efficiency, and personalization, improve production efficiency and quality, reduce production costs, and promote various industries towards intelligent manufacturing.

4. The advantages and challenges of intelligent mechanical manufacturing technology.

4.1 Advantages of intelligent mechanical manufacturing technology

(1) Improved production efficiency: Intelligent mechanical manufacturing technology can achieve digitalization, intelligence, and automation in production, greatly improving production efficiency. Digital processing technology can achieve digital, flexible, and high-precision processing of the production process, reducing production time and cost.

(2) Improved product quality: Intelligent mechanical manufacturing technology can achieve high-precision and high-quality production. Digital processing technology can improve machining accuracy and product quality. Intelligent detection technology can achieve automatic detection and quality control, ensuring product quality.

(3) Energy conservation and emissions reduction: Intelligent mechanical manufacturing technology can reduce manual operation, energy consumption, and emissions. Digital processing technology can optimize machining paths and reduce material waste. Intelligent detection technology can improve the accuracy and efficiency of quality inspection, reducing waste rates.

(4) Reduced production costs: Intelligent mechanical manufacturing technology can achieve automated production and digital processing, reducing labor and material costs. Intelligent detection technology can reduce waste and rework rates, reducing production costs.

(5) Enhanced production flexibility: Intelligent mechanical manufacturing technology can achieve digitalization, intelligence, and automation in production, improving production flexibility and adaptability. Digital processing technology can achieve flexible processing, adapting to production of different products and specifications.

In summary, intelligent mechanical manufacturing technology has significant advantages. It can improve production efficiency and product quality, reduce production costs, reduce energy consumption and emissions, enhance production flexibility and adaptability, and promote the development of the manufacturing industry towards high-end, intelligent, green, and sustainable development.

4.2 Challenges of intelligent machinery manufacturing technology.

With the development of intelligent machinery manufacturing technology, although it has brought many advantages, it also faces some challenges.

(1) High technical difficulty: Intelligent machinery manufacturing technology requires the application of multiple high-tech technologies, including artificial intelligence, machine vision, sensor technology, etc. The technical difficulty is relatively high. Continuous research and development and technical breakthroughs are needed to meet the needs of different industries and fields.

(2) Talent shortage: The intelligent machinery manufacturing industry requires a variety of technical talents, including mechanical engineers, electronic engineers, computer engineers, data analysts, etc. At present, there is a certain gap between talent supply and market demand, leading to talent shortage.

(3) High safety risks: Intelligent machinery manufacturing processes require digital and automated production, involving multiple risks such as personnel safety and data security. It is necessary to strengthen security precautions and measures to ensure production safety and data security.

(4) High maintenance cost: Intelligent machinery manufacturing technology requires maintenance of high-end equipment, software systems, etc., which incurs high maintenance costs. Especially in the process of use, malfunctions and damages may occur, requiring timely maintenance and repair, increasing costs and difficulty.

(5) Difficulty in integrating the industry chain: Intelligent machinery manufacturing technology

requires the integration of multiple technologies and industrial chains, including hardware, software, materials, etc. This requires companies to play a synergistic effect in the industrial chain, improve the efficiency and competitiveness of the entire industrial chain.

In summary, although intelligent machinery manufacturing technology has brought many advantages, it still faces challenges such as high technical difficulty, talent shortage, high safety risks, high maintenance costs, and difficulty in integrating the industry chain. In the future, continuous research and development and technical breakthroughs are needed, talent cultivation and introduction need to be strengthened, security management and maintenance need to be strengthened, and industrial chain integration and innovation need to be promoted to achieve the sustainable development of intelligent machinery manufacturing technology.

5. Conclusion

Currently, intelligent machinery manufacturing technology has been widely applied in many fields, such as automotive, machine tools, robots, and so on. Its advantages lie in improving production efficiency and quality, reducing production costs, and customizing production according to different needs. At the same time, with the continuous development and innovation of technology, intelligent machinery manufacturing technology will continue to develop towards intelligent, digital, flexible, and sustainable directions. Specifically, in the future, intelligent machinery manufacturing technology will pay more attention to the intelligence and automation of production. Through technologies such as big data and cloud computing, intelligent monitoring and control of the production process will be achieved. Meanwhile, flexible production will become a trend, making production more adaptable to market demands. Additionally, intelligent machinery manufacturing technology will make greater progress in the direction of sustainable development, such as green manufacturing and resource conservation, to achieve sustainable development goals. In summary, intelligent machinery manufacturing technology will play an increasingly important role in future development.

References

1. Zhang, X., & Qian, Y. (2021). Research on intelligent manufacturing technology of mechanical equipment based on industry 4.0. In 2021 IEEE 3rd International Conference on Electronics Technology (ICET) (pp. 1548-1551). IEEE.
2. Li, X., Li, Z., Li, L., Li, Q., & Li, Z. (2019). Design and implementation of intelligent manufacturing system for mechanical processing. Advances in Manufacturing, 7(1), 56-63.
3. Liu, J., & Wang, J. (2020). Research on intelligent manufacturing technology of mechanical equipment based on the integration of industrialization and informatization. In 2020 IEEE International Conference on Industrial Engineering and Engineering Management (IEEM) (pp. 1919-1923). IEEE.
4. Wang, C., & Zhou, L. (2019). Development of intelligent manufacturing technology based on industry 4.0. In 2019 IEEE 2nd Advanced Information Technology, Electronic and Automation Control Conference (IAEAC) (pp. 3026-3029). IEEE.

Preparation of MXene and its application and research progress in supercapacitors

Zhenkui Gao*

College of Textiles, Donghua University, Shanghai, China.

Abstract: As a new type of two-dimensional nanomaterial, two-dimensional transition metal carbon/nitride (MXene) has attracted much attention because of its high specific surface area, high conductivity and solvability. MXene is often used as a basic material to construct macroscopic composites. In recent years, it has been used to manufacture MXene structural-functional integrated fiber materials and mqd derivatives of MXene, which has a wide application prospect in multifunctional fibers, energy conversion and storage (ECS), sensors, catalysis and lubrication, electromagnetic shielding (EMI), supercapacitors (SCs) and so on. And has achieved some results. In this paper, the international research progress of MXene materials in recent years is reviewed, and its structure, properties and preparation methods are summarized. On this basis, its development in the field of energy storage (such as supercapacitors) is summarized, the shortcomings in this field are pointed out, and the future development trend is prospected.

Keywords: Two-dimensional material; MXene; Supercapacitor; Surface active group.

1. Introduction

The application of renewable energy has attracted more and more attention in the fields of portability and wearability, and in the case that fossil energy may be exhausted. However, sustainable renewable energy sources such as wind energy, hydropower energy and solar energy have the characteristics of fluctuation and discontinuity. At the same time, with the demand of the country, the development of new energy equipment such as electric vehicles is getting faster and faster, which requires the development of an energy storage device with small volume, large capacity, relatively large volume specific energy and high cycle stability. In recent years, supercapacitor materials with high energy density and high cycle stability have been widely used in the field of energy storage, which has become the key to solve this problem.

Supercapacitors (SCs) generally include electric double layer capacitors (EDLC) and Faraday Pseudo-Capacitors.[3][39]The purpose of high power density, stable cycle performance, wide working voltage range, low working temperature and environmental protection is finally achieved by using high surface area electrode materials and thin media. The most important step in the research and development of batteries is the selection of electrode materials. However, because the two-dimensional layered nanomaterial has achieved nanometer level (0-100 nm) to a certain extent, and its thickness is much smaller than the other two dimensions, it has large specific surface area, good mechanical properties and many surface active sites. At present, graphene, oxygen/hydroxide and carbon/nitride (MXene) have been widely used in two-dimensional systems.[38]. The development trend of high capacity and high power density batteries is to use two-dimensional materials with high specific surface area as battery electrode materials.

In 2011, Professor Naguib of Drexel University and his team developed a new type of two-dimensional layered material-MX gene. MXene is a new two-dimensional (2D) carbonitride compound with high conductivity (about 8000S/cm).[4][5][25]And has high specific capacity due to oxidation-reduction reaction between transition metal on its surface and electrolyte ions, and excellent electromagnetic shielding ability.

In this paper, the structure, properties and preparation methods of MXene are introduced in detail. Then, the progress and shortcomings in the research field of supercapacitors are emphatically summarized, and the future development of MXene is analyzed and prospected.

2. Introduction to MXene

MXene is a layered structure similar to graphene, with high specific surface area and good conductivity, and it is a new two-dimensional material with great application prospects.[25]. MXene is a transition metal carbide, nitride or carbonitride with $Mn+1\ XnTx$ (n=1,2,3,4) as the basic structural unit. X represents c and n, where m represents a pre-transition metal element; Tx is a kind of

* Corresponding author: 1246031275@qq.com

surface reactive end-capping groups, such as hydroxyl, hydroxyl, fluorine, chlorine and so on.[2].

Generally, in the process of preparing MXene, the A atom of MAX phase is selectively etched out, because the metal bonding strength of M-A of MAX species is weaker than the covalent bond of M-X, and the stability of M-X is higher than that of M-A, so MXene can be obtained by etching M-A.. The obtained MXene is divided into m layers and x layers. The MAX phase consists of more than 70 ternary metal carbides, nitrides and carbonitrides, and the general molecular formula is Mn+1AXn (n=1,2,3), with A as the main group (such as Al, Ga, In, Ti, Si, Ge, Sn, Pb, etc.). Compared with M-A, M-A is more metallic.[2][37].

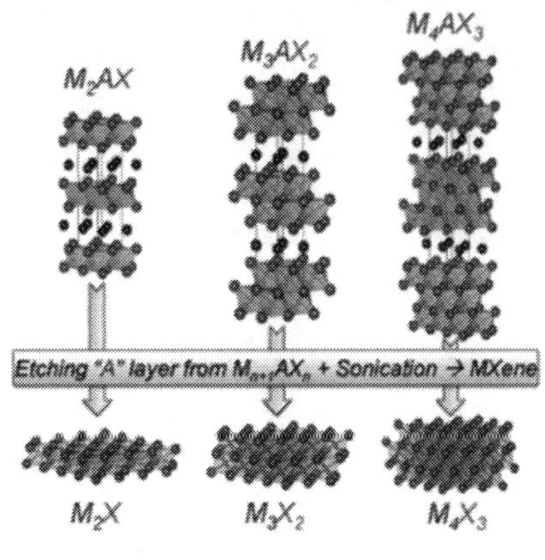

Fig. 1 structure of max phase and corresponding MXene[25]

At present, there are more than 60 kinds of MXene, more than 30 kinds have been synthesized, and more than 10 kinds can exist stably after calculation, such as Zr2C, Ti3N2, V4C3 and so on.[26]. MXene is a new type of ternary nitride. Up to now, only two materials, Ti2N and TiN3, have been reported.[2]. The M-A bond of ternary nitride MXene is more stable than that of ternary carbide, and it is not easy to be etched.[2]. Although M in all the above MXENes is the same metal, there are still 26 MXENes with high stability by density functional theory (DFT) calculation.[26]. The system consists of (TiV)2 C, (CrV)3C2, (CrV)3C2, (Cr2V)C2 and (V2Ti2)C3.[2].

3. Preparation technology of MXene

Most synthesis methods of MXene are chemical etching, which makes use of the differences in chemical activity and stability of raw material elements or structural units to carry out selective stripping.[26]. Because the covalent bond of M-X is stronger than the metal bond of M-A, under the conditions of strong acid, molten salt and high temperature, the M-A bond can be broken, leaving only the M-X bond, thus preparing two-dimensional materials of MXene. The high-temperature heating method, although it can break the M-A bond, is generally not used because the twinning process will form a 3D structure similar to rock.[4][5]. Chemical etching is the mainstream preparation method. However, at present, ammonium hydrogen fluoride, mixed solution of hydrochloric acid and lithium fluoride, hydrofluoric acid and other fluorides are mainly used in the market to prepare MXene thin films. These fluorides have strong electronic storage capacity in MXene thin films, and at the same time, the toxicity of fluoride itself will lead to safety problems in the preparation process. Therefore, it is urgent to develop a new method for preparing non-fluoride thin films. The following are several etching methods related to the preparation of MXene.

3.1 Etching based on HF and fluoride salt.

In 2011, Naguib et Al. etched Ti3AlC2 with HF to completely strip Al atoms in Ti3AlC2, and obtained two-dimensional Ti3C2 nanosheets.[4][29]. The preparation of HF and HF can be divided into the following three steps:

$$Ti_3AlC_2 + 3HF = AlF_3 + Ti_3C_2 + 3/2H_2 \quad (1)$$

$$Ti_3C_2 + 2H_2O = Ti_3C_2(OH)_2 + H_2 \quad (2)$$

$$Ti_3C_2 + 2HF = Ti_3C_2F_2 + H_2 \quad (3)$$

Among these processes, the mechanism of etching and exfoliation processes is illustrated in fig. 3.

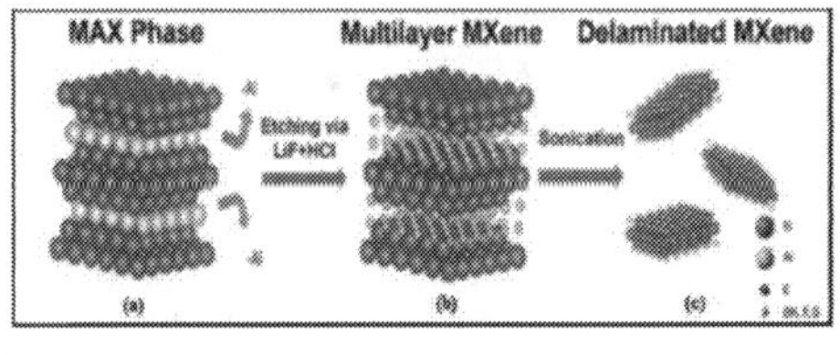

Fig. 2 Etching spalling mechanism diagram of Ti3AlC2 material (a)Ti3AlC2 structure diagram; (b) replacing Al atoms after acid etching with -OH groups; (c) hydrogen bonds between layers are broken under the action of ultrasonic waves, and the layers are separated.[29][32]

From the above equation, we can see that the surface of the two-dimensional nano-material MXene prepared by this method is rich in hydroxyl or fluorine groups, and the mechanical and thermodynamic stability of the material is improved by modifying its surface functional groups. At the same time, theoretically, the energy and width of the material can be adjusted by adjusting the surface functional groups of the material, so as to adjust the conductivity of the material.[26]. In addition, MXene materials with different properties can be prepared by adjusting the concentration, temperature and etching time of acidic solution. The experimental results show that the optimum preparation temperature of Ti3C2Tx MXene is around 55OC. During hydrofluoric acid etching, cations can play an insertion role, which increases the interlayer spacing of MXene films. In the preparation process, etching and stripping can be carried out simultaneously.

Ghidiu et al. etched with lithium fluoride and hydrochloric acid in addition to hydrofluoric acid, and finally obtained Ti3C2 similar to clay.[13][26]. Halim et al. grew Ti3AlC2 thin films by magnetron sputtering. However, the mixture of hydrogen fluoride and hydrogen chloride is essentially a strong acid turned into a weak acid, so hydrogen fluoride is finally at work.[12][26].

3.2 Other methods.

In addition to the above preparation methods, the latest research theory also uses chemical vapor deposition (CVD) to prepare nano-scale ultrafine structures. Xu et al. prepared large-area and high-quality two-dimensional AlphaMo2C single crystal and some transition metal carbides such as W2C and WC by chemical vapor deposition.[17][2]. Du et al. took the lead in putting forward a new idea of "topological transformation" in the world. Single-layer chalcogenides can be synthesized in large quantities by using non-Van der Waals systems (such as carbides, nitrides and carbon-nitrides (such as MAX phase)), which opens up a new idea for building MXene.[2][25].

4. Application of MXene in supercapacitors

As analyzed above, MXene materials have been widely used as electrode materials in supercapacitors because of their excellent conductivity and large specific surface area, and have a good development prospect. At present, the research direction of MXene for supercapacitors can be roughly summarized as the application of a single MXene electrode material in SCs. In view of the problem that MXene sheet structure is easy to accumulate, the MXene material is modified or compounded with other materials to make composite thin film materials. Of course, the application of MXene in hybrid supercapacitors is also developing rapidly.

4.1 Pure MXene electrode material.

Lukatskaya et al. discovered for the first time that cations (K+, Na+, Mg2+, etc.) in Ti3C2Tx MXene will be self-intercalated, but the specific mechanism is not clear.[36]. Lukatskaya et al. used dimethyl sulfoxide (DMSO) to separate Ti3C2, then used ultrasonic wave to form a thinner layer, then centrifuged, filtered and dried it, and finally made a "Ti3C: paper" electrode with a thickness of 2-20 microns. In KOH solution, under the condition of 2 mV/s, the capacity of this unbonded electrode can reach 340 F/cm3, and it will not produce any loss during 10000 times of charge and discharge.[14][26]. Ghidiu et al. etched Ti3AlC2 with lithium fluoride and hydrochloric acid to obtain slurry Ti3C2, and then rolled the slurry Ti3C2 into a thin sheet with a roller, thereby directly making the slurry Ti3C2 into an electrode. The electrode has a volume specific capacitance of 900 mV/cm3 at a scanning speed of 2 mV/sec in a sulfuric acid solution of 2 mV/sec, and a volume specific capacitance of 730 mV/cm3 at a scanning speed of 100 mV/sec. Ghidiu et al. determined the electrochemical properties of the prepared self-supporting Ti3C2 electrode material.[13][2]. As can be seen in figure 6. At 2 mV/s, the material has a specific capacity of 900 F/cm3, good rate characteristics and a cycle life of 10,000 times.

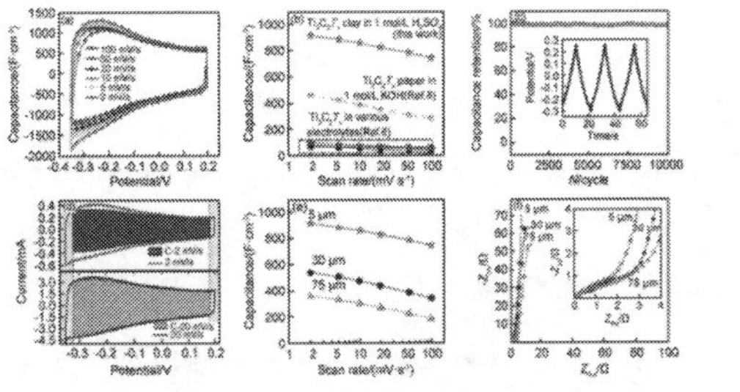

Fig. 3 electrochemical characteristics of ti3c2tx material (a) cyclic voltammetry (CV) curves of ti3c2tx at various scanning speeds; (b) The rate characteristics of Ti3C2Tx in various electrolytes; (c) The capacitance hypertension ratio of Ti3C2Tx during constant current charging/discharging is 10 A/g; (d) CV curves of Ti3C2Tx at scanning speeds of 2 mV/s and 20 mV/s; (e) Capacitor maintenance coefficient of Ti3C2Tx with different electrode thicknesses; (f) Impedance curves of Ti3C2Tx on electrodes with different thicknesses.[2]

MXene is a kind of supercapacitor material with great development potential, but its problems such as easy agglomeration, accumulation and oxidation seriously restrict its application in batteries. In addition, the surface functional groups of MXene will also affect the conductivity and cycle stability of the material.[35]. Based on this, this project intends to carry out related research from several aspects such as surface modification, structural regulation, ion/molecule insertion/doping, and antioxidant modification.

4.1.1 Structural control.

At present, the structural regulation of MXene is mainly realized by core-shell and porous methods. Xu et al. used the current collector Ni as the carrier, and loaded MXene nano-sheets on its surface to prepare a new type of wearable flexible fiber substrate. The substrate does not need any binder, and its specific capacity reaches 140 F/g in 1 mol/LKOH solution.[17][25]. Li et al. found that three-dimensional Ti3C2 aerogels have high specific surface area (176.3m2/g) and special pore structure. The electrode was tested in 1 mol/L KOH electrolyte with a scanning speed of 2 mV/s, and a high surface capacity of 1012.5 mF/cm2 was obtained.[16][25].

The above research results show that by adjusting the micro-morphology of MXene, the lamellar rearrangement of MXene can be effectively reduced, the specific surface area can be increased, the active sites can be increased, the ion transport channel can be shortened, and the ion intercalation/deintercalation can be accelerated, thus greatly improving the electrochemical performance of MXene, which is an important means to improve the rate performance of MXene.[25]

4.1.2 Intercalation Modification and Doping Modification.

Insertion modification of MXene usually does not change its surface functional groups, but only increases its interlayer distance; By doping modification, the functional groups on the surface of the material can be changed. Both methods can improve the related performance of MXene. By introducing cations, organic compounds, etc. between the layers of MXene, the interlayer spacing of MXene is increased, the active sites of MXene are increased, and its electrochemical, adsorption, mechanical and other properties are improved. Li et al. introduced Na+ ions into MXene (SI-Ti3C2 Tx), which increased the specific surface area of MXene, increased the surface active sites of MXene, and formed hydrogen bonds with ciprofloxacin (CPX), which was electrocatalytic decomposed into CO2 and H2O.[31]. The reaction process is shown in fig. 7. In the second cycle, the removal rate of chlorinated hydrocarbons can reach ninety-seven percent.

Doping MXene can effectively increase the interlayer distance and lay a foundation for its application in the fields of batteries and energy storage. However, the electrochemical performance of fluorine-terminated MXene prepared by existing etching methods in supercapacitors is not ideal. Therefore, in addition to intercalation modification, MXene also needs doping modification.

By introducing carbon, nitrogen, sulfur and other elements into MXene, and introducing chemical bonds into MXene, its conductivity is enhanced and its surface wettability is improved, thus increasing its reactive sites. At the same time, the collaborative optimization of MXene is realized by post-heat treatment; By doping transition metal elements in MXene, MXene has good catalytic performance for hydrogen evolution. As shown in fig. 8, LU et al. have fully studied the nitrogen doping mechanism of MXene.[25][31]

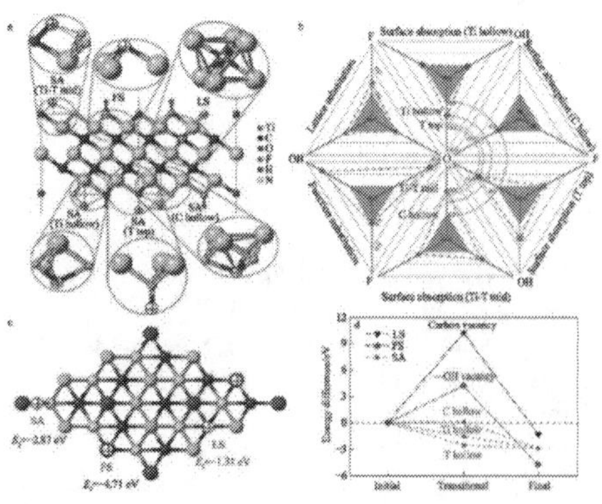

Fig. 4simulation of nitrogen dopant in ti3c2: atomic structure of Ti3C2Tx with all nitrogen dopants (a); Calculation of generating capacity (b) (shaded with positive and negative values); It has a nitrogen atom (top view) (c) in the most favorable position of the energy of Ti3C2 supermonomer; Calculation of Transition State Energy Levels (D)[31]

4.1.3 Anti-oxidation modification.

In the medium of water and air, MXene is easy to be oxidized by Ti to produce TiO2, which diffuses from outside to inside, resulting in the destruction of MXene structure. Therefore, in order to prevent the oxidative degradation of MXene, it is often necessary to conduct waterproof and insulation treatment. Taking MXene as the research object, MXene was treated with alkali by physical and chemical methods to improve the order of MXene crystals and reduce the content of functional groups such as hydroxyl and fluorine, and the MXene materials with high conductivity, high hydrophobicity and high conductivity were prepared. Chemical method can effectively isolate MXene from water by means of polyanion adsorption and polymer coating.[31]

4.2 MXene composite film material.

MXene is a new material which combines the advantages of many materials and has great potential.

4.2.1 And carbon composite

Graphene, carbon nanotubes (CNT), carbon nanofibers, carbon aerogels and other carbon materials are excellent conductive materials. They have excellent conductivity and are excellent conductive materials.[25]. Zhao et al. obtained the sandwich-shaped Ti3C2Tx/ carbon nanotube (CNT) composite electrode material by alternating suction filtration. It is in 1 mol/LMgSO4 aqueous electrolyte, at a scanning speed of 2 mv/s.[26]The volume specific capacitance is 390 f/cm3, and it is 280 F/cm3 at a scanning speed of 200 mv/s.[26]. In addition, Linetal's previous work also found that sandwich composite electrode materials have better electrochemical properties than random hybrid composite structures.

4.2.2 Compounding with metal oxide.

Transition metal oxide (TMO) is superior to carbon in theoretical capacity and safety, but it will have large volume expansion during charging and discharging, and its conductivity is low.[34]. However, if it is compounded with MXene, the performance of the two can be supplemented, thus improving the comprehensive properties of the material.[25]. Jiang loaded MnO2 on Ti3C2Tx by chemical deposition, and obtained MnO2/Ti3C2Tx composites.[25]The composite material has the highest specific capacity of 130.5 F/g at 0.2 A/g, which is equivalent to twice the same current density.[25]. Peng et al. used the strong electrostatic attraction between WO3 nanorods (WNR) and Ti3C2Tx (MXene) to compound WO3/MXen, so that it could obtain more than 5,000 cycles in the system of 1A/g current density and 0.5 mol/LH2SO4, and its specific capacity reached 297 f/g. Through the above research, we found that the addition of TMOs can greatly improve its specific capacity, and at the same time, the advantages of two-dimensional layered MXene and TMOs complement each other, which can ensure the volume stability of TMOs.[25]

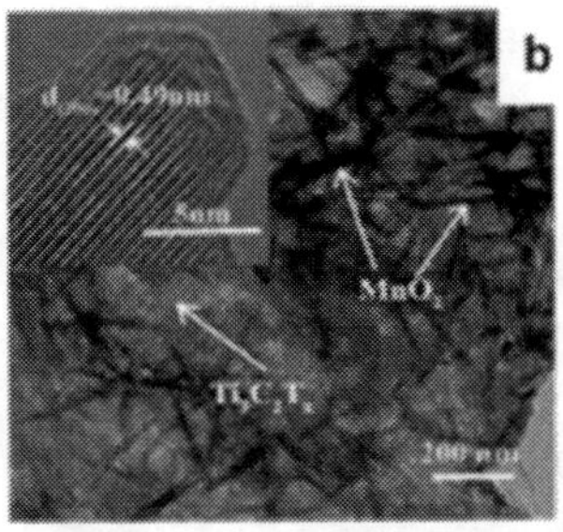

Fig. 5 TEM image of nano MnO2/Ti3C2Tx[25][33]

4.2.3 Compounding with polymer.

The introduction of polymers (such as polyaniline, polyaniline, polydimethyl acrylic acid, etc.) can effectively increase the interlayer spacing of MXene. As a carrier, it can effectively inhibit the stacking of layers, expose more surface active sites, shorten the ion diffusion channel of electrolyte, accelerate charge transfer, accelerate Ferrari reaction, and enhance its mechanical properties and stability.[25]. For example, Ling et al. used MXene with atomic layer thickness as the building block, and realized the controllable preparation of MXene/ polymer composite films for the first time by self-assembly chemistry.[26]. The Ti3C2Tx/PDA composite electrode prepared by Wang has a specific surface area capacity of 75 mF/cm2, and it can still maintain a specific surface area of 95.5% after 10,000 cycles.[25].

5. Conclusion and prospect

In this paper, firstly, the structure and properties of MXene materials are introduced, and the existing preparation methods are summarized in detail. At present, most of the preparation of MXene adopts chemical etching method, and fluorine-containing compounds account for the majority. However, there is a certain danger, and more efficient and safer etching methods in the future need to be further studied by scholars.
Secondly, the applications of various MXene in supercapacitors are reviewed. In the preparation process of monolayer two-dimensional nano-material MXene, there are usually defects such as easy accumulation, so it is necessary to improve its related properties by doping and intercalation. In addition, in order to solve the problem that Ti atoms on the surface of MXene are easy to be oxidized, the oxidation resistance of MXene is also improved from two aspects: physical alkaline annealing and chemical coating. Based on the comprehensive advantages of each material, composite MXene thin film material is the thriving development direction of supercapacitor pseudocapacitor in the future.
However, the research of MXene still faces the following problems: (1) More than 150 kinds of MAX phases are known, but only more than 30 kinds of MXene have been developed, and most of them are Ti3C2Tx limited by element A. The preparation and research of other kinds of MXene need further breakthrough. (2) The surface active groups of 2)MXene materials have great influence on the properties and electrochemical properties of the materials. However, after the surface functional groups are completely removed, the characteristics of this material are only theoretical speculation, and its actual characteristics need further study. (3) At present, the synthesis of MXene mostly needs to use some corrosive drugs, and its yield is low, which is easily disturbed by external conditions. How to realize efficient and safe large-scale production and industrialization of MXene is an urgent problem to be solved. (4) At present, the application of MXene material library in SCs is still in the primary stage, and it is urgent to develop the preparation technology, structural design and functional method of Ti3C2Tx electrode with high cost performance and high yield.[22].
However, based on the excellent performance of MXene materials and the continuous exploration of energy storage in recent years, the research of MXene materials in super capacitor (SCs) still has a very broad application prospect, and the country is also vigorously promoting the research and development of sustainable green pollution-free energy.

References

1. Winter M, Brodd R J. What are batteries, fuel cells, and
2. Dang Ali, Fang Chenglin, Zhao Wei, et al. Preparation of a new two-dimensional nano-material MXene and its application in energy storage field [J]. Materials Engineering, 2020, 48(4):1 4.
3. Classification and model naming method of supercapacitors: SJ/T 11625-2016[S]. 2016.
4. NAGUIBM , MOCHALINVNN, BARSOUMMW,,e tal. MXenes : a new family of two-dimensional materials [J]. Advanced Materials, 2014,26(7) :992-1005 .
5. NAGUIBM, KURTOGLUM, PRESSERV, etal. Two-dimensional nanocrystals produced by exfoliation of Ti3AlC2[J]. Advanced Materials, 2011, 23(37): 4248-4253.
6. HANM, YINX, LIX,etal. Laminated and two-dimensional
7. Carbon-supported microwave absorbers derived from MXenes [J] .
8. ACS Applied Materials & Interfaces, 2017 9(23):20038-20045.
9. ANASORIB, LUKATSKAYAMR, GOGOTSIY .2 D metal carbides and nitrides (MXenes) for energy storage[J].NatureReviews Materials,2017, 2:16098.
10. XUC,WANGL,LIUZ,etal. Large-area high-quality 2 Dul.
11. Trathin Mo2C superconducting crystals [J]. Nature Materials,2015,14(11):1135-1141.

12. HALEM J . LI, KATSKAYA M R . COOK K M . et al . Transparent conductive two . dimensional titanium carbide epitaxial thin films [J] . Chemistry ofMaterials, 2014, 26(71: 2374 . 2381 .
13. GHIDIUM, LUKATSKAYA M R . , ZHA0 M O, et al. Conductive Two. Dimensional Titanium Carbide Clay with Hi Vomitic Pacification [J]. Nature. 2014.516: 78.81.
14. LUKATSKAYA M R, MASHTALIII O.REN C E, et al . Cation intercalation and high volumetric capacitance of two . dimensional titanium carbide[J] . Science, 2013, 341: 1502—1505 .
15. XIE X, XUE Y, LI L, et al. Surface Al leached Ti3 AlC2 as a substitute for carbon for use as a catalyst support in a harsh corrosive electrochemical system[J]. Nanoscale, 2014, 6(19): 11035-11040.
16. LI M, LU J, LUO K, et al. Element replacement approach by reaction with lewis acidic molten salts to synthesize nanolaminated MAX phases and MXenes[J] . Journal of the American Chemical Society, 2019, 141(11): 4730-4737.
17. Xu C, Wang L, Liu Z, et al. Large-area high-quality 2D ultrathin Mo2C superconducting crystals[J]. Nature materials, 2015,14(11):1135-1141.
18. LIJB, LIUY, XU F, et al. K+ intercalation of NH4HF 2-exfoliated Ti3C 2 MXene as binder-free electrodes with high electrochemical capacitance[J] . Physica Status Solidi A-Applications and Materials Science, 2020, 217(8): 1900806.
19. LIUL, YING G B, ZHAO Y L, et al. Attapulgite-MXene hybrids with Ti3C 2Tx lamellae surface modified by attapulgite as a mechanical reinforcement for epoxy composites[J]. Polymers, 2021, 13(11): 1820-1820.
20. Li huapeng, Xu xusheng, bin bin, et al. controllable preparation of TiO _ 2/mxene nanocomposites and its application in photocatalysis and electrochemistry [J]. materials engineering, 2021, 49 (8): 54-62. doi:
21. Liang cheng, cheng qunfeng. research progress in preparation, properties and application of MXene fiber [J]. journal of composite materials, 202,239 (9): 4227-4243. doi: 10.13801/j.cnki.fhclxb. 20725.000000001006.
22. Niu Lili, Wang Pei, Zhang Li, et al. Progress in preparation of Ti3C2Tx MXene and its application in supercapacitors [J]. journal of synthetic crystals, 202,150

Nonlinear effects-based 1.7 μm fiber lasers: a review and prospect

Xinning Ma*

College of Optoelectronic Engineering, Changchun University of Science and Technology, Changchun, China

Abstract: The nonlinear effects in the fiber lasers have always been explored and studied as people are pursuing higher quality fiber lasers in different wavelengths for profound applications. In recent years, 1.7 μm band fiber lasers have received the tremendous attention due to their unique spectral properties in biological imaging, organic gases detection, material processing and other fields. In this paper, the research progress of nonlinear effects-based 1.7 μm fiber lasers is thoroughly reviewed. Meanwhile, the four nonlinear effects applied in 1.7 μm fiber lasers included stimulated Raman scattering (SRS), super-continuum (SC), four-wave mixing (FWM), soliton self-frequency shift (SSFS) are introduced, as well as the principle, characteristics and advantages of each method. In addition, the latest researches on the 1.7 μm fiber lasers based on the hybrid gain are summarized in detail. Finally, the conclusion included the obstacles and adversities is given and the future development tendency of nonlinear effects-based 1.7 μm fiber lasers is prospected.

Keywords: Fiber lasers; 1.7 μm band; nonlinear effects; hybrid gain

1. Introduction

Fiber lasers, which have the unique advantages of high output beam quality, good heat dissipation and stable operation, are enjoying their popularities in this decade, having become a significant potential laser light source and stood out among other lasers[1-3]. Fiber lasers are mainly divided into rare earth-doped and nonlinear-based. Although there are some highlights of rare-earth-doped fiber lasers[4-7], their poor tunability and low optical conversion efficiency limit their application in some specific fields[8-10]. In fiber lasers, the effect of nonlinear influences on the generation of laser light is complex. It will produce frequency conversion, wavelength change, and phase distortion[11]. Appropriate nonlinear effects can promote the generation and increase of laser light, increasing the power and efficiency of the laser.

Since nonlinear effects have a key impact on the development of ultrafast lasers, massive fiber lasers based on nonlinear effects in various bands have been continuously reported. With the development of short-wave infrared fiber lasers[12-14], 1.7 μm band has received the attention of many researchers because of its unique spectral properties and great application value in many fields[15-19]. This band is between the two absorption peaks of water molecules[20], where water molecules have low absorption, and is also at the absorption peak of lipids, making the 1.7 μm band pulsed laser light source an important tool for bioimaging applications[21, 22], which are widely applied to multi-photon fluorescence microscopy imaging of lipid-rich samples[23], lipid-targeted skin therapy[24] and optical coherence chromatography imaging[25]. In addition, the 1.7 μm band is located at the absorption peak of the carbon-hydrogen bond, which can be applied to the detection of organic gases [26], processing of polymers[27], etc. Therefore, 1.7 μm band fiber lasers have a very broad development prospect.

Tremendous studies related to 1.7 μm band rare earth-doped fiber lasers have been demonstrated these ten years. However, the most appropriate active gain fiber, the thulium doped fiber (TDF), has the disadvantages of limited emission cross-section and the large absorption cross-section at the 1.7 μm band, leading to the low performance of optical efficiency[10]. As mentioned before, if the nonlinear effects in optical fibers are exploited well, laser generation can be promoted and the performance of lasers can be improved. So nonlinear effect-based methods have been proposed to generate continuous/pulsed lasers in the 1.7 μm band. There are four main methods based on nonlinear effects: (1) SRS: Stimulated Raman Scattering (2) SC: super-continuum. (3) FWM: four-wave mixing (4) SSFS: soliton self-frequency shift[28-30]. Among them, SRS combined with solid-core fiber is mostly used to generate 1.7 μm continuous laser. In recent years, fiber gas Raman lasers (FGRLs) using SRS combined with hollow fibers have also been studied and reported to obtain 1.7 μm pulsed lasers[28], offering

* Corresponding author: 18852162188@163.com

a potential solution for generating pulsed fiber laser sources with high-power and high-efficiency performance in the 1.7 μm band[31]. Compared to SRS, the SC generated in the anomalous dispersion region is also one of the preferred methods to achieve the 1.7 μm band. Tunable pulsed light as a pumping source has a lower threshold, higher conversion efficiency, and easy control of the resulting broadband light source spectrum. FWM for fiber lasers can eliminate phase noise and changes in oscillation conditions between optical devices, making fiber lasers more stable and reducing problems such as wavelength fluctuations and frequency shifts. SSFS can increase the output power of fiber lasers, both continuous wave (CW) and pulsed lasers, due to the ability of optical solitons to eliminate dispersion effects through self-interaction, thereby reducing the broadband spread of the optical pulse in the fiber and increasing the power of the optical signal. However, laser generation based on fiber nonlinear effects will inevitably cause the broadening of the output spectrum. Complex device structures can also limit the development of Raman fiber lasers[29].

The article details the research progress and applications of 1.7 μm band fiber lasers based on nonlinear effects around the world in recent years. Four technical paths around the realization of nonlinear effect-based 1.7 μm-band fiber lasers are presented (SRS, SC, FWM, SSFS). Meanwhile, the characteristics of the different technical solutions are discussed and analyzed. In addition, a detailed summary of the recently emerged hybrid gain methods based on nonlinear effects and rare earth doping is elaborated. Combined with application requirements, the prospect of 1.7 μm band fiber optic light source is reasonably foreseen based on the current research status.

2. Four nonlinear effect-based methods and the related 1.7 μm fiber lasers

2.1 SRS-based implementation methods

SRS refers to the strong interaction between high-intensity laser light and matter molecules, so that the scattering process has the nature of stimulated emission. It displays threshold properties and can only be produced with appropriate pump power.

The Raman fiber laser (RFL) is a laser that operates on the SRS principle. It generally consists of a section of Raman fiber with high refractive index and a section of ordinary fiber with lower refractive index. SRS occurs when the high-power laser is injected into the Raman fiber, where a portion of the light is converted to Raman scattered light of lower frequency. The Raman scattered light will be reversed and put back into the normal fiber and interact with the original laser, thus causing feedback and laser amplification. Using the SRS process, the Raman fiber laser is capable of converting high frequency laser light into Raman scattered light at lower frequencies, resulting in second harmonic generation. Additionally, it can be used to generate lasers with multiple wavelengths, making itself have the advantages of flexible wavelength tunability and large output power.

SRS-based Raman fiber lasers are typically used to generate high power continuous and quasi-continuous pulsed lasers in the 1. 7 μm band, having tremendous potential. Since the first 1.7 μm band Raman fiber laser was reported by Alexander et al. from the University of Michigan in 2011[24], there has been a significance increase in the number of studies about RFL these 5 years. In 2018, the research group of Feng from Shanghai China, obtained a continuous-wave laser with a maximum power of 66.9 W at 1676 nm by pumping a section of Raman fiber with a home-made time-stable 1064 nm high-power ytterbium-doped fiber laser while cascading a random Raman laser and adjusting the pump laser wavelength[32]. Later, they replaced the pump source with the 1 μm band ASE source, the temporal stability of which is higher, and obtained a continuous laser with a maximum output power of 6.9 W at 1691.6 nm, increasing the spectral purity of the RFL[33]. The schematic diagram of ASE seed and the cascaded Raman random fiber laser (RRFL) is shown in Fig 1. To prevent lasing, isolators are spliced at both ends of an all-open cavity.

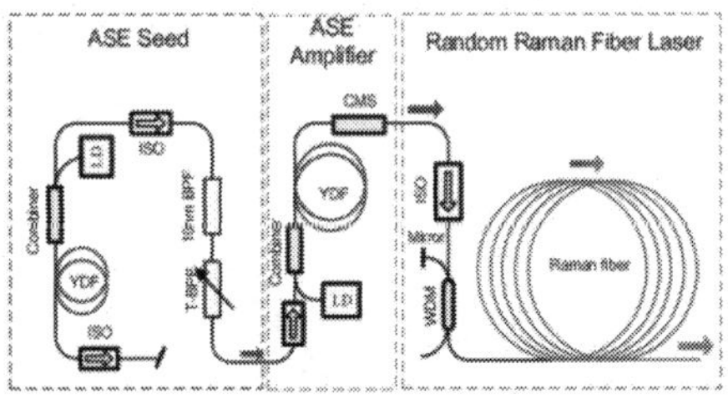

Figure. 1 Schematic diagram of the narrow linewidth Yb-doped ASE source and the cascaded Raman random fiber laser.[33]

In 2019, Thouroude et al. from France, reported a linearly-polarized, diffraction-limited and high-power 1. 7 μm RFL[34]. The schematic diagram of RFLS is shown in Figure 2. Two different configurations both use a single-mode germanium-doped silica PM fiber (PMSMF) as a Raman gain fiber. A monolithic all-fiber format laser was developed to generate the continuous-wave laser with over 5 W of optical power at 1679nm. In the same year, the Chinese research group of Zhou reported a tunable high-power 1. 7 μm RFL[35]. As shown in Fig 3, the overall structure of the experimental setup is presented as a half-cavity, aiming to reduce the threshold of SRS effect. A Raman fiber (RL) was pumped by a high-power ASE source with both wavelength and linewidth tunability. Figure 4 proves that significant temporal stability is the key to a more complete Raman conversion and higher spectral purity, as the result of which, they finally realized the spectral purity of 98.29%, the output power of 14W at 1.7 μm band.

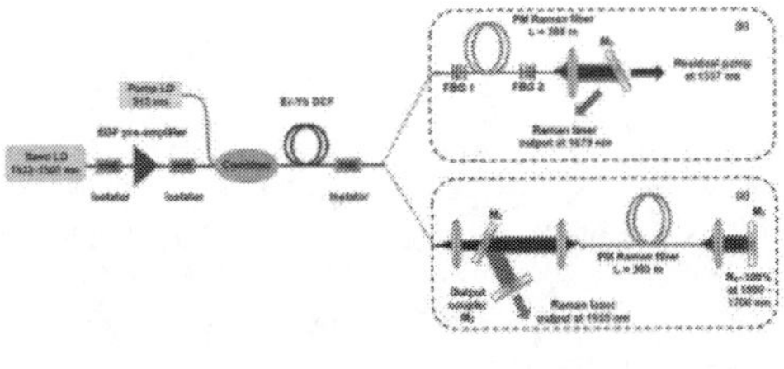

Figure 2. Schematic diagram of the linearly-polarized RFLs based either on (a) a free-running laser scheme with free-space coupling of the pump beam or on (b) an all-fiber-integrated Raman laser [34].

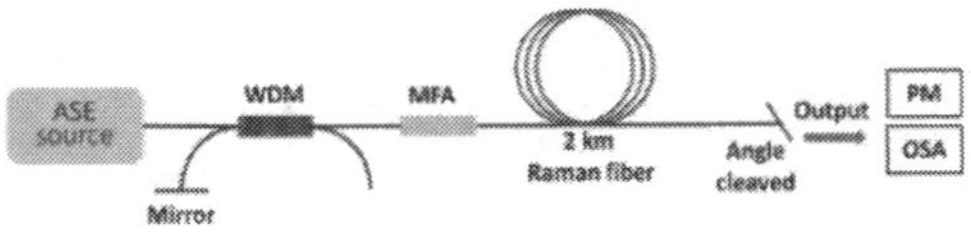

Figure 3. Experimental setup of the tunable cascaded RRFL[35].

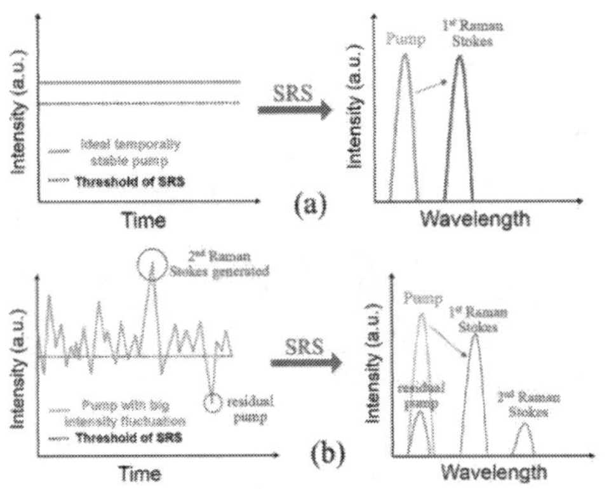

Figure 4. (a) The SRS effect pumped by an ideal temporally stable ASE source. (b) The SRS effect pumped by an ASE source with big intensity fluctuation[35].

In 2020, Zhang's research group of Changchun China, summarized their research results three years ago[36] and reported a 1.7 μm band tunable multiwavelength RFL[37]. On the basis of the previous research, they used the filtered ASE light source as the pump source, which was amplified by erbium-doped fiber amplifier for optical power, as shown in Fig 5. Highly nonlinear fiber (HNLF) and dispersion-shifted fiber (DSF) were used as nonlinear gain medium to obtain the 1.7 μm band gain spectrum, which was filtered by Sagnac ring filter. The threshold of stimulated Brillouin scattering (SBS) was increased by expanding the spectral line width of pump light. Later in 2020, Zhou et al. combined two different commercial communication fibers (CCF), successfully suppressing the spectral broadening near the fiber zero-dispersion wavelength (ZDW) in the cascaded RFL and obtaining a continuous wave Raman laser output at 1721 nm with a power of 2. 16 W, the result of which provides a significantly potential and cost-effective method to realizing high-power tunable fiber lasers[38]. In 2020, a high-power CW/pulsed 1.7 μm Raman fiber laser was reported by Grimes from Denmark[39]. A piece of RL made from the company OFS was pumped by a high-power 1117 nm fiber laser to achieve CW laser output power as high as 104W at 1692nm. By gain-switching, pulsed Raman laser was achieved with an average power of 23 W.

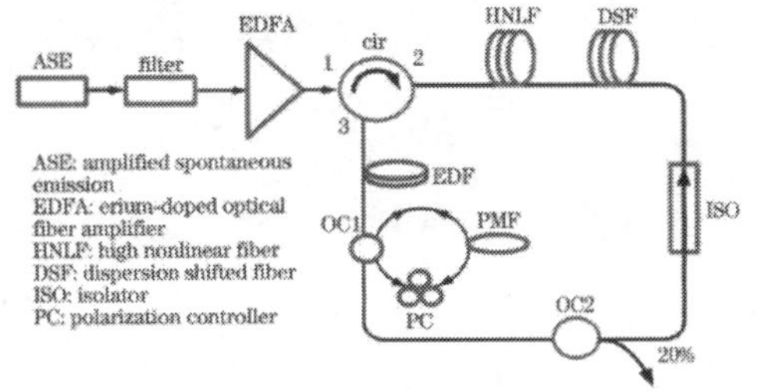

Figure 5. Experimental setup of 1.7 μm band tunable Raman fiber laser[37]

The two years from 2020 to 2021 witnessed the development of FGRLs. Wang's research group of China firstly reported a quasi-all-fiber structured pulse impulse 1.7 μm band fiber hydrogen Raman laser[40]. They used a tunable pulsed 1.5-μm-band fiber laser to pump a low-loss hollow-core fiber, realizing the nanosecond pulsed laser output with tunable wavelengths from 1687 to 1723 nm, 0.8 W average power. Later, they optimized the experimental parameters based on the guidance of the previously established numerical model[41], and used a higher power pump source and a shorter Hollow-core Photonic Crystal Fibers (HC-PCF) to achieve higher efficiency and Stokes power[31], shown in Fig 6, increasing the average output power of the pulsed laser with a central wavelength of 1705 nm to 3. 3 W. They finally achieved a pulsed fiber laser with a repetition frequency of 1.3 MHz, a pulse width of about 15 ns, and a slope efficiency of 84% as shown in Fig7. The output wavelength can be adjusted in the range of 1693~1705 nm.

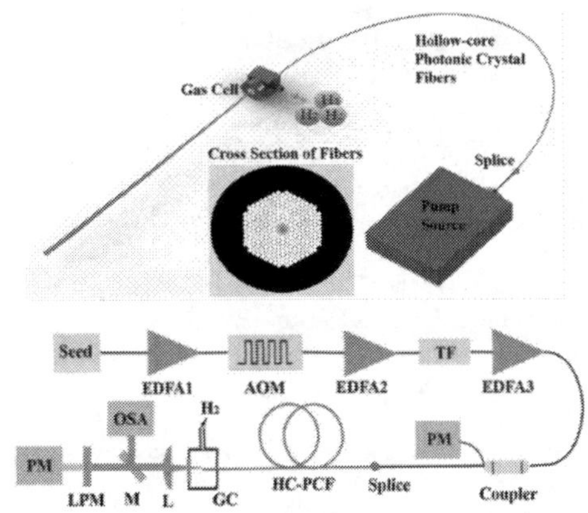

Figure 6. The graphical abstract of HC-PCF and the experimental setup of 1.7 μm band tunable Raman fiber laser[31]

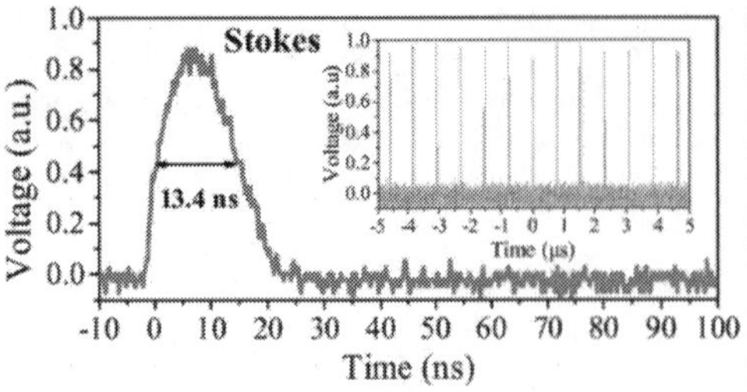

Figure 7. Stokes pulses when the pump power is maximum and repetition frequency is 1.3 MHz[31]

Subsequently, the group firstly reported an all-fiber structure of 1.7 μm band FGRL in 2021[42], the experimental setup of which is shown in Fig 8. By using 1.5 μm pulsed Erbium-doped fiber amplifier to pump 9m HC-PCF, they obtained the maximum output Stokes power of 1.63 W at 1.7 μm with the repetition frequency of 1 MHz and the corresponding optical–optical conversion efficiency of 58%. Later, they optimized the previous experimental structure[43] and systematically studied the relationship between pulse characteristics and Raman threshold, achieving a 1.7 μm pulsed fiber gas Raman laser oscillator by pure rotational stimulated Raman scattering of H2. The maximum Raman power of 1.5 W was obtained at 1693 nm at a repetition frequency of 6 MHz and pulse width of 30 ns[44].

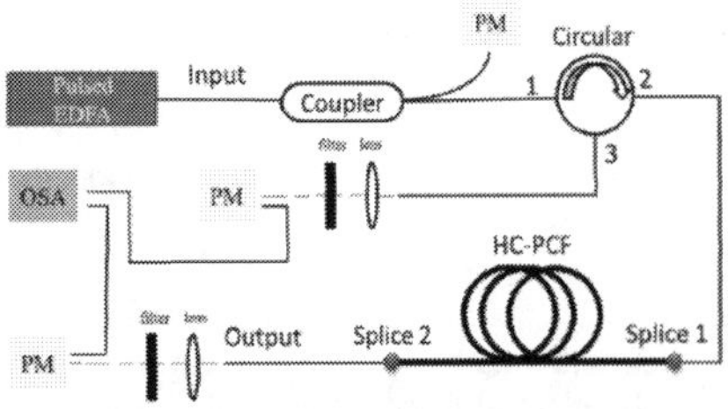

Figure 8. Experimental setup: EDFA: Erbium doped fiber amplifier, PM: power meter, OSA: optical spectrum analyzer[42].

2.2 SC-based implementation methods

The supercontinuum spectrum, which is produced by a fiber laser light source through nonlinear effects, has the high spectral density, with good spectral continuity and flatness. In fiber lasers, the SC can be generated by effects including self-phase modulation (SPM), cross-phase modulation (XPM), SRS, and FWM, according to the pump type and fiber dispersion condition and the laser output of a specific waveband can be obtained by matching the filter with the wave selection[28]. The technology of SC can offer high power, high spectral density, wide bandwidth, and modulation bandwidth advantages, which create favorable conditions for the development of fiber lasers operating in the 1.7 μm wavelength band.

In 2017, the research group of Chang from Germany[45] used a 1550 nm femtosecond fiber laser to pump a normal dispersion LMAF. A fiber-optic method of generating widely wavelength was applied to the fiber laser. After carrying a detailed numerical study of fiber-optic spectral broadening, the experimental setup was constructed, as shown in Fig 9. By using proper optical filters to select the leftmost or rightmost spectral lobes, they obtained the wavelength-tunable femtosecond pulsed laser output in the range of 1.35-1.7 μm with the maximum average power of 330 Mw, which features exceptional energy scalability and broad wavelength coverage.

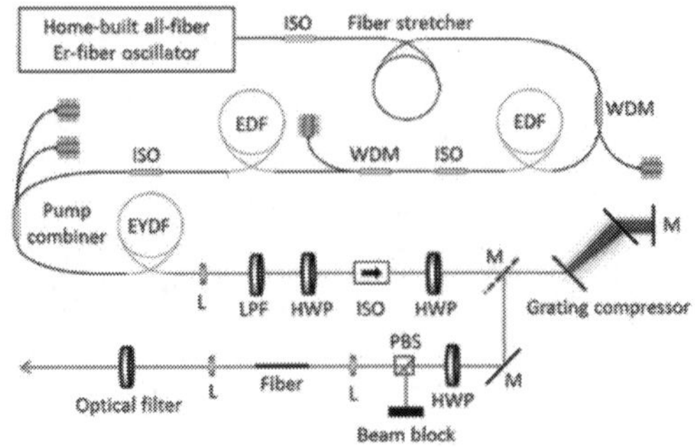

Figure 9. Experimental setup: WDM: wavelength-division multiplexing, EDF: erbium-doped fiber, EYDF: erbium ytterbium co-doped fiber, L: lens, M: mirror, ISO: isolator, HWP: half-wave plate, PBS: polarization beam splitter, LPF: long pass filter[45].

In 2018, Zhang's research group of Changchun China, reported a 1.7 μm band tunable picosecond-pulsed fiber light source based on SC and other nonlinear effects[46]. The experimental setup is shown in Fig 10. The pump source consists of a continuous wave laser and a cascaded intensity modulator (CIM) and a 1 km HNLF was employed as the gain medium. Through employing CIM in synchronization, the generation of repetition switched pulse trains with narrow pulse durations was achieved for pumping SC and investigating the output. By utilizing a wavelength division multiplexing (WDM) and a tunable filter for the filtration of SC, they obtained a single-pulse source with pulse duration of around 151 ps, a 200 MHz repetition rate, and a tuning range of 1660.1–1703.7 nm.

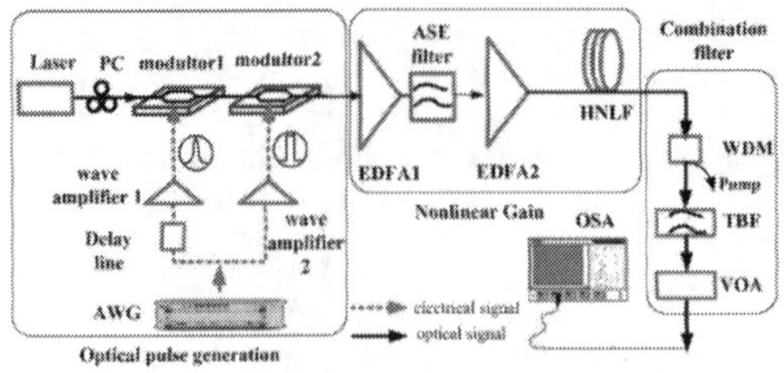

Figure 10. Experimental setup of the proposed picosecond-pulsed fiber light source[46]

In 2019, Zeng et al. reported a 1.7 μm femtosecond SC source[47]. They used an all-silica fiber thulium-doped oscillator and amplifier system to generate the low-seed pulse energy flat broad SC at a repetition frequency of 21.1 MHz. As shown in Fig 11, a 1565nm fiber laser serves as the pump source to backward pump Tm gain fiber. The output power of the 431 nm 20 dB spectral SC BW is equivalent to about 30% efficiency. This stable SC source covers a wavelength range from 1.7 μm to 2.33 μm, which is particularly attractive for gas spectroscopy that is interested in the vibration bands based on this window.

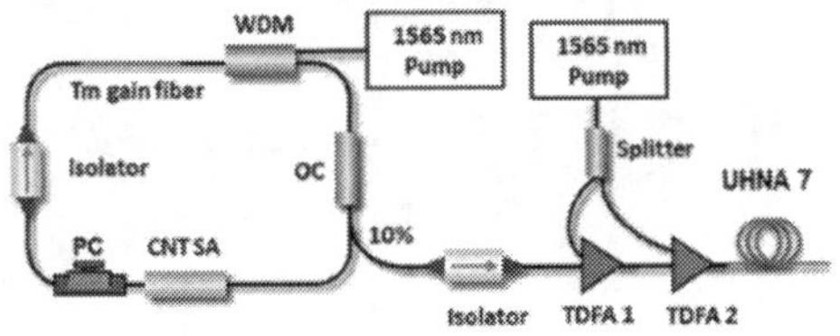

Figure 11. Experimental setup of the SC generation[47]

2.3 FWM-based implementation methods

Four-wave mixing in fiber lasers is the process of generating an output light wave by mixing three input light waves after they act nonlinearly in the fiber. As the light waves propagate through the fiber, they cause interactions between photons through nonlinear interactions, which leads to mixing between the light waves, creating a new frequency and wave vector of light waves.

A fiber optic parametric amplifier (FOPA)/oscillator (FOPO) based on simplex FWM is a device that utilizes the FWM effect to generate the interaction of pump and signal light in an optical fiber by controlling the power and wavelength difference of the input signal, thus enabling the amplification and modulation of optical signals as well as the generation of new frequency combinations. In fiber lasers, FOPA/FOPO can be used as optical amplifiers or oscillators to enhance the output power and spectral width of the laser signal. Meanwhile, the fiber parametric amplifier/oscillator based on the simplex FWM can also generate ultrashort pulses. Specifically, the signal light can generate SPM, XPM, and dispersion effects in the optical fiber through the action of nonlinear optical effects, which in turn create ultrashort pulses at the end of the fiber. In particular, the FOPO-based structure allows the locking and precise control of pulse repetition frequency through an optical feedback mechanism.

In 2018, Becheker et al. from France[48] reported a 1. 7 μm band FWM-based FOPO which can generate high-energy picosecond pulses, with the experimental structure shown in Fig.12. A silica-based DSF was applied as a gain medium, being pumped by a passively mode-locked Erbium-doped fiber laser. After numerical simulations, they found that the best results could be obtained using a 7-meter-long DSF. By optimizing the pump wavelength and the temporal dispersion tuning of the FOPO, they obtained a picosecond pulsed laser output with tunable wavelengths from 1617-1876 nm, and the pulse energies with the average power of 14.3 mW at the wavelength of 1668 nm for a pump wavelength of ~1566nm, as shown in the Fig 13. Later, they used the same experimental setup and increased the width of intra-cavity filter in pump source to optimize the parameter of chirped pulse (reached to 13 nJ, 35 ps)[49], achieving the picosecond ultrashort pulse with the average power of 204 mW.

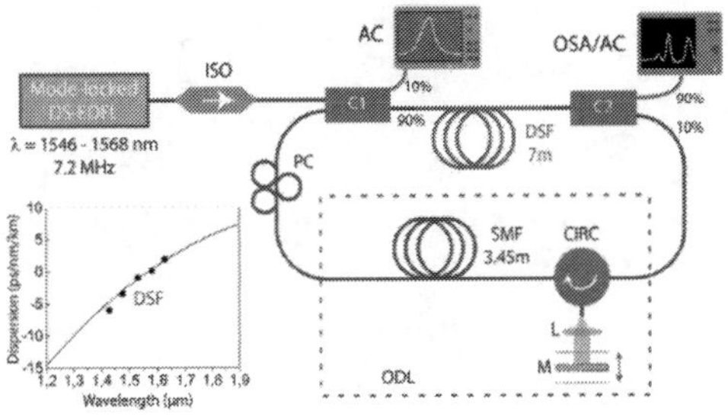

Figure 12. Experimental setup of the FOPO based on FWM[48]

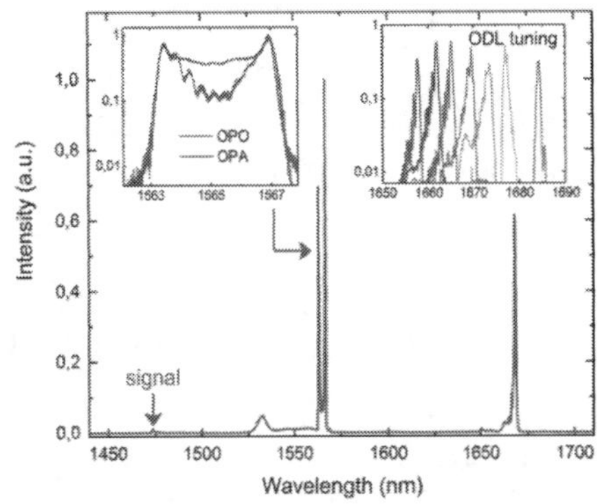

Figure 13. The output spectra at 1668 nm for a pump wavelength of ~1566nm[48]

However, the power scaling of the FOPO method is very challenging due to the nonlinearity of the fiber, leading to output power levels in the milliwatt range. To further increase the output power, an all-fiber high-power femtosecond fiber-optic parametric chirped pulse amplifier (FOPCPA) operating at 1.7 μm was reported by Qin et al. in 2020. By combining chirped pulse amplification with parametric interactions[50], they achieved high power amplification inside the fiber and obtained up-chirped pulses with the average power of 1.42 W. The schematic of the laser system with six stages is shown in Fig14. The output of the oscillator (stage 1) was amplified by the first stage pre-amplifier, a 1550nm EDFA (stage 2), and then was split into two branches for generating the SC signal (stage 3) and pump pulses (stage 4) with the proportions of 90% and 10%. In stage 5, the output pump pluses of the second pre-amplifier (stage 3) were amplified and launched into the WDM to combine with the signal from stage 2. A polarization maintaining (PM) DSF was pumped for the parametric amplification in stage 6. Additionally, the output pulses were compressed down to 450 fs.

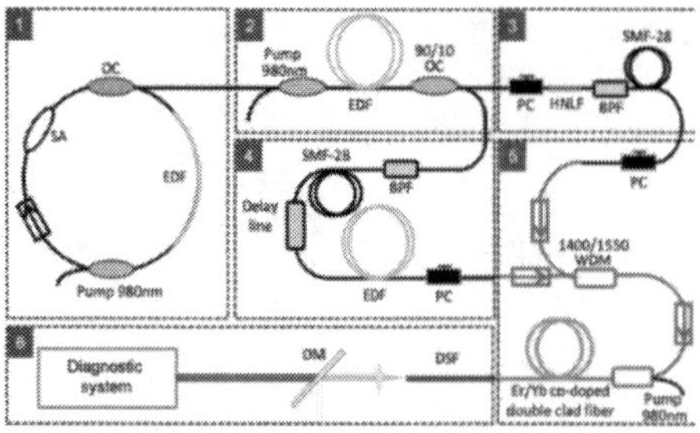

Figure 14. Experimental setup of the FOPCPA based on FWM[50]

2.4 SSFS-based implementation methods

In fiber lasers, the self-frequency shift of optical solitons is a phenomenon due to the interaction of optical solitons through non-linear effects, resulting in a red-shift or blue-shift of the frequency of optical solitons, referring to the automatic adjustment of the center frequency of the optical soliton waveform as the propagation distance increases. The dispersion effect in optical fibers usually causes light of different frequencies to travel at different speeds during propagation, which in turn causes distortion of the optical signal. The optical soliton is a nonlinear wave that can maintain its own shape and speed in optical signal transmission, and it can be transmitted over long distances in optical fiber without attenuation, realizing the single-mode optical fiber transmission and avoiding problems such as mode mismatch and multi-channel signal interference.

In 2017, the research group of Zhan from Shanghai China reported the generation of femtosecond ultrashort pulses operating at 1.7 μm wavelength[51]. As shown in Fig 15, a piece of DSF was pumped by a mode-locked fiber laser (MFL), and a variable optical attenuator was applied to change the power injected into the DSF. To create high-quality SSFS ultrashort pulses, they optimized the property of DSF including fiber length and ZWD (below1.55 μm), as well as an Er-doped MFL. As shown in Fig 16, they finally obtained sub-200 fs pulses with the maximum pulse energy and output power of 0.7 nJ and 26.8 mW at the pump power of 35.3 mW, conversion efficiency of over 76% and tunable wavelength from 1.70 to 1.74 μm, which had the advantages of economic and easy maintenance.

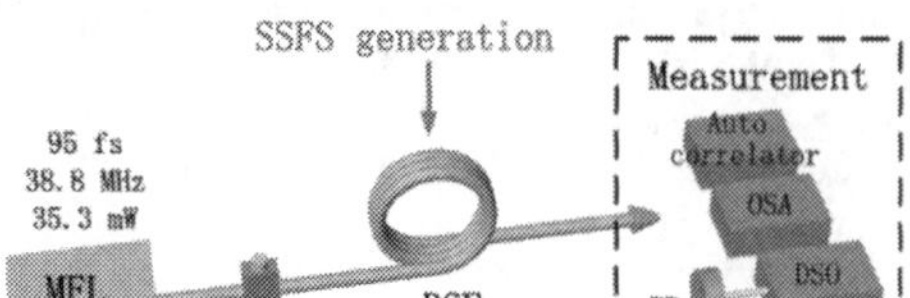

Figure 15. Experimental setup of the SSFS-based fiber laser source[51].

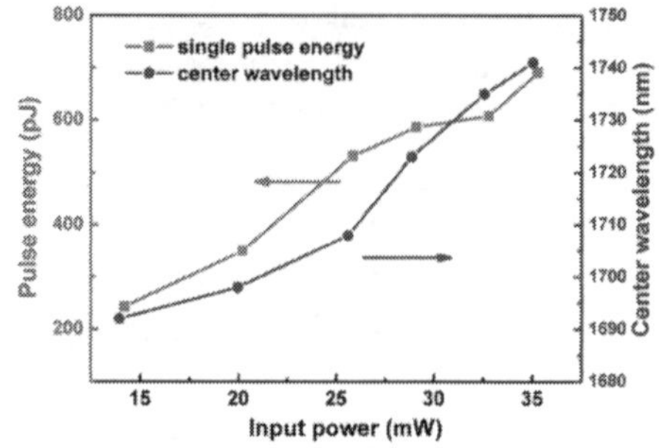

Figure 16. Single pulse energy and center wavelength versus the pump power[51].

In 2019, the research group of Zach from Germany[52] reported a versatile all-fiber wavelength-tunable ultrafast laser based on SSFS. A specially designed Er-doped polarization maintaining very large mode area fiber (PM VLMA) was applied in order to conquer the limitation that the pulse peak powers of mode-locked fiber oscillators would be limited to 1 kW by the tight mode confinement, increasing the tunable bandwidth and overall output power[52]. The experimental setup is shown in Fig 17. By using a 1480nm non-PM RFL as the pump source to pump PM VLMA, they realized the 120fs short laser pulses output with the maximum average power of 1.5 W at the tunable wavelength from 1620nm to 1990nm. Moreover, the all-fiber laser system provides the fantastic flexibility and high robustness for the application in multimodal imaging.

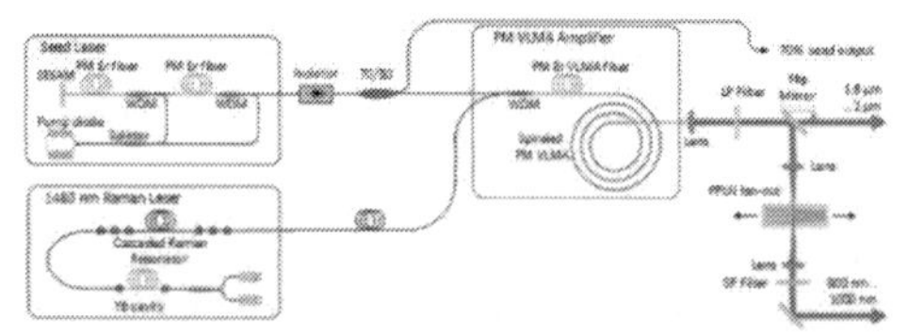

Figure 16. Laser setup showing an all-PM-fiber mode-locked seed laser with a pre-amplifier followed by a PM VLMA fiber amplifier core, pumped with a 1480 nm Raman laser[52]

3. 1.7 μm band fiber lasers based on hybrid gain

Whenever we discuss 1.7 μm fiber lasers, we always discuss rare earth doping methods separately from nonlinear effects-based methods. Few studies have combined and compared these two methods, using hybrid gain to produce higher quality laser beams. The weak points of the rare earth-doped fiber lasers including TDF have been discussed before. However, 1.7 μm band fiber lasers based on nonlinear effects also has the fly in the ointment[53]. Considering the tricky situation mentioned above, in 2022, Quan et al. from China groundbreakingly focused their research on 1.7 μm random fiber laser with hybrid gain[53]. Surprisingly, they emphatically compared the hybrid gain based laser with the Raman gain only structure and the TDF based active gain only one to highlight the advantages of the hybrid gain strategy, as shown in Fig 17. The structure of 1.7 μm RFL with hybrid gain is a common cavity with two backward pumped half-open cavities (Fig 16 (a)). A 3m long EYDF was pumped by a 976nm laser diode (the initial pump source), which was coupled by a signal/pump combiner. The generated 1565 nm lasing then played a role as the pump source to pump the SMF(1st order Raman gain) and the TDF[53]. It was the overlap of the Raman gain spectrum of SMF and the emission spectrum in the TDF that reduced the lasing threshold for the 1.7 μm lasing and realized the high performance output. As shown in Fig 18, when the pump power reaches 21.51 W, the maximum output power of the 1699nm random lasing is 2.48 W with the optical efficiency of 11.5%. As the comparations, the figure 16(b) and the figure 16(c) show the experimental schematic of a 1.7 μm RFL without the TDF only and a 1.7 μm RFL in a TDF based active gain only strategy respectively. The disadvantages of both structures were discussed and the conclusion was given that the combination of the two gain mechanisms could mitigate the limitations of each method, obtaining the high-efficient and low-threshold 1.7 μm band random lasing with 100% spectral purity.

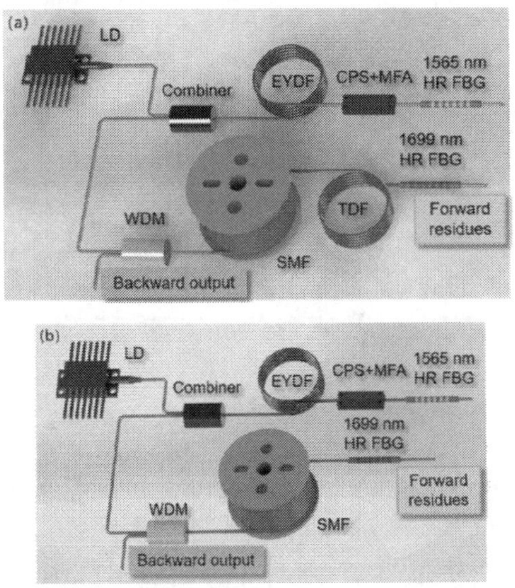

Figure 17. Comparison of three experimental setups: (a) Experimental setup of the 1.7 μm RFL with hybrid gain; (b) Experimental setup of the 1.7 μm RFL in a Raman gain only strategy; (c) Experimental setup of the 1.7 μm RFL in a TDF based active gain only strategy[53].

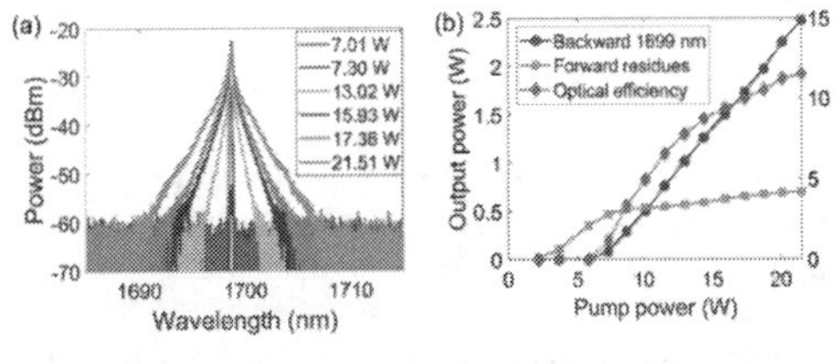

Figure 18. (a) Optical spectra of the 1699 nm RFL at different pump powers measured at the backward output; (b) Output power and optical efficiency versus the pump power[53].

4. Conclusion

Over the past few years, 1.7 μm band fiber lasers have received increasing attention and research interest due to their unique properties and potential applications , having a tendency to play the role as a springboard for launching into the imperative research fields with future technological breakthroughs including biological imaging, organic gases detection, material processing and atmospheric/underwater communication. Although the researchers from all over the world have joint their efforts to realize the marvelous boost in the performance of 1.7 μm band fiber lasers, there still some obstacles and adversities stuck in the way forward.

The nonlinear effects in the fiber lasers are always the inevitable research topics since the ultrafast fiber lasers entered onto the historical stage. The four nonlinear effects-based technologies applied in 1.7 μm band fiber lasers are mentioned in the paper, which are invented and created later than the rare earth-doped fiber lasers in 1.7 μm band, but are more potential to achieve greater lasing. So, the principle and characteristic of each method are elaborated and the advantages and weak points of each method are listed. Combined with the research progress of 1.7 μm band fiber lasers based on nonlinear effects in Table 1, we can draw the following conclusions by points.

(1) The stimulated Raman scattering technology has a significant positive impact on the implementation of 1.7 μm band fiber lasers, including extended wavelength range, improved efficiency and tunability. As seen in Table 1, the number of reports related to SRS is the most among four methods and the RFLs with solid-core fibers as the nonlinear medium are almost always used to generate 1.7 μm CW lasing, the maximum average power of which reaches to over 100 W. Although there are a few SRS-based reports of ultrashort pulse laser output in 1.7 μm band, the emergence of gas RFLs fills this gap, offering a promising method for achieving high performance (~60% conversion efficiency) 1.7 μm band pulse laser. However, the relatively higher lasing threshold, the complex optical structure design (non-all-fiber structure) and the relatively poor stability are stumbling stones which need to be removed by researchers.

(2) The super-continuum laser source should be obtained by combination of multiple nonlinear effects. From the only three SC-based reports listed in the Table 1, we can find that the fiber lasers have the wide spectral bandwidth

of 1350 nm-1700 nm or 1700 nm-2330 nm, but have lower average output power.

(3) The four-wave mixing technology is mainly applied in FOPA and FOPO, which can enhance the output power and spectral width of the laser signal with the wavelength tuning range, achieving picosecond ultrashort pulses. But the relatively lower nonlinearity coefficient of fibers will limit the realization of higher optical parametric amplification gain.

(4) The soliton self-frequency shift technology used in 1.7 μm band fiber lasers has the ability to achieve wavelength tunable femtosecond ultrashort pulses. However, for obtaining high average power, the smaller the fiber mode field area will lead to the lower the soliton pulse energy of the Raman soliton laser. For realizing flexible wavelength tuning, the changes in pump power will affect the stability of the average output power.

Table1 Research progress of 1.7 μm band fiber lasers based on nonlinear effects

Nonlinear effects	Year	Pump wavelength/nm	Output wavelength/nm	Output power/W	Efficiency/%	Pulse width	Line width/pm	Ref.
SRS	2018	1064	1676	66.9	42.4	CW	7670	[32]
SRS	2018	1062	1691	6.9	21	CW	4700	[33]
SRS	2019	1550	1655,1679	6.2,5	79,64	CW	5000,1600	[34]
SRS	2019	1055-1072	1695-1725	10-14.4	25.6	CW	$<1\times10^4$	[35]
SRS	2020	1527-1650	1653-1686	-4×10^{-4}	—	CW	80(3dB)	[37]
SRS	2020	1066	1721	2.16	6	CW	4240	[38]
SRS	2020	1117	1692	104	37	CW	2000	[39]
SRS	2020	1117	1692	23	8	100ms	—	[39]
SRS(gas)	2020	1550	1687-1723	0.8	60	10ns	<200	[40]
SRS(gas)	2021	1540-1550	1693-1705	3.3	60	13ns	<200	[31]
SRS(gas)	2021	1540-1550	1693-1705	1.63	58	10ns	<200	[42]
SRS(gas)	2021	1540	1693	1.5	54	30ns	<200	[44]
SC	2017	1550	1350-1700	0.23-0.33	10-20	50fs	—	[45]
SC	2018	1550	1660-1704	0.0021	—	755ps	—	[46]
SC	2019	1914	1700-2330	0.092	32	865fs	—	[47]
FWM	2018	1546-1568	1617-1876	0.0143	20	14ps	—	[48]
FWM	2018	1560	1620-1870	0.204	20	14.5ps	—	[49]
FWM	2020	1550	1700	1.42	20	450fs	—	[50]
SSFS	2017	1600	1700-1740	0.0268	80	196fs	—	[51]
SSFS	2019	1480	1620-1990	1.5	3	120fs	—	[52]
Hybird gain	2022	1565	1699	2.48	11.5	—	—	[53]

Additionally, one of the critical advancements is the development of the low threshold and high spectral purity 1.7 μm random fiber laser based on hybrid gain. The mechanism has been proven to provide better efficiency and performance compared to conventional TDF based laser system and the Raman gain only structure, providing a new scientific perspective for researchers and a significance method for the application in biological imaging and temporal ghost imaging.

For the future development trend of 1.7 μm fiber lasers, they are progressing toward all-fiber compact structures with higher power, more flexible tunability, higher efficiency, and narrower linewidths. The technologies based on nonlinear effects in 1.7 μm fiber lasers are just in tune with this trend, especially the SRS-based FGRL and the hybrid gain-based random fiber laser reported in the last year which illuminate the road forward and spark researchers' interests and passion for rising to the challenges. An increasing number of studies in the development of 1.7 μm band fiber lasers breathe confidence into us that we are embracing a promising a promising prospect.

References

1. C. Jauregui, J. Limpert, and A. Tünnermann, "High-power fibre lasers," Nature Photonics, vol. 7, no. 11, pp. 861-867, 2013.
2. Z. Liu, X. Jin, R. Su, P. Ma, and P. Zhou, "Development status of high power fiber lasers and their coherent beam combination," Science China Information Sciences, vol. 62, no. 4, 2019.
3. W. Shi, Q. Fang, X. Zhu, R. A. Norwood, and N. Peyghambarian, "Fiber lasers and their applications [Invited]," Appl Opt, vol. 53, no. 28, pp. 6554-68, Oct 1 2014.
4. E. M. Dianov, "Bismuth-doped optical fibers: a challenging active medium for near-IR lasers and optical amplifiers," Light-Science & Applications, vol. 1, May 2012, Art no. e12, doi: 10.1038/lsa.2012.12.
5. S. Firstov, S. Alyshev, K. Riumkin, M. Melkumov, O. Medvedkov, and E. Dianov, "Watt-level, continuous-wave bismuth-doped all-fiber laser operating at 1.7 μm," Opt. Lett., vol. 40, no. 18, pp. 4360-4363, 2015.

6. C. Li, C. Kong, and K. K. Wong, "High energy noise-like pulse generation from a mode-locked thulium-doped fiber laser at 1.7 μm," IEEE Photonics Journal, vol. 11, no. 6, pp. 1-6, 2019.
7. Y. Xiao, X. Xiao, L. Liu, and H. Guo, "Gain-switched watt-level thulium-doped fiber laser and amplifier operating at 1.7 μm," High Power Laser Science and Engineering, vol. 10, 2022.
8. S. K. Gupta, K. Sudarshan, and R. Kadam, "Optical nanomaterials with focus on rare earth doped oxide: A Review," Materials Today Communications, vol. 27, p. 102277, 2021.
9. T. Sun, X. Su, Y. Zhang, H. Zhang, and Y. Zheng, "Progress and summary of photodarkening in rare earth doped fiber," Applied Sciences, vol. 11, no. 21, p. 10386, 2021.
10. D. Creeden, B. R. Johnson, G. A. Rines, and S. D. Setzler, "High power resonant pumping of Tm-doped fiber amplifiers in core-and cladding-pumped configurations," Optics Express, vol. 22, no. 23, pp. 29067-29080, 2014.
11. P. Grelu and N. Akhmediev, "Dissipative solitons for mode-locked lasers," Nature Photonics, vol. 6, no. 2, pp. 84-92, 2012.
12. S. D. Jackson, "Towards high-power mid-infrared emission from a fibre laser," Nature Photonics, vol. 6, no. 7, pp. 423-431, 2012.
13. D. Kirsch, S. Chen, R. Sidharthan, Y. Chen, S. Yoo, and M. Chernysheva, "Short-wave IR ultrafast fiber laser systems: Current challenges and prospective applications," Journal of Applied Physics, vol. 128, no. 18, p. 180906, 2020.
14. M. R. Majewski, R. I. Woodward, and S. D. Jackson, "Dysprosium-doped ZBLAN fiber laser tunable from 2.8 mum to 3.4 mum, pumped at 1.7 mum," Opt Lett, vol. 43, no. 5, pp. 971-974, 2018.
15. J. X. P. Z.Y.Zhan, M.Liu, A.P.Luo, W.C.Xu, Z.C.Luo, "Recent progress of 1.7 μm ultrafast fiber lasers (Invited)," 2022.
16. J.-X. Chen et al., "1.7-μm dissipative soliton Tm-doped fiber laser," Photonics Research, vol. 9, no. 5, pp. 873-878, 2021.
17. S. Chen et al., "All-fiber short-wavelength tunable mode-locked fiber laser using normal dispersion thulium-doped fiber," Optics express, vol. 28, no. 12, pp. 17570-17580, 2020.
18. C. Li et al., "Fiber chirped pulse amplification of a short wavelength mode-locked thulium-doped fiber laser," Apl Photonics, vol. 2, no. 12, p. 121302, 2017.
19. J. Zhang et al., "Single-frequency 1.7-μm Tm-doped fiber laser with optical bistability of both power and longitudinal mode behavior," Optics Express, vol. 29, no. 14, pp. 21409-21417, 2021.
20. A. Bashkatov, E. Genina, V. Kochubey, and V. Tuchin, "Optical properties of the subcutaneous adipose tissue in the spectral range 400–2500 nm," Optics and spectroscopy, vol. 99, pp. 836-842, 2005.
21. H. Cheng et al., "Deep-brain 2-photon fluorescence microscopy in vivo excited at the 1700 nm window," Opt. Lett., vol. 44, no. 17, pp. 4432-4435, 2019.
22. N. G. Horton and C. Xu, "Dispersion compensation in three-photon fluorescence microscopy at 1,700 nm," Biomedical optics express, vol. 6, no. 4, pp. 1392-1397, 2015.
23. T. Hasegawa, I. Sogawa, and H. Suganuma, "A near infrared angioscope visualizing lipid within arterial vessel wall based on multi-spectral image in 1.7 μm wavelength band," in Endoscopic Microscopy VIII , vol. 8575: SPIE, pp. 25-32, 2013.
24. V. V. Alexander et al., "Photothermolysis of sebaceous glands in human skin ex vivo with a 1,708 nm Raman fiber laser and contact cooling," Lasers in surgery and medicine, vol. 43, no. 6, pp. 470-480, 2011.
25. M. Tanaka, M. Hirano, K. Murashima, H. Obi, R. Yamaguchi, and T. Hasegawa, "1.7-μm spectroscopic spectral-domain optical coherence tomography for imaging lipid distribution within blood vessel," Optics express, vol. 23, no. 5, pp. 6645-6655, 2015.
26. B. Li et al., "Development and measurement of a near-infrared CH4 detection system using 1.654 μm wavelength-modulated diode laser and open reflective gas sensing probe," Sensors and Actuators B: Chemical, vol. 225, pp. 188-198, 2016.
27. I. Mingareev, F. Weirauch, A. Olowinsky, L. Shah, P. Kadwani, and M. Richardson, "Welding of polymers using a 2 μm thulium fiber laser," Optics & Laser Technology, vol. 44, no. 7, pp. 2095-2099, 2012.
28. W. H. H.Li, Y.L.Cui, W.X.Peng, Z.F.Wei, , "Progress and Prospect of Fiber Lasers Operating at 1.7 μm Band," Laser & Optoelectronics Progress, vol. 59, no. 19, 2022.
29. P. Z. Y.Z.Ning, S.He, Q.Li, Y.L.Fan, Y.Liu, "Progress and applications of 1.7 μm waveband fiber laser," LASER TECHNOLOGY, vol. 47, no. 2, pp. 154-170, 2023.
30. P. Z. Y.Zhang, P.Liu, Q.L.Du, T.S.Wang, L.Z.Zhang, S.F.Tong, H.L.Jiang, "Fiber Light Source at 1.7 μm Wave band and Its Applications," Laser & Optoelectronics Progress, vol. 53, no. 9, p. 90002, 2016.
31. H. Li, W. Pei, W. Huang, M. Wang, and Z. Wang, "Highly Efficient Nanosecond 1.7 μm Fiber Gas Raman Laser by H2-Filled Hollow-Core Photonic Crystal Fibers," Crystals, vol. 11, no. 1, 2020.
32. L. Zhang, J. Dong, and Y. Feng, "High-Power and High-Order Random Raman Fiber Lasers," IEEE Journal of Selected Topics in Quantum Electronics, vol. 24, no. 3, pp. 1-6, 2018.
33. J. Dong et al., "High order cascaded Raman random fiber laser with high spectral purity," Optics Express, vol. 26, no. 5, pp. 5275-5280, 2018.

34. R. Thouroude et al., "Linearly-polarized high-power Raman fiber lasers near 1670 nm," Laser Physics Letters, vol. 16, no. 2, 2019.
35. Y. Zhang, J. X. Song, J. Ye, J. M. Xu, T. F. Yao, and P. Zhou, "Tunable random Raman fiber laser at 1.7 mu m region with high spectral purity," Optics Express, vol. 27, no. 20, pp. 28800-28807, Sep 2019.
36. P. Zhang et al., "1.7 μm band narrow-linewidth tunable Raman fiber lasers pumped by spectrum-sliced amplified spontaneous emission," Applied Optics, vol. 56, no. 35, pp. 9742-9748, 2017.
37. P. Z. Z.X.He, D.Wu, "1.7 μm Tunable Multi-Wavelength Raman Fiber Laser Based on Amplified Spontaneous Emission Pump," Laser & Optoelectronics Progress, vol. 57, no. 07, pp.234-241, 2020.
38. Y. Zhang et al., "Cascaded telecom fiber enabled high-order random fiber laser beyond zero-dispersion wavelength," Opt. Lett., vol. 45, no. 15, pp. 4180-4183, 2020.
39. A. Grimes et al., "Hundred-watt CW and Joule level pulsed output from Raman fiber laser in 1.7-μm band," in Fiber Lasers XVII: Technology and Systems, vol. 11260: SPIE, pp. 234-239, 2020.
40. W. Huang, Z. Li, Y. Cui, Z. Zhou, and Z. Wang, "Efficient, watt-level, tunable 1.7 μm fiber Raman laser in H2-filled hollow-core fibers," Opt. Lett., vol. 45, no. 2, pp. 475-478, 2020/01/15 2020.
41. H. Li, W. Huang, Y. Cui, Z. Zhou, and Z. Wang, "Pure rotational stimulated Raman scattering in H2-filled hollow-core photonic crystal fibers," Optics Express, vol. 28, no. 16, pp. 23881-23897, 2020.
42. W. Pei, H. Li, W. Huang, M. Wang, and Z. Wang, "All-Fiber Tunable Pulsed 1.7 μm Fiber Lasers Based on Stimulated Raman Scattering of Hydrogen Molecules in Hollow-Core Fibers," Molecules, vol. 26, no. 15, p. 4561, 2021.
43. H. Li et al., "All-fiber gas Raman laser oscillator," Opt. Lett., vol. 46, no. 20, pp. 5208-5211, 2021.
44. W. Pei, H. Li, W. Huang, M. Wang, and Z. Wang, "Pulsed fiber laser oscillator at 1.7 microm by stimulated Raman scattering in H(2)-filled hollow-core photonic crystal fibers," Opt Express, vol. 29, no. 21, pp. 33915-33925, Oct 11 2021.
45. H.-Y. Chung, W. Liu, Q. Cao, F. X. Kärtner, and G. Chang, "Er-fiber laser enabled, energy scalable femtosecond source tunable from 13 to 17 μm," Optics Express, vol. 25, no. 14, 2017.
46. P. Zhang et al., "1.7 μm tunable picosecond-pulsed fiber light source based on nonlinear effect combination using a cascaded intensity modulator," Laser Physics, vol. 28, no. 9, 2018.
47. J. Zeng, A. E. Akosman, and M. Y. Sander, "Supercontinuum Generation From a Thulium Ultrafast Fiber Laser in a High NA Silica Fiber," IEEE Photonics Technology Letters, vol. 31, no. 22, pp. 1787-1790, 2019.
48. R. Becheker et al., "High-energy dissipative soliton-driven fiber optical parametric oscillator emitting at 1.7 μm," Laser Physics Letters, vol. 15, no. 11, 2018.
49. M. Tang et al., "Low Noise High-Energy Dissipative Soliton Erbium Fiber Laser for Fiber Optical Parametric Oscillator Pumping," Applied Sciences, vol. 8, no. 11, p. 2161, 2018.
50. Y. Qin, O. Batjargal, B. Cromey, and K. Kieu, "All-fiber high-power 1700 nm femtosecond laser based on optical parametric chirped-pulse amplification," Opt Express, vol. 28, no. 2, pp. 2317-2325, Jan 20 2020.
51. X. Fang, Z. Wang, and L. Zhan, "Efficient generation of all-fiber femtosecond pulses at 1.7 μ m via soliton self-frequency shift," Optical Engineering, vol. 56, no. 4, 2017.
52. A. Zach, M. Mohseni, C. Polzer, J. W. Nicholson, and T. Hellerer, "All-fiber widely tunable ultrafast laser source for multimodal imaging in nonlinear microscopy," Opt. Lett., vol. 44 21, pp. 5218-5221, 2019.
53. X. Quan, R. Ma, H. Wu, Z. Yong Bai, D. Yuan Fan, and J. Liu, "Low threshold and high spectral purity 1.7 μm random fiber laser based on hybrid gain," Optics & Laser Technology, vol. 155, 2022.

Experimental study on the factors affecting the performance of wire-plate Biefeld-Brown thrust unit

Jinbao Li, Lin Lai*

School of aeronautics and astronautics, Sun Yat-sen University, Shenzhen 518107, China

Abstract. Biefeld-Brown effect (B-B effect for short) refers to the phenomenon that when a high voltage is applied to an asymmetric capacitor, the capacitor generates an additional force. In order to explore the influencing factors and series characteristics of the B-B effect of wire-plate asymmetric capacitors in atmospheric pressure, a set of experimental devices with adjustable parameters was designed. Based on this device, the effects of different wire-plate lengths, power supply types and electrode application modes on the thrust and efficiency of B-B effect are studied. At the same time, the optimal parameters are selected to study the series characteristics. The research shows that the thrust generated by B-B effect is directly proportional to the length of the wire plate. The thrust generated by applying positive emission voltage to the copper wire and grounding the aluminum plate is the largest, and the two-stage series connection can achieve multiple increase of the thrust, with better performance.

1. Introduction

Biefeld-Brown effect[1](hereinafter referred to as B-B effect) refers to the phenomenon that when a high voltage is applied to an asymmetric capacitor, the capacitor will produce an additional force. This phenomenon is named after its discoverers, physicists Paul Alfred Biefeld and Thomas Townsend Brown. Because it can directly convert electricity into thrust, it should have a broad application prospect in the aerospace field, but because of the lack of a reasonable and feasible scheme, it has not been really applied in engineering. In recent years, with the development of new structure, new scheme, high-performance integrated power supply and other technologies, the application of B-B effect in aerospace field has been paid more attention.
A series of researches have been made on Brown-B effect, and some patents have been applied [2-6]. One of the models is a pair of asymmetric arc electrodes with a dielectric rod placed between them. When high voltage is applied, a force pointing to the positive electrode will be generated. De Seversky[7]studied that the electrode structure area is 0.1 m2, the distance between electrodes is 5cm, and the thrust force is 50mN under 20kV voltage and 0.5 mA current, and pointed out that the thrust efficiency is related to atmospheric pressure and humidity. The early B-B effect is mostly used in the power demonstration device of the floating lift, and the thrust force level is mN[8-14], so the electric thrust conversion efficiency is low. Since then, Bedolla, Tajmar and Cabanas et al[15-18] have carried out a lot of theoretical analysis and experimental measurement research on the B-B effect based on the structure of the floating elevator. In 2018, Nature introduced the pure electric propulsion aircraft based on B-B effect in MIT atmosphere[19]. The aircraft adopted the structure of four levels in parallel up and down and two levels in series, with a wire board length of about 40m and a wingspan of about 5m, which produced a thrust of 3.2N and achieved an autonomous flight of 60m. Undefined Technology Company of the United States designed a new type of silent UAV prototype by using B-B effect[20,21]. The prototype realized the miniaturization and integration of power supply, and could take off vertically with its own power supply, at the same time, it significantly improved the electric-thrust conversion efficiency and thrust level of B-B effect.
Comparatively speaking, there are few researches on B-B effect in China. Jiang Xingliu of Beijing University of Aeronautics and Astronautics[22-26] studied the B-B effect, and thought it was related to vortex dynamics and torsion field theory. Wang Liying[27] made an experiment on the length and spacing of wire-plate structure, and simulated the model based on Maxwell software, and compared it with the experimental data. Cheng Xiangyu[28] made research and analysis on various electromagnetic propulsion devices, made theoretical calculation and related structural design of the B-B effect lift, and carried out experiments. Hefei Institute of Material Research, Chinese Academy of Sciences [28] increased the thrust of B-B effect to 5.2N through a large number of series and parallel connection methods of asymmetric electrodes with wire and plate layout, and made a certain breakthrough.
Throughout the research at home and abroad, there are relatively few researches on B-B effect at present, the experimental data are not comprehensive and detailed enough, the research on the influencing factors and laws

* Corresponding author: lailin@sysu.edu.cn

 https://doi.org/10.1051/matecconf/202338201029

of B-B effect is not deep enough, and there is a lack of research on the series structure that can obtain greater thrust.

In this paper, a new experimental device with adjustable structural parameters is designed. Under atmospheric pressure, based on the line-plate asymmetric capacitor, the influences of line-plate length, power supply type and electrode application mode on the thrust and efficiency of B-B effect are studied, and the reasons for the changes are analyzed. At the same time, the optimal structural parameters are selected for the series structure experimental study, which provides a reference for the subsequent research and design of B-B effect.

2. Experimental system and measurement method

2.1 Experimental system

The experimental system of B-B effect in this paper consists of a high-voltage power supply, a balance of measuring equipment, a self-designed line-plate B-B effect experimental device, etc., as shown in Figure 1. Among them, the high-voltage power supply is the DC high-voltage generator of Shanghai Shenggao Electric Technology (model ZJF-60kV/5mA), which outputs high voltage through the high-voltage cylinder and applies it to one pole of the line-plate asymmetric capacitor.

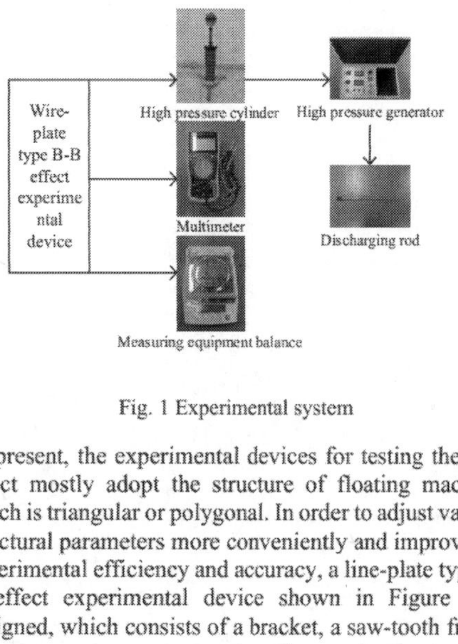

Fig. 1 Experimental system

At present, the experimental devices for testing the B-B effect mostly adopt the structure of floating machine, which is triangular or polygonal. In order to adjust various structural parameters more conveniently and improve the experimental efficiency and accuracy, a line-plate type B-B effect experimental device shown in Figure 2 is designed, which consists of a bracket, a saw-tooth frame, a copper wire component and an aluminum plate component. Among them, the copper wire component and the aluminum plate component are convenient to replace copper wires with different diameters and aluminum plates with different sizes, so that they are in the same plane, forming an asymmetric capacitor. At the same time, it is installed on the saw-tooth frame, and the distance between adjacent saw-teeth is 1cm, which is convenient to adjust the distance between the wire boards. The length of the saw tooth determines the length of the wire board. The saw-tooth frame is installed on the pallet and put on the balance of measuring equipment for experiment. All components are made of insulating materials, and the main dimensions of the device are shown in Figure 3.

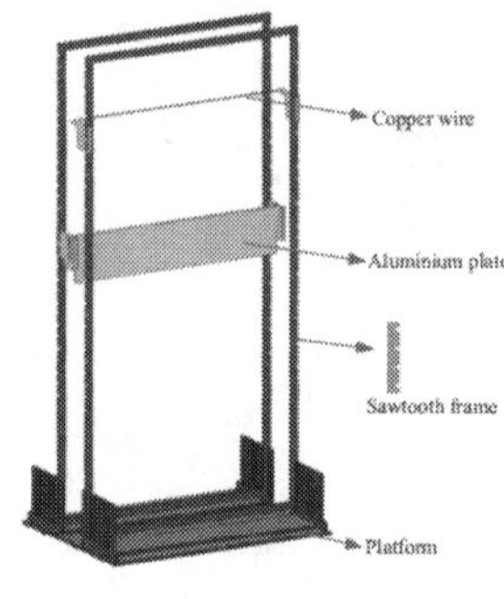

Fig. 2 Line-plate type B-B effect experimental device

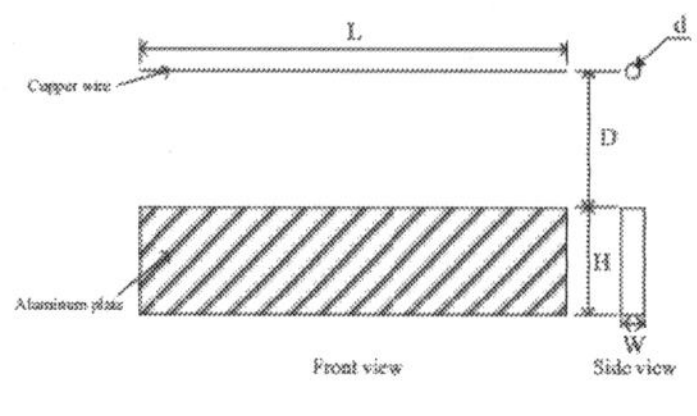

Fig. 3 Main dimensions of line-plate type B-B effect experimental device

2.2 Measurement method

The control box of DC generator adjusts the output voltage, and applies it to one pole (such as copper wire pole) of the experimental device through the high-voltage cylinder, while the other pole (such as aluminum plate pole) is grounded to form a potential difference. Micro-ammeter on the high-voltage cylinder displays the loop current in real time. The experimental device is placed on a measuring device balance (model JJ1023BF of Shuang Jie Balance, measurement accuracy is 0.1mg), and the thrust generated by B-B effect is the indicator change of the balance. Under atmospheric pressure, the B-B effect is mainly the ion wind effect. The high-voltage electrode ionizes the surrounding air to form electric charge, and the electric charge moves to drive the air to flow to form ion wind. In order to avoid the interference of ion wind and high-voltage environment on the measurement and reading of the balance, the measuring device balance is wrapped with aluminum foil and grounded together with the metal weighing plate. At the same time, a wind shield is added between the experimental device and the balance. After testing, the experimental device is suspended above the balance, and the applied voltage has no effect on the balance readings, so the balance readings are stable during the experiment. Fig. 4 is the circuit connection diagram of the experimental device.

Change the wire board length, power supply type and electrode application mode, etc. The voltage starts from 10kV, and the voltage and current and balance weight reduction indicator are recorded every 2kV until the breakdown voltage.

 https://doi.org/10.1051/matecconf/202338201029

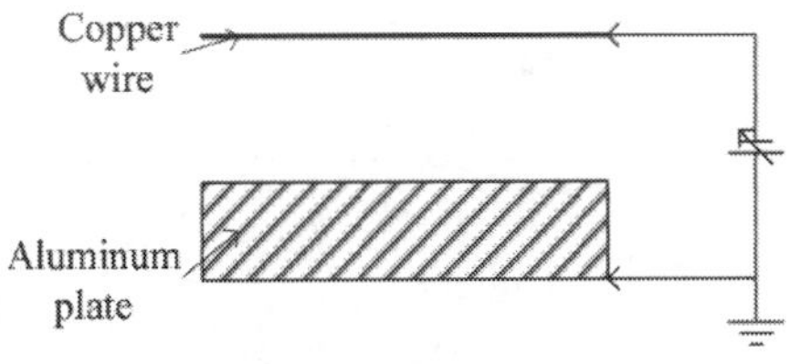

Fig. 4 Circuit connection mode 1

3. Experimental results and discussion

3.1 Characteristics of B-B Effect under Different Length of Wire Plate

The distance between fixed wires and boards is D=8cm, the diameter of copper wire is d=0.1mm, the width of aluminum plate is W=3mm, and the height is H=40mm. The circuit connection mode is connection 1 shown in Figure 4. The line length L varies between 10 and 20 cm, and every 2cm is a group. Fig. 5 shows the variation of thrust force of B-B effect with voltage under six groups of different wire board lengths. As shown in the figure, the force generated by the B-B effect is proportional to the voltage. Under the same voltage, the greater the length L of the wire board, the greater the thrust of the B-B effect. If the experimental phenomenon is analyzed from the angle of ion wind effect, ion wind effect refers to the strong electric field generated after the electrode is applied with high voltage, which ionizes the surrounding air and generates positive and negative charges. Under the action of electric field, the charges move and collide with air molecules to exchange momentum, and the air flow makes the asymmetric capacitor also have a "relative movement" and generate thrust. Therefore, the longer the length of the wire plate, the more the total charge generated by ionization, thus causing more air molecules to move, which makes the relative motion effect of the asymmetric capacitor stronger, and finally shows that the thrust of B-B effect increases.

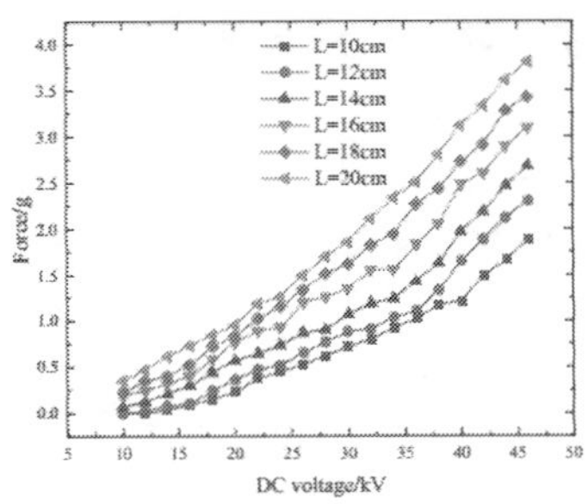

Fig. 5 Thrust force under different length of wire plate

3.2 Characteristics of B-B Effect under Different Power Supply Types and Electrode Application Methods

The above experiments are all based on the circuit connection method (A) shown in Figure 6. Different power emission polarities and electrode application modes also have influence on the B-B effect. There are two types of power sources: positive emission and negative emission; There are two ways to apply electrodes: copper wire to electrode, aluminum plate to ground, copper wire to electrode and aluminum plate to electrode, so there are four ways: (a) copper wire to negative high voltage, aluminum plate to ground; (b) Copper wire is grounded, and aluminum plate is connected with negative high voltage; (c) The copper wire is connected to positive voltage, and the aluminum plate is grounded; (d) Copper wire is grounded, and aluminum plate is connected to positive voltage.

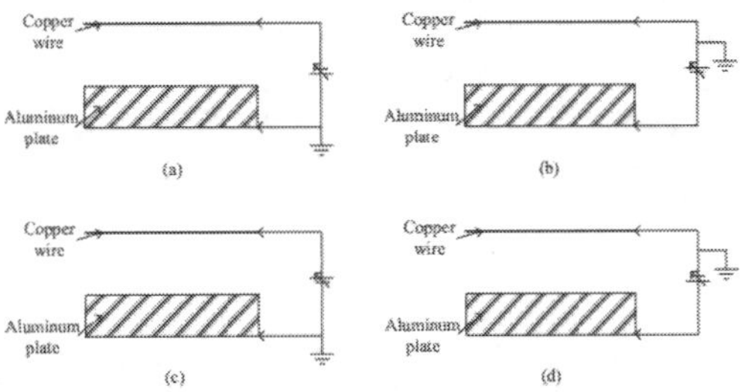

Fig. 6 Different power supply types and electrode application modes

Fig. 7 shows the thrust of B-B effect with voltage under different power supply types and electrode application modes. Under different forms, the thrust of B-B effect is still proportional to the voltage, and changing the polarity of the electrode will not change the thrust direction, but will only affect the thrust. Under the same voltage, the thrust generated by aluminum foil grounding and copper wire connecting to high voltage pole is greater, and the thrust generated by positive polarity transmitting power supply is greater than that of negative polarity transmitting power supply, so the B-B effect thrust generated by copper wire connecting to positive voltage and aluminum plate grounding in mode (c) is the largest.

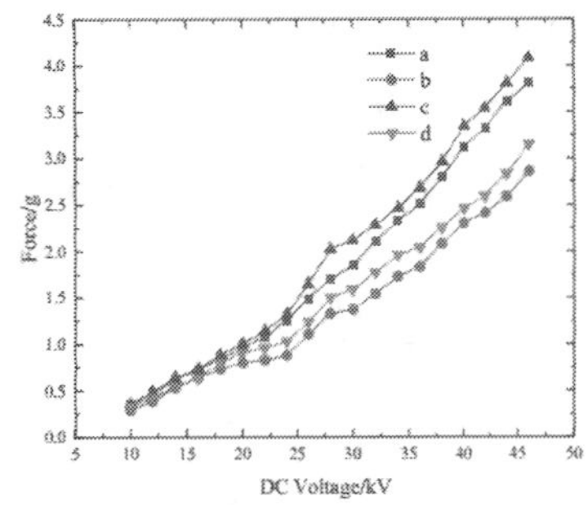

Fig. 7 Thrust force under different power supply types and electrode application modes

3.3 B-B effect characteristics of multistage series structure

In order to obtain better performance, to study the reason why MIT adopts the B-B effect of two stages in series, it is necessary to conduct experimental analysis on the multi-stage series structure. Select single-stage wire board length L=20cm, wire board spacing D=4cm, copper wire diameter d=0.1mm, aluminum plate height H=4cm, aluminum plate width W=15mm, power supply type and electrode application mode: the first-stage copper wire is connected to positive high voltage and aluminum foil is grounded. At the same time, in order to prevent breakdown, adjacent poles of different stages adopt the same polarity.

Fig. 8 is a schematic diagram of two-stage and three-stage series structure. Two-stage series connection is copper wire I-aluminum foil I-copper wire II-aluminum foil II, in which copper wire I and aluminum foil II are connected with positive high voltage, and aluminum foil I and copper wire II are grounded; The three-stage series connection is copper wire I-aluminum foil I-copper wire II-aluminum foil II-copper wire III-aluminum foil III, in which copper wire I, aluminum foil II and copper wire III are connected to positive high voltage, and aluminum foil I, copper wire II and aluminum foil III are grounded. The variables studied are the series distance D2 between single-stage line board modules and the series number. The series distance D2 is 2, 4, 6 and 8cm.

In addition to the thrust, the electrical efficiency and geometric efficiency are also analyzed. The electrical efficiency is defined as the ratio of thrust to consumed power, i.e. F/P, P=UI(U is the voltage, I is the corresponding current at this voltage). The higher the electrical efficiency, the less electric energy is needed to generate the same thrust. Geometric efficiency is defined as the ratio of thrust to geometric size, that is, F/(L*(D+H)). The higher the geometric efficiency, the smaller the device size will be when the same thrust is generated.

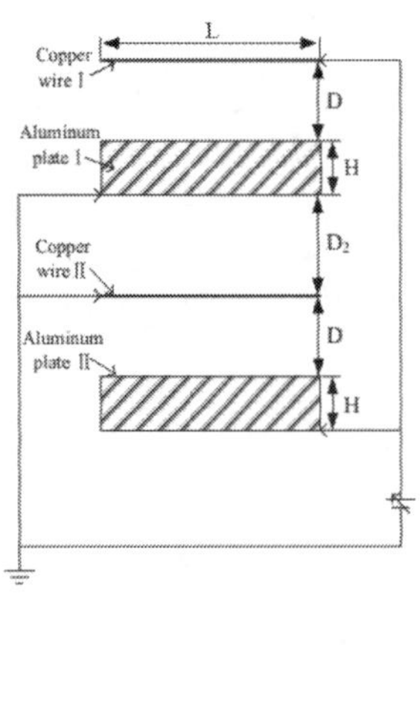

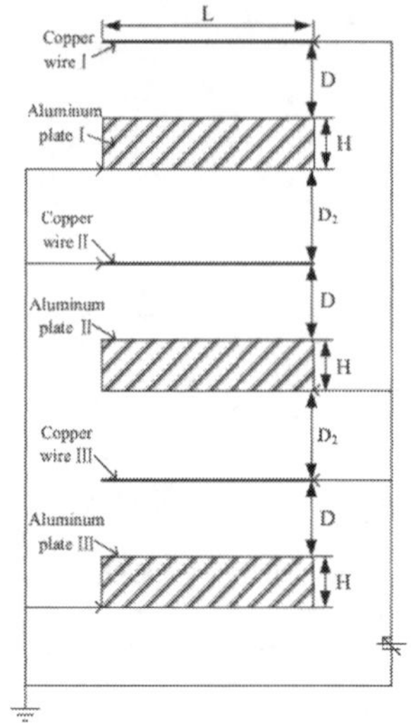

Fig. 8 Two-stage and three-stage series connection

Fig. 9 is a diagram of the thrust of the series structure B-B effect with voltage. In the series structure, the B-B effect thrust is still proportional to the voltage. Under the same voltage, the thrust generated by three-stage series is larger than that of two-stage series, and the single-stage thrust is the smallest, but the thrust generated by two-stage series exceeds twice that of single-stage, while the thrust generated by three-stage series is three times lower than that of single-stage, so the multiple increase is not realized. At the same time, the series distance D2 between single stages does not affect the thrust of B-B effect.

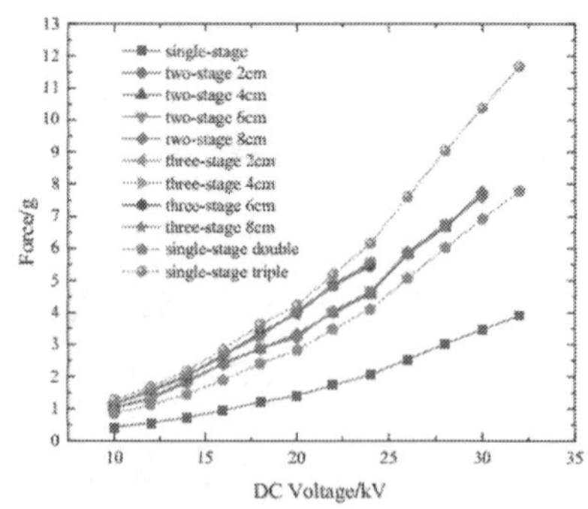

Fig. 9 Thrust force under series structure

Fig. 10 is a graph of the change of current in series structure with voltage. It can be seen that the current in the series structure is proportional to the voltage. Under the same voltage, the current consumed by the three-stage series is greater than that of the two-stage series, and at the same time it is greater than that of the single stage, that is, the more series stages, the greater the current consumed. Whether two or three stages are connected in series, the larger the series distance D2, the smaller the current consumed.

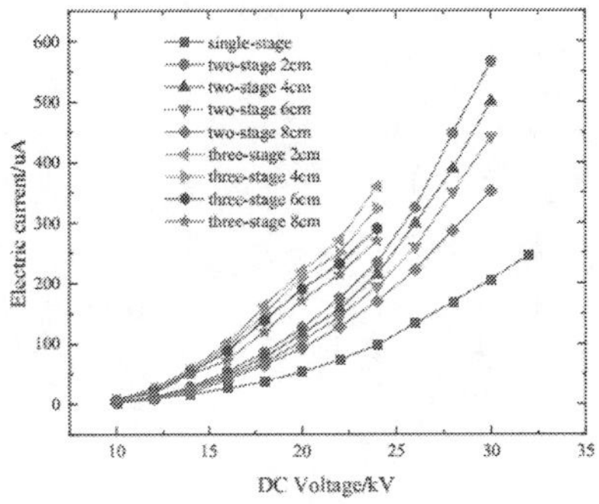

Fig. 10 Electric current under series structure

Fig. 11 is a graph showing the change of electrical efficiency of series structure with voltage. As can be seen from the figure, the electrical efficiency of the series structure is inversely related to the voltage. At a lower voltage (< 20kV), the electrical efficiency of two-stage series is greater than that of three-stage series, and it is basically greater than that of a single stage. Similarly, the electrical efficiency tends to be consistent at a high voltage. In the same voltage and series number, the greater the series distance D2, the greater the electrical efficiency.

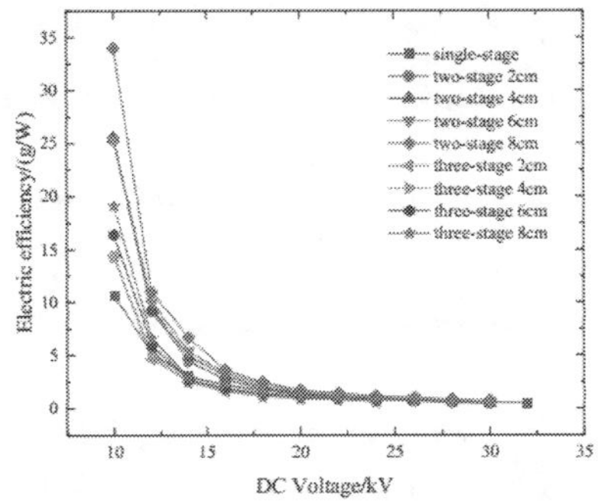

Fig. 11 Electric efficiency under series structure

Fig. 12 is a graph of the geometric efficiency of the series structure as a function of voltage. In the series structure, the geometric efficiency is still directly proportional to the voltage. Under the same voltage, the geometric efficiency of two-stage series is greater than that of three-stage series, but both are smaller than that of single stage, that is, the more series, the smaller the geometric efficiency. At the same time, in the same voltage and series number, the larger the series distance D2 is, the smaller the geometric efficiency is.

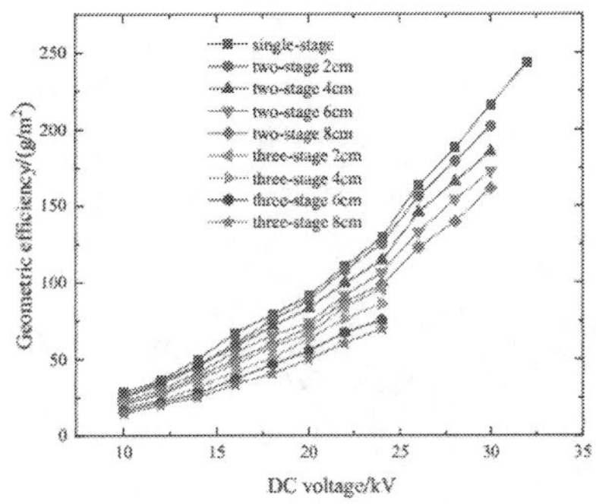

Fig. 12 Geometric efficiency under series structure

To sum up, the thrust generated by two-stage series connection achieves the multiple growth of a single stage, and at the same time, it consumes less current, and has higher electrical efficiency and geometric efficiency, which is more advantageous than three-stage series connection. Perhaps this is the reason why MIT ion propulsion aircraft adopts the front and back two-stage series connection structure. For the series distance D2, the distance will not affect the thrust, but the larger the distance, the smaller the current consumed, the greater the electrical efficiency and the smaller the geometric efficiency. Therefore, in the actual design of the aircraft based on the B-B effect, it is necessary to make a reasonable choice according to the application objectives whether to consume less energy or have a smaller and more compact structure.

4. Conclusion and prospect

In this paper, an experimental system of Biefeld-Brown effect is designed under atmospheric pressure. The length of the wire plate, the type of power supply and the way of applying electrodes are adjusted for experiments, and the characteristics of B-B effect under different parameters and series structures are studied. Research shows that:

(1)The Biefeld-Brown effect of wire plate is directly proportional to the length of wire plate. The longer the length of the wire, the greater the thrust generated.

(2)Different power supply types and electrode application methods will not change the thrust direction, but will only affect the thrust. The thrust generated by the positive emission power supply is larger than that generated by the negative emission power supply, and the thrust generated by the copper wire connected with high voltage and the aluminum plate grounded is larger and the electric efficiency is higher.

(3)The series structure can increase the thrust, and the two-stage series connection can realize the multiple increase of the thrust. At the same time, the two-stage series connection consumes less current and has higher electrical and geometric efficiency, which is better than the three-stage series connection. Series spacing does not affect thrust, but only electric efficiency and geometric efficiency.

The research object of this paper is the line-plate thrust unit, and the follow-up research will be carried out in the full-scale prototype. The optimized structure and series parameters obtained in this paper will provide an important reference for the next full-scale prototype research.

References

1. Thomas T B. How I Control Gravitation[J]. Science and Invention, 1929:374.
2. Thomas T B. A Method of and an Apparatus or Machine for Producing Force or Motion[P] .UK:300311 ,1928-11-15.
3. Thomas T B. Electrostatic motor[P] .US:1974483 ,1934-09-25.
4. Thomas T B. Electrokinetic apparatus[P] .US:2949550 ,1960-08-16.
5. Thomas T B. Electrical thrust producing device:[P] .US:2958790 ,1962-01-23.
6. Thomas T B. Electrokinetic apparatus[P] .US:3187206 ,1965-06-01.
7. SAP De. Ionocraft:[P] .US:3130945 ,1964-04-28.
8. Bahder T B, Fazi C. Force on an Asymmetric Capacitor[R]. Physics, 2002.
9. Tajmar M. Biefeld-Brown Effect: Misinterpretation of Corona Wind Phenomena[J]. Aiaa Journal, 2004,42(2):315-318.
10. Stephenson G V. The Biefeld Brown Effect and the global electric circuit[J]. Aip Conf Proc, 2005, 746(1): 1249-1255.
11. Musha T. Explanation of dynamical Biefeld-Brown effect from the standpoint of ZPF field [J]. Jbis-J Brit Interpla, 2008, 61(9): 379-384.
12. Wang W, Yang L J, Gao J, et al. Experimental study on the thrust and the ratio of thrust to power of multi-points/grid ionic wind exciter [J]. Acta Physica Sinica, 2013, 62(7): 292-298.
13. Einat M, Kalderon R. High efficiency Lifter based on the Biefeld-Brown effect [J]. Aip Adv, 2014, 4(7):201-220.
14. Wang W, Yang L J, Liu S, et al. Theoretical and experimental study of thrust produced by corona discharge exciter in wire-aluminum foil electrode configration [J]. Acta Physica Sinica, 2015, 64(10): 151-157.
15. Bedolla P O, Vorlaufer G , Sequard-Base P , et al. Altitude dependence of electrohydrodynamic flow in an electrostatic lifter[J]. Journal of Electrostatics, 2017, 87:32-44.
16. Tajmar M , Schreiber T . Put Strong Limits on All Proposed Theories so far Assessing Electrostatic Propulsion: Does a Charged High-Voltage Capacitor Produce Thrust?[J]. Journal of Electrostatics, 2020, 107:103477.
17. P. Huang, Z. -y. Xie, P. Liu, at el . Ionic Wind Simulation of Wire-Aluminum Foil Electrode Structure Lifter in 2-D Space[J] . Transactions on Plasma Science, 2022,50(3):566-573.
18. Cabanas M F, Gonzalez F P, Gonzalez A S, et al. Analysis of the Efficiency of the Electrohydrodynamic Propulsion Based on the Biefeld-Brown Effect for Manned and Unmanned Aircrafts [J]. Appl Sci-Basel, 2022, 12(6):512-521.
19. Xu H, He Y, Strobel K L, et al. Flight of an aeroplane with solid-state propulsion [J]. Nature, 2018, 563(7732): 532-535.
20. TECHNOLOGIES U. https://www.undefinedtechnologies.com/. 2020.
21. TECHNOLOGIES U. The US VTOL Ion Thruster Drone for Urban Cargo Delivery [M]. https://www.youtube.com/watch?v=uFtBdMm26Tk. 2020.
22. Lei Jinzhi, Jiang Xingliu. Kasimir effect and zero point energy extraction [J]. Science and Technology Herald, 1999(04):10-12.
23. Lei Jinzhi, Jiang Xingliu, Zhang Jiangang, et al. Electrochemical Anomalies and the Theory of Deflection Field [J]. Science and Technology Herald, 2000(06):3-5.
24. Jiang Xingliu, Yi Lizhi, Rui Liu, et al. Research on Corona Discharge Hoist [J]. Journal of Aviation, 2005(06):28-31.
25. Jiang Xingliu. The 21st century is the century of zero point energy [J]. Frontier Science, 2010,4(03):15-20.
26. Ren Junxue, Liu Yu, Jiang Xingliu, et al. Study on the lift mechanism and efficiency of electric floating machine [J]. journal of aerospace power, 2010,25(06):1395-1400.
27. Wang Liying. Experimental study of Biefeld-Brown effect in fluid [D]. Beijing: Beihang University, 2007.
28. Cheng Xiangyu. Exploration and pre-research of space electromagnetic propulsion system [D]. Hefei: University of Science and Technology of China, 2018.

https://doi.org/10.1051/matecconf/202338201030

Research on Anti-loosening Technology of Transmission Line Bolts

Qinghua Li*, Maohua Li, Binrong Zhu and Xiaoguang Hu

China Electric Power Research Institute Co., Ltd., Beijing, China

Abstract. Transmission lines are affected by the natural environment for a long time, and bolts are prone to frequent vibration, resulting in loosening or falling off. In this paper, the loose mechanism of transmission line threaded connection is analysed, and the transverse vibration comparative test of bolt locking method is carried out. A new type of locking nut has been developed, which has good locking performance and convenient installation. According to different operating environments, the principle of differential bolt locking is proposed.

1. Introduction

Loosening of bolts is a long-standing pain point in transmission lines, especially in region where bolts are prone to loosening such as galloping, strong wind, and ice-shedding. In 2021, there are more than 230,000 base towers in China to carry out the bolt tightening operation, and the high-altitude operation poses great personal safety hazards and high operation and maintenance costs. At present, the bolts of transmission towers mainly adopt anti-loosening methods such as spring washers, fastening nuts, double nuts, etc. The anti-loosening effect is greatly affected by the installation quality [2], and the anti-loosening durability is insufficient.

In the relevant standards for fasteners in the power industry, the technical requirements for locking nuts are currently not included, which restricts the improvement of the anti-loosening level of transmission line bolts. This paper analyzes the current transmission line bolt loosening mechanism, introduces a new type of transmission line anti-loosening nut, conducts a lateral vibration comparison test of anti-loosening performance, and puts forward differentiated anti-loosening suggestions for transmission line bolts.

2. Current status of bolt loosening in transmission lines

The threaded connection generally adopts a single-line common thread, and the thread lift angle is less than the equivalent friction angle of the screw pair, so the threaded connection meets the self-locking requirements. In addition, the friction force on the supporting surfaces such as the nut and the bolt head after tightening also has an anti-loosening effect, so the threaded connection will not automatically loosen when the static load and working temperature are small.

However, in the case of shock, vibration or variable load, or large temperature changes, the friction between the screw pairs may decrease or disappear instantaneously. After this phenomenon is repeated many times, the connection will be loosened [3-5]. The loosening of bolts results in a significant drop in connection strength, resulting in serious accidents such as missing tower members, connection failures, and structural damage. At the same time, due to the lack of automatic monitoring methods for bolt loosening, it is difficult to eliminate the hidden danger of bolt loosening in time, which directly threatens the safety of power grid operation.

* Corresponding author: epcri@qq.com

Figure 1. Loosening and falling off of transmission tower bolts.

The transmission line defect record table of each provincial company from January 2015 to April 2020 derived from the PMS system of SGCC shows that the proportion of bolt loosening defects in strong wind and galloping region exceeds 15%. Table 1 shows that the proportion of bolt loosening is relatively high provincial power company statistics.

Table 1. Statistics on loose bolts of transmission towers.

Provincial power company	Number of tower defects	Number of loose bolts	Bolt loose ratio
Tibet Electric Power Co.	1647	315	19.1%
Heilongjiang Electric Power Co.	10840	1920	17.7%
Liaoning Electric Power Co.	16399	2602	15.9%
Xinjiang Electric Power Co.	44669	5410	12.1%
Shandong Electric Power Co.	12561	1134	9.0%
Nei monggol Electric Power Co.	34876	3064	8.8%
Jiangsu Electric Power Co.	4266	353	8.3%
Jilin Electric Power Co.	35554	2799	7.9%
Shanxi Electric Power Co.	29492	2193	7.4%
Hebei Electric Power Co.	11532	828	7.2%

3. Commonly used bolt anti-loosening methods for transmission lines

3.1 Anti-loosening method of transmission tower bolts

The essence of the bolted connection is to prevent the relative movement between the bolt and the nut. There are many different kinds of anti-loosening bolts, nuts, gaskets and other anti-loosening types that are used in different industries. However, due to the characteristics of transmission line processing, installation, and field operation, and considering the economy and installation convenience, the current transmission line mainly use spring washers, fastening nuts, double nuts to fasten bolts.

(a)Spring washer (b)Tightening nut

(c) Double nuts (d) Thin nut

Figure 2. Anti-loosening methods for transmission lines.

The above-mentioned bolt anti-loosening methods are all friction-proof and need to ensure a certain bolt clamp force. However, due to the inability to ensure that the construction tightening torque of each bolt reaches the expected value during bolt installation, as well as the deformation of the bolt connection surface and long-term vibration, it is easy to cause the reduction or even disappearance of the bolt preload. The weakening or even disappearance of the clamp force of the bolted connection leads to the loosening of the bolts, mainly in the following situations:

(1) Creep occurs in the screw thread of bolted connection: the common problems of bolted connection, including thread plastic deformation and bolt rebound torque.

(2)Looseness caused by the collapse of the bearing surface: including the gap of multi-layer connectors, the gap of bolt holes, and the corrosion of the galvanized layer.

(3) Long-term exposure to vibration loads is the main cause of bolt loosening: including continuous and stable strong wind, galloping, and ice-shedding.

3.2 Bolt tightening torque

Check the standards for the design and construction of transmission lines, the relevant provisions of bolt tightening torque[6-7], the bolt tightening torque of each standard is slightly different, and the installation torque is mainly given according to the shear and tension bolts,

 https://doi.org/10.1051/matecconf/202338201030

respectively. Table 2 only contains the installation torques of M16, M20, and M24 bolts that are most commonly used in transmission lines.

For double nuts, the installation torque of the inner nut is implemented according to the standard, while the torque of the outer nut is mostly tightened in place, which is generally required to be installed according to half of the torque of the inner nut.

Table 2. Tightening torque of transmission tower bolts.

Bolt Load Type	Bolt Size	Bolt Strength Class	Tightening Torque (N.m)
Shear (connection of tower components)	M16	/	80
	M20	/	100
	M24	/	250
Tensile (steel pipe flange)	M16	6.8	80
	M20	6.8	160
	M24	6.8	280
	M24	8.8	380

3.3 Anti-loosening performance test

In order to test the anti-loosening performance of different anti-loosening methods, three commonly used bolts of 6.8M16, 6.8M20 and 8.8M24 on the transmission tower are selected, and the transverse vibration test is carried out according to the national standard [8]. The test conditions are as follows: the test frequency is 10Hz, and the no-load amplitude is ±1.6mm, ±1.9mm and ±2.0mm respectively. After installing the nut according to the torque required by the specification, record the initial clamp force of the bolt, and record the residual clamp force of the bolt during the test. Draw the change trend of the ratio of "residual clamp force (F) / initial clamp force (F_M)" of each group of test pieces. Figure 3(a)~(d) shows the trend diagram of M20 bolt. Table 3 lists the average value and standard deviation of the ratio of each group of test pieces. It can be seen from Figure 3 that the pre tightening force of bolts decreases rapidly after the vibration starts, and tends to be stable in the later stage. After a certain number of vibrations (N), the residual clamp force of bolts with different locking methods is small, and the locking performance is poor. The ratio of residual clamp force of double nut bolt is 20% - 30%, and the anti-loosening performance is relatively good.

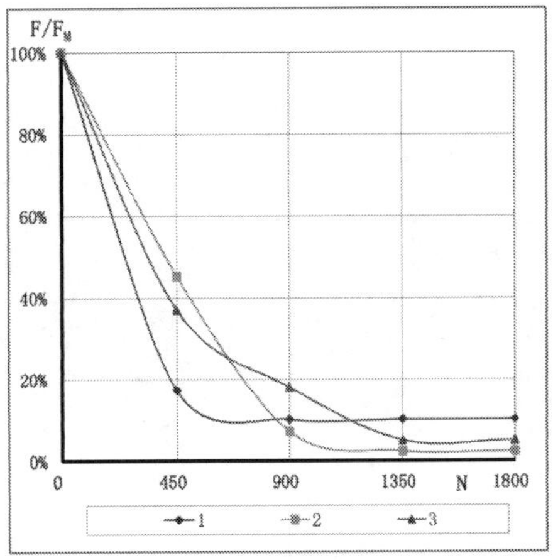

(a)Spring washer(M20)

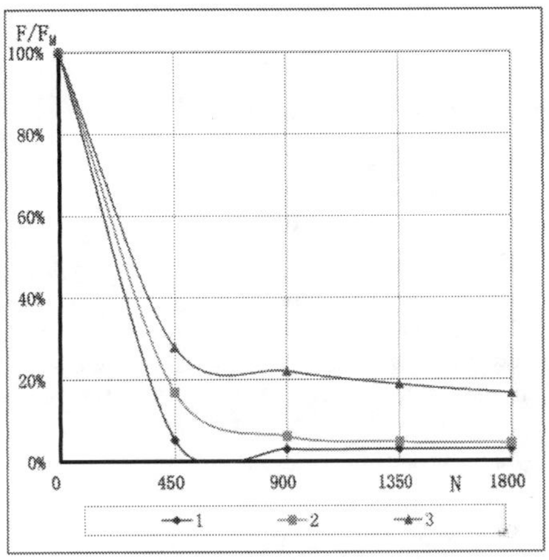

(b)Tightening nut(M20)

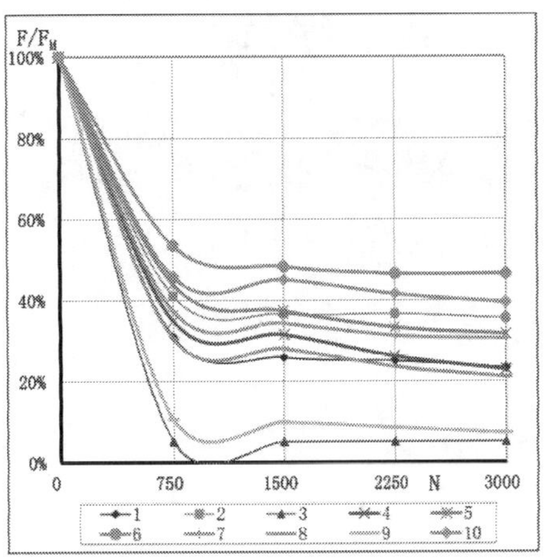

(c) Double nuts(M20)

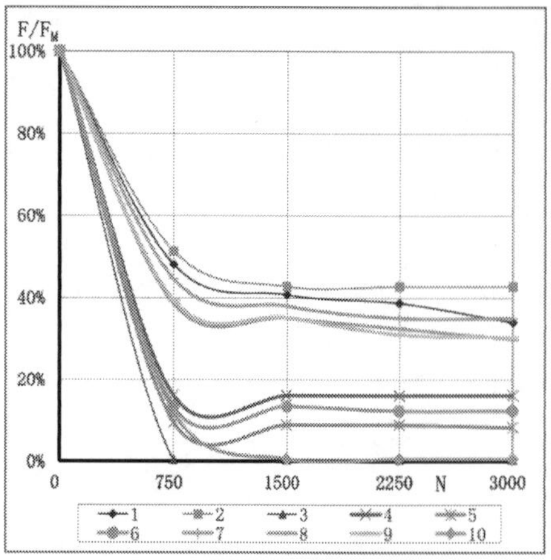

(d) Unequal thickness double nuts(M20)

Figure 3. Trend diagram of residual clamp force ratio of bolts

4. Integrated self-locking double nut

4.1 Basic structure

In response to the long-standing pain point problem of loose bolts in transmission lines, the project team developed an "integrated self-locking double nut", which applied the eccentric self-locking principle of the nut and the screw. Figure 4(a) is a standard height nut, suitable for new lines, Figure 4(b) is a thin nut, suitable for an in-transit line, and Figure 4(c) is a schematic diagram of the principle of the anti-loosening structure.

(a)Standard height style

(b)Thin height style

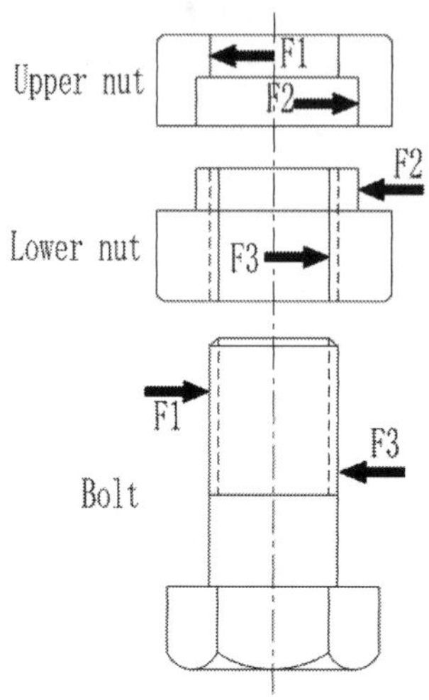

(c)Structure

Figure 4. Integrated self-locking double nut.

In the initial state of the integrated self-locking double nut, the upper and lower nut threaded holes are concentric; when the upper nut is locked, the upper nut rotates at a certain angle (60~90 degrees) relative to the lower nut, and causes the inner hole of the upper nut thread to be eccentric with respect to the lower nut thread. The upper nut and the lower nut will generate an interactive transverse force F2, thereby generating a transverse pressure F3 on the lower nut, and then a radial friction force F3 will be generated between the lower nut and the thread of the connecting bolt, and the friction force is the locking force that prevents the nut from loosening.

Since the lateral pressure F3 generated by the locking force is completely generated by the eccentric structure of the upper and lower double nuts, it is not affected by the installation torque. Even if the clamp force of the nut completely disappears, the lateral pressure F3 will not be weakened. Therefore, the locking force generated by the mechanical eccentric structure of the upper and lower double nuts can ensure the stability and durability of the anti-loosening performance.

4.2 Anti-loosening performance test

In order to evaluate the anti-loosening performance of the integrated self-locking double nut, an accelerated lateral vibration test was carried out and compared with the current anti-loosening methods of transmission lines. Figure 5 shows the change trend of the ratio of residual clamp force to initial clamp force.

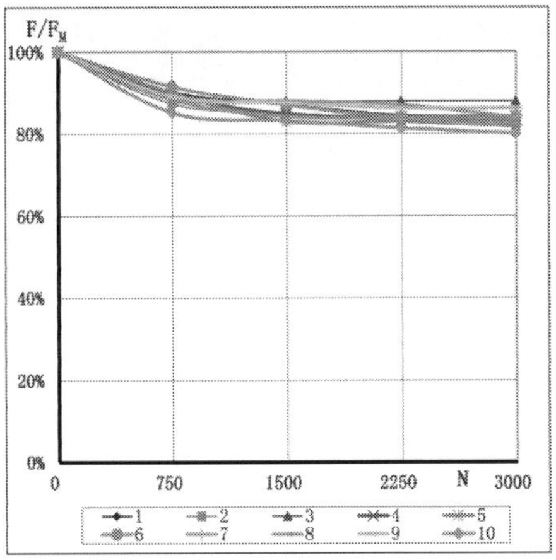

(a) 6.8M20

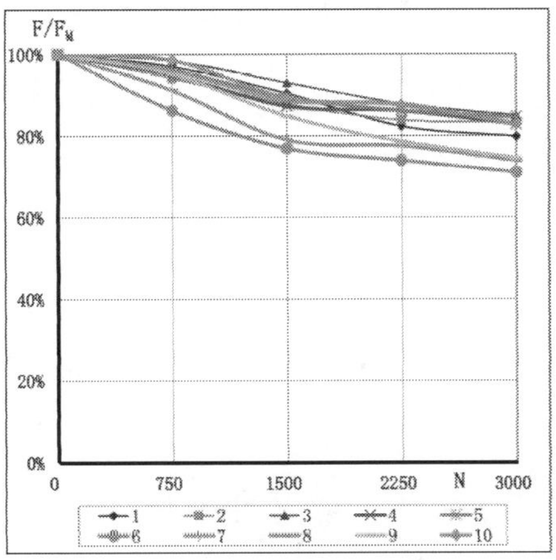

(b) 8.8M24

Figure 5. Trend diagram of residual clamp force ratio of integrated self-locking double nut.

The lateral vibration test shows that, compared with the double nut with the best anti-loosening performance of the current transmission line [9], the "residual clamp force ratio", which reflects the anti-loosening performance index, has increased from an average of 20 ~ 30% to about 80%, which is increased by more than 2 times, and the "standard deviation" reflecting the variability of bolt anti-loosening performance is reduced from an average of more than 10% to about 5%, indicating that the anti-loosening performance is more stable.

Table 3. Comparative test of anti-loosening performance of transmission line bolts.

No.	Bolt Type	Bolt Size	F/F_M	Standard Deviation	Number of Vibrations
1	Spring washer	6.8M16	16.7%	19.5%	1800
		6.8M20	5.9%	3.2%	
2	Tightening nut	6.8M20	24.2%	20.4%	3000
		8.8M24	8.1%	6.1%	
3	Double nuts	6.8M20	26.3%	12.5%	
		8.8M24	30.4%	11.8%	
4	Thin nut	6.8M20	21.0%	14.5%	
		8.8M24	12.3%	11.4%	
5	Integrated self-locking double nut	6.8M16	77.4%	7.4%	
		6.8M20	83.7%	2.1%	
		8.8M24	80.0%	4.8%	

5. Conclusion

The transmission lines are subjected to wind load for a long time, and the bolts are prone to frequent vibrations, resulting in loosening or falling off, which greatly weakens the bearing capacity of the members and reduces the overall bearing performance of the tower. Bolts should take corresponding anti-loosening methods.

Among the bolt anti-loosening methods commonly used in transmission lines at present, the anti-loosening effect of double nuts is relatively good, but the installation quality has a great influence on its anti-loosening performance, and the anti-loosening durability against long-term or strong vibration loads is still insufficient. The integrated self-locking double nut has excellent locking performance and convenient installation. It is recommended to be popularized and applied in transmission lines with high requirements for bolt locking in galloping, windy regions and no-man land.

Acknowledgments

This work was financially supported by "Research on the Loosening Mechanism and Anti-loose Technology of Bolts in Transmission Lines (GC83-21-007)" fund.

References

1. Yang Xiaohui,Lü Zhongbin,Han Baolin,et al.Antigalloping and locking performance test and analysis of joint bolts in transmission tower.Electric Power Construction,2014,35(4) 44-49 .
2. Qian Miao,Yu Hongyun,Wang Liang,et al.Torque Coefficient Determination and Analysis of Hot

Galvanized Bolts for Transmission Lines.ZHEJIANG ELECTRIC POWER, 2013, 32(6)17-19.

3. Wang Wei ,Xu Hao,Ma Yue,et al.Self loosening mechanism of bolted joints under vibration. Journal of Shock and Vibration, 2014,33(22)198-202.
4. Yang Fengli,Zhu Binrong,Xing Haijun.The slip characteristics and parametric study of bolted connections for transmission towers.Engineering Mechanics,2017,34(10)116-127.
5. Yang Fengli,Wang Xuming,Zhu Binrong.An experimental study on the construction torque and transversal vibration of bolted joints of transmission towers in low temperature regions with high temperature difference.JOURNAL OF VIBRATION AND SHOCK.2020,39(12)257-264.
6. General Administration of quality supervision, inspection and Quarantine of the people's Republic of China .Transverse vibration testing method for fasteners , Beijing: China Machinery Industry Federation,2012.
7. Ministry of housing and urban rural development of the people's Republic of China .Code for construction and acceptance of 110kV-750kV overhead transmission line .Beijing: China Planning Press,2014.
8. State Grid Corporation of China.Code for construction and acceptance of ±800kV overhead transmission line.Beijing: Science and Technology Department of SGCC,2018.
9. Yang Fengli,Li Zheng,Zhang Dachang, et al.Experimental study on the transversal vibration of double-nutbolted joints of transmission towers.Journal of Vibrationand Shock,2018,37(10) 164-171.

Experimental study on bearing capacity of corroded angle steel members of in-service transmission towers

Binrong Zhu [1,*], Yao Huang [1], Zheng Li [1], Bin Chen [2] and Hailin Liao [2]

[1] China electric power research institute, Beijing 100055, China
[2] Zhejiang Electric Power Co., Ltd. Shaoxing Power Supply Company, Shaoxing 312099, China

Abstract. The tower is the load-bearing structure of the overhead transmission line. Due to the complex operating environment and other factors, the transmission tower is prone to corrosion, which weakens the bearing capacity of the tower and affects the safe operation of the power grid. Through the investigation of the corrosion situation of the transmission towers in service, the angle steel members were removed from the in-service transmission towers, and the axial compression bearing capacity test was carried out. As the corrosion rate increases, the bearing capacity of components shows a decreasing trend, and due to the existence of corrosion, and at the same corrosion rate, due to the location of corrosion , The corrosion depth and corrosion area are different, and the axial compression bearing capacity of the components fluctuates greatly.

1. Introduction

The tower is the load-bearing structure of the overhead transmission line, and its reliable operation is crucial to the safety of the power system. Under the influence of multiple factors such as the complex operating environment of the transmission tower in service, its components are prone to corrosion, and the corrosion of the components will lead to the weakening of the bearing capacity of the tower structural components [1, 2], which will further affect the force safety of the entire structure, thereby affect the safe operation of the power grid.

2. Corrosion of transmission towers

According to the corrosion status of transmission towers, it can be divided into local corrosion and uniform corrosion[3]. Local corroded iron towers have galvanized layers on the surface of most components, and some local areas or individual components are corroded, as shown in Figure 1. Local corrosion of transmission towers mostly occurs in areas with weak anti-corrosion capabilities such as tower feet, tower connections, and areas where the tower has galvanized leakage or scars. Uniform corrosion mainly occurs in coastal areas with high salt spray concentration or industrial areas with severe acid rain. The galvanized layer of the transmission tower with uniform corrosion of the entire tower is completely corroded, and the base steel has been corroded, as shown in Figure 2, where the line is Chongqing The 220kV Tiancha Line[4] has been in operation for 48 years.

Figure 1. Local corrosion

Figure 2. Uniform corrosion.

* Corresponding author: amzhubr@163.com

https://doi.org/10.1051/matecconf/202338201031

3. Axial compression bearing capacity test

3.1 Test plan

Select the corrod angle steel members of the in-service transmission tower, there are four specifications:L75×8, L63×6, L50×5, L40×4. Each specification is divided into four categories according to the degree of corrosion: A, B, C and D. Category A is almost no corrosion; Category B is mildly corroded; Category C is more severely corroded; Category D is severely corroded. Three specimens of each specification and each degree of corrosion were taken for the axial compression bearing capacity test. The size and quantity of the specimens are shown in Table 1.

Table 1. Dimensions of test members.

Component size	Length (mm)	Degree of corrosion
L75×8	1500	A、B、C、D
L63×6	1300	A、B、C、D
L50×5	1200	A、B、C、D
L40×4	900	A、B、C、D

Before carrying out the axial bearing capacity test, it is necessary to measure the corrosion degree of the section of the member. Since all the axial compression members have little corrosion on the limb width edge, the edge corrosion (corrosion in the limb width direction) is ignored, and only the limb thickness direction is measured with a metal ultrasonic thickness gauge [5,6].

The axial compression test uses a hydraulic press to apply pressure, as shown in Figure 3. A dial indicator is used to measure the bending degree of the measuring member, and a plumb bob and a ruler are used to measure the axial displacement. The experimental process adopts one-time loading, preloading the component before the test, and then increasing the load step by step (by 25%, 50%, 70%, 100%, and then by 5%) until the component fails to bend, and can no longer be loaded.

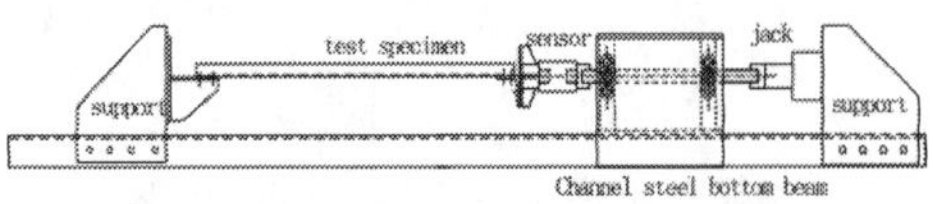

Figure 3. Test device

3.2 Test phenomenon

It can be seen from the test results that the corroded angle steel members basically suffer from bending failure, mainly bending in the direction of the middle minimum axis, as shown in Figure 4.

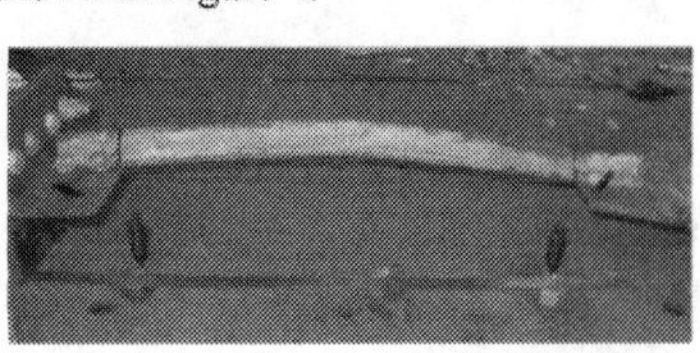

Figure 4. Axial compression member failure mode

3.3 Test data

Through the axial compression bearing capacity test of angle steel with different degrees of corrosion, the load-deflection curve of the corroded angle steel specimen is obtained, as shown in Figure 5-Figure 7. It can be seen from the figure that the load-deflection curve of the component is basically similar to that of the uncorroded component, and it can be divided into an ascending section, a strengthening section and a descending section.

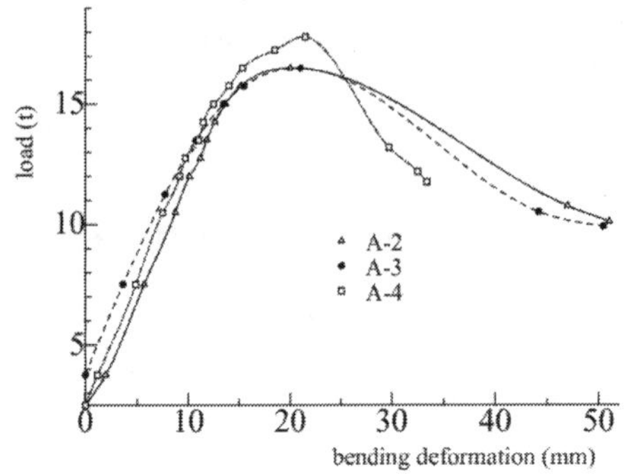

(a) Corrosion degree A

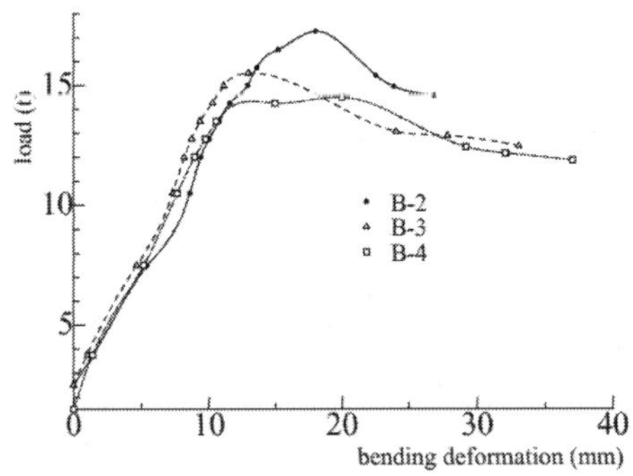

(b) Corrosion degree

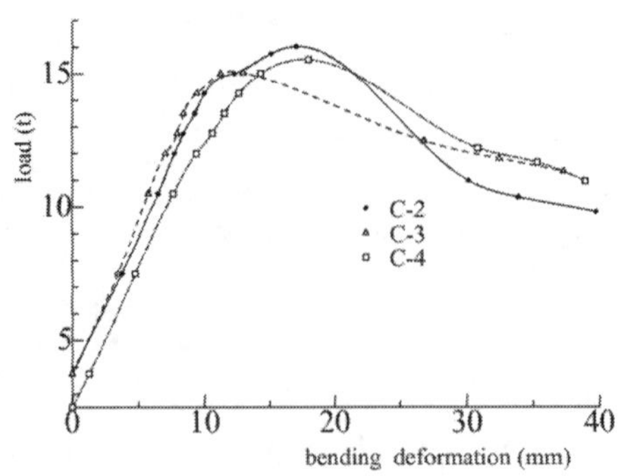

(c) Corrosion degree C

https://doi.org/10.1051/matecconf/202338201031

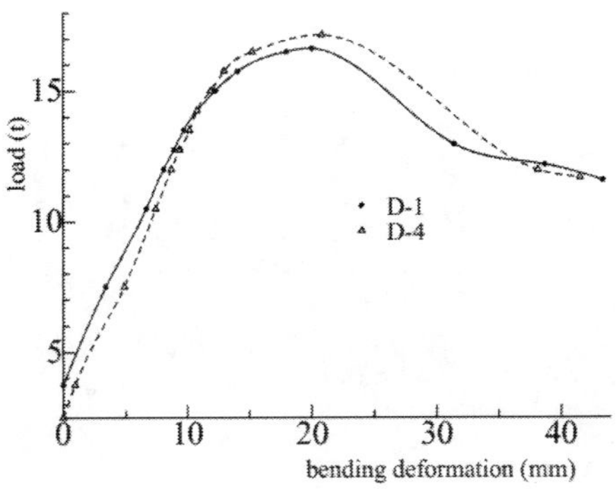

(d) Corrosion degree D

Fig.5 Load bending deformation diagram of L75×8 member

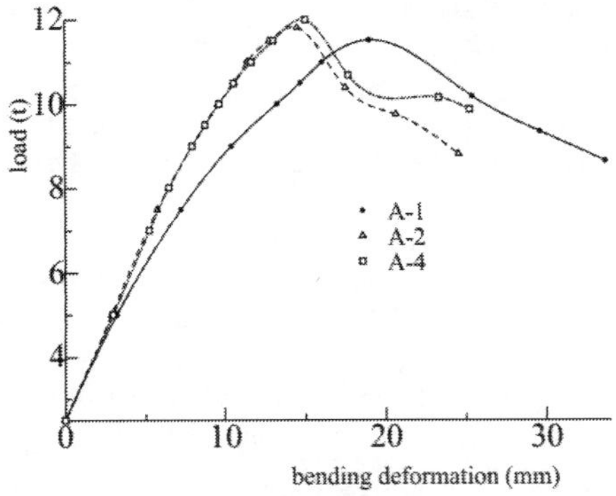

(a) Corrosion degree A

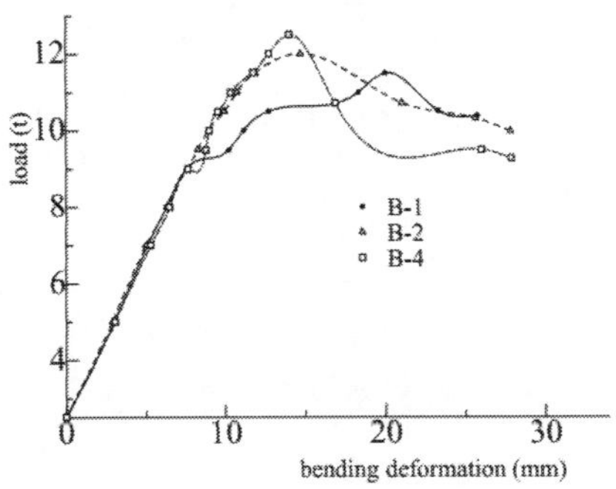

(b) Corrosion degree

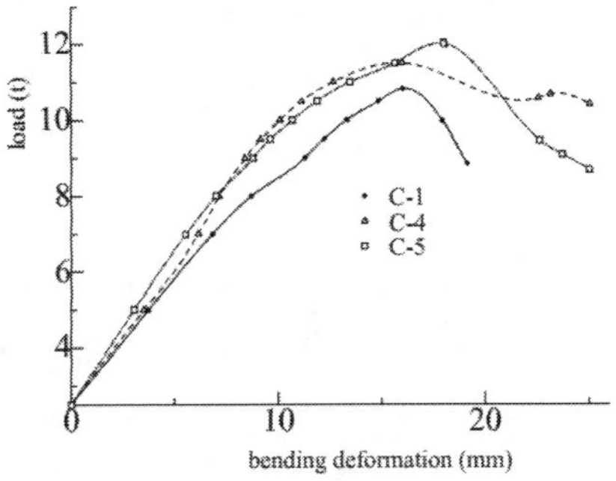

(c) Corrosion degree C

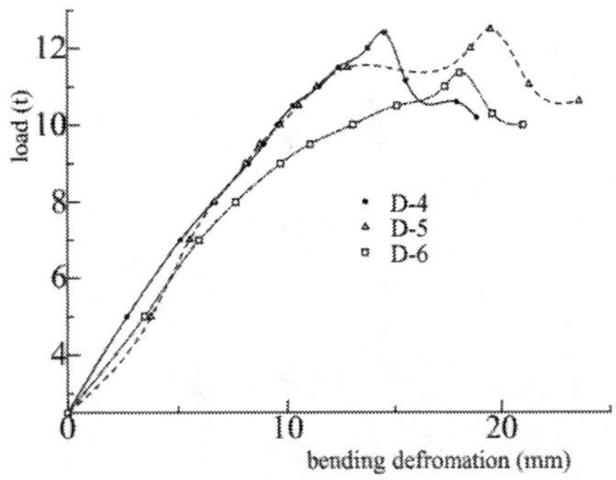

(d) Corrosion degree D

Fig.6 Load bending deformation diagram of L63×6 member

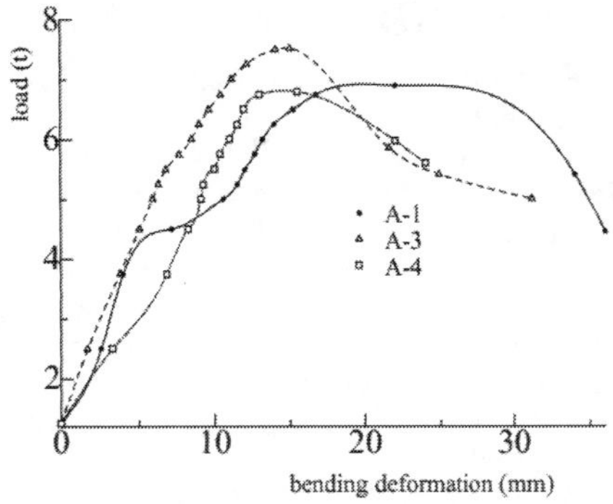

(a) Corrosion degree A

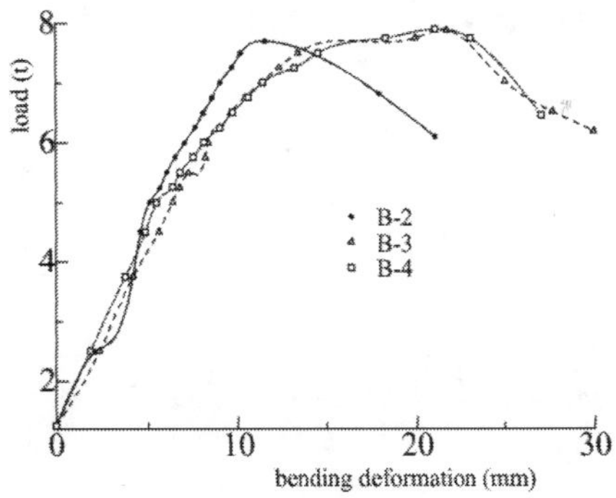

(b) Corrosion degree

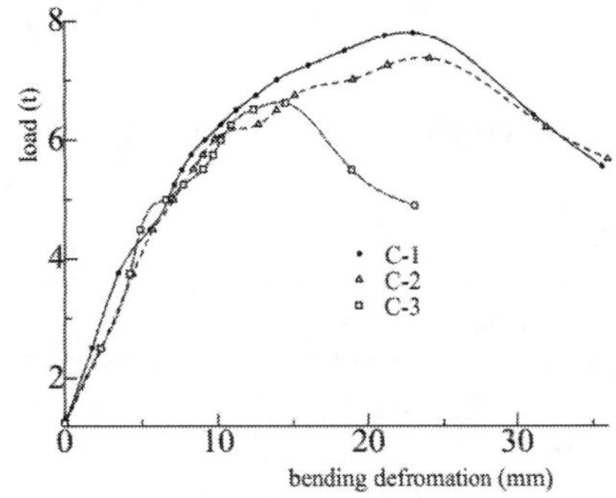

(c) Corrosion degree C

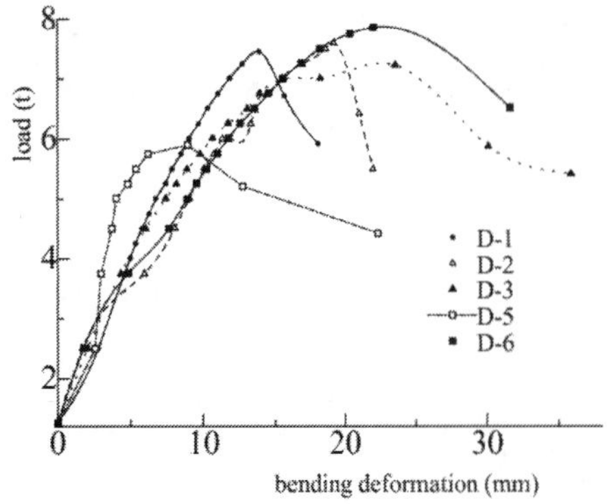

(d) Corrosion degree D

Fig.7 Load bending deformation diagram of L50×5 member

The bearing capacity drop percentage is defined as the ratio of the bearing capacity of the corroded angle steel member in the axial compression test to the full-section bearing capacity, and the relationship with the corrosion rate is drawn as shown in Figure 8. It can be seen that with the increase of the corrosion rate, the bearing capacity of the components shows a decreasing trend, and under the same corrosion rate, due to the differences in the corrosion position, corrosion depth and corrosion area, the axial compression bearing capacity of the components fluctuates greatly.

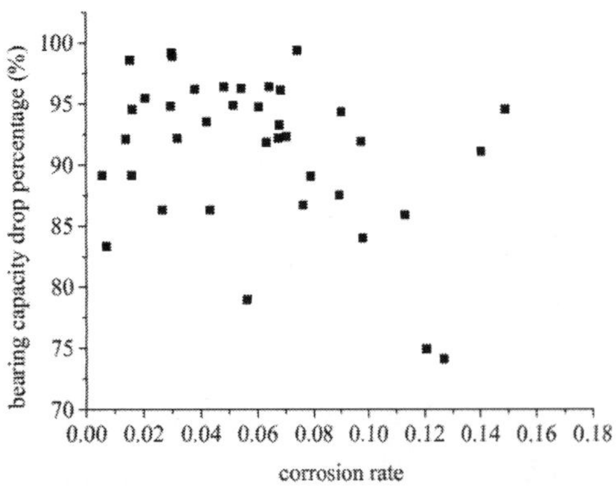

Fig.8 the ralationship between bearing capacity drop percentage and corrosion rate

4. Conclusion

The investigation of the current situation of corroded steel towers on transmission lines found that local corrosion of transmission towers mostly occurs in areas with weak anti-corrosion capabilities, such as the tower foot, the connection of tower parts, and the places where the tower has galvanized leakage or scars. The corroded components on the in-service transmission towers were sampled and the axial compression bearing capacity test was carried out. The results showed that with the increase of the corrosion rate, the bearing capacity of the components showed a decreasing trend. And under the same rust rate, due to the difference of rust position, rust depth and rust area, the fluctuation of the bearing capacity of the component's axial compression is large.

Acknowledgments

This work was financially supported by the State Grid Corporation of China. The project name is Research on edge intelligent perception and evaluation technology of health state of transmission tower structure. The project number is 5600-202019160A-0-0-00.

References

1. Shi Xianda, Chen Hao, Qiao Xin, etc. Analysis on corrosion of angle steel used for 500kV transmission tower[J]. Inner Mongolia petrochemical industry , 2021, 47(12): 38-41
2. Tan Jiancong, Long Zijun, Chen Jianping. etc. Analysis on corrosion of transmission tower and related factors in Guangdong coastal area. Mechanical & Electrical engineering technology, 2014, 43(11): 116-119.
3. Chen Yun, Yao Ningna, Xu Limin. etc. Discussion on rusting grade classification of transmission line steel tower. North China electric power, 2015, 4: 30-34.
4. Chang Hang. Corrosion and anti-corrosion research of transmission towers in Chongqing area[D]. Chongqing University, 2002.
5. Liu Ziyuan. Compressive bearing capacity of corroded single-angle connected with one leg[D]. China university of mining and technology, 2016.
6. Gao Chao, Liu Jianjun, Zheng Yichuan etc. Effects of corrosion on mechanical properties of transmission tower angle steel[J]. Corrosion & protection, 2020, 41(8): 32-38.

https://doi.org/10.1051/matecconf/202338201032

Study on looseness characteristics of bolted joints of transmission towers under lateral vibration

Binrong Zhu [1,*], Guo Huang [1], Zhigang Su [1], Bin Chen [2] and Hailin Liao [2]

[1] China electric power research institute, Beijing 100055, China;
[2] Zhejiang Electric Power Co., Ltd. Shaoxing Power Supply Company, Shaoxing 312099, China

Abstract. transmission towers have been subjected to alternating loads caused by the galloping of iced conductors for a long time, and their fastening bolts are easy to loosen or break, which seriously threatens the safe operation of transmission lines. Through the transverse vibration test of M20 bolts commonly used in transmission towers, the vibration frequency and amplitude are changed, and the change law of the preload of bolt joints is studied. The results show that with the increase of the initial preload, the decline speed of bolt clamping force decreases significantly; The clamping force drop curve of M20 bolt is fitted by shivakumar crews model, which is very consistent with the actual test curve.

1. Introduction

Bolt joint is the key component of the transmission tower member connection and force transmission. Its quality and fastening status not only affect the overall reliability of the tower, but also threaten the safe operation of the whole line. During the operation of the transmission tower, the long-term wind vibration and galloping of the conductor will loosen and fracture bolts of the cross arm[1], which will lead to the destruction of the cross arm, and in serious cases, the tower collapse accident will occur, bringing incalculable losses to the power grid. In the large-scale galloping accident of the power grid from the end of 2009 to the beginning of 2010, the loosening and falling off of the connecting bolts is one of the main reasons for the damage of the cross arm of the tower. The 28 damaged steel towers in Henan Province have suffered from bolt wear, loosening and falling off[2].

2. Dynamic test of bolted joints

2.1 Test plan

The M20 bolts are selected, which are widely used in transmission towers. The amplitude is 1.6mm, the vibration frequency is 4Hz, 3Hz, 2Hz, 1Hz, and 0.5Hz, and the pre-tightening torque is 100N·m, 120N·m, 140N·m, and 160N·m. There are 5 samples of each group of bolts, and the transverse vibration test is carried out on 100 bolts. The sample parameters of the test bolts are shown in Table 1. The bolt number Description: d0-t1-f4, the first letter is D, which means a single bolt; T refers to pre-tightening torque, t1=100N·m, t2=120N·m, t3=140N·m, t4=N·m; F represents the vibration frequency, and the following number represents the vibration frequency (Hz).

According to the test method in GB/T 10431-2008 "Transverse vibration testing method for fasteners"[3], the bolt connection vibration test is carried out on the Amboo transverse vibration test machine, the initial torque is read on the torque wrench, and the clamping force time history curve is collected during the test. Set the test conditions including amplitude, frequency, torque, etc., and then start the vibration. When the clamping force drops to a certain value, stop the test, and record the vibration times and vibration time[4-6].

* Corresponding author: amzhubr@163.com

https://doi.org/10.1051/matecconf/202338201032

Table 1. Bolt test sample parameters

number	No	Bolt Specifications	Pre-tightening torque(N·m)	frequency (Hz)	quantity
11	D0-T1-F4			4	5
12	D0-T1-F3			3	5
13	D0-T1-F2	M20	100	2	5
14	D0-T1-F1			1	5
15	D0-T1-F0.5			0.5	5
16	D0-T2-F4			4	5
17	D0-T2-F3			3	5
18	D0-T2-F2	M20	120	2	5
19	D0-T2-F1			1	5
20	D0-T2-F0.5			0.5	5
21	D0-T3-F4			4	5
22	D0-T3-F3			3	5
23	D0-T3-F2	M20	140	2	5
24	D0-T3-F1			1	5
25	D0-T3-F0.5			0.5	5
26	D0-T4-F4			4	5
27	D0-T4-F3			3	5
28	D0-T4-F2	M20	160	2	5
29	D0-T4-F1			1	5
30	D0-T4-F0.5			0.5	5

2.2 Test result

The clamping force time history curve of bolt transverse vibration test is shown in Figure 1. When the vibration frequency gradually increases from 0.5Hz to 4Hz, the percentage of residual clamping force first decreases and then increases after 240s. When the vibration frequency is 2~3Hz, the percentage of residual clamping force is the smallest, and the bolt loosens the fastest at this time.

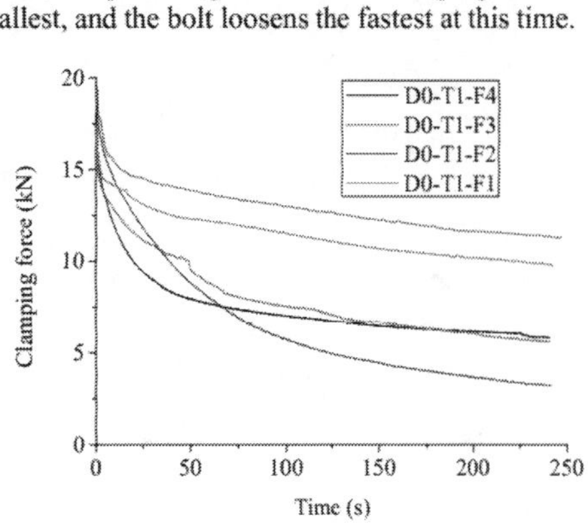

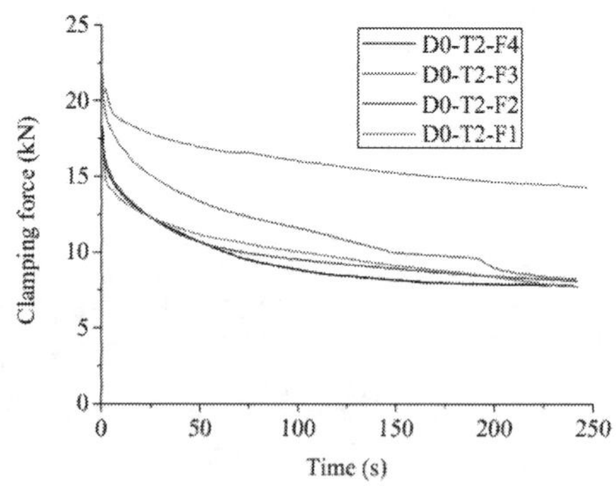

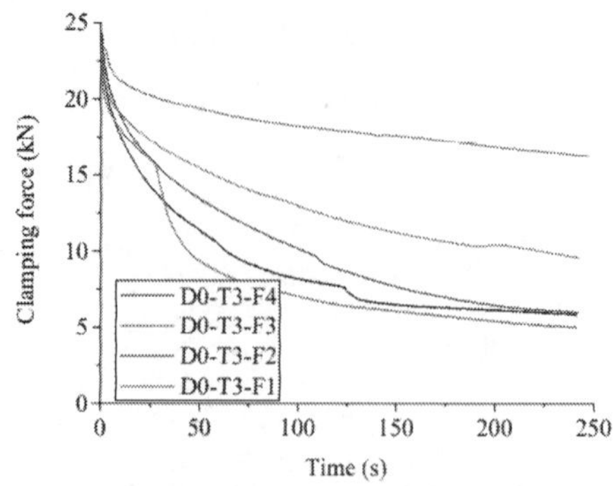

* Corresponding author: amzhubr@163.com

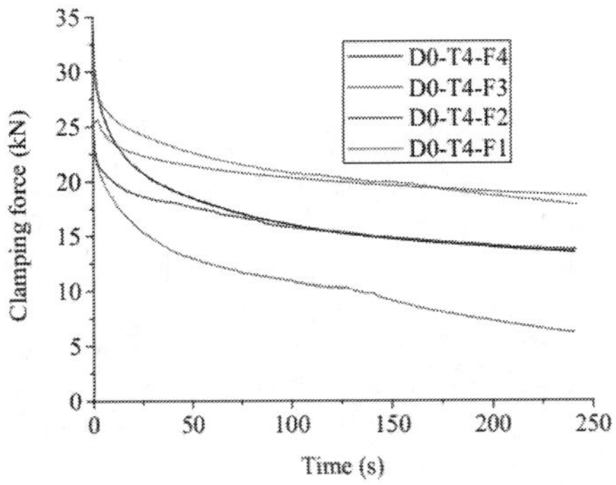

Fig. 1 Time history curve of clamping force

Under the same vibration frequency, the decline curve of clamping force with different initial pre-tightening force is shown in Figure 2. When the installation torque is from 100N·m gradually increases to 160N·m, the initial clamping force of the bolt is 17.77kN, 18.99kN, 23.96kN and 27.33kN respectively, and the residual clamping force also increases after 240s. It can be seen that with the increase of the initial pre-tightening force, the decline rate of the bolt clamping force decreases significantly. It can be seen that increasing the pre-tightening force can play a good role in preventing looseness, but in practice, bolts generally work only within the range of elastic deformation, and excessive pre-tightening force may cause other problems such as plastic yield and fatigue fracture of bolts.

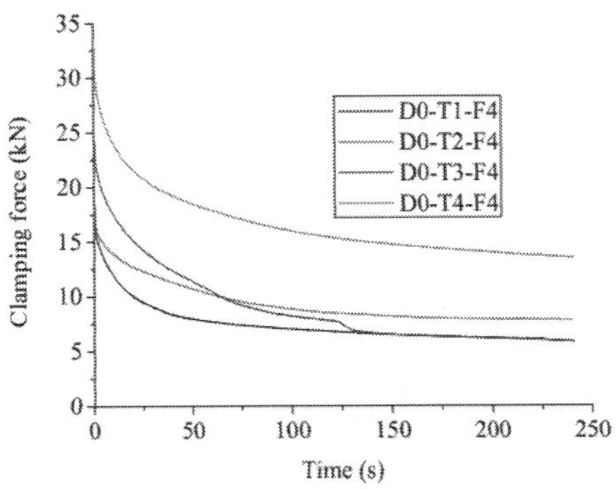

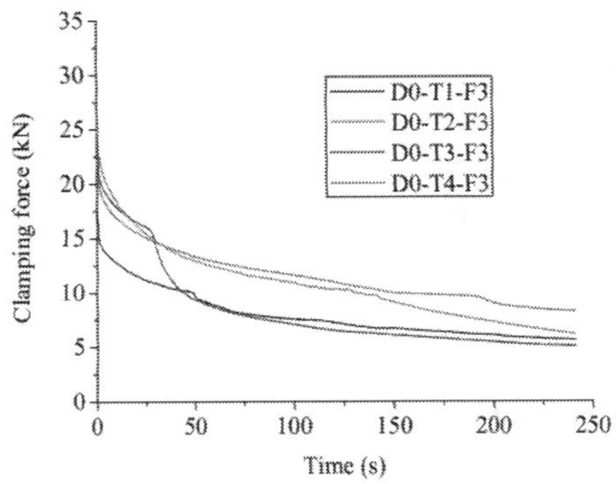

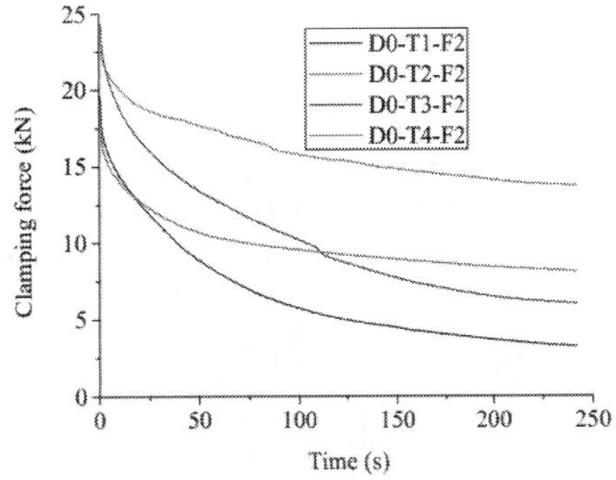

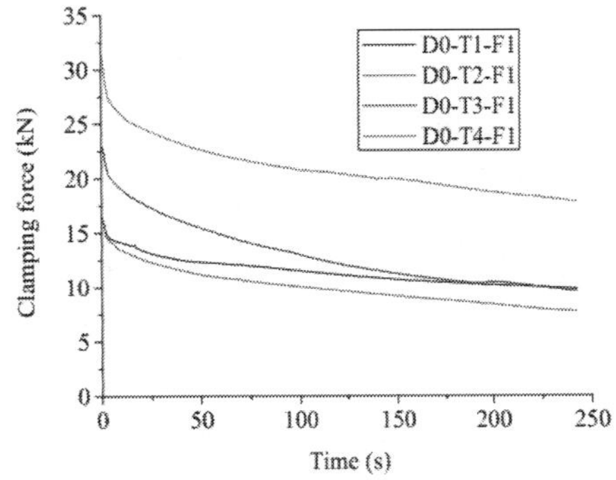

Fig. 2 Time history curve of M20 with different pre-tightening forces

2.3 Falling curve of clamping force of bolt joint

When the bolt connected structure bears high vibration load, the clamping force will drop, and the clamping force drop time history curve follows a similar change pattern. It can be seen from the above phenomenon that a general clamping force drop equation can be used to describe the clamping force drop law of bolted joints under vibration load. After years of research, scholars have proposed a variety of empirical formulas to describe the relaxation of bolt preload, such as Shivakumar Crews model[7], Hook Norton model[8]. Existing studies have proved that S-C model can describe the decline characteristics of bolt clamping force under vibration load.

The clamping force decline equation of Shivakumar Crews model is:

$$\frac{P}{P_0} = \frac{1}{1+Kt^n} \tag{1}$$

$$t = \frac{n}{f_b} \tag{2}$$

Where p is the current clamping force, P_0 is the initial clamping force, and t is the test duration; N is the number of vibrations, f_b is the vibration frequency.

In order to obtain the attenuation law of bolt clamping force in the process of transverse vibration, the test data are nonlinear fitted, and the fitting curve passes through the starting point. The clamping force drop curve and fitting curve of single nut M20 are shown in figure 3, and the clamping force drop formula is as follows:

$$\frac{P}{P_0} = \frac{1}{1+0.0603t^{0.4371}} \tag{3}$$

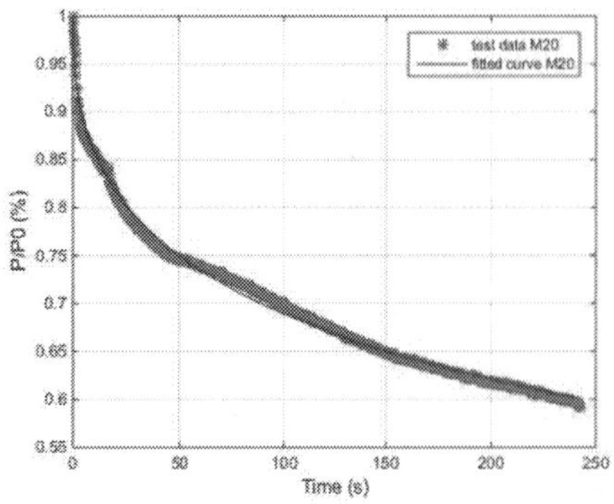

Fig. 3 falling curve of clamping force of single bolt M20

According to the comparison between the fitting curve and the test curve, the clamping force decline curve fitted by Shivakumar Crews model is very consistent with the actual test curve.

3. Conclusion

All manuscripts must be in English, also the table and figure texts, otherwise we cannot publish your paper. Please keep a second copy of your manuscript in your office.

(1) Through the transverse vibration test of transmission tower bolt joints, the test conclusion shows that when the vibration frequency is 2~3hz, the percentage of residual clamping force is the smallest, and the bolt loosens the fastest at this frequency; With the increase of the initial pre-tightening force, the decreasing speed of the bolt clamping force decreases significantly.

(2) The clamping force drop curve of M20 bolt is fitted by Shivakumar Crews model, which is very consistent with the actual test curve.

Acknowledgments

This work was financially supported by the State Grid Corporation of China. The project name is Research on edge intelligent perception and evaluation technology of health state of transmission tower structure. The project number is 5600-202019160A-0-0-00.

References

1. Chang Xingya. Research on loose characteristics of bolted joint used on transmission tower under the action of wind[D]. North China electric power university, 2015.
2. Zhang Tianguang, Sun Donghai, Ao Qingcheng, etc. Analysis of He' nan Grid's 500kV Yao-Shao lien gallop accident and its control. Electric power construction, 2009, 30(3): 24-27.
3. GB/T 10431-2008. Transverse vibration testing method for fasteners[s]. Beijing, 2008.
4. Zhang Ting, Jiang ZhaoXi, Feng Mei, etc. Shake proof and anti-looseness property comparison of fastener[J]. Metal products, 2012,38(1): 78-80.
5. Zhao Jianping, Zhang Yaobin, Wang Chunyao. Experimental Study on Lateral Vibration of Transmission Towers Using Bolt[J]. Journal of Xinjiang University (Natural Science Edition), 2018, 35(2): 243-247.
6. Zhang Pengcheng. Experimental Investigation on Self-loosening Mechanism for Bolted Joints under the Transverse Loading[D]. Zhejiang University of Technology, 2018.
7. Shivakumar K N, Crews J H. Bolt clampup relaxation in a graphite/epoxy laminate[M]//Long-Term Behavior of Composites. West Conshohocken:ASTM International, 1983:15-17.
8. Taira S. High temperature strength design theory of metal material [M]. Beijing: Science Press, 1983: 17-27.

Chemical composition and identification of glass products based on generalized linear model

Zhuo Pan *

College of Chemistry and Environment, Southwest Minzu University, Chengdu, 610041, China

Abstract. Weathering will cause the proportion of chemical composition of ancient glass products to change, thereby affecting the classification judgment of glass type. In order to reasonably analyze the composition content of ancient glass products, this paper provides a reference for studying the content change trend of glass weathering by exploring whether the appearance of glass cultural relics affects the weathering trend and the weathering composition law. According to the relationship between glass surface weathering and ornamentation, color and glass type, a linear regression model is constructed, and in order to solve the problem of heteroscedasticity in the perturbation term, the OLS robust + standard error method is used to improve the model parameters. The results show that the ornaments are B and C and are more likely to weather, high potassium glass is easy to weather, the color is green is mostly unweathered glass, and the color is black There is a tendency to change to weathered glass. The lower the content of potassium oxide, iron oxide and alumina in high potassium, the easier it is to weather, the higher the silica content, the easier it is to weather, and the regularity of weathering and component content of lead barium compared with high potassium is not obvious.

Keywords: linear regression model; glass weathering; high potassium; lead barium.

1. Background

In the ancient Silk Road, glass was introduced to China as an ornament, and China imitated its manufacturing technology and refined it from local materials, so the appearance of the glass manufactured was similar to that of glass products imported from outside, but the chemical composition was different [1]. The main component of glass is quartz sand, and silica is its main chemical component, accounting for most of the proportion. In the process of refining glass products, the melting temperature of quartz sand is too high, and co-solvents need to be added to reduce the melting temperature of quartz sand[2]. With different fluxes, the proportion of various chemical components of glass will also be different. For example, lead ore is a co-solvent of lead-barium glass, while the co-solvent of potassium glass is grass and wood ash. In ancient times, cultural relics glass was easily weathered due to the influence of the buried environment[3,4]. During the weathering process, the internal elements of the glass are exchanged with the environmental elements to change its composition ratio[5,6], which leads to an impact on the judgment of its category. Based on this difference in judgment, a regression model is established to analyze and identify glass products, and to find out whether there is a regularity of weathering component content. The chemical composition data is derived from Question C of the 2022 China National College Students Mathematical Modeling Competition.

2. Model establishment and solution

2.1 Establishment of linear regression model

Based on the chemical composition data of question C, the data of Form 1 was analyzed, and there were four indicators for each cultural relic in Form 1, namely whether the surface of the cultural relics was weathered, color, ornamentation, and type, and it was found that the four weathered lead barium numbered 19, 40, 48 and 58 did not detect color, considering that there were not many cultural relics missing from the data and there were errors in the data after interpolation, so these four data were discarded when analyzing the relationship between surface weathering and ornamentation, color and type. When seeking to explain the influence of three variables on surface weathering, the coefficient requirements are not large, and the OLS linear regression model is used to solve the problem. The model is as follows:

$$Weathering_i = \alpha + \sum \beta_{n_1}^{(1)} \times Color_{n_1} + \sum \beta_{n_2}^{(2)} \times Ornamontation_{n_2} + \sum \beta_{n_3}^{(3)} \times Type_{n_3} + \varepsilon_i$$

where $Weathering_i$ indicates whether the ith glass is weathered, $Weathering_i$ is a dichotomous variable,

* Corresponding author: panzhuo0809@163.com

glass weathering $Weathering_i$ = 1, vice versa 0. $Color_{n_1}$, $Oranmontation_{n_2}$, $Type_{n_3}$ are dummy variables of color, ornamentation, type, so n1=8, n2=3, n3=2. when the glass is a certain color, ornamentation, type, other dummy variables are 0. ε_i is the perturbation term, consider the multicollinearity problem, and then test to improve it.

2.2 Linear regression model solution and interpretation

Through Stata software to analyze the trend relationship between the weathering of the glass surface and its glass type, ornamentation and color, using the least squares estimation method, all the coefficients are estimated, and through the F test, the regression model is significant, the software automatically excludes complete multicollinearity, tests the significance of each regression coefficient, and uses the t-test to find that most of the coefficients are not significant, as shown in Table 1.

Table 1. Parameter estimation table

	Coef.	Std. Err.	t	P>|t|	[95%Conf.Interval]	
M_1	0	(omitted)				
M_2	1.018029	0.2093795	4.86	0	0.5957754	1.440284
M_3	0.2567611	0.1422772	1.8	0.078	-0.0301682	0.5436903
N_1	0	(omitted)				
N_2	-0.7973274	0.1470033	-5.42	0	-1.093788	-0.5008671
C_1	-0.3220914	0.3797931	-0.85	0.401	-1.088017	0.4438344
C_2	0.2324213	0.2901182	0.8	0.427	-0.3526579	0.8175004
C_3	0.0299077	0.338101	0.09	0.93	-0.6519379	0.7117533
C_4	0	(omitted)				
C_5	-0.1554248	0.3642688	-0.43	0.672	-0.8900428	0.5791933
C_6	-0.6554248	0.486468	-1.35	0.185	-1.636481	0.3256314
C_7	0.3806342	0.2953203	1.29	0.204	-0.2149359	0.9762043
C_8	0.6013363	0.379493	1.58	0.12	-0.1639843	1.366657
_cons	0.3986637	0.2733289	1.46	0.152	-0.1525567	0.949884

M_1, M_2, M_3 are dummy variables of ornamentation, N_1, N_2 are virtual variables of type, C_1... C_8 is the color dummy variable, the above is the variable calculated in practice, focusing only on the significant variables, M_2==B, M_3==C, N_2== high potassium.

Consider that the perturbation term may have heteroscedasticity, if there is heteroscedasticity, the hypothesis test fails, and the least squares estimation method is not the optimal linear unbiased measure, so it should be tested for BP. Using the BP test, the p-value is 0.012<0.5, and the null hypothesis is rejected with a 95% confidence level with heteroscedasticity.

In order to solve the problem of heteroscedasticity, the OLS robust and standard error method is used to give greater weight to the informative chemical composition, and Table 2.

Table 2. Parameter estimation table

	Coef.	Std. Err.	t	P>|t|	[95%Conf.Interval]	
M_1	0	(omitted)				
M_2	1.018029	0.0514293	19.79	0	0.9143124	1.121747
M_3	0.2567611	0.1252562	2.05	0.047	0.0041578	0.5093643
N_1	0	(omitted)				
N_2	1.018028	0.785933	-10.14	0	-0.9558268	-0.63882
C_1	-0.3220914	0.4539411	-0.71	0.482	-1.23755	1.593368
C_2	0.2324213	0.3450728	0.67	0.504	-0.4634843	0.9283268
C_3	0.0299077	0.3780666	0.08	0.937	-0.7325363	0.7923517
C_4	0	(omitted)				
C_5	-0.1554248	0.4376387	-0.36	0.724	-1.038007	0.7271578
C_6	-0.6554248	0.3362131	-1.95	0.058	-1.333463	0.0226136
C_7	0.3806342	0.3219399	1.18	0.244	-0.2686195	1.029888
C_8	0.6013363	0.3183382	1.89	0.066	-0.0406539	1.243326
_cons	0.3986637	0.3183382	1.25	0.217	-0.2433265	1.040654

Focus on significant variables, M_2==B, M_3==C, N_2== High Potassium, C_6== Green, C_8== Black. The final result can be obtained under the t-test, under the confidence level of 90%, the corresponding coefficient is significantly different from 0 when the ornament is B and C, the type is high potassium, the color is black or green, and the data analysis can obtain that the ornaments B and C are mostly weathered glass, B is the weathered glass The probability is much greater than C, that is, when the ornaments are B and C, the glass is more likely to weather. High potassium is easy to weather, the color is green is likely to be unweathered glass, when the color is black, there is a trend to weathered glass.

3. Statistical laws

3.1 Model establishment and solution of high potassium for dependent variables

Weathering is due to various chemical reactions between internal elements and external substances, resulting in changes in the proportion of the main components of glass. If the proportion of total cultural relics numbered 15 and 17 in Form 2 is less than 85%, it is invalid data, and it is discarded, and the component data column with many missing data is deleted. Because the production method of lead barium glass and high potassium glass and flux are different, whether the glass is weathered, the statistical weathering composition law can not describe lead barium and high potassium together, need to be divided into two categories, separate statistics. Form 2 is organized into two tables of glass type high potassium and lead barium, and first linear regression interpolation is performed using SPSS software to complete the data to obtain a reasonable value. To explore the law of the weathering content of cultural relics, first make statistical observations on the data. Table 3 shows the data description statistics of high potassium glass.

 https://doi.org/10.1051/matecconf/202338201033

Table 3. High potassium data descriptive statistics

N		Minimum value	Maximum value	Average	Standard deviation	Variance
SiO_2	17	59.01	96.77	76.4782	14.89435	221.842
K_2O	17	0.59	14.52	6.8100	4.98342	24.834
CaO	17	0.21	8.70	4.2506	3.10554	9.644
Al_2O_3	17	0.81	11.15	5.1747	3.12906	9.791
Fe_2O_3	17	0.08	6.04	1.4618	1.57042	2.466
CuO	17	0.47	5.09	2.2824	1.43473	2.058
P_2O_5	17	0.15	4.50	1.0594	1.28435	1.650
Valid sample	17					

High potassium glass has the largest silica content, and its maximum and minimum values differ greatly, and the maximum and minimum values of potassium oxide and alumina differ greatly.
Table4 shows the descriptive statistics of lead barium.

Table 4 . Lead barium data description statistics

N		Minimum value	Maximum value	Average	Standard deviation	Variance
SiO_2	48	3.72	75.51	39.2658	18.64039	347.464
Al_2O_3	48	0.45	14.34	3.7160	3.02143	9.129
CaO	48	0.37	6.40	2.2563	1.59302	2.538
CuO	48	0.11	10.57	1.7783	2.12098	4.499
PbO	48	9.30	70.21	33.4465	15.08981	227.702
BaO	48	1.29	35.45	10.1956	7.68756	59.099
P_2O_5	48	0.07	14.13	3.7665	3.90238	15.229
SrO	48	0.05	1.12	0.4052	0.22163	0.049
Valid sample	48					

The maximum and minimum values of silica and barium oxide content of lead barium glass are very different, the maximum and minimum content of lead oxide are very different, compared with high potassium glass, the average content of silica is much lower, and the average value of barium oxide and lead oxide is larger. The trend of difference between the two data is understood, and there is a general tendency for statistical regularity.
In order to know the influence trend of which chemical composition on whether high-potassium glass weathers, considering the problem of heteroscedasticity, the OLS standard and method are used to construct the model, and the model is as follows:

$$Weathering^{(1)} = \alpha^{(1)} + \sum b^{(m)} x_1^{(m)} + \varepsilon_i^{(1)}$$

$Weathering^{(1)}$ is a dichotomy variable of weathering in a high potassium type, with a value of 1 weathering for high potassium and unweathered for a value of 0, m = 7, m is the number of chemical components of high potassium after the adnexal treatment. $x_1^{(1)} \ldots x_1^{(7)}$ corresponds to the content of phosphorus pentoxide, copper oxide, iron oxide, alumina, calcium oxide, potassium oxide and silica, respectively.

The model passed the F-test to test the significance of each regression coefficient, and using the t-test, it was found that most of the coefficients were not significant.

Table 5. Parameter estimation table

	Coef.	Robust Std.Err.	t	P>|t|	[95%Conf.	Interval]
P_2O_5	0.0093493	0.1104601	0.08	0.934	-0.2405287	0.2592274
CuO	-0.0273651	0.0419187	-0.65	0.53	-0.1221917	0.0674615
Fe_2O_3	-0.1428314	0.0684786	-2.09	0.067	-0.2977408	0.012078
Al_2O_3	-0.0905678	0.0423343	-2.14	0.061	-0.1863345	0.005199
CaO	0.0280589	0.0804544	0.35	0.735	-0.1539416	0.2100594
K_2O	-0.1726901	0.0824662	-2.09	0.066	-0.3592416	0.0138614
SiO_2	-0.0462753	0.0244762	−1.89	0.091	-0.1016443	0.0090937
_cons	5.678746	2.352001	2.41	0.039	0.35815	10.99934

From Table 5, it can be obtained that under the t-test, the pre-coefficients of potassium oxide, iron oxide, silica, and alumina are significantly different from 0 under the t-test, and the four components are used to describe our law, so we can make the content plot as follows:

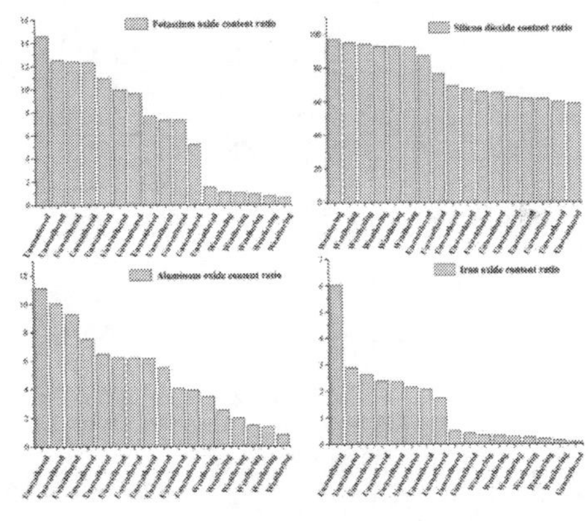

Figure 1. Trend of four components of high potassium versus whether weathering

From Figure 1, it can be seen that the proportion of four individual components has a clear trend with whether the glass is weathered, and the interval of the chemical composition when the glass is weathered can be obtained, and we analyze the proportions of the four components separately, the proportion range of potassium oxide, iron oxide, silica and alumina is 7.37~5.19, 1.74~0.5, 92.35~87.05, 4.06~3.93, these intervals.
It is the unweathered transition weathering interval, the lower the content of potassium oxide, iron oxide and alumina, the easier it is to weather, and the higher the silica content, the easier it is to weather.

3.2 The dependent variable is the model establishment and solution of lead barium

Establish the regression equation for lead barium:

$$Weathering^{(n)} = \alpha^{(2)} + \sum c^{(z)} x_2^{(z)} + \varepsilon_i^{(2)}$$

$Weathering^{(2)}$ is a dichotomy variable of weathering in the barium lead-type with a value of 1 weathering and a value of 0 unweathered barium, c=8, which is the chemical composition of lead barium after processing the data, $x_2^{(1)} \ldots x_2^{(8)}$ corresponds to the content of silica, calcium oxide, alumina, copper oxide, lead oxide, barium oxide, phosphorus pentoxide, strontium oxide, respectively.

The solution method is consistent with the linear model with high potassium, and the model passes the F-test.

Table 6. Parameter estimation table

	Coef.	Robust Std.Err.	t	P>|t|	[95%Conf.	Interval]
SiO_2	-0.6511925	0.0202499	-3.22	0.003	-0.1061518	-0.0242333
CaO	-0.2941218	0.0531655	-5.5	0	-0.4016593	-0.1865844
Al_2O_3	0.0509154	0.0234897	2.17	0.036	0.0034031	0.0984277
CuO	0.0240945	0.032123	0.75	0.458	-0.0408804	0.0890694
PbO	-0.0339802	0.0179748	-1.89	0.066	-0.0703376	0.0023772
BaO	-0.0791974	0.0324	-2.44	0.019	-0.1449445	-0.0134502
P_2O_5	0.112208	0.021263	0.35	0.729	-0.0537607	0.0762024
SrO	-0.6050173	0.2560187	-2.36	0.023	-1.122864	-0.0871706
cons	5.867445	1.914365	3.06	0.004	1.995276	9.73614

Table 6 analyzes that at the 90% confidence level, 6 components are significantly different from 0, and these 6 components are used to reflect the statistical law of whether lead barium weathers, and the 6 chemical components are shown below:

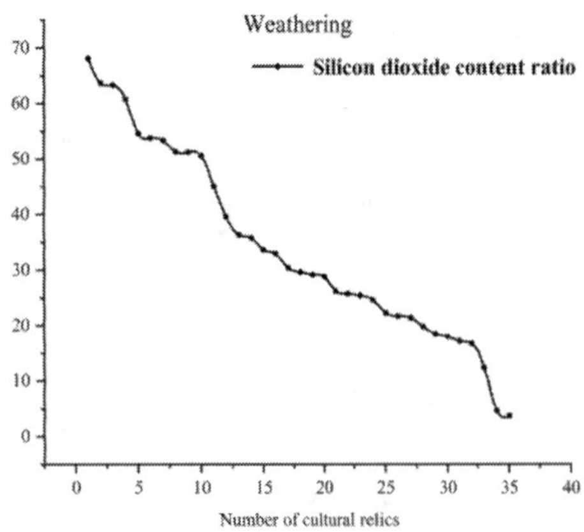

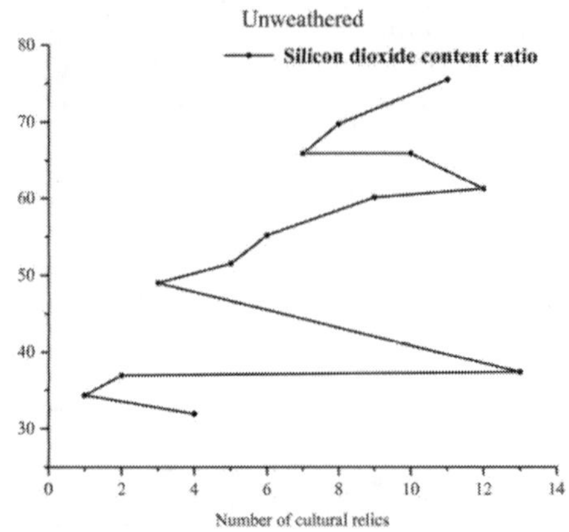

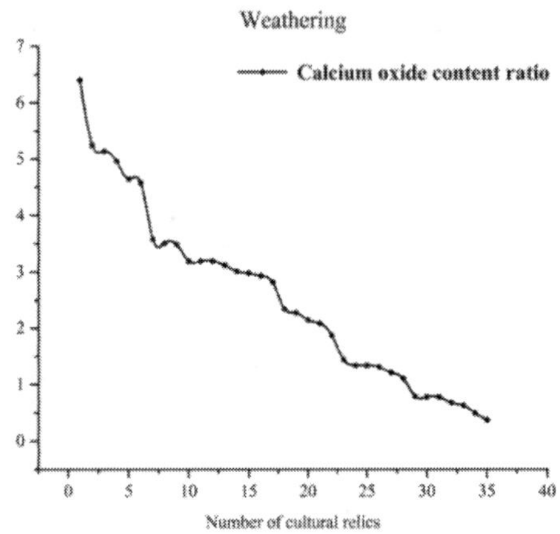

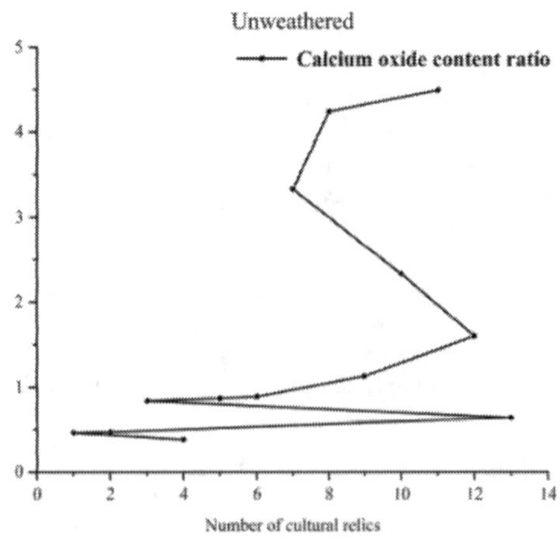

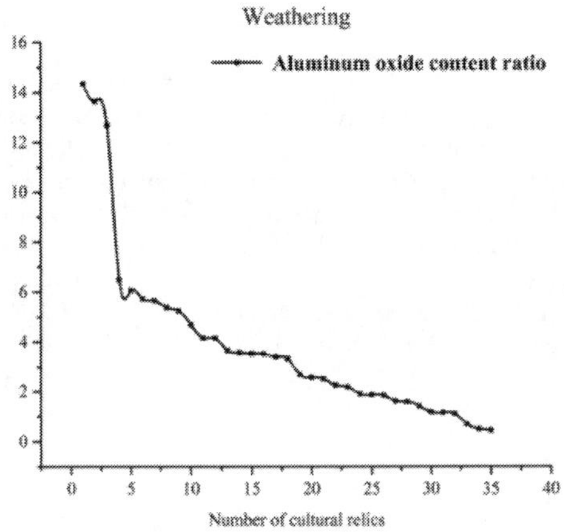
Weathering
Aluminum oxide content ratio
Number of cultural relics

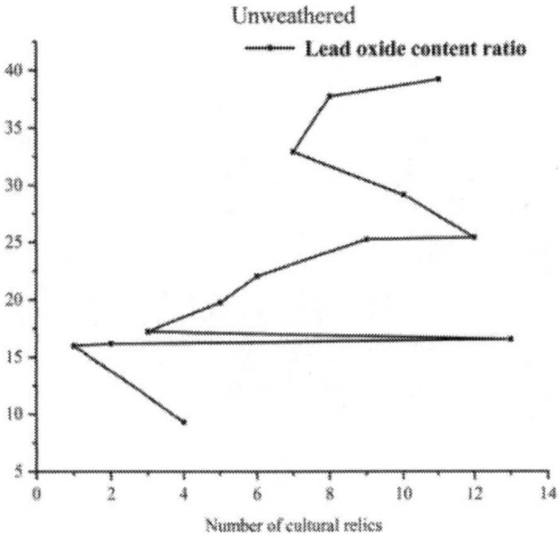
Unweathered
Lead oxide content ratio
Number of cultural relics

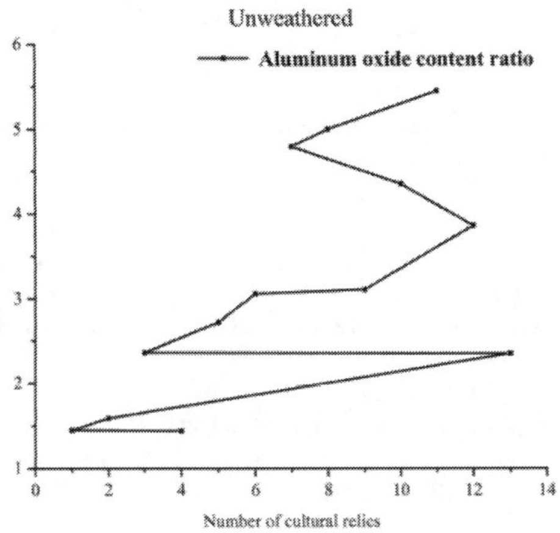
Unweathered
Aluminum oxide content ratio
Number of cultural relics

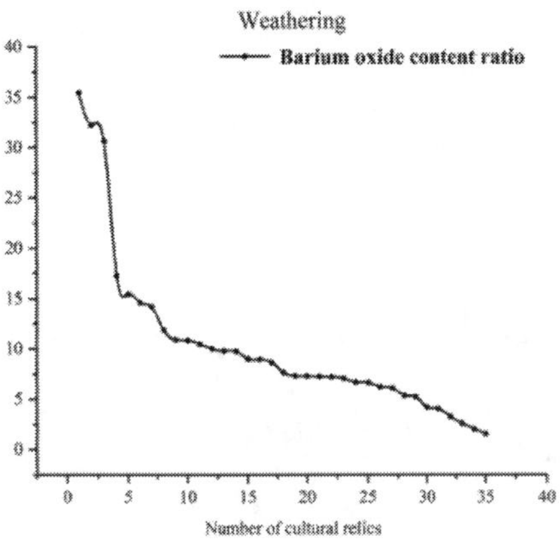
Weathering
Barium oxide content ratio
Number of cultural relics

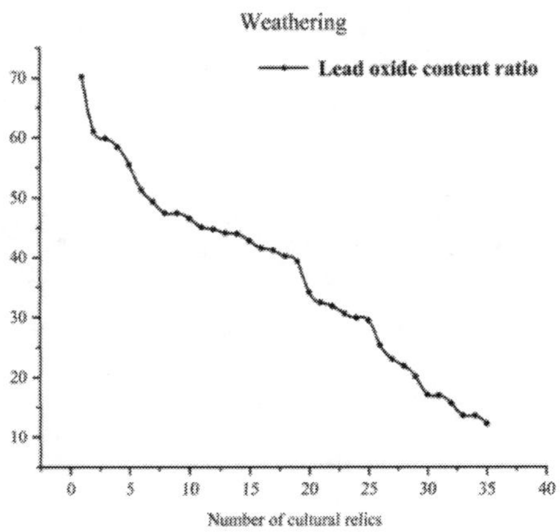
Weathering
Lead oxide content ratio
Number of cultural relics

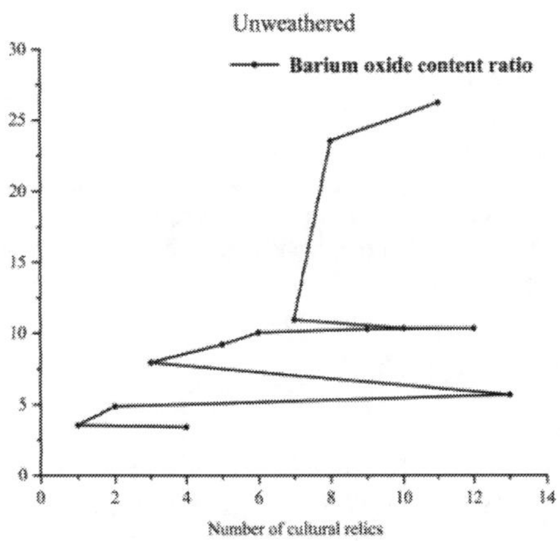
Unweathered
Barium oxide content ratio
Number of cultural relics

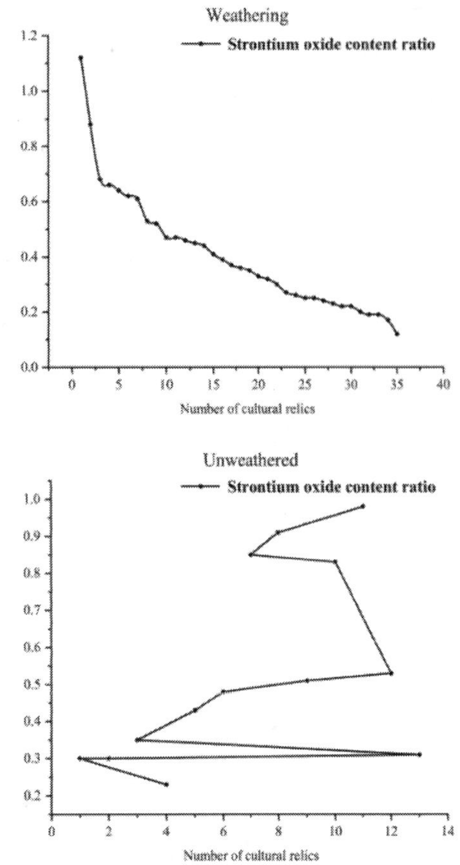

Figure 2. Lead barium chemical composition chart

The contents of silica, barium oxide and lead oxide in the barium component of lead can be obtained from Figure 2 in the weathering and unweathered content change range. Compared with the content of lead-barium composition and specific high potassium, there is no obvious statistical law of weathered and unweathered chemical composition content.

4. Conclusion

Based on the data in question C of the 2022 China Higher Education Society Cup National College Students Mathematical Modeling Competition, we established a regression model to obtain the relationship between the weathered and unweathered chemical components of the glass surface with ornamentation, color and glass type, and further based on the regression model, we found the statistical rules of chemical composition changes in high potassium and lead barium. The results of this paper are as follows:

(1) The ornamentation is B and C, the type is high potassium, and the color is black and green, these factors have a significant impact on the change of surface weathering. When the ornamentation is B and C, the glass is more weathered, high potassium is easy to weather, and the color is black and easy to weather.

(2) The proportion range of potassium oxide, iron oxide, silica and alumina components in high-potassium glass from unweathered to weathered is 7.37~5.19, 1.74~0.5, 92.35~87.05, 4.06~3.93, the lower the content, the easier it is to weather, and the higher the silica content, the easier it is to weather.

(3) The statistical law of lead barium glass compared with high potassium glass is not obvious.

References

1. J. Henderson,J. An,H. Ma.The Archaeometry and Archaeology of Ancient Chinese Glass: a Review[J].Archaeometry,2018,60(1).
2. Willem B. Stern.Phosphate: a neglected argument in studies of ancient glass technology[J].Swiss Journal of Geosciences, 2017,110(3).
3. Gentaz Lucile,Lombardo Tiziana,Loisel Claudine, Chabas Anne, Vallotto Marta.Early stage of weathering of medieval-like potash-lime model glass: evaluation of key factors.[J].Environmental science and pollution research international,2011,18(2).
4. Bettembourg J M. Composition et altération des verres de vitraux anciens[C]//Verres et réfractaires. Actes du IX colloque international du Corpus Vitrearum Medii Aevi, Paris, 8-12 September 1975. 1976, 30(1): 36-42.
5. Wang Chengyu,Tao Ying,Chen Min,Huang Ming.Weathering of Soda-Lime and Lead Glasses[J].Transactions of the Indian Ceramic Society,2014,50(6).
6. .SCIENTIFIC EXAMINATION AND CONSOLIDATION OF ANCIENT GLASS[J]. Studies in Conservation,2014,6(4).

Review of Recent Distillation Studies

Minghong Gao

Northwestern Polytechnic University, Shaanxi, China

Abstract. Knowledge distillation has gained a lot of interest in recent years because it allows for compressing a large deep neural network (teacher DNN) into a smaller DNN (student DNN), while maintaining its accuracy. Recent improvements have been made to knowledge distillation. One such improvement is the teaching assistant distillation method. This method involves introducing an intermediate "teaching assistant" model between the teacher and student. The teaching assistant is first trained to mimic the teacher, and then the student is trained to mimic the teaching assistant. This multi-step process can improve student performance. Another improvement to knowledge distillation is curriculum distillation. This method involves gradually training the student by exposing it to increasingly difficult concepts over time, similar to curriculum learning in humans. This process can help the student learn in a more stable and consistent manner. Finally, there is the mask distillation method. Here, the student is trained to specifically mimic the attention mechanisms learned by the teacher, not just the overall output of the teacher DNN. These improvements help to enhance the knowledge distillation process and enable the creation of more efficient DNNs.

Keywords: Knowledge distillation; Deep neural network; Distillation gap.

1. Introduction

Knowledge distillation [7] is a method of transferring the knowledge from a complex model, called the teacher, to a smaller and simpler model, called the student. In recent years, several variants of knowledge distillation have been proposed, including teaching assistant distillation, curriculum distillation, mask distillation, and decoupling distillation. This literature review summarizes the recent developments in these variants of knowledge distillation and discusses their strengths and limitations.

Knowledge distillation is a method of compressing a complex deep neural network (DNN) into a smaller and faster DNN while preserving its accuracy. The process of knowledge distillation involves training a smaller DNN, called the student, to imitate the predictions of a larger and more complex DNN, called the teacher. The student network is trained to produce similar results as the teacher network, but with fewer parameters and lower computational cost. Knowledge distillation has been widely used for model compression and acceleration, and has shown great promise in various applications[5, 8, 13, 18], such as computer vision and natural language processing.

In recent years, several variants of knowledge distillation[3, 4, 21] have been proposed and explored to improve the performance of knowledge distillation [6, 9]. One of these variants is teaching assistant distillation, which introduces an intermediate model, called the teaching assistant, between the teacher and the student. The teaching assistant is trained to mimic the behavior of the teacher, and the student is trained to imitate the outputs of the teaching assistant. This approach has been shown to provide better performance than traditional knowledge distillation, as it can better capture the knowledge learned by the teacher.

Another variant of knowledge distillation [10, 11] is curriculum distillation, which designs the learning process to follow a curriculum, similar to human education. The curriculum is designed to present easy examples first and gradually increase the difficulty of the examples as the student improves. This approach has been shown to provide better performance than traditional knowledge distillation, especially for tasks that require a lot of prior knowledge.

Mask distillation is a variant of knowledge distillation [12, 20] that focuses on transferring the attention mechanism learned by the teacher to the student. In mask distillation, the teacher is trained to produce a mask that indicates the importance of each input feature for a particular prediction. The student is then trained to imitate the predictions of the teacher while using the mask to weigh the importance of the input features. This approach has been shown to provide better performance than traditional knowledge distillation, as it can better capture the attention mechanism learned by the teacher.

Finally, decoupling distillation is a variant of knowledge distillation [1] that decouples the distillation loss from the task loss. In decoupling distillation, the student is trained to imitate the outputs of the teacher on a validation set, while being trained on the task loss on the training set. This approach has been shown to provide better performance than traditional knowledge distillation, as it can better balance the trade-off between preserving the knowledge of the teacher and adapting to the task.

In conclusion, knowledge distillation is a widely used method for compressing and accelerating deep neural networks

2. Teacher assistant

When knowledge distillation is performed, increasing the model size of the teacher network instead makes the knowledge distillation worse. The results are as follows: the student network is a 2-layer CNN, while the teacher network is a 4-10-layer CNN, respectively, and it can be seen from the results that continuing to increase the depth of the teacher network does not necessarily improve the performance of knowledge distillation.

To further prove that the gap does affect knowledge distillation, the authors conducted a small comparison experiment by fixing the teacher network as a 10-layer CNN and changing the number of layers of the student network, and the results are as follows.

Based on the above observations, it is not difficult to think that a medium-sized network (teacher assistant [16], TA) should be added between the large network and the small network, and the
teacher network should be used to distill the TA first, and then the student network should
be distilled by the TA. When TA was added, the distillation effect of the student network was significantly better than the direct distillation of the teacher network.

(NOKD in the table indicates no KD, BLKD indicates baseline KD, and TAKD indicates KD with the TA method proposed above) From the analysis of the following figure, we know that the TA distilled from the teacher network (KD-TA) is better than the TA trained directly (FS-TA).

The TA results obtained from multiple distillations are as follows, the more multiple stages the better the effect, but the more time and space consuming.

In conclusion, the TA approach used, with layers of distillation, where the next layer absorbs the knowledge of the previous layer and passes it to the next layer, is a good solution to the problem that an overly strong teacher network provides some knowledge beyond what the student network can learn, resulting in lower efficiency, and helps us to select a teacher network of the right size.

3. Inverse probability weighted distillation

Generally speaking, the knowledge embedded in categorical losses satisfies the assumption of independent homogeneous distribution, but the probability distribution of teachers'predictions is unbalanced across categories, and the presence of transfer gaps for both "strong" and "weak" teachers hinders the transfer ofknowledge in the tail categories of teachers'predictions. Transfer is hampered by the presence of transfer gaps. Therefore, the knowledge embedded in the teachers'predictions does not satisfy the assumption of independent homogeneous distribution across categories. The original KD ignores this transfer gap and assigns constant weights to categorization loss and distillation loss. Since teachers'knowledge is unbalanced, constant sample weights for distillation losses would be a bottleneck in knowledge transfer. Inverse probability weighted distillation (IPWD) [14] is a simple and effective way to supplement underweighted training samples in the machine domain. The propensity score of the machine domain is first estimated by comparing class-aware predictions with context-aware predictions . Then, IPWD uses the inverse probabilities as distillation loss sample weights to strengthen the weights of the underweighted samples. In this way, IPWD generates a pseudo-population of samples to handle unbalanced knowledge.

There is an imbalance problem in teachers'knowledge. CIFAR100 and ImageNet are used as examples, and the prediction probability sums of the teacher models for different categories in the training samples are counted. For example, a sample of a dog with a soft label of , then the corresponding values are accumulated in class and class, respectively. Figure 1 reflects the difference in distribution between ground-truth and teacher predictions, although the teacher model is trained on balanced data (blue dashed line), its prediction distribution is unbalanced across temperatures.

Also, the categories will be divided into 4 groups on CIFAR100 by teacher prediction probability and ranking, and knowledge distillation will be performed separately. As shown in Table 1, the KD achieved better performance in all subgroups compared to normal training. However, the first 25 categories brought much higher gains than the last 25 categories (mean 5.14% vs. 0.85%). This demonstrates that there is indeed a category imbalance in the knowledge implied by the teachers'predictions. Thus, the training samples in the human domain are no longer in the machine domain. Simply assuming that the training set is a perfect transfer set may lead to selection bias: samples matching "head" knowledge are over-represented, while samples matching "tail" knowledge are under-represented. This will inhibit the transfer of "tail" knowledge. Therefore, the inverse probability weighting (IPW) technique can be used to overcome the confusion effect caused by the transfer gap. In short, IPW is used to generate a pseudo-all sample, assigning larger weights to under-represented samples and smaller weights to over-represented samples to achieve de-biased distillation:

IPWD proposes an unsupervised way to estimate the propensity score in the machine domain , using the human domain using the CLS-trained classification head as a reference to compare with the output of the KD-trained classifier to determine whether the samples are underrepresented in the machine domain. The approach in this paper trains an additional classification head to calculate , and the propensity score is calculated as follows:

IPWD The final objective function can be expressed as

IPWD achieves the best performance in the vast majority of cases, and IPWD leads the other KD methods by a greater margin in heterogeneous distillation, with the following experimental results.

4. Early exit of KD

By comparing the training curves of ImageNet under distillation and without distillation for ResNet18, we find that KD is initially useful, but later becomes worse than scratch, regardless of whether the teacher network chooses ResNet34 or ResNet50. The following two plots directly negate the validity of KD, so is ImageNet a more difficult dataset to perform knowledge Is ImageNet therefore a more difficult dataset for knowledge distillation?
Based on the above observations, the ESKD method is used, and the results are as follows. At this
point, KD does work, but the degradation phenomenon still exists as the network depth increases. The training curve of ESKD [2] is as follows:
The teacher is not fully trained, and the performance of the incompletely trained teacher is the same as the small network. The experiment for the fully trained teacher is set to 200 epochs and 60 epochs will learn the rate (60/200), while the incompletely trained teacher follows the strategy of (15/50, 10/35), etc. The following figure shows the effect of the incompletely trained teacher used for distillation, and it can be seen that both are better than the fully trained teacher.
The effectiveness of ES teacher is also verified with different student and teacher networks in
the following figure:
The first method is to end the KD process early, while the second method is to actively reduce the capacity of the teacher network to match the student network, both of which improve on the original KD.

5. Decoupled knowledge distillation

Decoupled knowledge distillation (DKD) [17] is a new approach to knowledge distillation, which divides the traditional knowledge distillation loss into two parts: target class knowledge distillation (TCKD) and non-target class knowledge distillation (NCKD). The results show that TCKD conveys knowledge related to the difficulty of the training samples, while NCKD is important for Logit distillation.DKD is proposed as a more efficient and flexible method for implementing TCKD and NCKD, achieving comparable or better results than feature-based methods on image classifica- tion and target detection tasks. Decoupled knowledge distillation (DKD) reformulates knowledge distillation (KD) as a weighted sum of two parts, one related to the target class and the other unrelated to the target class. The reformulation uses binary probabilities and probabilities between non-target classes to separate predictions that are relevant and irrelevant to the target class. These two parts are named Target Class Knowledge Distillation (TCKD) and Non-Target Class Knowledge Distillation (NCKD), respectively. the weights of NCKD are coupled with the probabilities of the target classes. The reformulation stimulates the study of the individual effects of TCKD and NCKD, revealing the limitations of the classical coupling formulation. The role of TCKD and NCKD in logit distillation is that TCKD conveys knowledge related to the target class, while NCKD focuses on knowledge between non-target classes. The effectiveness of TCKD becomes apparent when the training data becomes challenging, while NCKD is crucial for logit distillation. However, the loss weight of NCKD is suppressed by the teacher's confident predictions. A study showed that better performance was obtained using NCKD on a good prediction sample, suggesting that the good prediction sample was more knowledgeable than the other samples. By considering TCKD and NCKD independently, decoupled knowledge distillation (DKD) is proposed to address these issues. The classical knowledge distillation (KD) approach has some limitations, i.e., the knowledge transfer between teacher and student models is coupled and cannot be balanced. Two types of knowledge transfer, difficulty-based and non-target-based, are considered critical. However, the latter is inhib- ited when teachers are confident in their predictions. To address this problem, researchers have proposed a new approach, called decoupled knowledge distillation (DKD), which considers each type of knowledge transfer independently and allows the weights to be adjusted.DKD provides an efficient and flexible method for logarithmic distillation, and Algorithm 1 provides the pseudo-code to implement it.
Experiments are described next for two tasks, image classification and object detection, using datasets such as CIFAR-100, ImageNet and MS-COCO. experiments explore the effectiveness of the DKD method, which improves the performance of knowledge distillation by decoupling certain factors. The results show that DKD can consistently improve the performance of teacher-student pairs with the same and different architectures, and obtain comparable or even better performance than feature-based distillation methods. DKD also improves target detection performance when combined with feature-based distillation methods.
Extensions to DKD, a logit distillation method for image classification and target detection tasks, are discussed next. These extensions include evaluating the training efficiency of DKD compared to other state-of-the-art methods, providing a new perspective on why bigger models are not always better teachers, evaluating the transferability of deep features learned by DKD, and showing visualizations that demonstrate the advantages of DKD. Experimental results show that DKD provides significant improvements on a variety of datasets and tasks.
In summary, this new approach to interpreting logit distillation decomposes the classical KD loss into target class knowledge distillation (TCKD) and non-target class knowledge distillation (NCKD). The limitations of the coupled knowledge decomposition formulation are discussed, and decoupled knowledge distillation (DKD) is proposed as a solution.DKD achieves significant improvements on image classification and target detection tasks on CIFAR-100, ImageNet, and MS-COCO datasets. However, DKD cannot outperform the state-of-the-art feature-based methods for the target detection task. This

paper provides guidance for tuning in supplements, but further research is needed.

5.1 Symmetric temperature scaling

The classical KD method uses the Kullback-Leibler scatter to minimize the difference between the output probabilities of teachers and students. However, recent studies have shown that more accurate teachers do not necessarily teach better, and the reasons for this remain unknown. In the following, a new asymmetric temperature scaling (ATS) method is proposed to improve the clarity of teacher-provided error class probabilities and make large teachers teach better. There are various approaches to knowledge distillation, which include transferring knowledge from complex teachers to smaller students through a combination of cross-entropy and distillation loss. Some studies have examined the dependence of knowledge transfer on student and teacher architecture and found that larger models are not always better teachers. Other work has focused on understanding the advantages of knowledge distillation from a principles-based perspective. The paper also explains the notation used, such as softmax functions, logarithms, and probabilities. In addition, a particular phenomenon where disadvantaged students are unable to fully imitate good teachers due to a mismatch of abilities is discussed and explained in detail. Knowledge distillation can be decomposed into three parts, namely, correct guidance, smooth regularization, and category discrimination. These three terms are measured quantitatively using the target category probability, the mean of the error category probability, and the variance of the error category probability. An appropriate temperature scaling method that incorporates the validity of these three terms simultaneously is presented here. Category defensibility is considered to be the basis of KD. A theoretical analysis of the application of temperature scaling to knowledge distillation is presented below. The softening probability vector used in KD is analyzed, showing that an increase in temperature leads to a more uniform probability distribution, and the concepts of intrinsic variance (IV) and derived variance (DV) are also introduced to analyze the class-discriminable terms. And further explains why larger teachers cannot teach better, showing that the effectiveness of KD is more related to differences between error classes than to all classes. A new temperature scaling method, the asymmetric temperature scale (ATS), is proposed here to make the distribution of error classes during distillation more discriminable. the ATS applies different temperatures to logits in correct and incorrect classes, allowing for more flexible and differentiated distillation instruction during instruction. This approach is particularly beneficial when teachers are overconfident, as it reduces the size of correct category logits, increases the diversity of error category logits, and ultimately increases the variance and accuracy of the derivations. Experiments using different datasets and network architectures are described here, including ResNet, WideResNet, ResNeXt, VGG, ShuffleNet, AlexNet, MobileNet, and DSCNN. the training setup mainly follows previous studies with slight modifications based on the dataset. The learning rate decayed every 30 cycles, with a batch size of 128 for CIFAR and 64 for the other datasets. The researchers investigated the phenomenon that having more accurate teacher models does not necessarily lead to better student model learning in knowledge distillation. They identified three factors of successful teachers:correct instruction, smooth regularization, and class differentiation. They propose a solution called asymmetric temperature scaling (ATS) to enhance the derived variance of large teachers to make their distillation labels more discriminatory when teaching. Experiments have shown that the method is effective. So far, researchers are studying why large teachers have difficulty teaching knowledge distillation (KD) effectively. They hope their research provides a new perspective on the KD field, but they do not anticipate any negative social impact from their work.

5.2 DIST

DIST method [24], i.e., better knowledge is extracted from stronger teachers. Existing methods perform poorly when the predictions of students and stronger teachers do not match exactly in KL dispersion. In this paper, we propose a correlation-based loss to explicitly capture the intrinsic inter- class relationships of teachers and extend this matching to the intra-class level. The method is simple and practical, and experiments show that it adapts well to various architectures, model sizes, and training strategies, consistently maintaining state-of-the-art performance on image classification, target detection, and semantic segmentation tasks. The most commonly used approach is to use KL scatter to match probabilistic prediction scores between teachers and students. However, this approach can be challenging when teacher and student models differ significantly in size or training strategy. Here, a new method, DIST, is proposed to extract intra-class relationships using Pearson correlation coefficients instead of KL scatter to further improve performance. This method outperforms the common KD and state-of-the-art KD methods for various tasks such as image classification, target detection, and semantic segmentation. The concept of knowledge distillation (KD), which involves transferring knowledge from a pre-trained teacher model to a student model by minimizing the difference between prediction scores, is discussed here, highlighting the importance of balancing raw classification loss and KD loss in training students, and also discussing the challenges of using stronger teachers in KD, which may lead to greater differences between teachers and students and make it difficult to achieve exact matching. Therefore, an easy way of matching teacher-student predictions is proposed here to solve this problem. Here the "Distillation from a Stronger Teacher" (DIST) approach can be used to improve knowledge distillation in deep learning. It proposes a loose matching prediction based on the relative rank between teachers and students rather than the exact probability value, and introduces inter-class and intra-class relationship loss to transfer the relationships of multiple classes and similarities of multiple instances to one class per instance, respectively. The method achieves significant improvements in the baseline settings for image classification tasks such as CIFAR-100 and ImageNet. Experimental setups and results for image classification using the DIST method are described here, which outperforms previous knowledge distillation (KD) methods to handle the large differences between teacher and student models. The method is also shown to be effective on stronger training strategies and larger models, such as the state-of-the-art swing - transformer. results on the CIFAR-100 dataset show that DIST outperforms feature-based distillation methods. The next describes experiments performed on the MS COCO object detection dataset, using DIST as an additional supervision on the final prediction of classes. The results show that DIST achieves competitive results on the COCO validation set and significantly outperforms vanilla KD. by combining DIST with mimic, state-of-the-art KD methods designed for object detection can be surpassed. The latter experiments on two different neural network models (DeepLabV3 and PSPNet) are used to perform semantic segmentation on the cityscape dataset. The proposed DIST method is applied to the prediction of classification heads using the ResNet-101 backbone teacher of DeepLabV3. The results show that DIST outperforms existing knowledge distillation methods on the semantic segmentation task. Two types of relations are also proposed:interclass and intraclass, and experiments show that both of them outperform vanilla knowledge distillation. The proposed DIST significantly outperformed vanilla knowledge distillation without ground-truth labels when students were trained only on knowledge distillation loss. In summary, the new knowledge distillation method, DIST, aims to improve knowledge distillation from a stronger teacher's perspective. In this paper, we address the problem of differences between students and teachers and propose relationship-based loss to solve this problem.The DIST method is simple and effective in dealing with strong teachers and outperforms the state-of-the-art KD method in various benchmarking tasks including object detection and semantic segmentation.

5.3 Curriculum distillation

The classical KD method, which minimizes the loss of kl dispersion between two predictions obtained from the teacher/student model, has a fixed temperature in the softmax layer. A new approach, called Course Temperature for Knowledge Distillation (CTKD), is proposed here, which improves the performance ofknowledge distillation by gradually increasing the learning difficulty of students through dynamic and learnable temperatures. The loss of distillation between the instructor and the student is maximized by learning the temperature in a reverse gradient during the student's training. The method is easy to implement and achieves overall improvement at negligible additional computational cost. The concepts of two domains, course learning and knowledge distillation, are discussed here. Course learning is a training strategy that organizes learning tasks into incremental difficulty and has applications in several fields such as computer vision and natural language processing. Knowledge distillation is a method for transferring knowledge from a pre-trained teacher model to a smaller student model. Existing approaches have been designed to transfer knowledge in various forms, including logic-based, representation-based, and relational-based approaches. Temperature hyperparameters control the difficulty level of the distillation process, with lower values concentrated on the maximum logarithm and higher values flattening the distribution. Most of the work fixes temperature as a hyperparameter, but recently MKD proposes to learn temperature through meta-learning. The proposed CTKD method is more efficient than the MKD method and can be seamlessly integrated into the existing KD framework. The technique of knowledge distillation course temperature (CTKD) consists of a learnable temperature module to predict the suitable temperature for distillation and a gradient inversion layer to invert the gradient of the temperature module during back propagation. The technique uses an easy-to-hard curriculum that progressively increases the learning difficulty for students. A method called adversarial distillation is involved here, which involves optimizing the student model to minimize task-specific losses and distillation losses. Distillation losses are calculated using a dynamic temperature value called tau. The authors suggest using an adversarial approach to learn the optimal temperature value where the temperature module is optimized in the opposite direction of the student to maximize the distillation loss. The optimization process consists of solving two subproblems alternately using stochastic gradient descent and using a nonparametric gradient inversion layer to implement the adversarial process. The course temperature approach organizes the distillation task from easy to difficult, gradually increasing the magnitude of the loss relative to the temperature. This is inspired by course learning, and students learn better when tasks are organized in a meaningful order, starting with basic concepts and gradually presenting more advanced ones. A cosine table is used to gradually increase , and a learnable temperature module is introduced to predict temperature values, either globally or by instance. The sub-instance temperature method has a better representation capability and thus better distillation performance. The researchers evaluated their CTKD technique on various neural networks (VGG, ResNet, Wide ResNet, ShuffleNet, and MobileNet) and compared it with other distillation frameworks, including vanilla KD, PKT, SP, VID, CRD, SRRL, and DKD. using a standard experimental setup. Evaluation was performed on the CIFAR-100, ImageNet-2012 and MS-COCO datasets. The results are reported as the mean (standard deviation) of three experiments. Here the results of the proposed CTKD method are applied to various tasks such as CIFAR-100 classification, ImageNet image classification and MS-COCO object detection. The method is shown to significantly improve the performance of different student networks and its effectiveness is demonstrated by loss curves, feature separability and learning curves. The method can be applied as a plug-in to existing distillation devices and can improve their performance without increasing the computational cost. The results show that CTKD can further improve the detection performance in MS-COCO target detection. The study used an ablation study on CIFAR-100 on the effectiveness of distillation method hyperparameters and components to assess the effectiveness of course parameters, strategies, and counteracting temperature and course distillation. The results showed that a smooth increase in task difficulty and the use of a cosine course strategy worked best. Learning the temperature parameters in an adversarial manner also improved distillation performance, and the pairing of adversarial temperature with course distillation was more effective than for a single element. Course temperature for knowledge distillation (CTKD) utilizes a dynamic, learnable temperature to organize knowledge distillation tasks from easy to difficult. The temperature is learned in an adversarial manner during student training to maximize distillation losses. CTKD can be easily integrated into existing knowledge distillation frameworks and leads to general improvements at negligible additional computational cost.

5.4 Masked generative distillation

The Masked Generative Distillation (MGD) method [23] guides the student's feature recovery by masking random pixels and forcing it to generate the teacher's full features through a simple block.MGD is a general feature-based distillation method that can be used for different tasks. It is demonstrated here that MGD achieves excellent improvements on various models for a wide range of datasets such as image classification, target detection, semantic segmentation, and instance segmentation. This paper provides impressive results, such as improving the ImageNet top-1 accuracy of ResNet-18 from 69.90% to 71.69%.

Masked Generative Distillation (MGD) involves students generating the teacher's features using the teacher's masked features instead of mimicking it directly. This approach improves the repre- sentation of student features and has been shown to bring considerable improvements for a variety of tasks, including image classification, object detection, semantic segmentation, and instance segmentation. The method is simple to use and has only two hyperparameters. The effectiveness of the method on different datasets has been verified through a large number of experiments.

Different approaches to knowledge distillation in machine learning are discussed here, where knowledge is transferred to a student model using a teacher model. The different approaches include extracting information from intermediate layers, attention transfer and contrast learning, and specific applications of knowledge distillation in object detection and semantic segmentation are also mentioned, where the challenge is to determine where to extract information from because of the imbalance between foreground and background. MGD generates the teacher's feature map by using a random mask to overlay the student's feature map and then tries to generate the teacher's feature map using the projector layer with the left side pixels. The proposed MGD distillation loss formulation is designed to make the students generate the teacher's features instead of imitating them.MGD can be easily applied to a variety of tasks and is effective in both classification and intensive prediction tasks.The total loss of MGD is a combination of the original loss and distillation loss.

The next experiments use the Masked Generation Distillation (MGD) feature-based distillation method for various tasks including classification, object detection, semantic segmentation, and instance segmentation. For the classification task, the method is evaluated on the ImageNet dataset and distillation loss is computed on the last feature map of the backbone. The results show that MGD provides a significant improvement in accuracy compared to other feature-based and logarithm- based knowledge distillation methods. The hyperparameters used are and the model is trained for 100 epochs using the SGD optimizer. The experiments were conducted using Pytorch-based MMClassification and MMRazor. Experiments on the COCO2017 dataset for target detection and instance segmentation are described next. Various distillation methods are compared and the results are evaluated for average precision. The authors used the MMDetection framework with an inheritance strategy to initialize the students with the teacher's parameters. The results show that their proposed method, Masked Generation Distillation (MGD), outperforms other state-of-the-art methods in terms of target detection and instance segmentation. The ResNet-50-based retinal net and SOLO models achieved significant improvements with MGD, improving by 3.6 Boundingbox mAP and 3.1 Mask mAP, respectively, on the COCO dataset.The following describes a study on semantic segmentation in which the researchers evaluated their method using the CityScapes dataset. They trained a teacher model (PspNet-Res101) and two student models (PspNet-Res18 and deeplabp3 - res18) and used distillation to transfer knowledge from the teacher to the student. The results showed that their method outperformed state-of-the-art distillation methods, with both homogeneous and non-homogeneous distillation resulting in significant improvements for students. The researchers used MGD, a feature-based distillation method, and combined it with a logit-based distillation method to further improve the results. The model was evaluated using the mIoU metric, and experiments were conducted using MMSegmentation. Researchers are comparing two approaches to extracting knowledge from teacher networks and student networks. One method directly mimics the teacher's feature map, while the other uses Masked Generation Distillation (MGD), which forces students to use their masked features to generate the teacher's full feature map. The researchers found that students achieved better performance and accuracy with MGD even when the teacher was themselves, while the improvement from direct imitation was negligible. The researchers also visualized training loss curves for both methods, showing that the students'feature maps gained stronger representation using MGD. At the time of the study, random channels in the masked global distillation (MGD) image classification method were masked instead of spatial pixels. They found that this method improved the performance of the student model. They also investigated the effect of using different teachers on knowledge distillation and found that stronger teachers with similar architectures were better suited for feature-based distillation, while teachers with high accuracy but different architectures were not as effective. The researchers used a generative block called MGD to recover features using two convolutional layers and one activation layer, ReLU. they explored the effects of different compositions of generative blocks and chose the architecture with two convolutional layers and one activation layer. The method can also be applied to other stages of the model, and refining deeper stages is more beneficial for students. The current study is insensitive to the hyperparameter alpha, but when the lambda is less than 0.5, the students have higher performance with larger ratios, while when the lambda is too large, the semantic information on the left side is too poor to generate a complete feature map. In summary, the masked generation distillation (MGD) method allows students to use the teacher's masked features to generate the teacher's features instead of directly imitating the teacher's features. The algorithm has enhanced representational capabilities and can be applied to various tasks such as image classification, target detection, semantic segmentation, and instance segmentation. Numerous experiments including on various models and datasets have demonstrated the simplicity and effectiveness of the method.

6. Simple knowledge distillation

The use of Simple Knowledge Distillation (SimKD) allows the compression of powerful teacher models into lightweight student models without sacrificing performance. This is achieved by reusing the discriminative classifier of the teacher model for student inference and training the student encoder by feature alignment with a single loss. A projector was also developed to help match the student encoder with the teacher classifier, making the technique applicable to a variety of teacher and student architectures. Experiments show that with the addition of the projector, the technique achieves state-of-the-art results at the cost of a reduced compression ratio. Vanilla KD aligns the logit or class projections of the two models, but the performance gap between the original teacher model and the refined student model is still large. Various methods have been proposed to overcome this problem, but they require detailed knowledge representation and optimized hyperparameters. The SimKD technique is proposed, which trains the student model by feature alignment in the previous layer of the classifier and directly replicates the teacher classifier for student inference. On a standard benchmark dataset, SimKD outperforms all state-of-the-art methods compared with various combinations of teacher-student architectures. The concept of knowledge distillation (KD), a technique for compressing knowledge from a powerful teacher model into a smaller student model, is discussed here. Transferred knowledge, usually in the form of soft targets, is thought to capture the relationships between different categories and serve as effective regularization during student training. Feature distillation is a common solution to prevent performance degradation in teacher-to-student compression, which utilizes more information from the middle layer of the teacher model. One of the proposed methods, SimKD, is related to hypothetical transfer learning (HTL) and aims to close the performance gap between teachers and students on the same dataset. Here, by analyzing the limitations of the existing method and focusing on its improvement, a new method, logit distillation, is proposed, and the knowledge distillation loss function is re-represented as a weighted sum of two components, target class knowledge distillation (TCKD) and non-target class knowledge distillation (NCKD). The effectiveness of each component is investigated and compared with vanilla knowledge distillation. A new technique called simple knowledge distillation (SimKD) is discussed here, which allows for state-of-the-art results without the need for complex knowledge representations or hyperparameters.SimKD involves student inference using a pre-trained teacher classifier, which eliminates the need for labeling information and makes feature alignment the only source of generated gradients. The discriminative information contained in the teacher classifier is considered important here, but has been neglected in the knowledge distillation (KD) literature. It is assumed that there is ability-invariant information in the data that can be easily obtained across different models, while the teacher model contains additional necessary ability-specific information that is difficult to obtain for a simple student model. Empirical evidence is provided here to support this hypothesis and to show that SimKD greatly mitigates the performance degradation in teacher-student compression.

6.1 Summary

In this paper, we reviewed recent developments in knowledge distillation[13, 15, 19, 22], a tech- nique for compressing complex deep neural networks (DNNs) into smaller and faster DNNs while preserving accuracy. We discussed four variants of knowledge distillation, including teaching assistant distillation, curriculum distillation, mask distillation, and decoupling distillation. Teach- ing assistant distillation introduces an intermediate model between the teacher and the student, while curriculum distillation designs the learning process to follow a curriculum. Mask distillation focuses on transferring the attention mechanism learned by the teacher, and decoupling distillation decouples the distillation loss from the task loss. These variants of knowledge distillation have shown promise in improving the performance of knowledge distillation, making it a valuable tool for model compression and acceleration.

References

1. Kunlong Chen, Liu Yang, Yitian Chen, Kunjin Chen, Yidan Xu, and Lujun Li. GP-NAS-ensemble: a model for the NAS Performance Prediction. In CVPRW.(2022)
2. Jang Hyun Cho and Bharath Hariharan. On the efficacy of knowledge distillation. In ICCV.(2019)

3. Peijie Dong, Lujun Li, and Zimian Wei. DisWOT: Student Architecture Search for Distillation WithOut Training In CVPR.(2023)
4. Peijie Dong, Xin Niu, Lujun Li, Zhiliang Tian, Xiaodong Wang, Zimian Wei, Hengyue Pan, and Dongsheng Li.RD-NAS: Enhancing One-shot Supernet Ranking Ability via Ranking Distillation from Zero-cost Proxies. arXiv preprint arXiv:2301.09850 (2023).
5. Peijie Dong, Xin Niu, Lujun Li, Linzhen Xie, Wenbin Zou, Tian Ye, Zimian Wei, and Hengyue Pan. Prior-Guided One-shot Neural Architecture Search. arXiv preprint arXiv:2206. 13329 (2022).
6. Peijie Dong, Xin Niu, Zhiliang Tian, Lujun Li, Xiaodong Wang, Zimian Wei, Hengyue Pan, and Dongsheng Li. Progressive Meta-Pooling Learning for Lightweight Image Classification Model. arXiv preprint arXiv:2301. 10038 (2023).
7. Geoffrey Hinton, Oriol Vinyals, and Jeff Dean. Distilling the Knowledge in a Neural Network. arXiv preprint arXiv:1503.02531 (2015).
8. Yiming Hu, Xingang Wang, Lujun Li, and Qingyi Gu. Improving one-shot NAS with shrinking-and-expanding supernet. Pattern Recognition (2021).
9. Lujun Li. Self-Regulated Feature Learning via Teacher-free Feature Distillation. In ECCV.(2022)
10. Lujun Li and Zhe Jin. Shadow Knowledge Distillation: Bridging Offline and Online Knowledge Transfer In NeuIPS.(2022)
11. Lujun Li, Liang Shiuan-Ni, Ya Yang, and Zhe Jin. Boosting Online Feature Transfer via Separable Feature Fusion In IJCNN.(2022)
12. Lujun Li, Liang Shiuan-Ni, Ya Yang, and Zhe Jin. Teacher-free Distillation via Regularizing Intermediate Representation. In IJCNN.(2022)
13. Lujun Li, Yikai Wang, Anbang Yao, Yi Qian, Xiao Zhou, and Ke He. Explicit Connection Distillation. In ICLR. (2020)
14. Li Liu, Qinwen Huang, Sihao Lin, Hongwei Xie, Bing Wang, Xiaojun Chang, and Xiao-Xue Liang. Exploring Inter-Channel Correlation for Diversity-preserved Knowledge Distillation. ICCV (2021).
15. Yifan Liu, Changyong Shun, Jingdong Wang, and Chunhua Shen. Structured knowledge distillation for dense prediction. arXiv preprint, arXiv:1903.04197 (2019).
16. Seyed Iman Mirzadeh, Mehrdad Farajtabar, Ang Li, Nir Levine, Akihiro Matsukawa, and Hassan Ghasemzadeh. Improved knowledge distillation via teacher assistant. In AAAI.(2020)
17. Wonpyo Park, Yan Lu, Minsu Cho, and Dongju Kim. Relational Knowledge Distillation. In CVPR.(2019)
18. Jie Qin, Jie Wu, Xuefeng Xiao, Lujun Li, and Xingang Wang. Activation Modulation and Recalibration Scheme for Weakly Supervised Semantic Segmentation. In AAAI.(2022)
19. Tao Wang, Li Yuan, Xiaopeng Zhang, and Jiashi Feng. Distilling object detectors with fine-grained feature imitation. In CVPR.(2019)
20. Zimian Wei, Hengyue Pan, Lujun Li Li, Menglong Lu, Xin Niu, Peijie Dong, and Dongsheng Li. ConvFormer: Closing the Gap Between CNN and Vision Transformers. arXiv preprint arXiv:2209.07738 (2022).
21. Liu Xiaolong, Li Lujun, Li Chao, and Anbang Yao. NORM: Knowledge Distillation via N-to-One Representation Matching. In ICLR.(2023)
22. Guodong Xu, Ziwei Liu, Xiaoxiao Li, and Chen Change Loy. Knowledge distillation meets self-supervision. In ECCV.(2020)
23. Kaiyu Yue, Jiangfan Deng, and Feng Zhou. Matching Guided Distillation. ArXiv (2020).
24. Sukmin Yun,Joon-Seok Park, Kimin Lee, andJinwoo Shin. Regularizing Class-Wise Predictions via Self-Knowledge Distillation. In CVPR.(2020)

MATEC Web of Conferences **382**, 01035 (2023) https://doi.org/10.1051/matecconf/202338201035
AMME 2023

Influence of Material Characteristics of Hoppers on Particle Flow Behavior

Panpan Zhang [1, 2], Guoqing Huang [1, 2], Qingwei Chen [1]

[1] China Waterborne Transport Research Institute, Beijing,100088, China.
[2] Key Laboratory of Logistics Equipment & Control Engineering, Beijing 100088, China.

Abstract: In the previous study, different structures of hoppers were proposed. And the dust suppression performance of hoppers were evaluated. In this paper, different materials of hoppers were selected.And influence of different material on particle behaviors were studied based on the discrete element method (DEM). Results shows that the wall material with low collision recovery coefficient for the hopper has positive effect on suppressing dust diffusion.and the wall material with high static friction coefficient showed little effect on the particle flow state through the hopper.

1. Introduction

In the previous research ,in order to access the particle behavior of different hoppers[1-2], different structures of hoppers were designed, and the discrete element method was used via the software package EDEM to evaluate the particle behaviors. The hopper shown in Figure 1 is the same as the one used in ports. Unlike the common hopper, a deflection cone as shown in Figure 4 was set in the hopper to change the flow of materials. Besides,In recent years, researchers and some companies focus on the study of particle behavior and dust suppression [5-7].

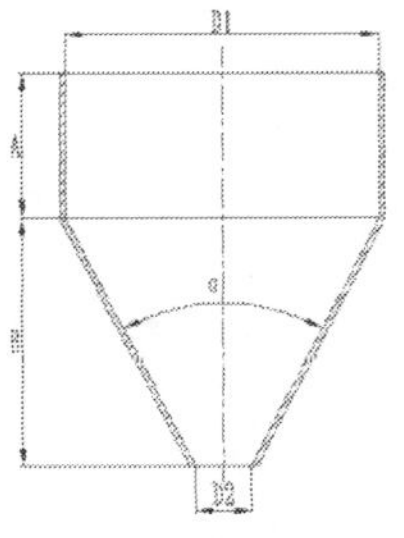

Fig. 1. The hopper

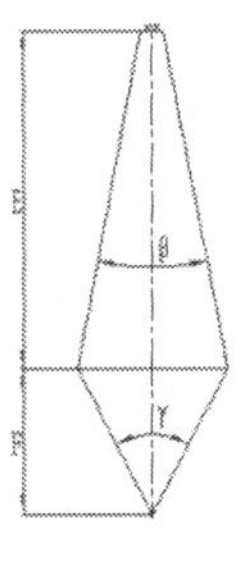

Fig. 2. The deflection cone

Results shows that structure of the hopper greatly affected the particle behavior. After adding the deflection cone in the hopper, the trajectory of particle material can be effectively restrained and the dust diffusion can be reduced. The annular channel formed by the deflection cone and the inner wall of the hopper has a guiding effect on the particle movement as shown in Figure 3.

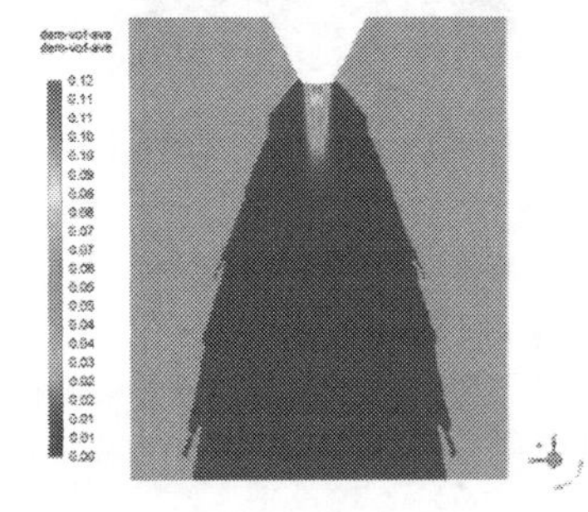

(a)the hopper without the deflection cone

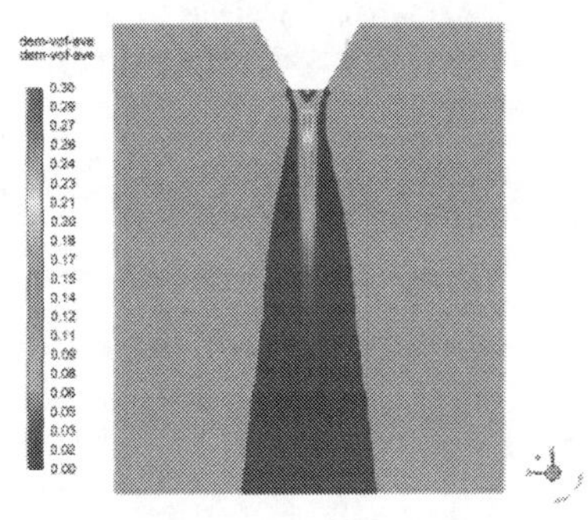

(b)the hopper with the deflection cone

Fig. 3. The percentage of dust particle for different hoppers

https://doi.org/10.1051/matecconf/202338201035

Based on previous study, different materials of hoppers were selected to investigate particle flow behavior in this paper. For different materials, different recovery coefficient and different static friction coefficient were chosen as main factors and influence on the particle behavior were studied based on the discrete element method (DEM).

2. Simulation setting

Different materials can be selected as the inner wall of dust suppression hopper. Therefore, it is necessary to study influence of different materials on the particle flow of the hopper. In this research, it is assumed that two kinds of wall materials have different collision recovery coefficients or static friction coefficients compared with the original materials.
1) Different impact recovery coefficient material: impact recovery coefficient of material A is 0.72, impact recovery coefficient of material B is 0.1
2) Different static friction coefficient material: static friction coefficient of material A is 0.2, and static friction coefficient of material B is 0.01.
The hopper without the deflection cone was selected in this study. The material flow rate of 100 mm dust suppression hopper was 10 t/h.
Calculation domain of the hopper was set as a cylinder under the outlet of the hopper with 600mm in height and 600mm in diameter as shown in Figure 4. During the simulation, Hertz-Mindlin non-slip model was adopted.

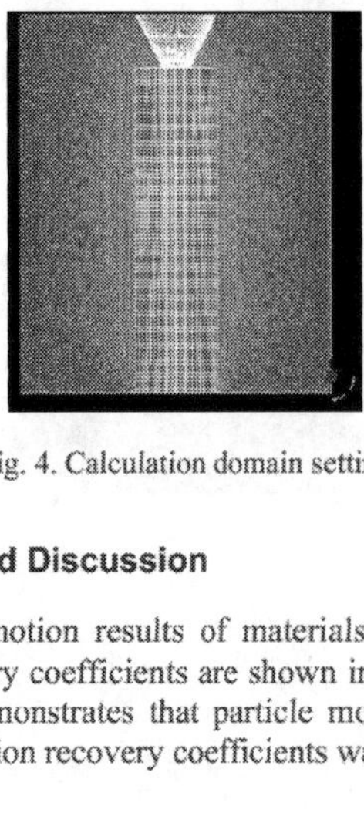

Fig. 4. Calculation domain setting

3 Results and Discussion

The particle motion results of materials with different impact recovery coefficients are shown in Fig.5. and the simulation demonstrates that particle motion difference between collision recovery coefficients was obvious.

(a) collision recovery coefficient of 0.72

(b) collision recovery coefficient of 0.1

Fig. 5. Particle motionof materials with different static friction coefficients

In order to carry out further evaluation, particle volume fractions of materials with different collision recovery coefficients were shown in Fig.6 and Fig.7 and it shows that the wall material with low collision recovery coefficient has obvious effect on suppressing dust diffusion, because the low collision recovery coefficient can weaken the rebound of particles falling behind the wall, so that the particles can form a more consistent wall-attached flow, and thus it generated a beam closure effect on the material flow.

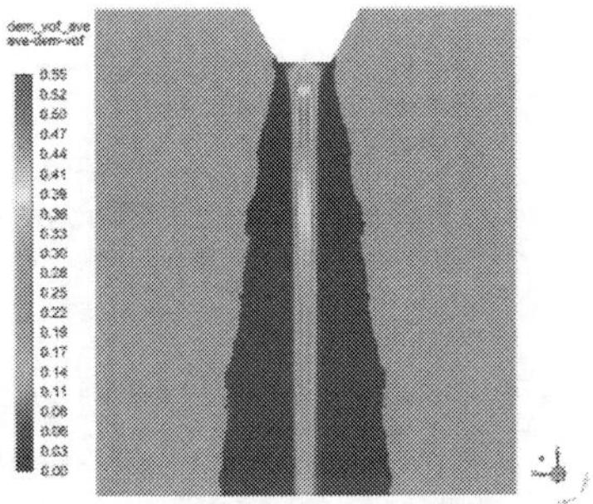

(a) collision recovery coefficient of 0.72

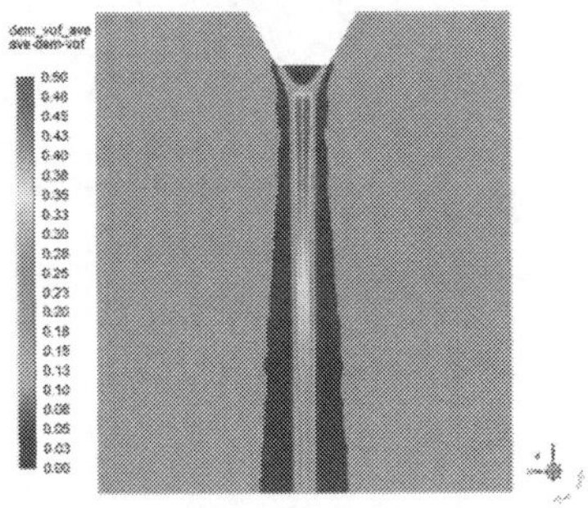

(b) collision recovery coefficient of 0.1

Fig. 6. Particle volume fractions of materials with different collision recovery coefficients

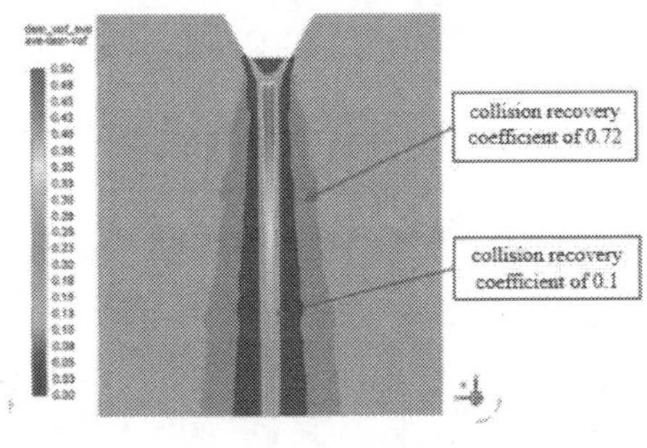

Fig. 7. Particle diffusion range of materials with different collision recovery coefficients

Besides, different static friction coefficients of materials were taken into consideration.and the simulation results of materials with different static friction coefficients are shown in the Fig. 8. and it shows that particle motion difference between two static coefficients was little. For further evaluation, particle volume fractions of materials with different collision recovery coefficients was shown in Fig.9 and Fig.10 and it is shown that the similar trend to the particle motion occurred. In this case, hopper materials of the static friction coefficients has little effect on the particle flow state, and the particle volume fractions range for material with a higher static friction coefficient was just slightly larger under low friction coefficient.

(a) static friction coefficient of 0.2

(b) static friction coefficient of 0.01

Fig. 8. Particle motion state of materials with different static friction coefficient

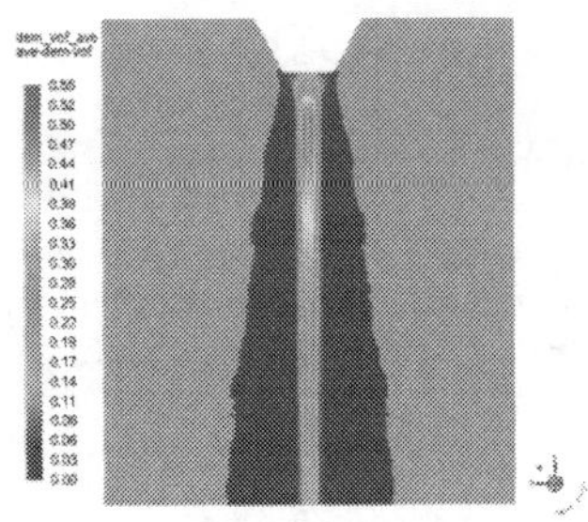

(a)static friction coefficient of 0.2

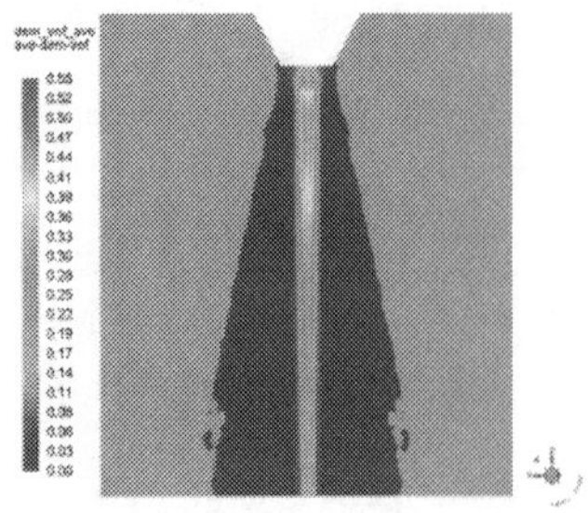

(b)static friction coefficient of 0.01

Fig. 9. Particle volume fraction distribution of materials with different static friction coefficients

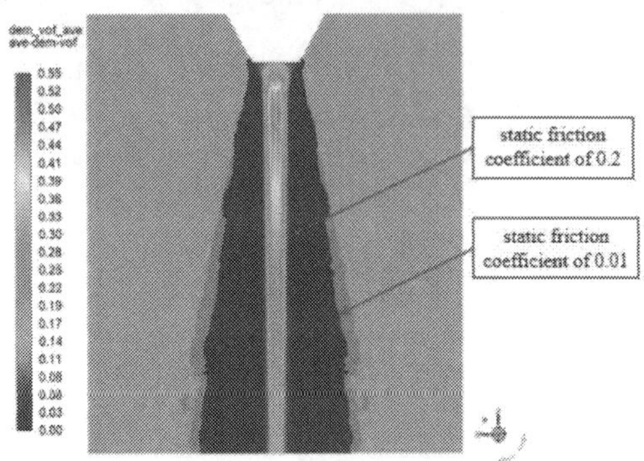

Fig. 10. Particle diffusion range of different static friction coefficient materials

3. Summary

In the previous study, different structures of hoppers were proposed and performance on particle flow were evaluated. In this paper, influence of material of hopper on particle flow behaviors were studied. And the dust suppression performance of hoppers were evaluated. In this paper, different materials of hoppers were selected and influence of different material characteristic on the particle behavior were studied based on the discrete element method (DEM). it reveals that low collision recovery coefficient of the hopper has positive effect on suppressing dust diffusion.and the wall material with high static friction coefficient showed little effect on the particle flow state through the hopper. In future, more factors that may affect particle flow behaviors of hoppers will be taken into account and provide further instruction for hopper design.

References

1. Panpan Zhang, Simulation of Particle Behavior for Different Hoppers, Advances in Engineering Research, Vol.125(2017):228-231.
2. Panpan Zhang,Particle Behavior Investigation for Different Structures of Hoppers,MATEC Web of Conferences 353, 01015 (2021)
3. Ogata K., Funatsu K., Tomita Y.Experimental investigation of a free falling powder jet and the air entrainment . Powder Technology 2001, (115): 90-95 .
4. Wypych P., Cook D., Cooper P. Controlling dust emissions and explosion hazards in powder handling plants [J]. Chemical Engineering and processing.2005, 44:323-326.
5. Liu Z.Q., Ma L.Y., Zhao X.Y. . Experimental Study the affection of the Dust Emission on Sand Stockpile by the Height of Windbreak Wall and Wind Speed. Applied Mechanics and Materials, 2014, 823:989-994.
6. Ai, Jun, Jian-Fei Chen, J. Michael Rotter, and Jin Y Ooi. Assessment of rolling resistance models in discrete element simulations. Powder Technology 206 (2011): 269-282.
7. Baran, Oleh, Alfred DeGennaro, Enrique Rame, and Allen Wilkinson. DEM Simulation of a Schulze Ring Shear Tester. Proceedings of the 6th International Conference on Micromechanics of Granular Media. 2009. 409-412.

Research on privacy protection of intelligent construction

Jingshi Shang [1], Xiuwen Li [1] and Qingyong Wang [2]

[1] China Institute of Water Resources and Hydropower Research, Beijing 100048, China
[2] Xinjiang Erqisi River Investment and Development (Group) Co., Ltd, Urumqi 830000, China.

Abstract. With the application of new generation intelligent technologies such as big data, cloud computing and numerical simulation in engineering, the level of intelligent construction has been gradually improved. In the development process of intelligent construction, it faces many problems, such as new technology, safety, human recognition and so on. This paper studies the privacy protection of intelligent construction, analyzes the causes of privacy disclosure on the basis of information classification, and then puts forward privacy protection methods. The research shows that the privacy security of intelligent construction needs to clarify the scope and formulate effective protection strategies.

Keywords: Intelligent construction, privacy protection, information classification, privacy disclosure.

1. Introduction

Since the beginning of this century, China's water conservancy and hydropower construction has developed vigorously, and the intelligent construction technology represented by intelligent dam [1] has been widely used. Xiluodu hydropower station has created a precedent for intelligent construction of high arch dam. Baihetan Hydropower Station has built an intelligent construction technology system for key construction processes and business processes based on the theory of closed-loop control of comprehensive perception, real analysis and real-time control [2~5].

The development of intelligent construction is inseparable from the support of a new generation of technology. At the same time, like many new things, it also faces many new problems. In the information age, privacy security has become a big problem. Intelligent construction and its participants need to put forward reasonable suggestions on how to protect privacy. This paper studies from the perspective of privacy protection of intelligent construction, and the research results can provide suggestions for the development of intelligent construction of hydropower projects.

2. Question raising

The basic explanation of privacy is information that is not open to the public. The development of modern science and technology has not only provided many conveniences for human society, but also caused a series of problems, of which privacy disclosure is one of the more prominent. With the strengthening of people's self ideology and the relative lag of the legal environmental protection of social privacy, privacy protection has gradually attracted attention [6].

The technologies applied in intelligent construction include: structural safety and health monitoring technology, construction environment awareness technology, personnel safety and health monitoring technology, etc. Among them, the construction environment perception technology analyzes and identifies the surrounding environment, determines the location, matches the perception, and makes real-time prediction and early warning. The personnel safety and health monitoring technology monitors the physiological indicators of construction personnel, warns and guides their construction behaviors, and ensures their safety and health. This involves the privacy protection of personnel. In addition, from the macro level, all or part of the engineering information of intelligent construction is confidential, which should also belong to the category of privacy.

Therefore, the privacy protection of intelligent construction includes not only the pure personnel information, but also the intermediate links of intelligent construction, scientific research achievements and other information. This paper first classifies the information of intelligent construction, then analyzes the possible links of privacy problems, and finally puts forward the methods of privacy protection.

3. Intelligent construction information classification

Information in the process of intelligent construction can be roughly divided into four categories. (1) Survey and design information: mainly including geological survey, geological survey during construction, preliminary design, design data during construction, etc; (2) Intelligent construction information: including construction progress, scheme, quality, safety and other information; (3) Operation and maintenance: including patrol inspection, dispatching and other information in the intelligent operation stage after the completion of project construction; (4) Feedback evaluation: including calculation and evaluation of project safety and quality during construction and use. With a wide range of information sources, a large number of people and miscellaneous information, there is a risk of privacy disclosure at the source.

Information collection methods mainly include manual collection, monitoring equipment collection, image collection, and self information feedback of construction equipment. Due to the diversity of collection devices and processing methods, there is also a risk of information and privacy disclosure in the collection process.

The transmission modes of intelligent construction information mainly include optical fiber, Internet, wireless 4g/5g, Beidou, etc. With the development of information technology, there is a certain risk of privacy disclosure due to insufficient information protection.

The objects of intelligent construction information application can be roughly divided into nine categories. There are overlapping demands for information among various units, the division of application personnel is not clear, and there is a certain risk of privacy disclosure in the process of information transmission and application.

Table 1 Intelligent construction information application classification

Unit category	demand	application
Employer	Construction information, design information, quality information and other comprehensive information.	Comprehensive management of project progress, quality, safety and contract.
Design unit	Engineering geology, construction information, monitoring information, scientific research data, etc.	Preliminary design, design during construction and technical changes.
construction control unit	Construction information, supervision process information, design information, etc.	Construction process supervision, quality, safety, contract supervision, etc.
Construction unit	Engineering geological information, construction plan and progress information, etc.	Complete construction tasks as required.
Quality supervision unit	Construction process quality information, quality objectives, etc.	The whole process quality supervision shall be conducted for all parties involved in the construction.
Scientific research institutions	Geological, design, construction and other information required for scientific research projects.	Technical breakthrough and support for engineering problems.
Auxiliary support unit	Supply and demand information, production and living supporting information, etc.	Assist in the completion of various works other than project implementation.
Superior competent unit	Project construction, design, scientific research and other important information.	Project decision-making and supervision.
government sector	The owner reports the project quality, progress, safety and other information.	Policy formulation and supervision.

4. Analysis on privacy leakage of intelligent construction

The reasons for smart building privacy disclosure can be summarized as management, technology, personnel, unclear boundaries and other reasons. (1) Management reason: there is no clear information protection workflow, measures and personnel during the implementation of intelligent construction. (2) Technical reasons: it mainly focuses on the transmission and timeliness of information in the process of intelligent construction, and ignores the security of methods and methods in the process of transmission. For example, many projects now transmit data through the establishment of wechat groups, which itself has great security risks. (3) Personnel negligence: the main reason is that the personnel do not have a strong sense of privacy protection. There is a large amount of engineering information on the mobile phone. Due to the current low security performance of the mobile phone, it is easy to become the source of privacy disclosure. In addition, short videos, circles of friends and other channels may become a way of privacy disclosure. (4) Unclear boundary: the main reason is that there is no clear industry standard to restrict what is engineering information privacy, and we do not know how to protect privacy during the implementation of intelligent construction.

5. Intelligent construction privacy protection method

By combing the application scenarios of intelligent construction data information, formulate targeted privacy protection strategies. In each application scenario of intelligent construction, the data collection and storage devices are complex, and the data processing capacity and protection capacity vary greatly. According to the characteristics of the collection devices and the data information involved (such as Internet of things data, personnel information, engineering construction data, equipment operation data, etc.), the data are classified, and different levels of privacy protection strategies are developed to prevent privacy disclosure and data abuse [7].

At the management level, attention should be paid to the definition of privacy permissions, such as granting corresponding permissions to relevant units and personnel according to the classification and application ways of intelligent construction engineering information. Second, in terms of storage, different storage methods are formulated according to the importance level of information, and information managers. Third, when using, it is necessary to leave traces of information such as the way of use and personnel to ensure that they are well documented.

Privacy disclosure risk involves the whole life cycle from information collection, information transmission, information storage, information processing, information use to information destruction. Among them, information collection and information transmission interface is the entrance of intelligent construction information platform, and its security is directly related to the operation safety of the whole intelligent construction project. Information collection needs to identify the types of information involved to ensure that the project and personal privacy are not disclosed; In the information transmission phase, it is necessary to use encryption, signature and other authentication mechanisms to manage the security of the transmitted information; The collected information needs to be stored on the information platform before calculation and processing, which puts forward requirements for data storage environment, disaster recovery and backup capability, encryption measures, etc. In the information processing and use stage, it is necessary to ensure the security of the whole process and prevent data from being tampered with and leaked. At the last stage of the production process, the information needs to be encrypted and stored or destroyed according to the security level to ensure that important sensitive information is not disclosed.

According to the personal information protection law, the collection, processing, use and other activities of personal information need to comply with the relevant provisions of the protection law. Personal consent, desensitization and encrypted storage shall be obtained when collecting and processing biometric, financial account, whereabouts and other personal sensitive information. Improve data display and desensitization standards, and try to use source desensitization to protect personal privacy.

6. Conclusion

This paper studies the privacy protection in the frontier of intelligent construction, and the main conclusions are as follows:

(1) By analyzing the relevant technical fields of intelligent construction, this paper puts forward the possible leakage ways involved in intelligent construction, and then puts forward the research ideas of problems;

(2) From the perspective of intelligent construction information classification, reasons for privacy disclosure and protection methods, this paper analyzes the problems of intelligent construction privacy protection, and puts forward reasonable suggestions.

With the development and improvement of intelligent construction technology, privacy protection will also attract the attention of researchers, and continue to put forward reasonable methods and suggestions to promote the safe development of intelligent construction.

Acknowledgements

Foundation of State Key Laboratory of Simulation and Regulation of Water Cycle in River Basin (IWHR-SKL-KF202112).

References

1. Q.B. Li, P. Lin: Demonstration on Intelligent Dam, Journal of Hydropower Engineering, Vol. 33 (2014) No.1, p.139-146.
2. P. Lin, R.N. Wang: Intelligent construction safety culture connotative development - from concept to action, Project Mnagement Rview, (2021) No.5, p.46-50.
3. P. Lin, Z.Y. Ning and M. Li: Study on prototype intelligent control test of cooling pipeline for a super-high arch dam, SHUILI XUEBAO, Vol. 52 (2021) No.7,p. 819-828.

https://doi.org/10.1051/matecconf/202338201036

4. Y.S. Tan, Q.X. Fan and Z.L Wang: Intelligent construction methods for the Baihetan super high arch dam, Journal of Tsinghua University (Sci&Technol), Vol. 61 (2021) No. 7, p. 694-704.
5. Q.X. Fan, P. Lin and P.C. Wei: Closed-loop control theory of intelligent construction, Journal of Tsinghua University (Sci&Technol), Vol. 61 (2021) No. 7, p. 660-670.
6. Y.Y. Zhang: Criminal law path choice for privacy protection in the era of big data, Journal of Shenyang University (Social Science), Vol. 23 (2021) No. 7, p. 625-630.
7. X. Zhang: Research on railway data security and privacy protection management strategy, Railway computer application, Vol. 30 (2021) No.11, p. 43-46.

https://doi.org/10.1051/matecconf/202338201037

Study on the Measuring and Reading Mode of High Precision Thermometer in Hydraulic Engineering

Shisen Zhen [1], Hao Wu [2]

[1]Water Conservancy Bureau, Shouguang 262700, China
[2]China Institute of Water Resources and Hydropower Research, Beijing 100048, China.

Abstract. After the operation of large water conservancy projects, water temperature changes will have a significant impact on the surrounding ecological environment. In order to monitor the water temperature, a water temperature monitoring system covering the whole flow channel has been established in China's giant hydropower stations. In order to accurately measure the water temperature in the water level fluctuation area, a ± 0.1 ℃ water temperature gauge is selected as the front sensing device. Compared with traditional thermometers, high-precision thermometers are rarely used in projects. After investigation and comparison, platinum resistance thermometer has been applied in water conservancy projects. In view of the fact that there is no special manual measuring and reading equipment for the platinum resistance thermometer before the automation is connected during the construction period, a four wire temperature measuring and reading mode is proposed, which is applicable to the platinum resistance thermometer by using the existing hydraulic bridge, and the rationality of the measuring and reading mode is verified through the comparison of indoor tests.

Keywords. Water conservancy projec, temperature monitoring, high precision thermometer, manual reading.

1. Introduction

In the context of the current great protection of the Yangtze River basin [1], in order to mitigate the adverse water temperature impact caused by the construction of the Project, the hydropower stations in the Yangtze River basin are equipped with water temperature monitoring systems in the reservoir area, in front of the dam and at the downstream of the dam. During the operation period, the water temperature in the river reach is comprehensively and systematically monitored. The online monitoring system of water temperature includes the online monitoring system of water temperature along the reservoir area, the online monitoring system of vertical water temperature at the water intake, the online monitoring system of water temperature in the unit channel and the online monitoring system of water temperature under the dam. According to the prediction results of the water temperature prediction model in the reservoir area of the power station and the vertical stratification characteristics of the water temperature, the water pressure resistance and accuracy of the thermometer are reasonably selected. The water temperature gauge with an accuracy of ± 0.3 ℃ is proposed to be selected in the area with relatively stable water flow and temperature below the elevation of 700m, and the water temperature gauge with an accuracy of ± 0.1 ℃ is proposed to be selected in the area with major changes in the water level above the elevation of 700m. At present, the temperature sensors commonly used in the safety monitoring of water conservancy projects are mostly resistance thermometers and vibrating wire thermometers, which meet the accuracy of ± 0.3 ℃. Compared with the traditional thermometer, the water temperature gauge with ± 0.1 ℃ accuracy is less used in the safety monitoring of water conservancy projects. This paper defines it as a high-precision thermometer, and explores the selection of high-precision thermometer in the water temperature monitoring system of Baihetan Hydropower Station, the measurement and reading mode during construction and the application effect.

2. Thermometer selection

2.1 Introduction to Common Thermometers

Temperature sensors are embedded in or on hydraulic structures or other geotechnical structures for a long time to measure the temperature inside the concrete of hydraulic structures, water temperature in the reservoir area, ambient air temperature, etc. The temperature sensors commonly used in water conservancy projects are mostly resistance thermometers. The Inspection and Test Procedures for Dam Safety Monitoring Instruments (SL530-2012) also stipulates the inspection and test methods and procedures for three types of thermometers, namely copper resistance thermometer, thermistor thermometer and vibrating wire thermometer. Copper resistance thermometer is a thermometer that uses copper resistance to work in a linear relationship with temperature within a certain temperature range. The thermistor thermometer uses the semiconductor thermistor as the sensor. With the characteristics that the number of carriers in the semiconductor increases drastically with temperature, the more carriers, the stronger the conductivity, and the smaller the resistivity, the temperature is measured by measuring its resistance value. It is just opposite to the temperature change characteristic of copper resistance thermometer. When the ambient temperature changes, the vibrating wire thermometer transmits to the internal inductor through the sensor. Because the temperature has a linear relationship with the change of the inductor, the inductor expands or contracts, causing the natural frequency of the vibrating wire to change. The current ambient temperature change can be calculated by the secondary instrument exciting the vibrating wire through the coil and receiving the digital signal.

These three types of thermometers have been used in water conservancy projects for a long time with good results, and portable and mature secondary instruments are used together during the construction period. Table 1 lists the relevant indicators of typical thermometers.

Table 1 Comparison Table of Typical Thermometer Parameters

Type	Copper resistance	Thermistor	Vibrating wire
Measuring range	-30℃ ~70℃	-25℃~80℃	-20℃ ~80℃
Measurement accuracy	±0.3℃	±0.5℃	≤0.03℃/F
Resolution	/	0.1℃	≤ 0.03%F.S
Measurement and reading method	Hydraulic bridge	Vibrating wire reader with temperature measurement	Vibrating wire reader

2.2 Selection of high-precision thermometer

Through the manufacturer's investigation and comparison of different types of thermometers, platinum is an ideal thermometer material, which is easy to be purified, processed, and has good physical and chemical stability. The thermometer with platinum resistance as the sensor has excellent temperature measurement performance, such as wide temperature measurement range, stable performance, and good repeatability, and can meet the ± 0.1 ℃ accuracy requirements of the project. From the perspective of application, standard platinum resistance thermometers are often used as verification instruments for various standard thermometers and precision temperature measuring instruments, and are widely used in high-precision temperature control systems in high-tech, industrial production and scientific research fields.

The platinum resistance thermometer used in this project has the characteristics of good long-term stability, stainless steel structure suitable for various harsh environments, good waterproof performance, long service life, high sensitivity, etc. See Table 2 for technical parameters.

Table 2 Instrument Technical Parameters

Type	Parameters
Measuring range	-30°C～+70°C
Measurement accuracy	±0.1°C
Sensitivity	0.1°C
Stability	≤0.1%/Y
Withstand voltage	1500V
Insulation resistance	≥100MΩ
Resistance	100Ω@0°C

3. Study on the measurement and reading mode during construction

3.1 Measurement and reading method

In water conservancy projects, the connecting cable of monitoring instruments is usually long, up to thousands. The platinum resistance thermometer has small resistance and high sensitivity. For example, the thermal resistance of the platinum resistance is 0.379 Ω/℃ at 100 ℃. If the resistance of the connecting cable is 2 Ω, the error will reach 5.3 ℃. Therefore, when measuring and reading the resistance of the sensor normally, the measurement error caused by the resistance of the cable led by the instrument must be eliminated [2].

As mentioned in the above section, platinum resistance thermometers are commonly used in precision industrial systems. The user manual of the selected thermometer gives the access and reading methods of the automatic acquisition system. There is no method for accurate measurement and reading without access to automation during the construction period, and there is no matching portable measurement and reading instrument. From the perspective of instrument principle, it is obvious that high-precision thermometers cannot use vibrating wire readout or the accompanying temperature module of vibrating wire readout to measure and read. Therefore, starting from the platinum resistance measurement principle in the industrial field, the feasibility of using the mature copper resistance thermometer that is also used to measure resistance for reference is analyzed.

There are three international measurement methods for thermal resistance leads [3], namely two-wire system, three wire system and four wire system. Two wire system is a way to connect a wire at both ends of the thermal resistance to lead out the resistance signal. This lead wire method is very simple, but because there must be lead wire resistance r in the connecting wire, the size of r is related to the material and length of the wire, so this lead wire method is only applicable to the occasions with low measurement accuracy. The three wire system is a way to connect one lead at one end of the root of the thermal resistance and two leads at the other end. This method is usually used with the bridge, which can better eliminate the influence of lead resistance. It is the most commonly used method in industrial process control. The four wire system is a way to connect two conductors at the root and both ends of the thermal resistor. Two leads provide the thermal resistor with constant current I, convert R into voltage signal U, and then lead U to the secondary instrument through the other two leads. It can be seen that this lead way can completely eliminate the influence of lead resistance, and is mainly used for high-precision temperature detection.

The thermometer selected is a four core cable, so it is proposed to use the four wire system method of double arm bridge for measurement and reading. There are mature bridge products in the field of safety monitoring of water conservancy projects, which are mainly used for the measurement of resistance ratio, resistance value, reverse resistance ratio and core wire resistance of differential resistance instruments, and can eliminate the influence of long cable resistance and core wire resistance variation on the measured value. The differential resistance instrument uses a five core hydraulic cable, and the existing bridge products are designed according to this. The measured resistance value can be obtained by matching the color of the five terminal posts of the bridge with the five core hydraulic cable core. Starting from the measuring principle and the principle of instrument and equipment, the four wire method is used to measure the platinum resistance thermometer with the existing bridge products.

The wiring equivalent diagram of four wire temperature measurement is shown in Figure 1, where R_t is the thermal resistance and R_{Li} is the equivalent resistance of conductor L_i. Apply exciting current I to the thermal resistance through wires L_1 and L_2, measure the potential V_3 and V_4, connect wires L_3 and L_4 to the high input impedance circuit, and make $I_{L3}=0$ and $I_{L4}=0$. Therefore, V_4-V_3 is equal to the voltage at both ends of the thermal resistance, and then use formula (1) to obtain the thermal resistance value [4].

$$\left(V_4 - V_3\right) / I = R_t \quad (1)$$

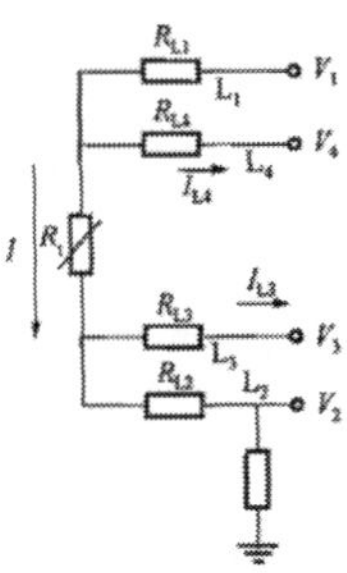

Fig. 1 Equivalent schematic diagram of four wire system temperature measurement

By analyzing the internal structure of the existing bridge products, the connection mode of the four core measuring and reading cable of high-precision thermometer using the four wire system method is shown in Figure 3. The measured temperature can be calculated using the temperature resistance characteristics of the platinum resistance, and the calculation formula is Formula (2).

$$R_t = R_0\left[1 + aT - bT^2 - cT^3\left(T - 100\right)\right] \quad (2)$$

Where, R_t is the resistance value when the temperature is T, R_0 is the resistance value thermal resistance when the temperature is 0 ℃, and a, b, c are coefficients. Platinum resistance is used for the instrument as TCR=0.003851 in accordance with IEC751 standard. See Table 3 for coefficient values.

Table 3 Coefficient value table when TCR=0.003851

Temperature	a	b	c
T<0℃	3.90802×10^{-3}	5.80195×10^{-7}	4.27351×10^{-12}
T≥0℃	3.90802×10^{-3}	5.80195×10^{-7}	0

At present, there is little practice of measuring and reading high-precision (platinum resistance) thermometers by hydraulic bridges in literature and work. The validity of the wiring mode of measuring and reading high-precision (platinum resistance) thermometers by hydraulic bridges will be tested and verified below.

3.2 Test verification

3.2.1 Instrument and equipment

The equipment used in the test includes thermostatic water tank (the maximum temperature difference in the working area shall not exceed ± 0.1 ℃), second-class standard mercury thermometer, electric bridge, multimeter and other secondary instruments. The thermostatic water tank is used to provide a standard temperature environment for the instrument. The platinum resistance thermometer and mercury thermometer are all immersed in the thermostatic water tank. The mercury thermometer is used to correct the temperature environment. The instrument is measured by an electric bridge and a multimeter. The two secondary instruments represent the four wire system and the two wire system respectively. The measured results are used for comparison.

3.2.2 Test result

The test was carried out in a standard laboratory in strict accordance with the Inspection and Test Procedures for Dam Safety Monitoring Instruments (SL530-2012). At 0 ℃, 30 ℃ and 60 ℃ three temperature points in the thermostatic water tank, the thermometer is connected to 3m cable, 50m cable and 100m cable, and the resistance is measured with an electric bridge and multimeter respectively, and then the temperature value is calculated. See Table 4 for the test results.

Table 4 Temperature measurement

Standard temperature (℃)		0	30	60
With 3m long cable	Bridge measurement (℃)	-0.03	29.96	59.96
	universal meter measurement (℃)	0.84	30.84	60.84
With 50m long cable	Bridge measurement (℃)	0.04	30.02	60.04
	universal meter measurement (℃)	5.37	35.39	65.46
With 100m long cable	Bridge measurement (℃)	-0.01	30.02	60.04
	universal meter measurement (℃)	10.76	40.88	71.00

The results show that:
(1) The multimeter is directly used to measure with the two-wire wiring method. With the increase of the instrument wiring length, the error of the measurement results will increase accordingly, because the two-wire system cannot eliminate the cable error.
(2) The bridge measurement is conducted by using the four wire wiring method. The error at each test temperature point is small. Under the laboratory conditions, the maximum error is ± 0.04 ℃, which is not affected by the instrument wiring length. This shows that the precision of the platinum resistance thermometer selected meets the engineering needs, the measuring instrument selected is appropriate, and the four wire system wiring mode studied is correct.

4. Conclusion

Research methods such as extensive investigation and indoor test are adopted to explore the selection of high-precision thermometers in the water temperature monitoring system of large hydropower stations and the measurement and reading methods during construction. The main conclusions are as follows:
(1) Platinum resistance thermometer has good performance, its accuracy is superior to the commonly used hydraulic monitoring thermometer, its accuracy meets the requirements of engineering design, and the instrument selection is accurate.
(2) Through the indoor comparison test, the platinum resistance thermometer can be manually measured and read by the existing hydraulic bridge according to the four wire system described in the article. The measurement and reading results are far better than the two wire system, which can eliminate the error caused by connecting cables.

Acknowledgements

Foundation of State Key Laboratory of Simulation and Regulation of Water Cycle in River Basin (IWHR-SKL-KF202112).

References

1. Q.X. Fan. Water resources utilization based on sustainable ecosystems. Practice of Jingsha river hydropower development, Journal of Hydroelectric Engineering, vol. 29 (2010) No.4, p.1-5+13.
2. Y. Zhang. Accurate measurement of industrial thermal resistance, Metrology & Measurement Technique, (2004) No. 4, p.6-7.
3. M. Zhou, X.S. Ge. Research on measuring method of three wire and four wire thermal resistance, Industrial Metrology, vol. 25 (2015) No.3, p.33-37.
4. X.W. Zhang, K. Zhang. Application of high precision digital multimeter in platinum thermistor temperature measurement, China Metrology, (2005) No. 10, p.71-72.

https://doi.org/10.1051/matecconf/202338201038

Key Influencing Factors Research of Domestic and Import Aviation Lubricating Oil Foam Characteristics

Yuxiong Geng, Xiaoai Wang, Yongtao Wang, Yitian Zheng, Jun Ma *

Air Force Logistics Academy, Jiangsu, Xuzhou, China

Abstract. The foam characteristics plays a key role towards the application of aviation apparatus, which also affected by several influencing factors. This paper focuses on researching the key factors of the foam characteristics of the domestic and imported aviation lubricating oils, by comparing a certain type of domestic aviation lubricating oil or imported aviation lubricating oil's foam characteristics at different oxidation temperature or oxidation length by virtue of utilizing aviation lubricating oil oxidation experimental device. The influencing factors that impact foam characteristics were analyzed to provide important theoretic and experimental support.

Keywords. Aviation Lubricating Oil; Foam Characteristics; Influencing Factors; Oxidation temperature; Oxidation time

1. Introduction

With the continuous improvement of the performance of new aviation turbine engines, studies on key factors influencing foam characteristics of domestic and imported aviation lubricating oil are attracting more researchers and funding. Due to the increasing rotation speed, working power and load, important parameters such as the turbine front temperature, pressure, supercharging ratio and thrust-weight ratio are constant increasing for better performances, which makes aviation lubricating oil subject to more intense oscillation and agitation. This puts forward higher requirements for foam characteristics of aviation lubricating oil. As aviation lubricating oil inevitably mixes with air during work, bubbles are formed under the action of interfacial tension, which will not only significantly increase the foam volume of aviation lubricating oil and seriously affect the lubrication condition of aviation engine, but also cause the oil to have obvious physical and chemical properties decay under the action of high temperature. Make its lubrication wear reduction, cooling and heat dissipation, clean dispersion, sealing and plugging, corrosion and rust prevention functions are adversely affected. Existing studies [2] have found that the amount of foam produced by the oxidation of gasoline oil at 150 ℃ increases significantly. The results show that the increase of oxidation temperature will lead to the increase of foam volume and the decrease of foam resistance. The addition method [3] and the blending method [4] may also lead to the uneven dispersion of the anti-foaming agent, and the settlement of the anti-foaming agent after standing, leading to the deterioration of the anti-foaming performance of the oil, or even the unqualified phenomenon. In addition, the selection and cleaning of gas diffuser also have an important influence on the measurement of foam characteristics. It is found that for some oil products with good foam characteristics, the aperture size of the diffuser has little influence on the test results. However, for oil products with certain foaming properties, the smaller the aperture, the larger the test data [5]. At the same time, the gas content of oil [6], inorganic salts [7], silicon [8] and laboratory conditions [9] will also affect the foam characteristics of lubricating oil.

This paper intends to conduct a comparative study on the foam characteristics of a certain type of domestic and imported aviation lubricating oil, investigate the change law of the foam characteristics of this type of domestic and imported aviation lubricating oil under different oxidation conditions, and study the key influencing factors of the foam characteristics of domestic and imported aviation lubricating oil, so as to provide a beneficial practical basis for solving the problem of engine performance limitation caused by the increase of aviation lubricating oil foam.

2. Experimental Section

The two types of aviation lubricating oil studied in this paper are purchased from Henan Aviation Materials Technology Co., LTD., in which the operating temperature of domestic aviation lubricating oil is -40 ℃ to 200 ℃, and that of imported oil is -40 ℃ to 175 ℃. In order to further study the key influencing factors of foam characteristics of this type of domestic and imported

* Corresponding author: manudt@163.com

https://doi.org/10.1051/matecconf/202338201038

aviation lubricating oil, This type of domestic and imported aviation lubricating oil was oxidized at 175 °C, 200 °C and 230 °C respectively by using the aviation lubricating oil oxidation test device shown in Figure 1.
The specific experimental steps are as follows: 200 mL oil sample is added into the clean and dry oxidation tube, and the dry air flow rate is 10 L/h. After the oil bath at the corresponding temperature (175 °C, 200 °C, 230 °C), the oxidation tube is cooled to room temperature, and the simulated oxidized oil sample is prepared. Table 1 shows the preparation conditions of aviation lubricating oil samples at 175 °C, 200 °C and 230 °C for 0.5 to 8 hours:

Table 1. Preparation scheme of simulated oxidized oil sample for aviation lubricating oil.

Oxidizing temperature	Oxidation time
175 oC	0.5h, 1h, 4h, 8h
200 oC	0.5h, 1h, 4h, 8h
230 oC	0.5h, 1h, 4h, 8h

The oxidized oil samples of the domestic and imported aviation lubricating oil were prepared by oxidation reaction. According to GB/T12579-2002 standard, the foam characteristics of the two types of aviation lubricating oil before and after oxidation were measured at 24 °C。

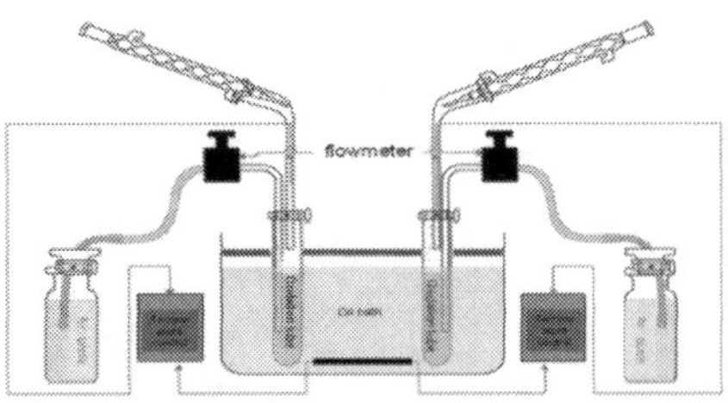

Figure 1. Oxidation test device for aviation lubricating oil.

3. Restle and discussion

3.1 Foam characteristics of a domestic aviation lubricating oil after oxidation at 175 o C

Oil samples of this type of domestic aviation lubricating oil oxidized for 0.5h, 1h, 2h, 4h and 8h were prepared at 175 °C by simulated oxidation test. The foam characteristics of the new oil and the oxidized oil samples were determined respectively, and the influence of oxidation time factors on the performance of this type of domestic aviation lubricating oil at 175 °C was analyzed.

Figure 2. Oil samples of a certain domestic aviation lubricating oil oxidized for 8h, 4h, 1h and 0.5h at 175 oC.

Table 2. Foam volume of domestic aviation lubricating oil (oxidized at 175 oC

Simulated oil sample	Total volume (mL)	Foam volume (mL)
New oil	215	55
0.5h	220	60
1h	295	180
4h	335	215
8h	345	220

Firstly, the foam volume of the new domestic aviation lubricating oil was obtained through foam characteristic test. Table 2 shows the foam volume measurement results of the simulated oil sample. The experimental results show that the foam volume of this type of domestic aviation lubricating oil is 60mL when oxidized for 0.5h at 175°C, and 215 and 220mL after oxidizing for 4 and 8h, respectively. The foam volume of this type of domestic aviation lubricating oil is about 4 times higher than that of the oxidized oil sample at 0.5h. In general, with the increase of oxidation time of the domestic aviation lubricating oil, the foam volume increases significantly and the foam characteristics become worse.

3.2 Foam characteristics of a certain domestic aviation lubricating oil after oxidation at 200°C

After oxidation at 200 °C, the color of this domestic aviation lubricating oil gradually deepened, but the oil did not precipitate, indicating that there was no obvious abnormality in the appearance of this oil at 200°C. The foam volume of this oil will be determined in accordance with GB/T12579-2002 standard.

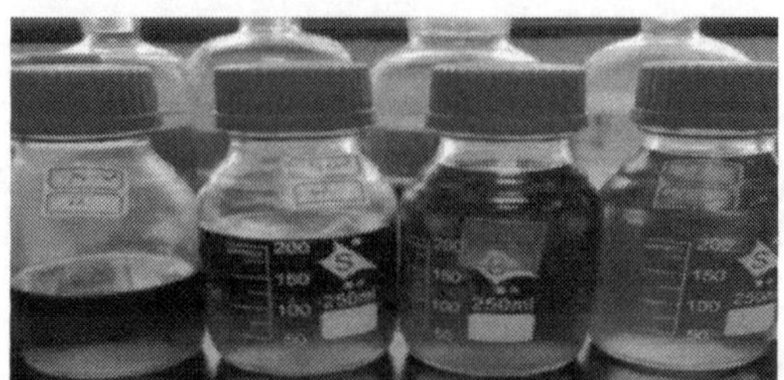

Figure 3. Oil samples of a certain domestic aviation lubricating oil oxidized for 8h, 4h, 1h and 0.5h at 200°C

https://doi.org/10.1051/matecconf/202338201038

The foam characteristics of the domestic aviation lubricating oil samples oxidized for 0.5h, 1h, 4h and 8h at 200 ℃ were tested. It was found that the foam characteristics changed significantly with the oxidation time.

Table 3. Foam volume of domestic aviation lubricating oil (oxidized at 200℃)

Simulated oil sample	Total volume (mL)	Foam volume (mL)
New oil	215	55
0.5h	215	50
1h	360	245
4h	400	280
8h	440	345

Table 3 shows the measurement results of foam volume of simulated oil sample. Measured results show that the foam volume of this type of domestic aviation lubricating oil is 50mL when oxidized for 0.5h at 200℃, and the foam volume corresponding to oxidized oil for 1h is 245mL. The foam volume is 390% higher than that of oxidized oil sample for 0.5h, and the foam volume corresponding to oxidized oil sample for 4h is 280mL. Compared with the oxidized oil sample at 0.5h, the foam volume increased by 460%, the foam volume corresponding to the oxidized oil sample at 8h was 345mL, and the foam volume increased by 590% compared with that of the oxidized oil sample at 0.5h. It can be seen that the foam volume corresponding to the oxidized oil for 0.5h was basically the same as that of the new oil at 200℃. However, after oxidizing for 1h, the foam resistance deteriorates and the foam volume increases significantly. The foam properties of the oil sample oxidized for 8h were the worst, and the foam volume of the oil sample oxidized for 0.5h increased by 590% compared with that of the oil sample oxidized for 0.5h.

3.3 Foam characteristics of a certain domestic aviation lubricating oil after oxidation at 230℃

After oxidation at 230℃, the appearance color of oxidized oil sample shown in Figure 4 changes greatly, and there is no precipitated substance. Table 4 shows the foam volume measurement results of 230℃ simulated oxidized oil sample of this domestic aviation lubricating oil. The measured results show that the foam volume of this type of domestic aviation lubricating oil is 110mL when oxidized for 0.5h and 30mL when oxidized for 4h and 8h at 230℃. At 230℃, the foam characteristics of oxidized oil sample decreased instead of rising, and its foam volume was even smaller than that of new oil, which was different from the previous foam volume test results measured at 175 ℃ and 200℃.

Figure 4. Oil samples of a certain domestic aviation lubricating oil oxidized for 8h, 4h, 1h and 0.5h at 230℃

Table 4. Foam volume of domestic aviation lubricating oil (230℃)

Simulated oil sample	Total volume (mL)	Foam volume (mL)
New oil	215	55
0.5h	225	110
1h	220	60
4h	210	30
8h	210	30

3.4 Foam characteristics of a certain imported aviation lubricating oil after oxidation at 175℃, 200℃ and 230℃

The oil samples of this type of imported aviation lubricating oil oxidized for 0.5h, 1h, 4h and 8h were prepared by simulated oxidation tests at 175 ℃, 200 ℃ and 230℃ respectively. The foam characteristics of the imported aviation lubricating oil were determined respectively and the influence of oxidation factors on its performance was analyzed.

Table 5. Foam volume of imported aviation lubricating oil (oxidized at 175℃, 200℃ and 230℃)

Simulated oil sample	Total volume (mL)			Foam volume (mL)		
	175℃	200℃	230℃	175℃	200℃	230℃
New oil	205	205	205	10	10	10
0.5h	205	205	205	10	10	10
1h	205	215	205	10	20	10
4h	205	210	215	10	10	15
8h	210	210	215	15	10	15

Table 5 shows the foam volume measurement results of simulated oil samples of imported aviation lubricating oil at 175, 200 and 230℃. The measured results show that the foam volume of this type of imported aviation lubricating oil is less than 20mL when oxidized for 8 hours at 175 ℃, 200 ℃ and 230℃, and the foam characteristics are very excellent.

Specific view, oxidative 0.5 hours after the simulation sample of bubble volume is almost the same with the new oil, that import aviation lubricating oil bubble features at the beginning of the oxidation with oxidation temperature insensitive. With the increase of oxidation time, the foam increases slightly, indicating that the oxidation has an effect on the oil composition, and then affects the foam characteristics. In general, the foam characteristics of imported aviation lubricating oil are significantly better than those of domestic aviation lubricating oil. Under the same oxidation decay condition, the foam volume of imported aviation lubricating oil is about 1/10 to 1/20 of domestic. At the same time, even when oxidized at 230℃, the foam volume of imported aviation lubricating oil is still below 15mL, which basically conforms to the foam characteristic standard of this new imported aviation lubricating oil, indicating that its foam characteristics still have great potential at higher oxidation temperature.

3.5 Analysis on key influencing factors of foam characteristics of domestic and imported aviation lubricating oil

According to the above experimental studies, the foam characteristics of imported aviation lubricating oil are better, and the foam volume of domestic aviation lubricating oil is still lower than 30mL (24℃) under relatively strict experimental conditions (8h oxidation at 230℃). By contrast, the foam characteristics of domestic aviation lubricating oil are worse, and the foam volume of domestic aviation lubricating oil is as high as 345mL (24℃) after 8h oxidation at 230℃. It shows that domestic aviation lubricating oil is greatly affected by experimental factors, especially oxidation temperature has a significant effect on foam characteristics of domestic aviation lubricating oil. The foam volume of oil samples oxidized at 200℃ for 8h is significantly larger than that of oil samples oxidized at 230℃ for 8h, indicating that the foam characteristics of this type of domestic aviation lubricating oil decrease significantly with higher oxidation temperature. In addition, the foam volume of oxidized oil samples at 175℃ and 200℃ increases with the increase of oxidation time, suggesting that oxidation time is also a key factor affecting the foam characteristics of this type of domestic aviation oil.

In addition, it was found that the foam volume of the aviation lubricating oil decreased significantly after oxidation at 230℃, and even the foam volume of the aviation lubricating oil after oxidation for 8 hours was smaller than that of the new oil. It is found in the research literature that a variety of domestic and imported aviation lubricating oil products will undergo physicochemical property jump at 230℃ [10]. The viscosity decreased significantly and acid value increased significantly. In order to investigate the reasons for the abnormal changes, the PAO base oil of this domestic aviation lubricating oil was oxidized and its kinematic viscosity changes at 40 and 100℃ were measured, as shown in Figure 5.

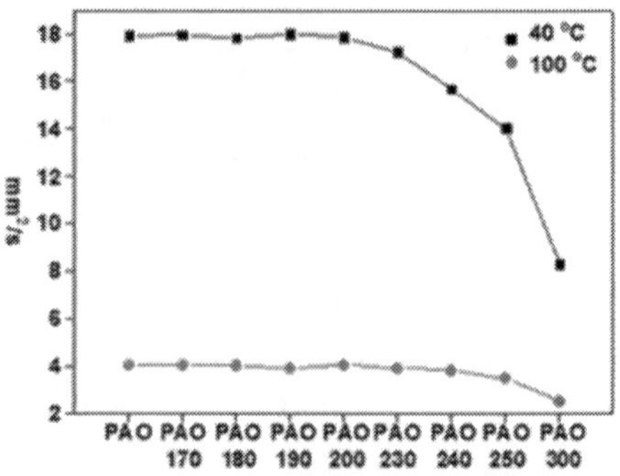

Figure 5. Movement viscosity of oil samples in PAO high-temperature oxidation tests at 40 ℃ and 100℃

The experimental results showed that for PAO oxidized oil samples, the motion viscosity at 100℃ was relatively flat, while the motion viscosity at 40℃ was more obvious, and the reference value was higher than that at 100℃. Therefore, the PAO high-temperature reoxidation test oil samples at 40℃ were used as the reference for analysis. The kinematic viscosity of PAO oxidized oil at 40℃ did not change significantly when the oxidation temperature was lower than 230℃. However, if the oxidation test temperature continued to rise, the kinematic viscosity would change significantly, which can be clearly seen from FIG. 5: The motion viscosity of PAO oxidized oil starts to change gently at 40 ℃ , and decreases significantly from 230℃ to 230℃. With the increasing of the oxidation temperature, the curve presents a steep slope and the viscosity gradually increases. Therefore, it is concluded that the molecular structure of the oil sample changes little when the oxidation temperature is below 230℃, and the degree of pyrolysis reaction is shallow. When the oxidation temperature exceeds the temperature, the degree of pyrolysis deepens abruptly. The foam characteristics of lubricating oil are related to the viscosity of oil. When the oxidation temperature of oil is above 230℃, the viscosity decreases and the molecular structure is damaged obviously, which affects the foaming properties of oil and reduces the foam volume. However, when the oxidation temperature is below 230℃, the decrease of foam volume does not indicate that the quality of the oil becomes better, and the decrease of its viscosity may lead to the significant deterioration of the service performance of the oil.

4. Conclusion

In this paper, the variation law of foam characteristics of domestic and imported aviation lubricating oil is studied. It is found that domestic aviation lubricating oil is greatly affected by oxidation time and temperature, while imported aviation lubricating oil is not sensitive. At low oxidation temperature (175℃), the foam volume of domestic aviation lubricating oil showed a significant

https://doi.org/10.1051/matecconf/202338201038

increase trend, and the trend expanded further at 200℃, but when further increased to 230 ℃ , the viscosity decreased significantly, resulting in a significant decrease in foam volume, indicating that oxidation temperature, oxidation time and kinematic viscosity are the key factors affecting the foam characteristics of domestic aviation lubricating oil.

References

1. Li Yong. Research on Fault Diagnosis of an aeroengine Lubrication System [D]. Wuhan: Wuhan University of Technology. 2010.
2. OSTERLAND S, MULLER L, WEBER J. Influence of Air Dissolved in Hydraulic Oil on Cavitation Erosion[J]. International Journal of Fluid Power, 2021,22(3): 373-392.
3. Andrenko P, Rogovyi A, Hrechka I, Khovanskyi S, Svynarenko M. The Influence of the Gas Content in the Working Fluid on Parameters of the the Hydraulic Motor's Axial Piston[M]. London: Oxford Press, 2021.
4. YUAN X, WANG C, ZHU X, ZHANG L. Research on Theoretical Model of Dynamic Bulk Modulus of Elasticity of Gas-liquid Mixed Fluid[J]. Journal of Mechanical Engineering, 2020,56(4): 209.
5. Bureček A, Hružík L, Vašina M. Determination of Undissolved Air Content in Oil by Means of Compression Method[J]. Strojniški vestnik - Journal of Mechanical Engineering, 2015,46(3): 56-61.
6. EDVARDSSON J, RAUTANEN P, LITTORIN A, et. al. Deactivation and coke formation on palladium and platinum catalysts in vegetable oil hydrogenation[J]. Journal of the American Oil Chemists' Society, 2001,78(3): 319-327.
7. Zhang Jie, Meng Xuemei, Lei Ling. Effect of Blending Parameters on High Temperature Foam Characteristics of Gasoline Oil [J]. Synthetic Lubricating Materials, 2017, 44(2): 10-11 .
8. Effect of Ma Zhansheng Antifoaming Agent on Physicochemical Properties of lubricating oil [J]. Chemical Management,2016(32):120-121.
9. Cen Zhiyuan, Chen Bingyao, Yang Shanjie et al. Discussion on Anti-foaming agent in Lubricating Oil [J]. Shandong Industrial Technology, 2019.06(6):32-33.
10. Influence of gas diffuser on Determination of Foam Characteristics of lubricating oil [J]. Guangdong Chemical Industry, 2020,47(9): 38-40.
11. Chen Chao, Gong Guofang, Jin Yaolan et al. Design and Experimental Study of Hydraulic Oil Vacuum Degassing Device [J]. Chinese Hydraulics & Pneumatics,2008,(2):88-91.
12. Li Hongqiang, Zhang Wen, Tian Chengtao et al. Effect of inorganic salts on the characteristics of gas-liquid two-phase foam [J]. Comprehensive Utilization of mineral resources, 2021, (1): 204-210.
13. Zhu Yaan, Wang Zhuo, Tao Chunsheng, et al. Effect of Silicon Content in Engine Oil on Foam Characteristics [J]. Lubricating Oil, 2017, 32(3): 35-38.
14. Wu Rina, Zhang Zhiyong, Chen Jie. Evaluation of Uncertainty in Determination of Foam Characteristics of Lubricating Oil [J]. Inner Mongolia Electric Power Technology, 2016, 34(5):6-13.
15. Peng Xiancai, Fei Yiwei, Wu Nan et al. Comparative analysis of oxidation decay Properties of Two aviation lubricating base Oils under Metal Catalysis [J]. Lubrication and sealing, 2017, 42(8):71-81.

Vibrational analysis of stiffened double-layered plate in vacuum

Peng Wang [1], Zhao Li [1], Hao Zhang [1], Jiandong Chen[2], Wei Li[3], Min Liu [1,*]

[1] Wuhan Second Ship Design and Research Institute, Wuhan 430205, China
[2] Ship Development and Design Center, Wuhan, 430064, China
[3] School of Naval architecture and Ocean Engineering, Huazhong University of Science and Technology, Wuhan 430074, China

Abstract: Stiffened plate has widely application in engineering, stiffened double-layered plate is one typical structure. In this work, research object is the stiffened double-layered plate, modal and forced vibration responses are studied through experimental work and numerical simulation with FEM software, and comparative analysis is also conducted. Numerical results show that, experimental results match well with numerical simulation results, the maximum difference for first four modal frequencies are less than 5%, so numerical simulation is verified to be correct. Conclusions here will help to support the further design and application of stiffened double-layered plate in engineering area.

1. Introduction

Stiffened plate has widely application in engineering, since it has good mechanic performance and quite simple structure. Stiffened double-layered plate is one typical structures of stiffened plate, which have received great attention from researchers.

Zhang et.al.[1] studied the vibration of double-layered plate and coupled plate with rectangular cross section. Reference [2] established a new model using effective density and modulus of Helmholtz resonator for noise reduction and energy absorption of double-layered plate. Xin et.al.[3-6] applied modal expansion method to describe the distribution of sound pressure, and analyzed the vibro-acoustics performance of double-layered plates with simply-supported and clamped boundary condition. It's noted that available work about double-layered plate mainly focus on sound transmission loss and vibro-acoustics characteristics. But it should be pointed out that vibration performance of double-layered plate is the foundation for further design (*i.e.*, vibration and noise reduction) and application in engineering. So it's important to get clear the vibration of double-layered plate. In this work, stiffened double-layered plate is the research object, and modal performance and forced vibration is studied by conducting experimental work and numerical simulation.

2. Vibration test for stiffened double-layered plate in vacuum

2.1 Test model

The test model is shown in Fig.1, length and width are both 690mm. model consists of upper plate ring-stiffeners and lower plate. There 25 stiffeners distributed between upper and lower plate. Thicknesses for the upper and lower plates are both 2mm. The thickness for stiffeners is also 2mm, and the height of frame is 18.8mm. Four large I-shape beams with many holes are welded to the bottom edges of the lower plate, which will help to hang the model.

Fig.1 Test model

* Corresponding author: 549931866@qq.com

https://doi.org/10.1051/matecconf/202338201039

2.2 Measurement point

169 points are marked and well-distributed on the surface of upper plate, as is shown in Fig.2. During the experiment, point load will be conducted by knock the load point with force hammer. Two points are selected to measure the vibration of plate with acceleration sensors, and the locations are shown in Fig.2.

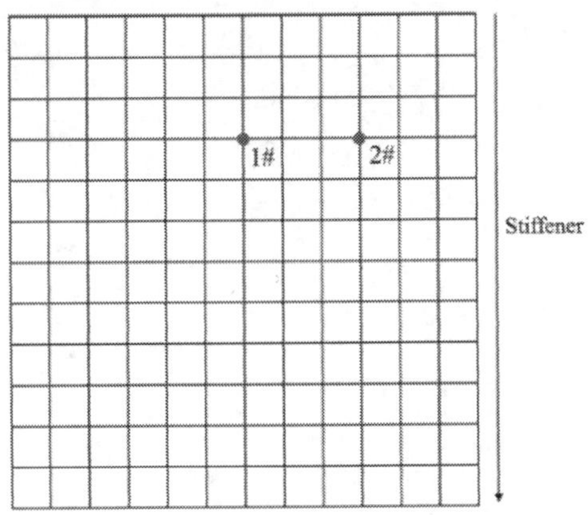

Fig.2 measurement point

2.3 Calculation of mean square velocity level

Based on the reciprocity of transfer function of two different measure points, the mean square velocity of plate could be obtained by conducting normalization of test data at each measurement point. The mean square velocity level is defined as

$$< v^2 >= \frac{1}{S}\iint_S |V|^2 ds$$

$$l = 10\log_{10}\left(< v^2 >/v_0^2\right)$$

Where, v is the normalized square velocity, S is the area of upper surface, V is the distribution function of normal velocity, v0 is reference velocity, $v_0 = 1\times10^{-9} m/s$.

3. Result discussions

3.1 Modal performance

Numerical simulation is conducted for modal performance of experimental model with ANSYS, and modal frequencies are compared with those from experimental work, as is shown in table 1, mode shape from experimental work are shown in Fig.3-5.

It's noted that modal frequencies from numerical simulation match well with that from experimental work, and the maximum difference is less than 5% for the first three modes.

For mode shapes of stiffened double-layered plate, the first mode has one half-wave in two directions; the second mode has two half-waves in one direction and only one in the other direction; the third mode has three half-waves in one direction and still one in the other direction. This phenomenon is quite similar to that of a single rectangular plate.

Table 1 Comparative analysis of model frequencies of stiffened double-layered plate

Mode order	experiment/Hz	simulation/Hz	difference
1	306.3	291.0	-4.9%
2	434.4	430.3	-0.9%
3	565.6	564.7	-0.2%

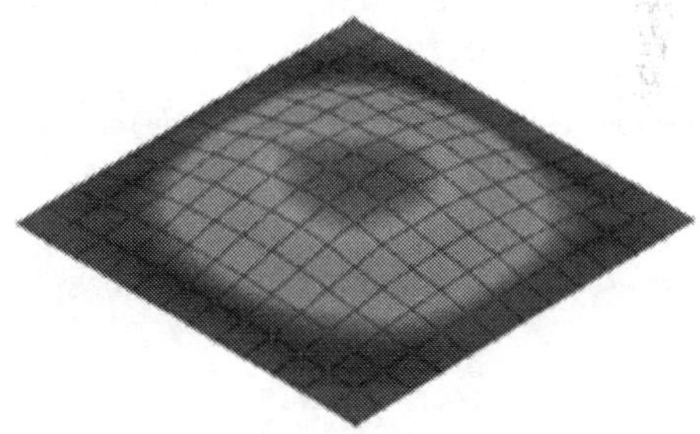

(a) first mode;

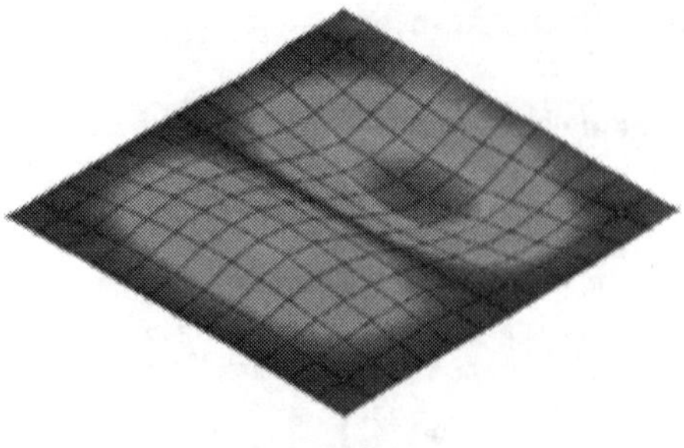

(b) second mode;

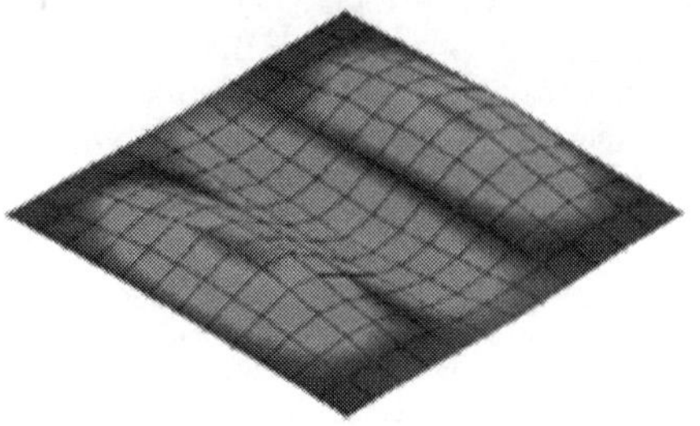

(c) third mode

Fig.3 Mode shape of stiffened double-layered plate

3.2 Forced vibration

Distributions of mean square velocity versus exciting frequency in case of point load at JL1 and Jl2 are plotted in Fig.6-7, and numerical results are also given for comparative analysis.

It's noted that, distribution of mean square velocity from experimental work and numerical simulation match well, especially in the frequency range <1000Hz, which indicates the validation of numerical results.

Besides, with comparative analysis on mean square velocity in case of JL1 and JL2,its notes that the change of point load location will lead to obvious change of peak amplitude or new appearance of peak frequency. This is due to complex distribution of different mode or new excitation of a certain structure mode.

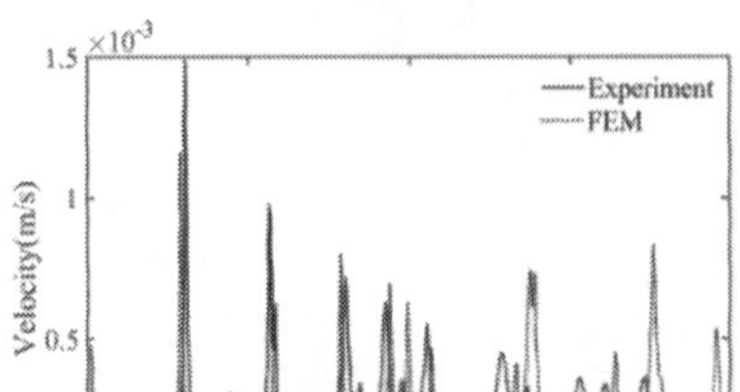

(a) JL1;

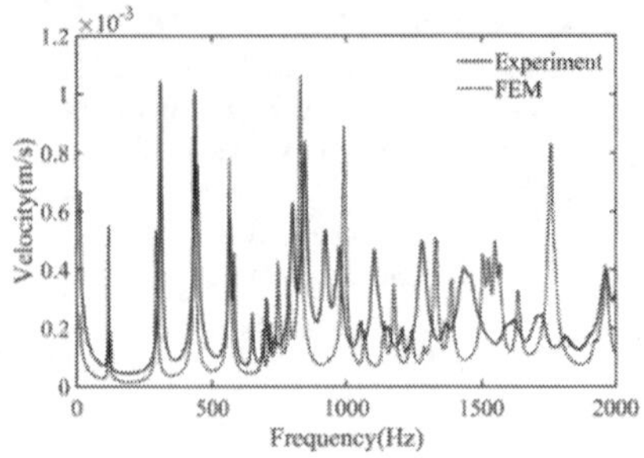

(b) JL2

Fig.4 Comparative analysis of mean square velocity from experiment and FEM in case of different load point

4. Conclusion

In this work, research object is the stiffened double-layered plate, modal and forced vibration responses are studied through experimental work and numerical simulation with FEM software, and comparative analysis is also conducted. Numerical results show that, experimental results match well with numerical simulation results, the maximum difference for first four modal frequencies are less than 5%, so numerical simulation is verified to be correct.

For mode shape of stiffened double-layered plate, the firsts one has one half-wave in two directions; the second on has two half-waves in one direction and one half-wave in the other direction; the third one has three half-waves in one direction and still one half-wave in the other direction. This is quite similar to modal performance of single plate.

Besides, the change of point load location will lead to change of peak amplitude or new peak frequencies, which is due to complex contribution of each mode to total vibration response.

Conclusions here will lay foundation for the analysis of vibration and sound radiation performance of double-layered structures further design and application of stiffened double-layered plate in engineering area.

Acknowledgements

We sincerely thank the financial support from the Hubei Provincial Natural Science Foundation (No. ZRMS2020002231) and the Young Top-notch Talent Cultivation Program of Hubei Province.

References

1. Zhang C Y, Jin G Y, Ye T G, et al. Harmonic response analysis of coupled plate structures using the dynamic stiffness method [J] . Thin- Walled Structures, 2018, 127: 402-415.
2. F. L, H. H, W. G. Broadband low-frequency sound transmissionloss improvement of double walls with Helmholtz resonators[J]. Journal of Sound and Vibration, 2020,476.

https://doi.org/10.1051/matecconf/202338201039

3. Xin F X, Lu T J, Chen Q. Vibro-acoustic behavior of clamp mounted double-panel partition with enclosure air cavity[J]. Journal of the Acoustical Society of America, 2008, 124(6): 3604-3612.
4. Xin F X, Lu T J. Analytical and experimental investigation on transmission loss of clamped double panels-Implication of boundary effects [J]. Journal of the Acoustical Society of America, 2009, 125(3): 1506-1517.
5. Xin F X, Lu T J, Chen C Q. Sound Transmission through Simply Supported Finite Double-Panel Partitions with Enclosed Air Cavity [J]. Journal of Vibration and Acoustics, 2010, 132(1): 011008: 11001-11011.
6. Xin F X, Lu T J, Analytical modeling of sound transmission through clamped triple-panel partition separated by enclosed air cavities [J]. European Journal of Mechanics A/Solids, 2011, 30: 770-782.

MATEC Web of Conferences **382**, 01040 (2023) https://doi.org/10.1051/matecconf/202338201040
AMME 2023

Experimental Platform Design for Particle Behavior Investigation

Panpan Zhang[1,2], Siyang Zhou[1], Qingwei Chen[1,*]

[1]China Waterborne Transport Research Institute, Beijing,100088, China.
[2]Key Laboratory of Logistics Equipment & Control Engineering, Beijing 100088, China.

Abstract. The hopper is a kind of important equipment in bulk handling process. In this paper, an experimental platform was designed. The experimental platform was mainly composed of a feeding head, a vertical conveying screw machine, a horizontal conveying screw and a lifting device, a supporting structure, the visual experimental system,the electrical system and the acquisition system. the particle behavior for hoppers can be assessed by the experimental platform. And the visual experimental system will help to investigate the real particle flow behavior and will provide instructions for hopper design in future

1. Introduction

In the process of industrial production, transportation and unloading of bulk materials such as sand, coal and food particles, etc it can be seen that hoppers are widely applied. In bulk handling process of port, performance of the hopper has a significant influence on the site environment. Therefore, dust suppression of hoppers is the prone point during operation management as shown in Figure 1. in order to enhance the performance of hoppers, the particle behavior of material in the hopper is a key in the development of hoppers. In recent years, researchers and some companies focus on the study of particle behavior and dust suppression [1-3]. Figure 2 is an experimental platform was designed by Liu Z. Q to carry out research on flow field characteristics of different geometric parameters for the discharging hopper

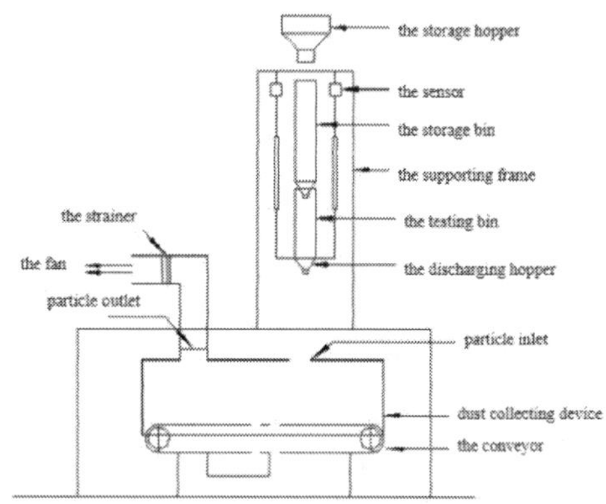

Figure 1. The diagram of the experimental platform desigened by Liu Z. Q

During the process of industrial production, transportation and unloading of bulk materials (such as sand, coal and food particles, etc), it can be seen everywhere that free-fall motion of granular flow; in the process of free-fall motion, due to the impact of gravity, buoyancy, the frictional resistance between particles, and the force ambient air and particles, some fine particles are attracted to flow field of particle beam by ambient air due to the unbalance force, then a particle stream was formed. In the process of particle streamfree-fall motion, the dust escape into the surrounding environment when particle stream confronted at the bottom of the receiving device due to the influence of gravity of material itself, which caused air pollution as well as a huge damage on human health.

In this paper, in order to study particle behavior during the transport process, an experimental platform was put forward. The composition of the experimental platform was introduced. And with help of the experimental platform, design of hoppers will be more scientific and efficient.

* Corresponding author: chenqw@wti.ac.cn

https://doi.org/10.1051/matecconf/202338201040

2. Overall composition of the system

This system is mainly composed of a feeding head, a vertical conveying screw machine, a horizontal conveying screw, a lifting device, a supporting structure, the visual experimental hopper,the electrical system and the acquisition system.

When working, the vertical conveying screw machine and horizontal conveying screw machine are first started, and then the feeding head is driven. Then the feed screw machine is inserted into the heap to extract materials, unloaded into the horizontal conveying screw machine, and then transported to the rear test system.

According to the test requirements, the material flow can be provided for the later test project to complete the test research, and the related performance parameters of the feeding screw machine can be studied separately.

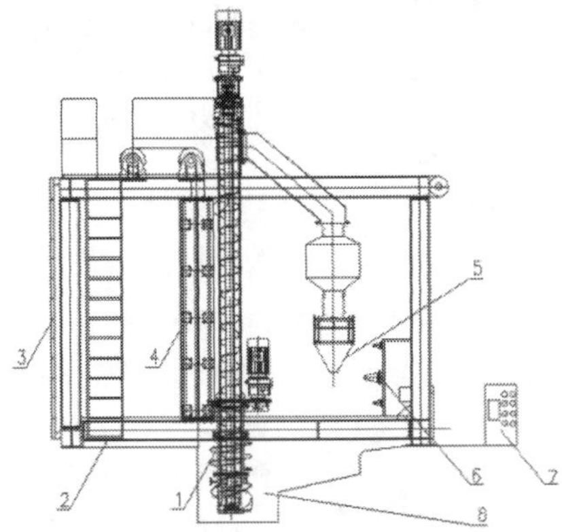

1-a feeding head, 2-a supporting structure,, 3-a vertical conveying screw machine,, 4- a lifting device,5-the visual experimental hopper,7- the acquisition system,8-the electrical system,9- a horizontal conveying screw

Figure 2 Overall structure of the experimental platform

3. System configuration and functions

3.1 The feeding head

The feeding head is mainly composed of the internal spiral and the external spiral. Driven by variable frequency speed regulation, the feeding capacity can be adjusted to the experiment setting value. The experimental study of different materials and different structural forms can be realized by setting the removable structure. different spiral feeding head for future spiral feed device test provides an interface. The specific structure form is shown in ' Figure 3 spiral feeding head '.

Figure 3. The feeding head

The main technical parameters of the feeding head were shown in Table 1.

Table 1 Main parameters of the feeding head

Basic parameters	Productivity	100t/h	Driving unit	Motor type	Y15-4P
	Material	Plastic particles		Motor drive power	15KW
	Density	0.61t/m³		Motor speed	1460r/min
Spiral parameters	Outer spiral pitch	200mm		Reducer type	GRF97
	Internal diameter of outer spiral blade	406mm		Transmission speed ratio of reducer	10.83
	External Diameter of External Spiral Blade	636mm		Output speed of reducer	135r/min

3.2 The vertical conveying screw machine

The vertical conveying screw machine mainly lifts the material extracted from the feeding head to a certain height, this is convenient to feed material to the horizontal conveying screw machine. The feeding capacity can be adjusted by using frequency conversion drive. The specific structure is shown in Fig. 4 vertical screw conveyor.

Figure 4 Vertical conveying screw machine

The main technical parameters of the vertical screw conveyor are shown in Table 2.

Table 2 Main technical parameters of the vertical screw conveyor

Table 2 Main parameters of vertical screw conveyorBasic parameters	Productivity	100t/h	Driving unit	Motor type	Y15-4P
	Material	Plastic particles		Motor drive power	15KW
	Density	0.61t/m³		Motor speed	1460r/min
Spiral parameters	outer spiral pitch	240mm		Reducer type	GRF97
	Internal diameter of spiral blade	133mm		Transmission speed ratio of reducer	4.5
	External diameter of Spiral Blade	301mm		Output speed of reducer	327r/min

3.3 The horizontal screw conveyor

The horizontal conveying screw machine mainly delivers the material discharged by the screw machine to the rear test system. The structural form is shown in Fig. 5 horizontal screw conveyor.

Figure 5 Horizontal conveying screw machine

The main technical parameters of the horizontal screw conveyor are shown in Table 3.

Table 3 Main parameters table of horizontal screw conveyor

Basic parameters	Productivity	110t/h	Driving unit	Motor type	Y22-4P
	Material	Plastic particles		Motor drive power	22KW
	Density	0.61t/m³		Motor speed	1460r/min
Spiral parameters	Spiral pitch	360mm		Reducer type	GRF97
	Internal diameter of outer spiral blade	159mm		Transmission speed ratio of reducer	4.5
	External Diameter of External Spiral Blade	475mm		Output speed of reducer	327r/min

3.4 The lifting device

The lifting device is mainly composed of a lifting support and a lifting drive mechanism. The lifting bracket is used to support the screw machine. The lifting action can be realized in the sliding track. In order to realize the feeding height change of the vertical conveying screw machine. a wire rope drum device was adopted for the lifting mechanism. The structural form of lifting bracketshown in Fig. 6 is the lifting device.

Figure 6 The lifting device

The main technical parameters of the lifting device are shown in Table 4.

Table 4 Main parameters table of the lifting device

Framework parameter	Model of guide wheels	6309-2Z	Driving unit	Motor type	Y5.5-4P-M1
	number of Guide wheels	20		Motor drive power	5.5KW
	Spacing of Guide wheels	700mm		Motor speed	1460r/min
steel wire rope parameters	Type	6x19-¢ 10		Reducer type	GS100
	Number of wire rope winding	6x19		Transmission speed ratio of reducer	212.81
	Length of wire rope	30m		Output speed of reducer	6.8r/min

3.5 The support structure

The supporting device is mainly composed of supporting beams, maintenance platform, ladders and railing, which is used to support and install the equipment in the higher level.

The form of supporting structure is shown in Fig. 7.

Figure 7 The support structure

3.6 The visaual hopper system

In order to obtain the flowing behavior of particles in the hopper, transparent hoppers were chosen in the platform design. The hoppers including the discharging hopper and the storage hopper are made from PMMA, which is a kind of transparent polymer as shown in Figure 8. The transparent structure can provide sight for the acquisition system.

Figure 8 The Support structure

3.7 The acquisition system

The acquisition system is mainly used for collecting and processing the experiment data. In the experimental platform, a high-speed camera was introduced to get the actual particle behavior through the visual hopper system as shown in Figure 9. And the dust concentration detector as shown in Figure 10 was chosen to investigate the dust generated during the experiment.

Figure9 The high-speed camera

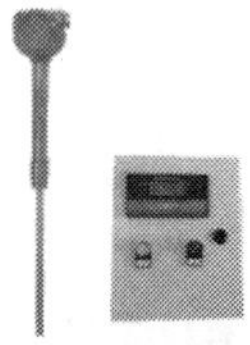

Figure10 The dust concentration detector

4. Summary

In this paper, a novel experimental platform was proposed to study particle behavior during the transport process. the composition was introduced. the characteristics for the experimental platform was that the transparent hopper with the The high-speed camera was adopted to investigate the real particle flow behavior. Through the transparent hoppers, the real flow behavior can be obtained and will provide direct instructions for efficient dust-control hopper design and development.

References

1. Liu Z.Q.Ma L.Y.Zhao X.Y.Experimental Study the affection of the Dust Emissionon Sand Stockpile by the Height of Windbreak Wall and Wind Speed. Applied Mechanics and Materials,2014,823:989-994.
2. Panpan Zhang, Qingwei Chen,Particle Behavior Investigation for Different Structures of Hoppers, 4th International Conference on Advances in Materials, Machinery, Electronics (AMME 2021),353:01015
3. Panpan Zhang, Simulation of Particle Behavior for Different Hoppers, Advances in Engineering Research, Vol.125(2017):228-231.
4. Ai, Jun, Jian-Fei Chen, J. Michael Rotter, and Jin Y Ooi. Assessment of rolling resistance models in discrete element simulations. Powder Technology 206 (2011): 269-282.
5. Baran, Oleh, Alfred DeGennaro, Enrique Rame, and Allen Wilkinson. DEM Simulation of a Schulze Ring Shear Tester. Proceedings of the 6th International Conference on Micromechanics of Granular Media. 2009. 409-412.
6. Ansart R.,and Ryck A.Dust emission in powder handling:Free falling particle plume characterization Chemical Engineering Journal 2009,152:415–420

Simulation analysis of internal and external faults in transformer area

Peixian Bai, Yixuan Jia

Electrical Engineering and its Automation of North China Electric Power University), Baoding 071003, Hebei Province, China

Abstract: Under different types of internal and external faults of transformers, electrical quantities such as voltage, differential current and transformer oil flow rate change accordingly. In this paper, the variation law is studied, and the variation law of differential current and other electrical quantities under different types of faults and abnormal conditions inside and outside the transformer area is verified by simulation. The characteristics of faults inside and outside the area are different, and the faults inside and outside the transformer area are judged. At the same time, ANSYS Maxwell 3D, ANSYS Fluent and other software are used to model and simulate the transformer, and the oil flow rate changes under different operating conditions such as normal operation, external fault and internal fault of the transformer are analyzed.

Key words: Transformer, internal and external fault, electrical parameters, oil flow rate, simulation.

1. Introduction

Power transformer is the main equipment in power system and plays an important role in the safe operation of power system. When a fault occurs, the voltage, differential current and other electrical quantities and transformer oil flow rate will change accordingly under different types of internal and external faults. Good simulation verification is carried out to analyze the electrical quantities and oil flow rate changes under different faults, and more effective measures can be taken to protect when actual faults occur.

2. Electrical quantity simulation analysis

2.1 Transformer internal and external fault line model

2.1.1 Outside fault model

In this paper, PSCAD / EMTDC simulation software is used to establish the simulation model of lines with transformers. By setting different types of faults (single-phase ground short circuit, two-phase ground short circuit, two-phase short circuit, three-phase short circuit and three-phase ground short circuit), the characteristics of various faults are further studied.
The external fault simulation model is shown in Figure 1.1 :

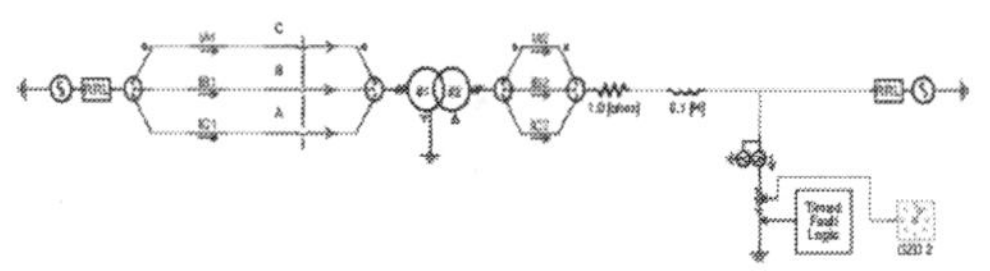

Figure 1.1 Outside fault simulation model

The two AC power supplies with phase voltages of 220 kV and 35 kV are connected by transformers and concentrated impedance lines. The transformer adopts the UMEC model, the capacity is 360MVA, the connection mode is YN / d11, and the ratio is 220kV / 35kV. The current transformer is a model based on Jiles-Aherton (JA) hysteresis theory. ABC is a three-phase circuit breaker ; gZD2 is the fault type controller ; timed Fault Logic is a fault timing controller module. This module is used to provide electrical quantities (single-phase ground short circuit, two-phase ground short circuit, two-phase short circuit, three-phase short circuit and three-phase ground short circuit) under different types of faults.

2.1.2 Internal fault model

Transformer internal faults include internal single winding inter-turn short circuit, two winding inter-turn short circuit, turn-to-ground short circuit and transformer outlet short circuit. When the inter-turn and turn-to-ground faults occur at different positions of the internal winding, the primary current and short-circuit turn current of the transformer are different. Generally speaking, the current is smaller when the short-circuit turns are located

at both ends of the winding, while the current is larger when it is located in the middle of the winding [1].
Based on the single-phase three-winding transformer, this paper constructs a model suitable for transformer internal fault analysis. As shown in Figure 1.2, the two windings of the secondary side (winding 1 : Y-y, winding 2 : Z-z) are in series to simulate the secondary side winding of the dual-winding transformer, that is, y is connected with Z. When faults such as inter-turn short circuit occur, the third winding can be short-circuited. Setting different turns ratio of winding 2 to winding 3 can simulate internal faults at different positions.

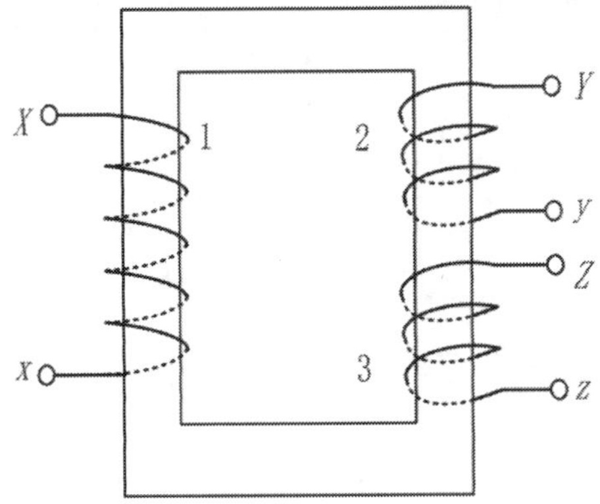

Fig.1.2 Three-winding transformer winding connection diagram

Based on the above model, the proposed model for transformer internal fault simulation is shown in Figure 1.3.

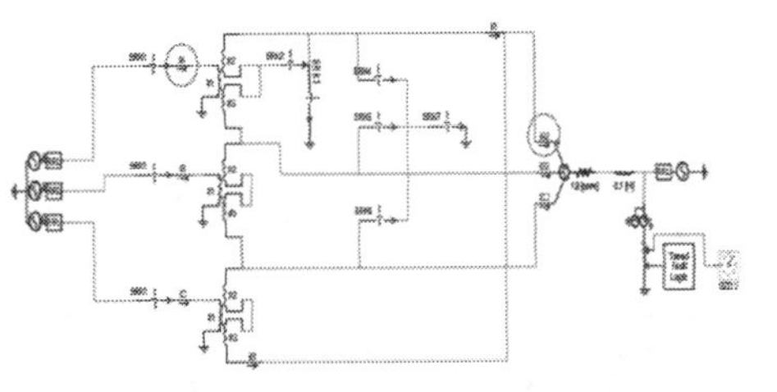

Figure 1.3 Internal fault simulation model

In order to ensure the three-phase balance, the three-winding transformer is selected, and the winding 2 and winding 3 in series are equivalent to the secondary winding of single-phase double-winding transformer[2]. In the above model, the transformation ratio of the equivalent single-phase dual-winding transformer is 220kV / 35kV, the connection mode is YN / d11, and the parameter settings of the power supply and other components are consistent with the external fault model. In Figure 1.3, BRK1 ~ BRK7 are control switches, which can be realized by controlling BRK1 ~ BRK7 when simulating different types of internal faults. As shown in Table 1.1. During normal operation, BRK1 is closed, other switches are disconnected, and BRK1 is always in a closed state.

Table 1.1 Implementation method of transformer internal fault

Switch state (closed)	fault type
BRK2	A phase inter-turn short circuit
BRK3	A phase turn-to-ground short circuit
BRK4、BRK5	AB two-phase inter-turn short circuit
BRK4、BRK5、BRK6	ABC three-phase inter-turn short circuit
BRK4、BRK5、BRK7	AB two-phase turn-to-ground short circuit
BRK4、BRK5、BRK6、BRK7	ABC three-phase turn-to-ground short circuit

2.2 Internal and external fault simulation

2.2.1 Electrical quantity change research

This section mainly verifies the variation law of electrical quantities after various faults occur to ensure the correctness of the built transformer model.
(1) Normal operation
When the line is running normally, the line current waveform, three-phase ground voltage and transformer differential protection differential current waveform are shown in Fig.1.4.

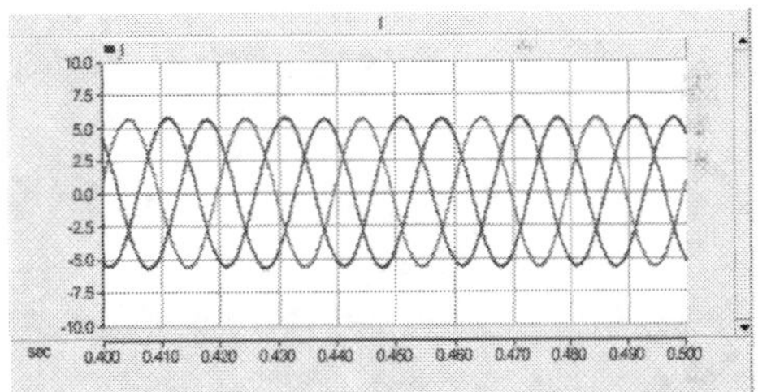

(a) three-phase current

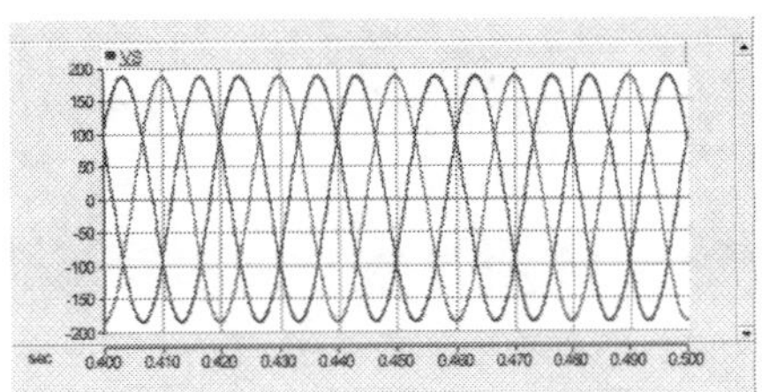

(b) three-phase turn-to-ground voltage

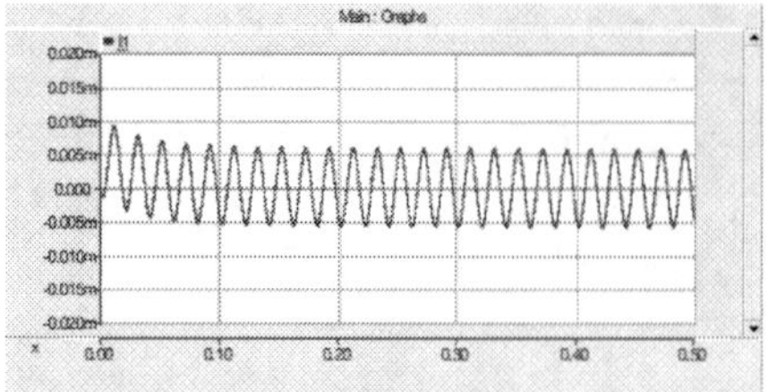

(c) differential current

Figure 1.4 Voltage and current waveform during normal operation

It can be seen from Fig.1.4 that when the simulation model is in normal operation, the three-phase current and ground voltage waveform are sinusoidal, and the phase difference between each two phases is 120 °, and the differential current value is very small.

(2) External fault

1) Setting A ground phase fault outside the transformer fault model area, the fault point F ground current, ground voltage and transformer differential protection differential current waveform are shown in Figure 1.5.

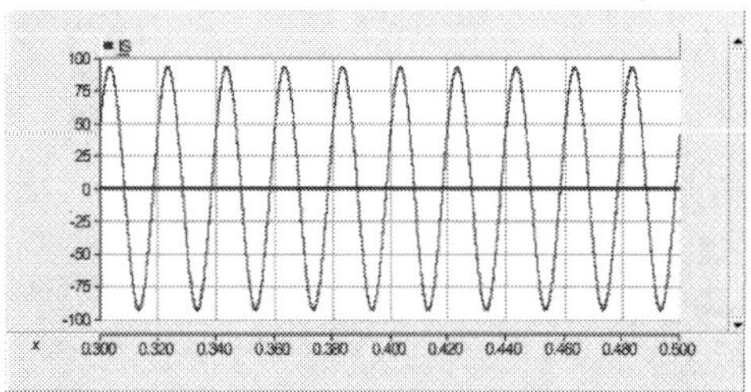

(a) Fault point F to ground current

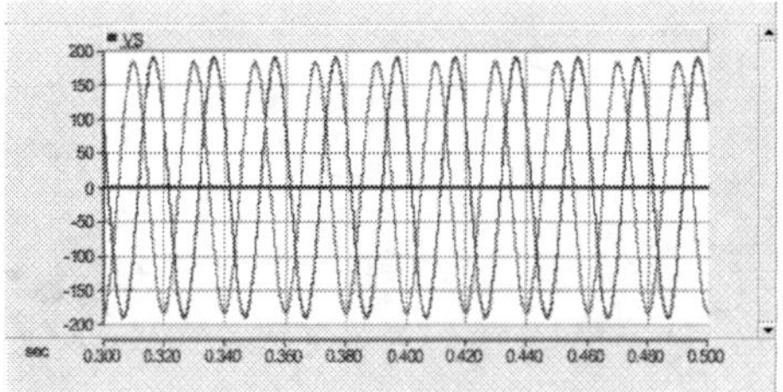

(b) Fault point F to ground voltage

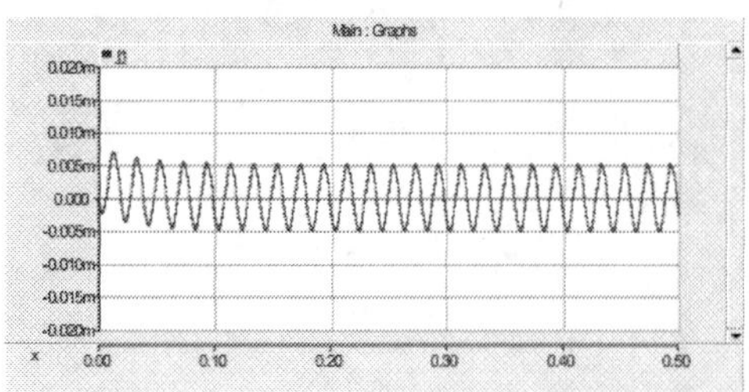

(c) Differential current

Fig.1.5 A phase fault to ground current and voltage waveform

As shown in Fig.1.5, in the ground current of fault point F, the amplitude of phase A current is not 0, and the ground current of the other two phases is 0. In the ground voltage of fault point F, the two-phase voltage of B and C is the same as that of normal operation, and the relative ground voltage of A is 0. The differential current value of differential protection is close to 0. It is known from the above that when the A-phase fault occurs, the voltage, current and differential current conform to the theoretical boundary conditions when the single-phase fault occurs, and the transformer protection does not operate [3].

2) Setting A, B two-phase short-circuit fault outside the transformer fault model area, the fault point F ground current, ground voltage and transformer differential protection differential current waveform is shown in Figure 1.6.

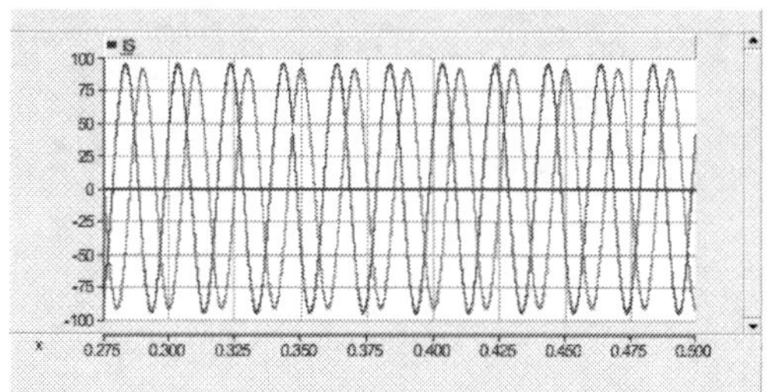

(a) Fault point F to ground current

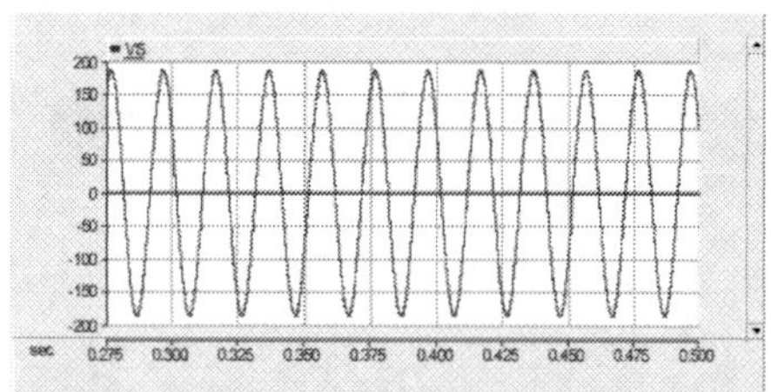

(b)Fault point F to ground voltage

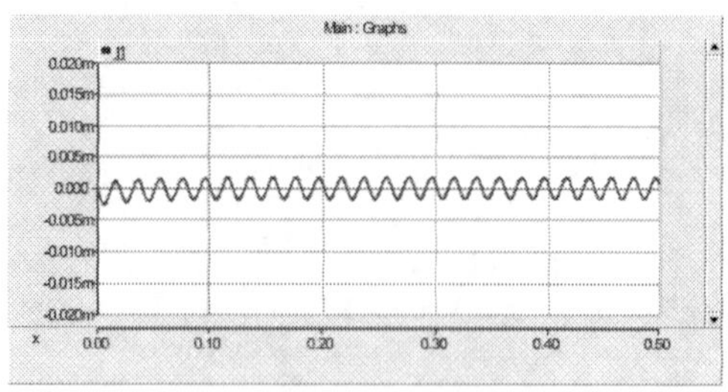

(c) Differential current

Fig.1.6 Two-phase fault-to-ground current and voltage waveform

In the fault point F to ground current, shown in Fig.1.6, the amplitude of A and B two-phase current to ground current is not 0, and the phase difference is 120 °, C relative to ground current is 0 ; in the ground voltage of the fault point F, the C two-phase voltage is the same as the ground voltage during normal operation, and the A and B two-phase ground voltage is 0 ; the differential current value of differential protection is close to 0. From

the above, it is known that when A and B phase faults occur, the voltage, current and differential current conform to the theoretical boundary conditions of single-phase fault, and the transformer protection does not act.
3) When three-phase fault symmetrical fault occurs outside the transformer fault model area, the fault point F ground current, ground voltage and transformer differential protection differential current waveform are shown in Figure 1.7.

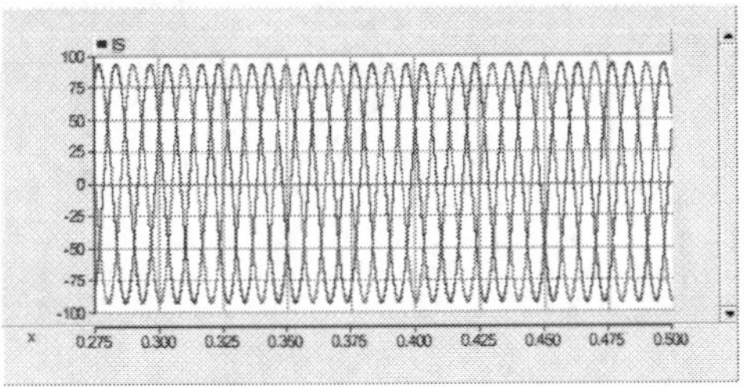

(a) Fault point F to ground current

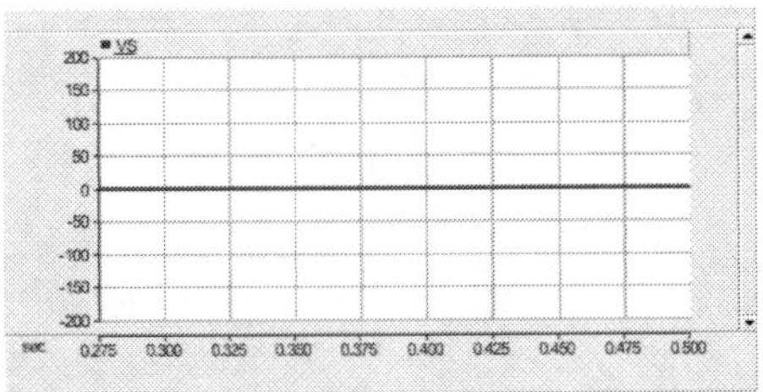

(b) Fault point F to ground voltage

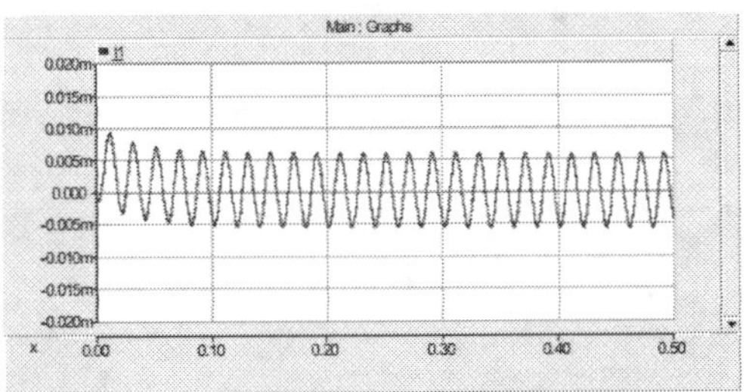

(c) Differential current

Fig.1.7 Three-phase symmetrical fault-to-ground current and voltage waveform

As shown in Fig.1.7, in the fault point F to ground current, the amplitude of A, B and C three-phase current to ground current is not 0, and the phase difference is 120 °. In the ground voltage of fault point F, the three-phase ground voltage of A, B and C is 0 ; the differential current value of differential protection is close to 0. From the above knowledge when A, B, C three-phase fault voltage, current and differential current in line with the theory of single-phase fault boundary conditions, transformer protection does not act.
According to the above analysis, the simulation results of the transformer external fault model established in this paper are in line with the theory under various external faults.

(3) Internal fault
1) When phase A inter-turn short circuit occurs in the transformer fault model [4], the differential current waveform of transformer differential protection is shown in Fig.1.8.

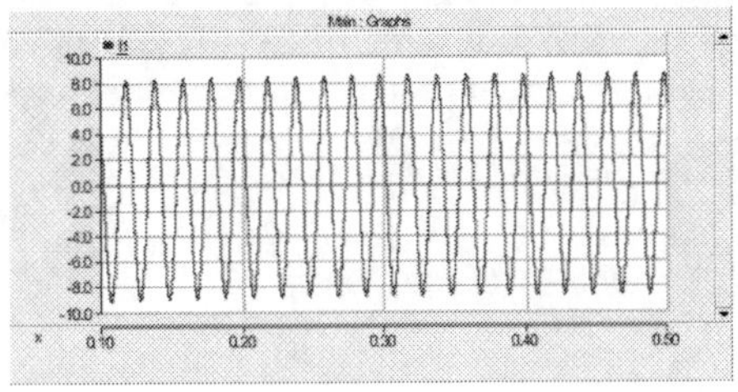

Fig.1.8 Differential current waveform while A phase inter-turn short circuit

Figure 1.8 shows that when a phase A inter-turn short circuit occurs in the area, the differential current is significantly larger than that of various faults outside the area, and the differential protection operates at this time.
2) The differential current waveform of transformer differential protection is shown in Fig.1.9 when the B-phase short circuit occurs in the transformer fault model.

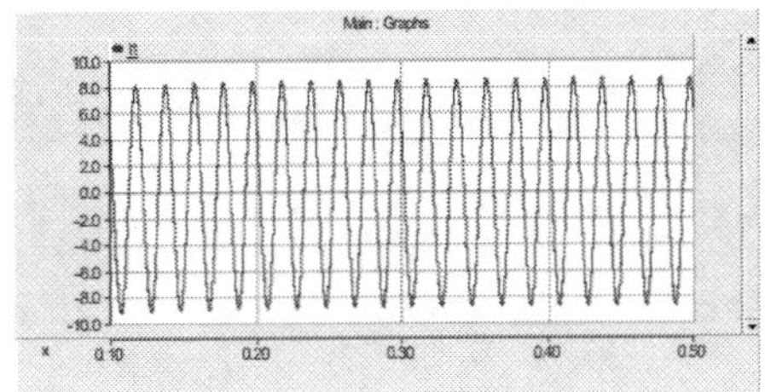

Fig.1.9 B phase turn-to-ground short circuit differential current waveform

As shown in Fig.3.9, when the phase B turn-to-ground short circuit occurs in the region, the differential current is similar to that of the phase A inter-turn short circuit in the region, which is significantly larger than that of various faults outside the region, and the transformer protection act this time.
3) When the ABC three-phase inter-turn short circuit occurs in the transformer fault model, the differential current waveform of the transformer differential protection is shown in Fig.1.10.

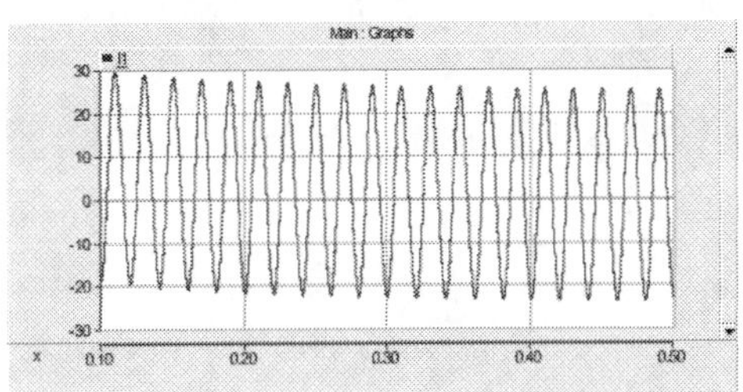

Fig.1.10 ABC three-phase inter-turn short circuit differential current waveform

Figure 1.10 shows that when the ABC three-phase inter-turn short circuit occurs in the area, the differential current increases compared with the A-phase inter-turn short circuit and the B-phase turn-to-ground short circuit in the area, and is larger than the various faults outside the area. At this time, the differential protection operates.
According to the above analysis, the simulation results of the transformer internal fault model established in this paper meet the requirements of transformer protection under various internal faults.

2.2.2 Comparative study of models

The external fault simulation model established in section 1.1 is consistent with the internal fault simulation model except for the different transformer components [5].In order to ensure the accuracy of the follow-up research, the two are compared for consistency.
Due to the YN / d11 connection mode adopted by both the external fault simulation model and the internal fault simulation model, the phase difference between the two sides of the transformer should be 30 ° in theory, and the simulation verification is shown in Figure 1.11.

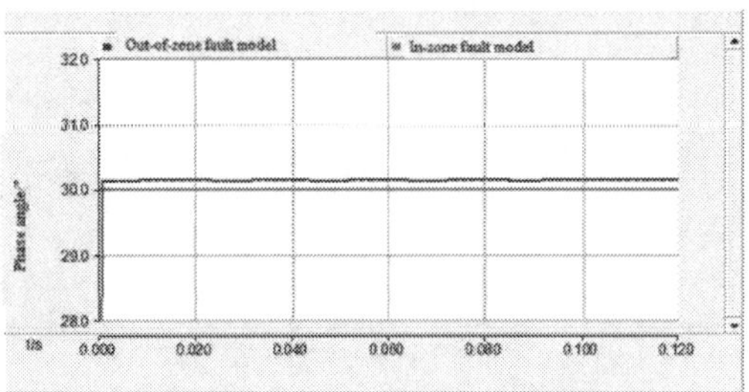

Fig.1.11 Phase difference on both sides of transformer

According to Fig.1.11, the phase difference between the two sides of the transformer in the external fault model is 30.09 °, and the phase difference between the two sides of the transformer in the internal fault simulation model is 29.99 °. Then, the error is

$$\Delta\varphi=\frac{30.09°\text{-}29.99°}{30°}\times100\%=0.33\%$$

, within the allowable range of error, it is considered that the external fault simulation model is consistent with the internal fault simulation model.
The phase A current comparison between the external fault simulation model and the internal fault simulation model under the same working conditions is shown in Fig.1.12 and Fig.1.13.

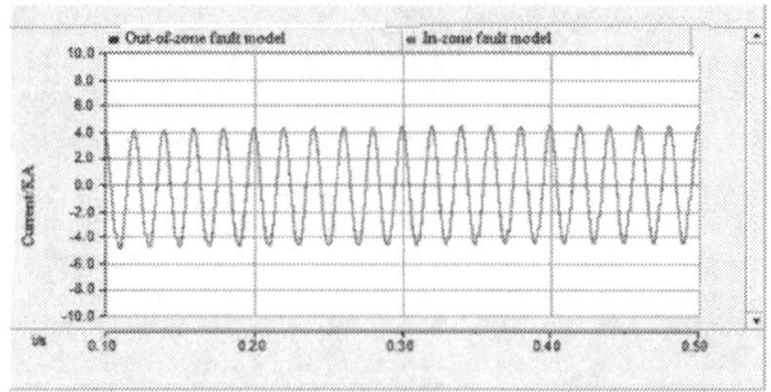

Figure 1.12 Phase A current of transformer primary side under normal working conditions

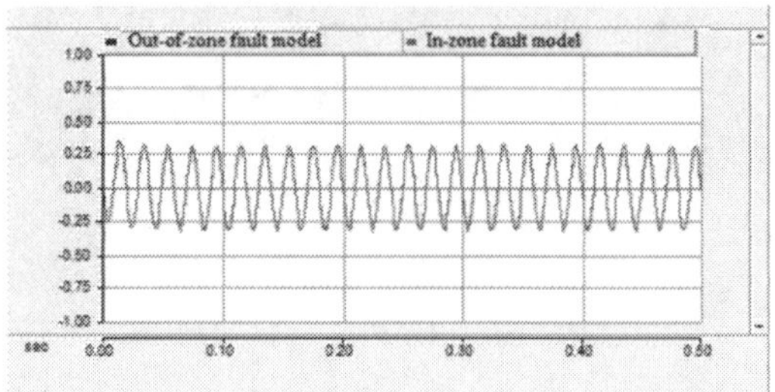

Fig.1.13 A phase grounding short circuit transformer primary side A phase current

The comparison of phase A current amplitude between the transformer external fault simulation model and the internal fault simulation model under different types of faults is shown in Table 1.2.

Table 1.2 A phase current amplitude comparison

Fault type	External fault model (kA)	Internal fault model (kA)
Normal operation	0.4974	0.4977
A phase grounding short circuit	0.4600	0.4605
B, C two-phase grounding short circuit	2.1491	2.1429
A, B, C three-phase grounding short circuit	4.8164	4.7923
B, C two-phase short circuit	2.1858	2.1760

The normal operation differential current of the external fault simulation model and the internal fault simulation model of the transformer is shown in Figure 1.14, and the results of the two simulation models are basically the same.

https://doi.org/10.1051/matecconf/202338201041

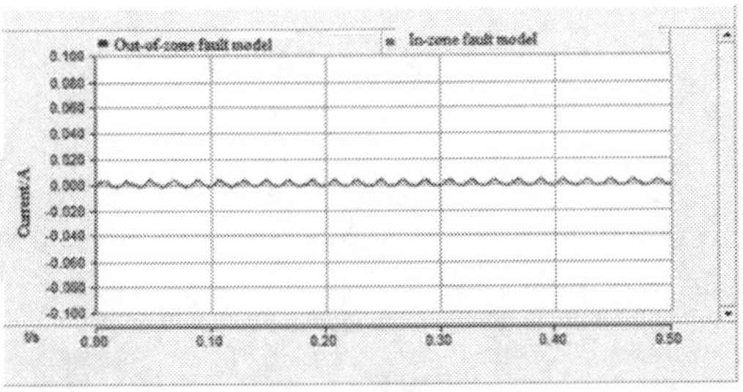

Fig.1.14 Differential current during normal operation of the line

When the A-phase short circuit occurs in the line, the differential current of the two simulation models of the transformer is shown in Figure 1.15, and the results of the two simulation models are basically the same.

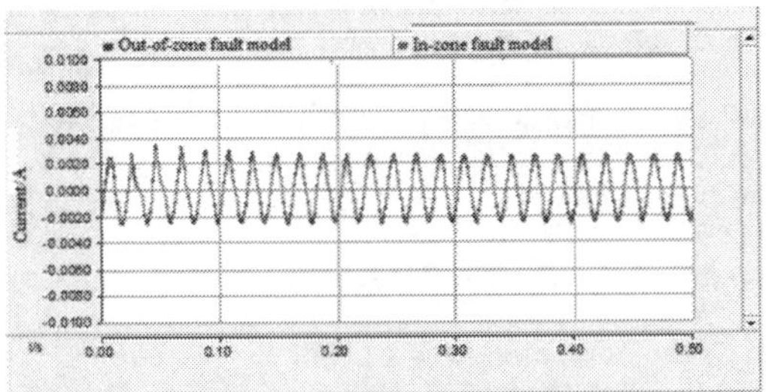

Fig.1.15 Differential current in A-phase fault

By comparing the A-phase current and differential current waveform of the two transformer simulations under different working conditions, the specific values of each fault are listed in the table for comparison [6]. Finally, it can be concluded that the transformer external fault simulation model and the internal fault simulation model are consistent within the allowable error range and can be replaced by each other. Therefore, the internal fault model is selected for convenience in the later simulation.

3. Oil flow simulation analysis

3.1 Transformer modeling

Taking 5000kVA, 35 / 10.5kV three-phase oil-immersed transformer as the object, the three-dimensional simulation is carried out to analyze the magnetic field, force vibration and oil flow rate change of the transformer core and winding. The finite element analysis method and the finite volume method are used for analysis and calculation. The transformer model parameters are as follows.

Table 2.1 Structural parameters of 35kV oil-immersed transformer

Transformer parameters	figure
High voltage winding height	733mm
Inner radius of high voltage winding	267.5mm
Outer radius of high voltage winding	315.5mm
High voltage winding width	3.4mm
Low voltage winding height	726mm
Inner radius of low voltage winding	195.5mm
Outer radius of low voltage winding	248mm
Low voltage winding width	3.2mm
Fuel tank size	2210*980*1860mm

The core of the simplified model of the transformer is silicon steel sheet, the permeability follows the B-H curve, the winding is copper wire, and there are 11 arc cooling tubes on both sides. In the fluid simulation, the inlet velocity is used to simulate the heat dissipation device. The resulting three-dimensional model of the transformer is shown in Fig.2.1. In order to simulate the different operating conditions of the transformer, the electromagnetic field simulation is connected to the external circuit, the primary side is connected to the voltage source, and the secondary side is connected to the load. The trigger switch is set to simulate the normal operation of the transformer, the external single-phase ground short-circuit fault and the internal one-phase coil short-circuit fault. The external circuit is shown in Figure 2.2.

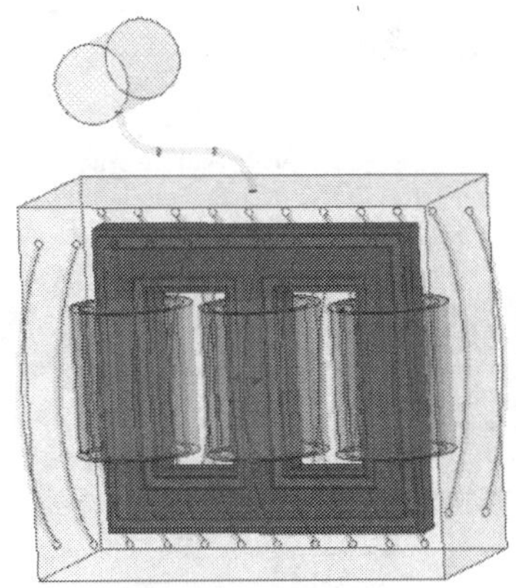

Fig 2.1 Simplified transformer model

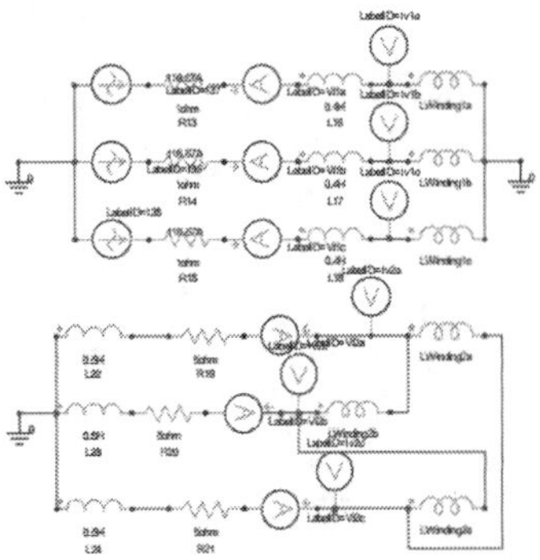

Figure 2.2 External circuit diagram

3.2 Electromagnetic field simulation

According to the above model, the electromagnetic field simulation calculation is carried out by using ANSYS Maxwell software, and the electromagnetic field distribution and the electromagnetic stress change of the coil under different operating conditions such as normal transformer, external single-phase ground short circuit and internal one-phase short circuit are obtained. When the simulation time is set to 1 s, the step size is 0.002 s, and 0.495 s (the maximum current of phase A), the trigger switch is used to simulate various faults.

Figure 2.3 is the magnetic field intensity distribution map when the transformer is 0.2 s. At this time, the transformer is in normal operation, and the transient current of phase A is 0 A. The simulation results show that the magnetic field intensity is mainly distributed with B and C phases. Figure 2.4 shows the distribution of magnetic induction intensity at the same time. The magnetic induction intensity at the upper and lower positions of the transformer is higher, and the magnetic induction intensity at the middle position is lower. The magnetic induction is mainly distributed in B and C phases. The simulation results are in line with the basic law of electromagnetic field [7].

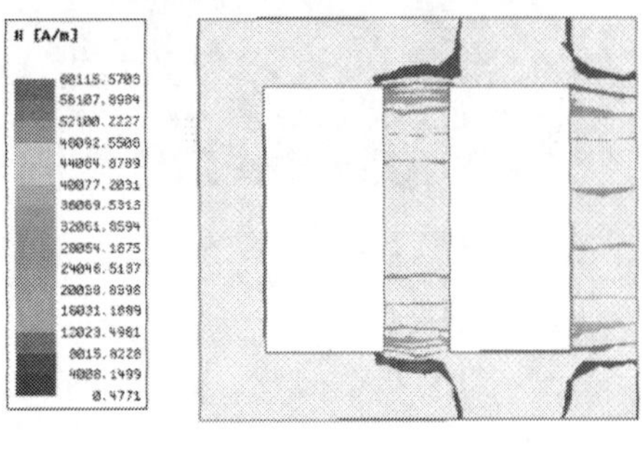

Fig 2.3 Magnetic field intensity distribution of transformer (0.2s)

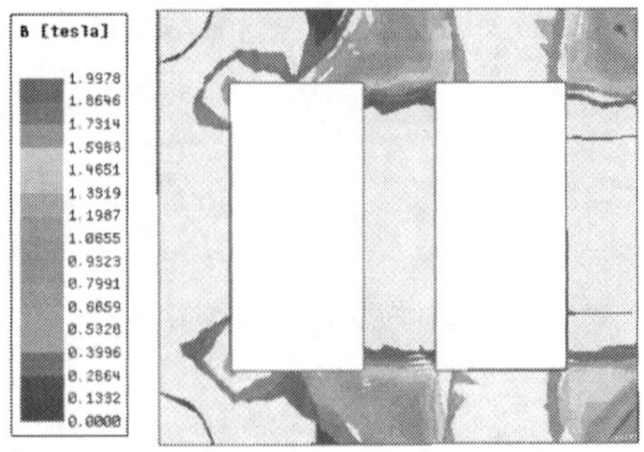

Fig.2.4 Distribution diagram of transformer magnetic induction intensity (0.2s)

When the phase A ground fault occurs outside the area, the coil is subjected to electromagnetic stress changes as shown in Figure 2.5 and Figure 2.6.

Figure 2.5 shows the stress change curve of the three-phase coil on the primary side of the transformer. Before 0.495 s, the transformer is in normal operation, and the coil force changes periodically. After the fault, the coil force increases to 77 N, and then the force begins to decay. Figure 2.6 shows the stress change curve of the three-phase coil on the secondary side of the transformer, and the change trend is similar to that on the primary side. The maximum force of the coil is 135 N.

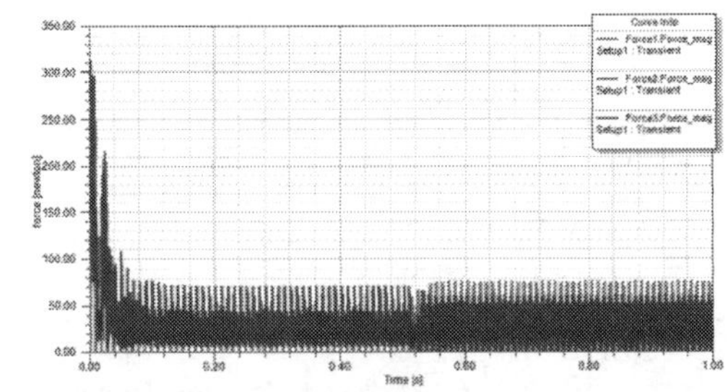

Fig. 2.5 Force change of three-phase coil on primary side of transformer

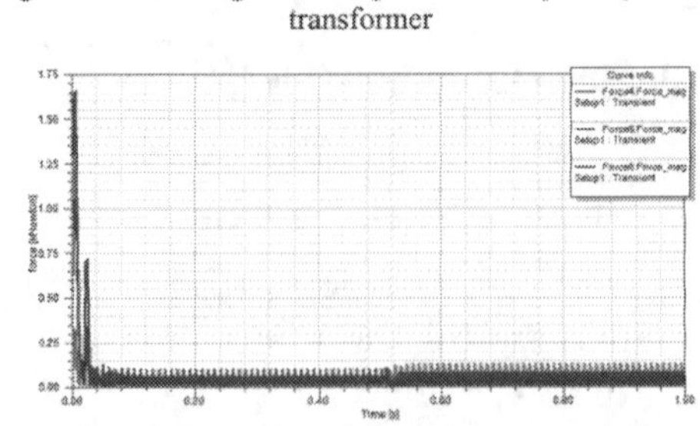

Figure 2.6 Force change of three-phase coil on secondary side of transformer

When a phase A short circuit fault occurs in the primary side of the transformer, the coil is subjected to electromagnetic stress changes as shown in Figure 2.7 and Figure 2.8.

Figure 2.7 is the stress change curve of the three-phase coil on the primary side of the transformer, and Figure 2.8 is the stress change curve of the three-phase coil on the secondary side of the transformer. The change trend of the

https://doi.org/10.1051/matecconf/202338201041

coil stress after 0.495 s fault is similar to that of the external fault. However, compared with the external fault, the coil force is greater at this time. The maximum force of the primary side coil reaches 370 N, and the secondary side reaches 340 N ; in addition, the transient process of the fault in the area is longer, the force change is more complex, and the change period is longer.

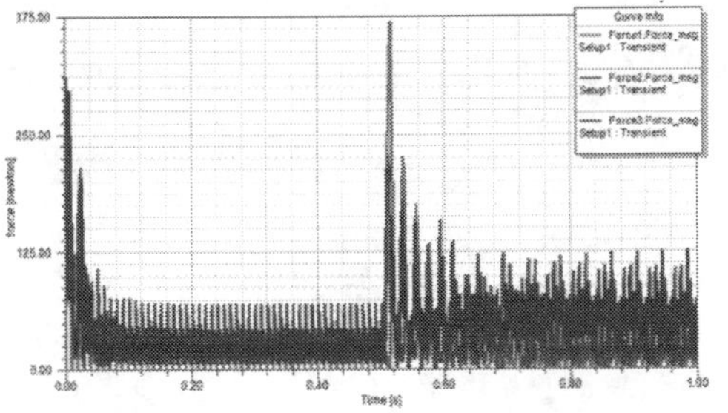

Fig.2.7 Stress change of three-phase coil on primary side of transformer

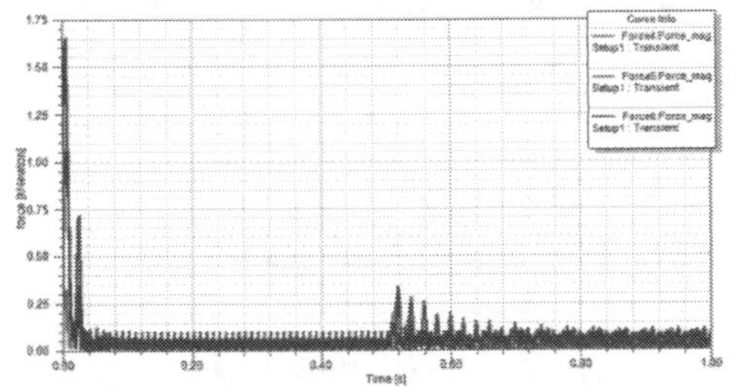

Fig.2.8 Stress change of three-phase coil on secondary side of transformer

3.3 Simulation of transformer oil flow field

The change of electromagnetic stress on the transformer coil will cause the vibration of the coil [8], which will cause the change of transformer oil flow rate. Taking the electromagnetic stress of the coil as the load, the curve of the outlet velocity change under different operating conditions of the transformer is calculated by using computational fluid dynamics. The simulation time is 0.1 s. During normal operation, the electromagnetic stress after 0.3-0.4 s of electromagnetic field simulation is selected as the load. When the fault occurs, the electromagnetic stress within 0.1 s after the electromagnetic field simulation fault is selected as the load. The initial flow rate is 0 m / s. The flow field simulation results at 300 K and 0.004 s are shown in Fig.2.9.

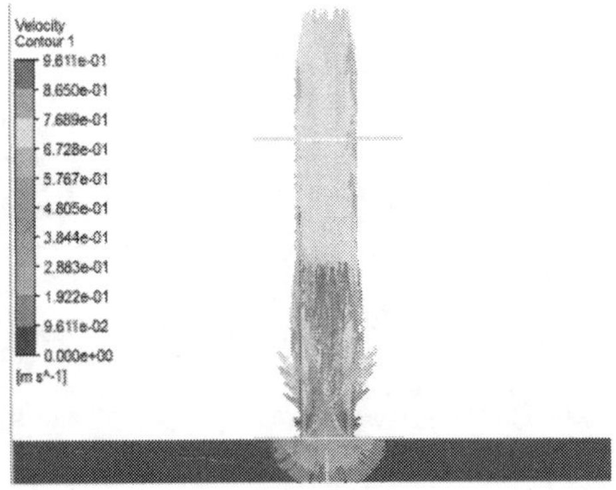

Fig.2.9 Outlet flow velocity vector diagram at normal operation (0.004s)

As shown in Fig.2.9 above, a flow velocity monitoring point is selected in the outlet center to measure the change of outlet flow velocity within 0.1 s, and the curve is drawn as shown in Fig.2.10.

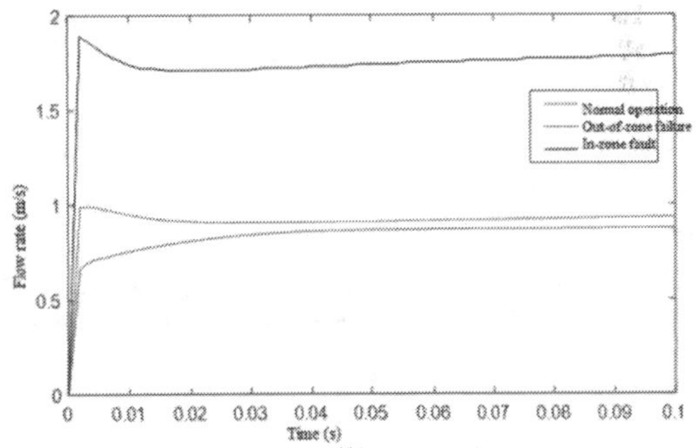

Figure 2.10 Outlet flow velocity curve

The initial flow velocity of the transformer in the diagram is 0 m / s. After the electromagnetic force load is applied, the flow velocity increases rapidly to the maximum value. Under different operating conditions, the flow velocity is also different due to the difference in the force change of the coil. During normal operation, due to the periodic change of coil force, the flow rate tends to be stable after increasing. When the fault occurs, the force of the coil is attenuated, and the flow rate is also attenuated after reaching the maximum value.

3.4 Results analysis

According to the simulation results of electromagnetic field and flow field [9-10], the simulation results of transformer under different operating conditions are shown in table 2.2.

 https://doi.org/10.1051/matecconf/202338201041

Table 2.2 Simulation data of transformer under different operating conditions

Transformer status	The maximum coil electromagnetic force (N)		Maximum outlet flow velocity (m/s)	Maximum outlet pressure (Pa)
	Primary side	Secondary side		
Normal operation	71	68	0.88	70500
External fault	73	117	1.00	71400
Internal fault	370	345	1.89	138000

From the above simulation analysis, it can be seen that the electromagnetic force received by the coil during the internal one-phase short-circuit fault of the transformer is much larger than that during normal operation and external single-phase ground short-circuit, and the resulting transformer outlet flow rate and outlet pressure are also greater. In addition, according to the changes of outlet flow velocity and pressure under different operating conditions, combined with the changes of related electrical quantities, it is helpful to identify the fault location and type, and avoid the influence of misoperation by locking when an external fault occurs.

4. Summary

In this paper, according to the transformer in different types of internal and external faults, voltage, differential current and other electrical quantities and transformer oil flow rate will produce corresponding changes, using PSCAD / EMTDC simulation software to establish a simulation model with transformer lines, using ANSYS Maxwell software for electromagnetic field simulation calculation, the relationship between electrical quantity and oil flow rate under different faults inside and outside the area is obtained. In this way, according to the changes of outlet flow velocity and pressure under different operating conditions, combined with the changes of related electrical quantities, it is helpful to identify the fault location and type, and better design the gas relay, so that it can avoid the influence of misoperation by blocking when an external fault occurs.

References

1. Yang Ling. A new method for waveform identification to prevent misoperation of transformers with faults outside the differential protection zone [D]. Tianjin: Tianjin University. 2005.
2. Yu Hua , Yang Gang , Wang Tianzheng , et al . Test and simulation analysis of transformer over-excitation Protection[J] , Transformer , 2015 , 52(11): 51-55(in Chinese) .
3. Yan Chenguang, Zhang Baohui, Hao Zhiguo, et al . Modeling and simulation of pressure characteristics of power transformer tanks' internal faults[J] . Proceedings of the CSEE , 2014 , 34(1) : 179-185(in Chinese) .
4. Yan Chenguang , Hao Zhiguo , Zhang Baohui . Power transformer digital protection based on internal pressure characteristics[J] . Proceedings of the CSEE , 2015 , 35(24) : 6405-6412(in Chinese) .
5. Tu Weimin , Cui Guomin , Li Yu , et al . Internal streaming simulation of oil-immersed transformers andon-line fault prediction method[J] . Power System Technology , 2010 , 34(5): 195-200(in Chinese) .
6. Liao Weihan , Guo Chuangxin , Jin Yu , et al . Oil-immersed transformer fault diagnosis method based on four-stage preprocessing and GBDT[J] . Power System Technology , 2019 , 43(6) : 2195-2203(in Chinese) .
7. Wang Longbo. Transformer oil flow analysis fault determination method [J]. Heilongjiang Electric Power, 2002,24(4): 273-276.
8. Huang Limei , Wang Wu , Lin Qiongbin , et al . Fault diagnosis method of converter based on kernel density estimation classifier[J] . Power System Technology , 2019 , 43(6) : 2204-2210(in Chinese) .
9. Pu Tianjiao , Li Ye , Chen Naishi , et al . Key technology and research framework for optimal operation control of hybrid AC/DC system based on power electronic transformer[J] . Power System Technology, 2018, 42(9): 2752-2759(in Chinese) .
10. Sun Qiu, Fan Chunju, Xu Feng, et al . Research on characteristics of pole-to-ground fault in mv power electronic transformer DC port[J] . Power System Technology, 2018, 42(11): 3616-3622(in Chinese) .

A Method for Monitoring Operation Status of Oil-immersed Transformer Based on the Baffle Angle Characteristic of Gas Relay

Peixian Bai, Yiling Xie, Longjiang Dou
Electrical Engineering and its Automation of North China Electric Power University), Baoding 071003, Hebei Province, China

Abstract: A gas relay in the transformer, an important non-electricity protection equipment, only triggers the stem leaf spring for the serious gas alarm when the baffle rotates to a certain angle, which cannot judge the transformer fault in the event of heavy gas before running state and fault development process, this paper puts forward a kind of based on gas relay panel corner of transformer running state monitoring methods. The partial structure of the gas relay is improved, and the baffle corner signal is extracted by installing a Hall sensor on the back of the baffle. On this basis, this paper develops a test stand capable of simulating the baffle characteristics of gas relays at different fault energies. Subsequently, the baffle angle variation law under different fault energies is studied, and the influence of pipe pressure and pipe flow rate on the baffle angle is given. The results show that, based on this improved gas relay, the real-time baffle angle information can reflect the internal operating state of the transformer and provide an effective method for transformer operating condition monitoring.

Key words: Transformer; gas relay; baffle angle; state monitoring

1. Introduction

The existing protection methods for oil-immersed transformers include differential protection [1], over-excitation protection [2], zero sequence current protection [3], gas protection [4], etc. The literature [5-6] proposes a protection method based on pressure characteristics by studying the pressure distribution characteristics inside the oil tank during transformer faults, in which the gas relay is a common non-electric protection method used for transformers, installed in the oil storage cabinet of the transformer and It is installed in the pipeline between the transformer's oil storage cabinet and the oil tank, using the internal fault of the transformer to cause the oil to decompose and produce gas or cause an oil flow surge, which causes the contact of the gas relay to operate, connects the designated control circuit and sends a timely signal to warn or activate the protection element to automatically remove the transformer [7-8].

The gas relay is a mechanical protection, the baffle in the process of displacement from 0 to the rectification value, the electromagnetic contact is jumped from a fully separated state to a fully closed state, the electrical signal of its exit is similar to the condition of 0 -1 level, that is, whether the heavy gas action. The current field installed gas relay heavy gas action rectification value is selected based on test results and operational experience provided by the former Soviet Union in the 1950s. The literature [9] proposes to adjust the heavy gas protection flow rate rectification value for the transformer in the case based on simulation, combined with field experience. The literature [10] proposes to increase the protection delay to avoid heavy gas misoperation, but all of these methods only consider the 2 state points of heavy gas not issuing an alarm signal and the flow rate value reaching the action setting value of heavy gas issuing an alarm signal. It is only through the heavy gas alarm that the transformer box is judged whether a serious fault has occurred, but ignores the entire process of the gas relay heavy gas baffle from sensing to flow rate to trigger protection, and cannot determine the operating state of the transformer box and the fault development process before the fault occurs to the heavy gas alarm [11-14]. In order to sense the operating state and fault development process of the transformer before the occurrence of a heavy gas fault, this paper makes a first theoretical investigation into the action that the gas relay will produce under transient oil flow, and it is feasible to provide a theoretical basis for improving the structure of the gas relay section through theoretical analysis [15].

2. The construction and principle of Gas relay

The QJ gas relay is widely used and its internal structure is shown in Figure 1. The QJ gas relay uses an open cup structure for light gas and a baffled magnetic contact structure for heavy gas. In normal operation, the relay is filled with oil and the float is immersed in the oil and in the upward floating position. When a serious fault occurs inside the transformer, it produces a strong gas, the pressure in the oil tank instantaneous sudden increase, resulting in a large oil flow to the direction of the oil pillow impact, because the oil flow impact baffle, baffle to overcome the resistance of the spring, drive the magnet to the direction of the reed contact, so that the reed contact closed, turn on the tripping circuit, so that the circuit breaker tripped, that is, heavy gas protection [16].

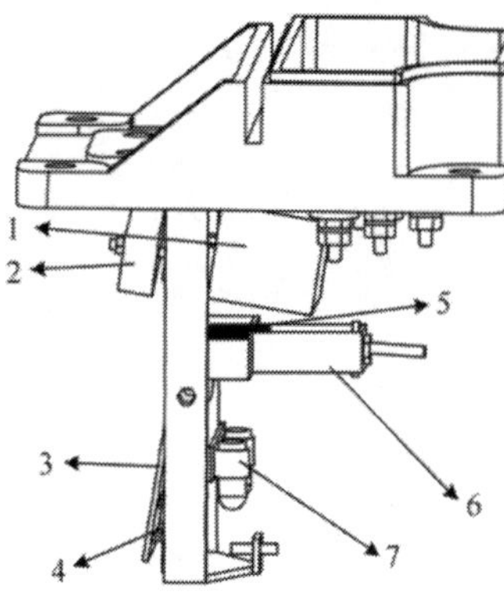

1-Open cups 2-Heavy hammers 3-Retainer 4-Magnets 5-Spring 6-Adjustment levers 7-Reed contacts

Fig. 1 Internal structure of QJ type gas relay

3. Theoretical basis for gas relay operation

3.1 Fundamentals of fluid dynamics theory

The study of fluid motion in pipes and the dynamic analysis of baffles requires a combination of physical conservation laws, of which the basic ones are: the law of conservation of mass, the law of conservation of momentum and the law of conservation of energy. The control equations are the mathematical description of these conservation laws.

3.1.1 Conservation of mass equation

In fluid motion, the law of immortality of matter also applies. This law can be expressed as follows: the addition of mass per unit time in a fluid micro-element is equal to the net mass flowing into that micro-element in the same time interval. This law leads to the equation of mass conservation.

$$\frac{\partial \rho}{\partial t}+\frac{\partial(\rho u)}{\partial x}+\frac{\partial(\rho v)}{\partial y}+\frac{\partial(\rho w)}{\partial z}=0 \quad (2\text{-}1)$$

Introduction of vector notation $div(a)=\frac{\partial a_x}{\partial x}+\frac{\partial a_y}{\partial y}+\frac{\partial a_z}{\partial z}$,equation (2-1) can be written as:

$$\frac{\partial \rho}{\partial t}+div(\rho u)=0 \quad (2\text{-}2)$$

Some of the literature uses ∇ to denote scatter, i.e. $\nabla \cdot a = div(a)=\frac{\partial a_x}{\partial x}+\frac{\partial a_y}{\partial y}+\frac{\partial a_z}{\partial z}$, equation (2-2) can be written as:

$$\frac{\partial \rho}{\partial t}+\nabla \cdot (\rho u)=0 \quad (2\text{-}3)$$

In equations (2-1) to (2-2), is the density, t is time, u is the velocity vector, and u, v and w are the components of the velocity vector u in the x, y and z directions.

Given above is the mass conservation equation for a transient three-dimensional compressible fluid. If the fluid is incompressible and the density p is a constant, equation (2-1) may become:

$$\frac{\partial u}{\partial x}+\frac{\partial v}{\partial y}+\frac{\partial w}{\partial z}=0 \quad (2\text{-}4)$$

If the flow is in a steady state, the density does not vary with time and equation (2-1) can be changed to:

$$\frac{\partial(\rho u)}{\partial x}+\frac{\partial(\rho v)}{\partial y}+\frac{\partial(\rho w)}{\partial z}=0 \quad (2\text{-}5)$$

3.1.2 Conservation of momentum equations

In practical engineering flow systems sometimes only the total combined forces are of interest, but not the distribution of the forces, so the law of conservation of momentum must be used to solve for this. The law is expressed as follows: the rate of change of the momentum of a fluid in a micro-element with respect to time is equal to the sum of the various external forces acting on the micro-element. This law is in fact Newton's second law. According to this law, the equations for the conservation of momentum in the x, y and z directions can be deduced.

$$\frac{\partial(\rho u)}{\partial t}+div(\rho u \mu)=div(\mu \cdot grad \cdot u)-\frac{\partial p}{\partial x}+S \quad (2\text{-}6)$$

$$\frac{\partial(\rho v)}{\partial t}+div(\rho v \mu)=div(\mu \cdot grad \cdot v)-\frac{\partial p}{\partial y}+S_v \quad (2\text{-}7)$$

$$\frac{\partial(\rho w)}{\partial t}+div(\rho w \mu)=div(\mu \cdot grad \cdot w)-\frac{\partial p}{\partial z}+S_w \quad (2\text{-}8)$$

In general , S_x , S_y and S_z are small quantities, and for incompressible fluids where the viscosity is a constant, $S_x = S_y = S_z = 0$.

3.1.3 Conservation of energy equation

When a flow system needs to take into account temperature and energy changes, then a fundamental equation is also needed, namely the equation of conservation of energy. It is the fundamental law that must be satisfied by a flow system containing heat exchange. For this law it can be expressed as follows: the rate of change of the total energy in the micro-element is equal to the net heat flux per unit time transferred to the

micro-element plus the work done on the micro-element by the external forces (physical and surface forces). This law is actually the first law of thermodynamics.
The energy E of a fluid is usually the internal energy i, the kinetic energy $K=\frac{1}{2}\left(u^2+v^2+w^2\right)$ and the sum of the potential energy P, we can create an energy equation for the total energy E. However, the energy conservation equation obtained in this way is not very useful and the change in kinetic energy is usually deducted from it. This gives us the conservation equation for the internal energy i. And we know that there is a relationship between the internal energy i and the temperature T, i.e. $i=c_P T$, where c_p is the specific heat capacity. This gives us the equation for the conservation of energy with temperature T as the variable.

$$\frac{\partial(\rho T)}{\partial t}+div(\rho\mu T)=div\left(\frac{k}{c_p}\cdot gradT\right)+S_T \qquad (2\text{-}12)$$

where c_p is the specific heat capacity, T is the temperature, k is the heat transfer coefficient of the fluid, S_T is the internal heat source of the fluid due to the viscous action of the fluid mechanical energy into heat energy fraction, referred to as S_T is the viscous dissipation term.

3.1.4 A generic form of the control equation

In order to facilitate the analysis of each of the above control equations and the use of the same program to solve each control equation when writing a computer program, a general form of each of the basic control equations is established. As can be seen from (2-2), (2-6) and (2-12), although the dependent variables in these equations are different, they all reflect the conservation of physical quantities per unit of time per unit of volume. If ϕ is used to denote the generic variable, then all of the above control equations can be expressed in the following generic form:

$$\frac{\partial(\rho\phi)}{\partial t}+div(\rho\mu\phi)=div(\Gamma\cdot grad\phi)+S \qquad (2\text{-}13)$$

where ϕ is a multi-generic variable that can represent the solution variables u, v, w, T, etc., Γ is the generalised diffusion coefficient, and S is the generalised source term. The terms in Eq. (2-13) are, in order, the transient, convective, diffusive and source terms. For a particular equation, ϕ , Γ and S have a specific form and Table 2-1 gives the correspondence between the three symbols and each characteristic equation.

Table 2-1 Specific forms of the symbols in the generic control equations

Formula	ϕ	Γ	S
Continuity equation	1	0	0
X-Momentum equation	u	μ	$-(\partial p/\partial x)+S_x$
y-Momentum equation	v	μ	$-(\partial p/\partial y)+S_y$
z-Momentum equation	w	μ	$-(\partial p/\partial z)+S_z$
Energy equation	i	$\frac{k}{c}$	S_T

Numerical simulation methods for turbulent flows
In nature, we often encounter fluid flow can be broadly divided into three types: laminar flow, transitional flow, turbulent flow. When the fluid flow velocity is very small and for laminar flow, the layers and layers do not mix with each other called laminar flow; with the flow velocity gradually increased, the fluid flow line began to appear wave-like oscillation, and the frequency and amplitude of oscillation with the increase in flow velocity and increase, the flow condition is called transitional flow; when the flow velocity increases to a large, the flow line is no longer clearly identifiable, there are many small swirls in the flow field, the phenomenon of sliding and mixing between adjacent layers, the flow condition is called turbulent. At this time, the flow condition is called turbulence.
When the fluid movement, commonly used Reynolds number Re of the fluid flow state to determine. Engineering general round pipe when the Reynolds number Re< 2320 for the laminar flow state, Re = 2320 ~ 4000 for the transition state, Re ⩾ 4000 for the turbulent state. For a circular cross-section pipe of diameter d, the Reynolds number calculation formula is as follows: $Re=vd/\upsilon$, where υ is the kinematic viscosity, d is the diameter of the pipe, v is the liquid flow rate. In this paper, in the numerical simulation calculation process 25# transformer oil density $\rho=0.895\text{kg}/m^3$, kinematic viscosity $\upsilon=9.6\times10^{-6}m^2/s$. According to the standard DL/T540-94 in the 500kv forced oil circulation transformer gas relay flow velocity range 1.3m/s~1.4m/s, the characteristic length is taken to connect the pipe diameter ϕ 80mm, corresponding to the range of Reynolds number 10833.33~ 11666.67, are greater than the internal flow critical Reynolds number $Re_{cr}=2320$, belong to turbulent flow, so this paper in the numerical simulation calculation of the oil flow state according to turbulent treatment.
Turbulence is the most common and complex flow phenomenon in nature, and many problems have not yet been effectively solved. There are currently two main approaches to numerical simulation of turbulence problems: direct numerical simulation and non-direct numerical simulation. The direct numerical simulation method is the direct use of the instantaneous Navier-Stokes equations to solve turbulence problems. In theory, relatively accurate results can be obtained using this method, but in practice, the resolution of the computational grid has to be high enough to simulate the smallest scale vortices, and as turbulence is a multi-scale

irregular flow, information on all scales of flow has to be obtained for accurate computation, thus the number of grid nodes in the whole computational domain is very large in magnitude and the computing performance of the computer is also very demanding. Based on the current level of development of computer technology, direct numerical simulations can only be used to calculate simple turbulent motions with low Reynolds numbers, and cannot yet be used to calculate complex turbulent motions. Instead of solving the problem directly by numerical calculation, the non-direct simulation method is to build a turbulence model and to simplify the calculation model to a certain degree of approximation, and then carry out the calculation. Non-direct simulation methods can be divided into large eddy simulation, Reynolds averaging and statistical averaging methods. In contrast, the non-direct simulation method requires relatively low computer performance and is relatively fast, making it the most common numerical simulation method in use today.

Modern engineering numerical calculations are more often carried out by solving the time-homogenised Navier-Stokes equations (RANS method). Since the time-homogenised set of Navier-Stokes equations is not closed, additional equations need to be introduced to close the set of equations, which leads to the concept of a turbulence model. The so-called turbulence model is a theory based on theoretical knowledge of turbulence, experimental data and experience to create expressions or equations for the Reynolds stress terms in the Reynolds mean equations of motion and to make reasonable modifications to some of the terms in the Reynolds stress equations so that the set of mean Reynolds equations for turbulence is closed. Turbulence models can be classified into Reynolds stress models and vortex viscosity models depending on the assumptions made or the treatment given to Reynolds stress.

(1)Reynolds Stress Model In the Reynolds Stress Model approach, equations representing Reynolds stress are constructed directly. Typically, the Reynolds stress equation is in differential form and is referred to as a Reynolds stress equation model. If the differential form of the Reynolds stress equation is reduced to the form of an algebraic equation, the model is referred to as an algebraic stress equation model.

(2)Vortex viscosity model In the vortex viscosity model approach, instead of dealing directly with the Reynolds stress term, a vortex viscosity factor is introduced so that the following relationship is established between the Reynolds stress of the turbulent flow and the mean velocity gradient in the turbulent field, i.e:

$$-\rho\overline{u_i'u_j'} = \mu_t\left(\frac{\partial u_i}{\partial x_j}+\frac{\partial u_j}{\partial x_i}\right)-\frac{2}{3}\left(\rho k+\mu_t\frac{\partial u_i}{\partial x_i}\right)\delta_{ij} \qquad (2\text{-}14)$$

where μ_i is the time-averaged velocity; μ_t is the turbulent viscosity, δ is a function of the spatial coordinates and depends on the flow state rather than the physical parameters; and k is the turbulent kinetic energy, which is expressed as follows:

$$k = \frac{\overline{u_i'u_i'}}{2} = \frac{1}{2}(\overline{u'^2}+\overline{v'^2}+\overline{w'^2}) \qquad (2\text{-}15)$$

The vortex viscosity model is based on Boussinesq's vortex viscosity hypothesis. In terms of the number of equations given, the vortex viscosity model can be divided into a zero equation model, a one equation model and a two equation model. In contrast, the two-equation model is a significant improvement over the zero-equation and one-equation models and has been most widely tested and successfully applied in scientific research and engineering practice.

The two-equation model, known as the standard k- ε model, is formed by introducing an equation for the turbulent dissipation rate ε on top of the equation for the turbulent kinetic energy k. In the model, ε , which represents the turbulent dissipation rate, is defined as:

$$\varepsilon = \frac{\mu}{\rho}\left(\frac{\partial u_i'}{\partial x_k}\right)\left(\frac{\partial u_i'}{\partial x_k}\right) \qquad (2\text{-}16)$$

The turbulent viscosity class can be expressed as a function of μ_t and ε , i.e:

$$\mu_t = \rho C_\mu \frac{k^2}{\varepsilon} \qquad (2\text{-}17)$$

where C_μ is an empirical constant.//

However, the standard $k-\varepsilon$ model produces some distortion when used for strong cyclonic flows, curved wall flows or curved flow lines. The reason for this is that in the standard $k-\varepsilon$ model the viscosity coefficient μ_t is assumed to be the same for each component of the Reynolds stress, i.e. it is assumed to be an isotropic scalar. In the case of curved flow lines, however, turbulence is anisotropic and μ_t should be an anisotropic tensor. In order to compensate for the shortcomings of the standard $k-\varepsilon$ model, the RNG $k-\varepsilon$ model was developed through extensive practice and refinement.

In the RNG $k-\varepsilon$ model, these small-scale motions are systematically removed from the control equations by reflecting the small-scale effects in the large scale motions and the modified viscosity terms. The resulting equations and equations are as follows:

$$\frac{\partial k}{\partial t}+\frac{\partial(u_j k)}{\partial x_j} = \frac{1}{\rho}\frac{\partial}{\partial x_j}\left[\left(\mu+\frac{\mu_t}{\sigma_{\varepsilon RNG}}\right)\frac{\partial k}{\partial x_j}\right]+\frac{1}{\rho}P_K-\varepsilon \qquad (2\text{-}18)$$

$$\frac{\partial k}{\partial t}+\frac{\partial(u_j k)}{\partial x_j} = \frac{1}{\rho}\frac{\partial}{\partial x_j}\left[\left(\mu+\frac{\mu_t}{\sigma_{\varepsilon RNG}}\right)\frac{\partial k}{\partial x_j}\right]+\frac{1}{\rho}\frac{\varepsilon}{k}C_{\varepsilon 1RNG}P_K-C_{\varepsilon 2RNG}\frac{\varepsilon^2}{k} \qquad (2\text{-}19)$$

where: P_K - the term for the generation of the speculative flow energy k due to the average velocity gradient, has the following equation:

$$P_k = \mu_t\left\{2\left[\left(\frac{\partial u}{\partial x}\right)^2+\left(\frac{\partial v}{\partial y}\right)^2+\left(\frac{\partial w}{\partial z}\right)^2\right]+\left(\frac{\partial u}{\partial y}+\frac{\partial v}{\partial x}\right)^2+\left(\frac{\partial u}{\partial z}+\frac{\partial w}{\partial x}\right)^2+\left(\frac{\partial v}{\partial z}+\frac{\partial w}{\partial y}\right)^2\right\} \qquad (2\text{-}20)$$

σ_{kRNG} 、 $\sigma_{\varepsilon RNG}$ -RNG $K-\varepsilon$ are empirical constants for turbulence models;

$C_{\varepsilon 1RNG}$ 、 $C_{\varepsilon 2RNG}$ -RNG $K-\varepsilon$ are turbulence model correction factors;

$$C_{\varepsilon 1RNG} = 1.42 - \frac{\eta(1-\eta/4.38)}{(1+\beta_{RNG}\eta^3)}, \qquad (2\text{-}21)$$
$$\eta = (P_K / \rho\varepsilon C_{\mu RNG})^{1/2} 。$$

The RNG $K-\varepsilon$ turbulence model is a turbulence model used more frequently in flow field simulation calculations. It has the following main advantages: firstly, it improves the calculation accuracy by correcting the turbulent viscosity and taking into account the turbulent swirl; secondly, the RNG theory provides an analytical formula that takes into account the viscosity of low Reynolds number flows, and this formula can correctly solve the flow problems in the near-wall region. Therefore, the RNG $K-\varepsilon$ model can better simulate rotational and bending flows. In this paper, the internal flow channel structure of the gas relay is complex and the liquid flow process is prone to vortex and other phenomena. In order to accurately deal with the resulting high strain on the fluid and the severe bending of the flow line, the RNG $K-\varepsilon$ turbulence model is used in this paper.

3.2 Theoretical analysis of the force conditions on the baffle (main body of the gas relay action)

3.2.1 Static analysis of baffles

The baffle in this paper is based on the QJ80 gas relay manufactured by Shenyang Yamin. The movement of the baffle around the rotating axis in this model can be considered as a rigid body rotating on a fixed axis and is subject to three types of moments: a heavy moment, a spring tensioning moment and an impact moment. This is shown in Figures 2-1a and 2-1b.

Fig. a QJ80 gas relay flap

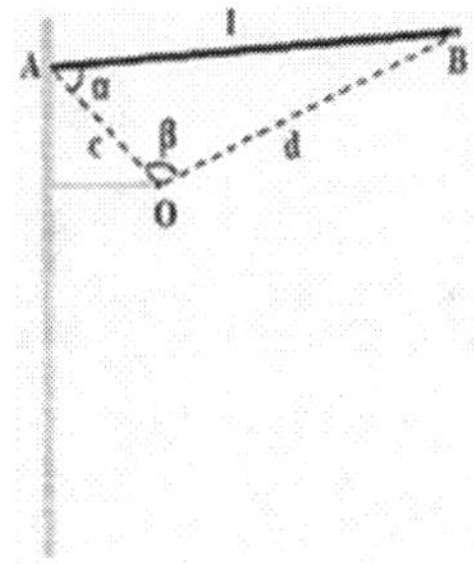

Fig. b Force analysis diagram of the baffle

Figure 2-1 Force analysis diagram of the baffle

(1) Calculation of the gravitational moment Mg

The centre of gravity of the baffle is just below the axis of rotation, so the weight moments are:

$$Mg = mg \cdot h \cdot l \sin\theta \qquad (2\text{-}22)$$

$$hl = h * \cos\theta = 0.36 * \cos\theta \qquad (2\text{-}23)$$

$$Mg = mg \cdot hl \cdot \sin\theta = 0.3(\qquad (2\text{-}24)$$

In the above equation: m is the mass of the baffle, initial value m=38g=0.038kg; h is the vertical distance from the centre of mass of the baffle to the axis of rotation, initial value h=0.36m; h1 is the vertical distance from the centre of mass of the baffle to the axis of rotation; θ is the angle of rotation of the baffle; g is the acceleration of gravity, initial value is taken as g=9.8N/kg.

(2) Calculation of the Spring tensioning torque Ms

A sketch of the calculation of the spring pair is shown in Figure 7-1. In the diagram: the line AB represents the length of the spring after stretching, the size of l, can be changed with the rotation of the baffle, let the initial length be l_1, the amount of change is Δl; c, d is the length geometry, α, β are the angle geometries. Let the direction of incoming flow be horizontal to the right, the initial position of the baffle is perpendicular to the velocity of the oil flow and it can rotate counterclockwise around point O under the action of the oil flow.

The expressions (2-25), (2-26) and (2-27) for the spring tensioning torque are:

$$M_S = C_1(1 - \frac{C_2}{\sqrt{C_3 - \cos(\theta+\theta_1)}})\sin(\theta+\theta_1) \qquad (2\text{-}25)$$

$$\theta_1 = \arccos(\frac{c^2+d^2-l_1^2}{2cd}) \qquad (2\text{-}26)$$

$$C_1 = \mathrm{ckd}, C_2 = \frac{\mathrm{l_1 - \Delta l}}{\sqrt{\mathrm{2cd}}}, C_3 = \frac{\mathrm{c^2+d^2}}{\mathrm{2cd}} \qquad (2\text{-}27)$$

The spring tensioning moments corresponding to several special angle values of 0°, 5°, 10°, 15°, 20° and 25° are calculated by the above equation and are 0.0778, 0.0809, 0.0830, 0.0841, 0.0842 and 0.0832 respectively.

According to the formula (2-24) Mg= mg, h1, sin θ=0.36*mg, sin θ.cos θ , and the formula Mf= Ms+Mg,, the weight moment Mg and impact moment Mf of the baffle at 0°, 5°, 10°, 15°, 20° and 25° are calculated as shown in Table 2-2 below:

Table 2-2 Impact moments at different angles

θ	0	5	10	15	20	25
Mg(N.m)	0	0.0116	0.0229	0.0335	0.0431	0.0514
Ms(N.m)	0.0778	0.0809	0.0830	0.0841	0.0842	0.0832
Mf(N.m)	0.0778	0.0925	0.1059	0.1176	0.1273	0.1346

Using the software, these points are fitted to a curve as shown in Figure 2-2 below to reflect the relationship between the baffle angle and the weight moment M_g. Figure 2-3 shows the relationship between the baffle angle of rotation and the spring tensioning moment M_S and the impact moment M_f.

The above data results are analysed without taking into account the influence of the complex flow field around the oil flow and baffles of the pipeline. The results are intended as qualitative conclusions only and the trends can be used as a reference for the results of the simulation.

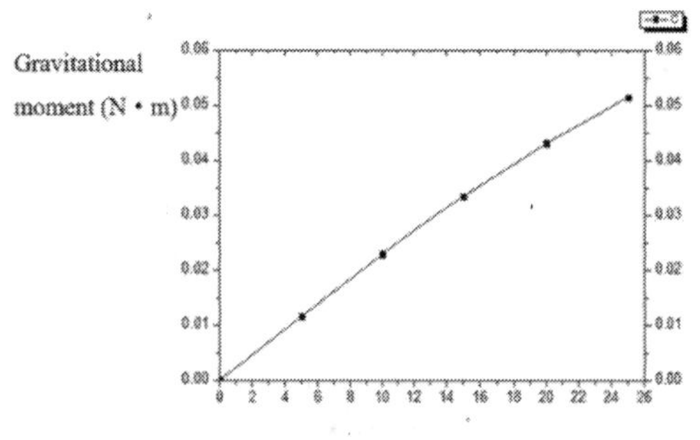

Figure 2-2 Relationship between baffle angle and gravitational moment Mg

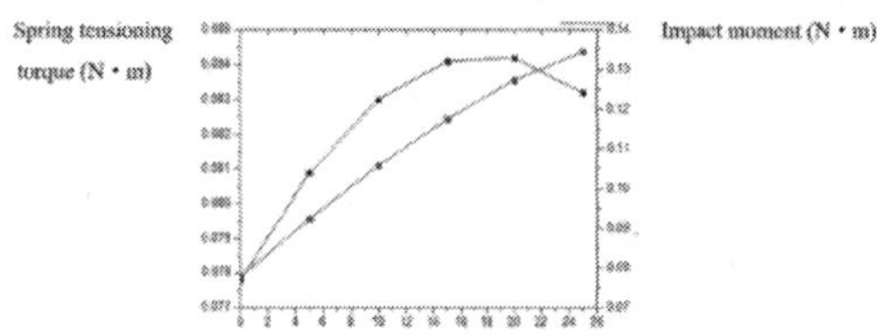

Figure 2-3 Relationship between turning angle and Ms and Mf

As can be seen from Figures 2-2 and 2-3, as the baffle angle increases, the baffle weight moment M_g, the spring tensioning moment M_S and the impact moment m all increase. The trend of the data shows that the larger the impact moment, the larger the baffle angle occurs when the oil flow surges and impacts the baffle.

Based on the processing of the data in Table 2-1, the following data were obtained as shown in Table 2-2:

Table 2-2 Variations of the three moments for equal angular variations

Θ (°)	Δθ (°)	ΔMg(N.m)	ΔMs(N.m)	Δ*Mf*(N.m)
0-5	5	0.0116	0.0031	0.0147
5-10	5	0.0113	0.0021	0.0134
10-15	5	0.0106	0.0011	0.0117
15-20	5	0.0096	0.0001	0.0097
20-25	5	0.0083	-0.001	0.0073

As can be seen from Table 2-2, the corresponding ΔMg, ΔMs and ΔMf (N.m) gradually decrease at two adjacent corner spacings $\Delta\theta$ = 5°. This data shows that at the beginning, when the baffle angle changes by 5°, the impact moment changes more, and the further the baffle angle continues to change by 5°, the impact moment required gradually decreases, which is related to the change in the acceleration of the fluid flow rate.

3.2.2 *Analysis of baffle dynamics*

(1)Impact moment Mf

The oil flow in the relay is turbulent and the flow field around the baffle is complex, with the impact of the oil flow on the front of the baffle and the adsorption of the low pressure vortex zone at the back of the baffle, making the total moment of oil flow impossible to solve accurately from theoretical calculations. Fluent software can calculate the transient flow field and then find the total moment of oil flow on the baffle under this transient flow field. The moments on the computed surface area are extracted by the macro function Compute_ Force _And _Moment. The total impact moment on the baffle is calculated and returned by specifying the inertia and axis of rotation of the baffle in the parameters of the macro function.

The simulation flow diagram for the oil flow impact baffle is shown in Figure 2-4.

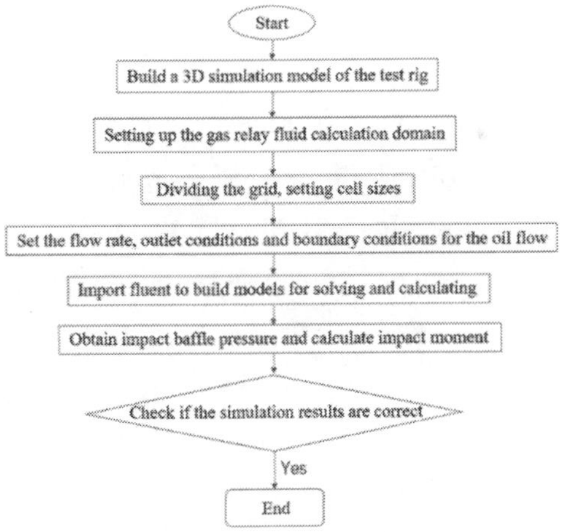

Figure 2-4 Simulation flow diagram of oil flow impact baffle

(2)Total torque *Mt*

$$Mt = Mf - Ms - Mg \quad (2\text{-}28)$$

The angular acceleration Wn is obtained by dividing Mt by the inertia of the baffle along the axis of rotation I. For programming purposes, the angular velocity Wn at the current moment is expressed in differential format.

$$Wn = Wn-1 + (Mt/I)\Delta t \quad (2\text{-}29)$$

In equation (2-29):Wn-1 is the angular velocity at the previous moment; Δt is the time step. Then the instantaneous linear velocity of the sensor is also the instantaneous flow velocity of the impact baffle.

$$Vn = Wn \,.\, R \quad (2\text{-}30)$$

In equation (2-30) R is the radius of the sensor's rotating shaft. The above is a forward derivation of the equation for the vertical displacement h of the gas relay baffle movement versus the velocity Wn or Vn, we are inverting and verifying the movement characteristics of the baffle by means of the sensor. Compared to the static analysis, the impact moment Mf in the kinetic analysis is closer to the real situation and the theoretical calculation results are more accurate.

3.3 Discrete inverse algorithm for transient oil flow (steady state calculation method)

In order to quantitatively study the specific relationship between the baffle under the action of oil flow impact, firstly by the dimensional analysis, the dimensionless number related to the resistance moment is obtained, and then the specific relationship of the dimensionless number is obtained by the method of fluent simulation test, the advantage of this method is that the combination of static analysis and dynamics analysis, is the simulation calculation results are more realistic.

3.3.1 Dimensionless relations for resistance moments

When studying the specific relationships that affect the resistance torque of a relay baffle using dimensional analysis, a necessary condition is to find all the factors that affect the resistance torque. As the baffle is impacted by the oil flow, the first consideration is the components of liquid dynamics: density ρ and velocity v. Secondly, the influence of liquid viscosity is considered, i.e. the dynamic viscosity μ . Finally, consider the geometric elements that affect the drag moment coefficient: the baffle area, the pipe diameter and the baffle axis and its centre of gravity, the eccentricity of the formation of ε . The above factors are unified into the following equation:

$$M = f(\rho, u, \mu, A, d, \varepsilon) \quad (2\text{-}31)$$

where the quantitative equation for each variable is expressed as:

$$\begin{aligned} &\dim M = ML^2T^{-2} \\ &\dim\rho = ML^{-3} \\ &\dim u = LT^{-1} \\ &\dim\mu = ML^{-1}T^{-1} \qquad (2\text{-}32) \\ &\dim d = L \\ &\dim A = L^2 \\ &\dim\varepsilon = L \end{aligned}$$

Choosing u, d and ρ among them as independent variables, the above equation is written in the following scale using the π theorem:

Table 2-4 Matrix of Quantities

	ρ	u	d	M	μ	A	ε	θ
M	1	0	0	1	1	0	0	0
L	-3	1	1	2	-1	2	1	0
T	0	-1	0	-2	-1	0	0	0

A primary transformation of the matrix gives:

Table 2-5 Table of transformed magnitude matrices

	ρ	u	d	M	μ	A	ε	θ
M	1	0	0	1	1	0	0	0
L	0	1	0	2	1	0	0	0
T	0	0	0	3	1	2	1	0

From this we have the dimensionless numbers $\pi_1 = \frac{M}{\rho u^2 d^3}$, $\pi_2 = \frac{\mu}{\rho u d}$, $\pi_3 = \frac{A}{d^2}$, $\pi_4 = \frac{\varepsilon}{d}$ and $\pi_5 = \theta$ which represent the drag coefficient, the reciprocal of the Reynolds number, the dimensionless area and the dimensionless bias canton distance respectively. This leads to:

$$F(\pi_1, \pi_2, \pi_3, \pi_4, \pi_5) = 0 \quad (2\text{-}32)$$

$$\frac{M}{\rho u^2 d^3} = f(\frac{\mu}{\rho u d}, \frac{A}{d^2}, \frac{\varepsilon}{d}, \theta) \quad (2\text{-}33)$$

As this paper is only for the domestic QJ-80 relay to study the calculation, without considering the baffle and the change of eccentricity, the above equation can be reduced to:

$$\frac{M}{\rho u^2 d^3} = g(\frac{\rho u d}{\mu}, \theta) \quad (2\text{-}34)$$

After defining the resistance moment coefficient $C_m = \frac{M}{\rho u^2 d^3}$, the Reynolds number $Re = \frac{\rho u d}{\mu}$, the above equation can be expressed as:

$$C_m = g(\mathrm{Re}, \theta) \quad (2\text{-}35)$$

3.3.2 Results of fluent simulations

In order to obtain the variation of the baffle moment coefficient C_m with the incoming flow velocity θ and the opening degree q, different velocities and different openings are combined into different operating conditions for calculation.

The velocity ranges from 0.25m/s to 1.5m/s, with one operating point taken every 2.5°, for a total of 6; the baffle angle ranges from 0° to 25°, with one operating point taken every 2.5°, for a total of 11. This makes a total of 11*6 operating points, and for each case a steady state analysis is carried out to extract the drag coefficients summarised in Table 2-6 below.

Table 2-6 Summary of resistance coefficients for 66 operating conditions calculated by fluent steady state

(Transverse gauge speed m/s, longitudinal gauge opening °)

	0.25	0.5	0.75	1	1.25	1.5
0	0.1228101	0.1181215	0.11625	0.11538	0.11479	0.11454
2.5	0.116469	0.1117809	0.11019	0.10959	0.10922	0.10891
5	0.114862	0.1103615	0.10856	0.10775	0.10731	0.10704
7.5	0.114862	0.10994	0.10793	0.10696	0.10659	0.10649
10	0.1133553	0.10861	0.10667	0.10561	0.10523	0.10519
12.5	0.1102737	0.10577	0.10383	0.10302	0.10278	0.10266
15	0.108205	0.10391	0.10197	0.10104	0.10062	0.10058
17.5	0.1058175	0.10185	0.100139	0.09916	0.09887	0.09869
20	0.1040161	0.09934	0.09743	0.09641	0.09584	0.09532
22.5	0.1010543	0.09717	0.09547	0.09469	0.09469	0.09385
25	0.0951355	0.09111	0.08929	0.08837	0.08801	0.08766

The following equation was fitted in MATLAB using non-linear regression:

$$C_m = c_{00} + c_{10}\mathrm{Re} + c_{01}\theta + c_{20}\mathrm{Re}^2 + c_{02}\theta^2 + c_{30}\mathrm{Re}^3 + c_{21}\mathrm{Re}^2\theta + c_{12}\mathrm{Re}\theta^2 + c_{03}\theta^3 + c_{40}\mathrm{Re}^4 + c_{31}\mathrm{Re}^3\theta + c_{22}\mathrm{Re}^2\theta^2 + c_{13}\mathrm{Re}\theta^3 \quad (2\text{-}36)$$

The coefficients are as follows (95% confidence intervals in parentheses):

$$c_{00} = 0.1312(0.1249, 0.1374)$$
$$c_{10} = -6.111e^{-006}(-1.045e^{-005}, -1.769e^{-006})$$
$$c_{01} = -0.09418(-0.1366, -0.05179)$$
$$c_{20} = 8.812e^{-010}(-1.418e^{-010}, 1.904e^{-009})$$
$$c_{11} = 2.022e^{-007}(-1.203e^{-005}, 1.244e^{-005})$$
$$c_{02} = 0.3071(0.1198, 0.4944)$$
$$c_{30} = -5.883e^{-014}(-1.576e^{-013}, 3.994e^{-014})$$
$$c_{21} = -1.037e^{-010}(-1.479e^{-009}, 1.272e^{-009})$$
$$c_{12} = 2.66e^{-006}(-2.636e^{-005}, 3.168e^{-005})$$
$$c_{03} = -0.5423(-0.8128, -0.2717)$$
$$c_{40} = 1.48e^{-018}(-1.868e^{-018}, 4.829e^{-018})$$
$$c_{31} = 8.538e^{-015}(-4.844e^{-014}, 6.551e^{-014})$$
$$c_{22} = -2.797e^{-010}(-1.563e^{-009}, 1.004e^{-009})$$
$$c_{13} = 2.513e^{-006}(-3.084e^{-005}, 3.586e^{-005})$$

Quality indicator of the fitted formula; SSE (sum of squares of error terms) 4.651e-005

R-square (R-squared): 0.9884

Adjusted R-square (adjusted R-square): 0.9856

RMSE (root mean squared error): 0.0009458

The above indicators reflect the very good quality of the regression fit. The fitted images are shown in 2-5.

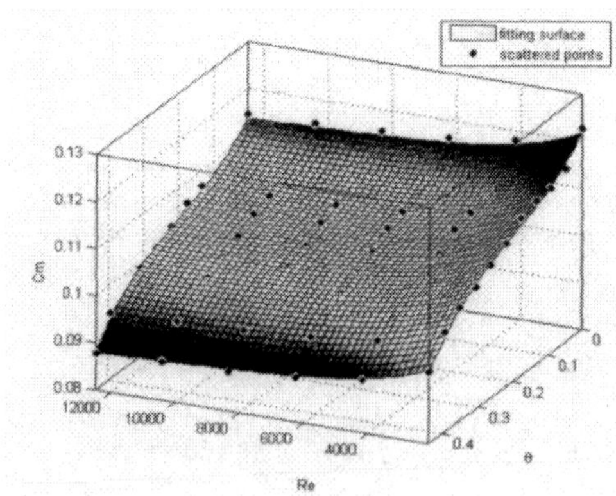

Figure 2-5 Three-dimensional image of the moment coefficient with C_m as a function of Reynolds number Re and θ

The resistive moment coefficient in equation (2-36) is defined as follows:

$$M_f = C_m(u, \theta)\rho u^2 d^3 \quad (2\text{-}37)$$

The moment of resistance of the spring is abbreviated by the equation (2-38):

$$M_s = f(\theta) \quad (2\text{-}38)$$

Assuming that the water flow moment is just sufficient to open the baffle to the alarm position, the condition (2-39) is satisfied.

$$M_f = M_s \quad (2\text{-}39)$$

The speed adjustment value u_{set} is obtained when the angle θ reaches the alarm angle $\theta_{\max}$ by combining equations (2-37) to (2-39).

$$C_{m}(u_{set},\theta_{\max})\rho u_{set}^{2}d^{2} = f(\theta_{\max}) \qquad (2\text{-}40)$$

With the help of the mathematical software Matlab, the value of u_{set} in the implicit function equation is solved numerically, i.e. the value of the gas relay speed is obtained. In this paper, from the above data, substituted in (2-40), we get u_{set} = 1.268m/s.

Based on the above principle, taking the domestic Shenyang explosion-proof QJ1-80 baffle type 80 calibre gas relay 500KV forced external circulator as an example, its rectification value range is 1.3~1.4m/s, the initial rectification value is set at 1.25m/s. Using the operation calculation described in Figure 2-6, the final rectification value is stabilised at about 0.821m/s after multiple sets of calculations, each set of 10 iterations.

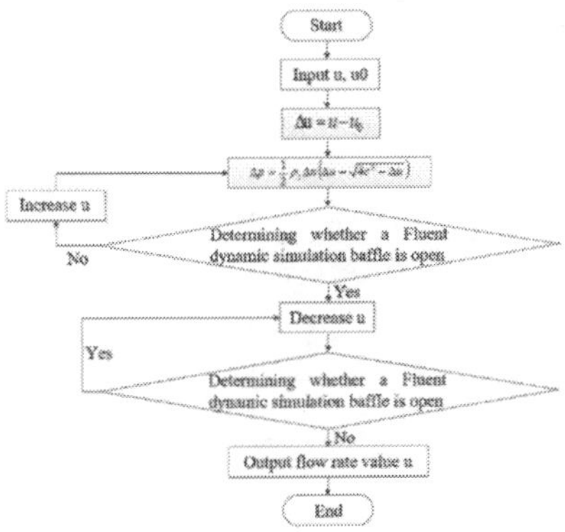

Figure 2-6 Flow diagram for calculating flow velocity by inverse calculation of baffle opening and closing angle

Detailed data are shown in Tables 2-6.

Table 2-6 Summary results of back-calculated flow rate values

Serial number	Back-calculated flow rate values (rectified values)m/s
1	0.817218470
2	0.825025491
3	0.820613796
4	0.822892186
5	0.824611742
6	0.818956810
7	0.823747245
8	0.817953860
9	0.818978072
10	0.816125413
Average value	0.820702309

As can be seen, the value of the calibration is much lower when the transient shock effect is taken into account. In essence, this corresponds to a steady state velocity of 1.268m/s calibrated for the gas relay, which is equivalent to a transient value of 0.821m/s when the shock effect is taken into account, so it can be seen that the inertial force of the transient effect cannot be ignored and this section provides a theoretical basis for the calculation of transient flow velocity values for the gas relay.

4. Summary

This paper analyses the theory involved in the operation of a gas relay under transient oil flow from the conservation of mass equation, conservation of momentum equation, conservation of energy equation and control equation in fluid dynamics. Secondly, this paper analyses the turbulence values of gas relays. The paper analyses the weight moment, spring tension moment and impact moment of the baffle plate at rest in conjunction with the main body of the gas relay action - the baffle plate, the dimensionless relationship of the resistance moment of the baffle plate under transient oil flow and analyses fluent simulations. These analyses are consistent with the actual gas relay operation under transient oil flow.

References

1. Weng Hanli, Liu Hua, Lin Xiangning, et al . A novel criterion of the transformer differential protection based on the Hausdorff distance algorithm[J] . Proceedings of the CSEE, 2018, 38(2): 475-483(inChinese) .
2. Yu Hua , Yang Gang , Wang Tianzheng , et al . Test and simulation analysis of transformer over-excitation Protection[J] , Transformer , 2015 , 52(11): 51-55(in Chinese) .
3. Fang Yudong , Xu Xidong , Zhu Bingquan . Influence of transformer inrush on zero-sequence current protection[J] . Electric Power Automation Equipment , 2008 , 28(9) : 115-118(in Chinese) .
4. Zhang Yunfeng . Reason and measure about transform gas profect motion[J] . Power System Technology, 2008, 32(S2): 300-302(in Chinese) .
5. Yan Chenguang, Zhang Baohui, Hao Zhiguo, et al . Modeling and simulation of pressure characteristics of power transformer tanks' internal faults[J] . Proceedings of the CSEE , 2014 , 34(1) : 179-185(in Chinese) .
6. Yan Chenguang , Hao Zhiguo , Zhang Baohui . Power transformer digital protection based on internal pressure characteristics[J] . Proceedings of the CSEE , 2015 , 35(24) : 6405-6412(in Chinese) .
7. Guan Deqiu, Cao Yaping . The analysis of reason of act of light gas relay in transformer[J] . High

Voltage Engineering, 1996, 21(1): 79-80(in Chinese).

8. Zhao Youyang. The countermeasure for frequently operation of gas relay of on load tap changer[J]. High Voltage Engineering, 2004, 30(S1): 130-131(in Chinese).
9. Yang Xian, Ke Chunjun, Zhou Dan. Research on prevention strategy and its application for heavy gas protection malfunction of 500kv shell-type transformers[J]. Transformer, 2015, 52(3): 60-65(in Chinese).
10. Tu Weimin, Cui Guomin, Li Yu, et al. Internal streaming simulation of oil-immersed transformers andon-line fault prediction method[J]. Power System Technology, 2010, 34(5): 195-200(in Chinese).
11. Liao Weihan, Guo Chuangxin, Jin Yu, et al. Oil-immersed transformer fault diagnosis method based on four-stage preprocessing and GBDT[J]. Power System Technology, 2019, 43(6): 2195-2203(in Chinese).
12. Huang Limei, Wang Wu, Lin Qiongbin, et al. Fault diagnosis method of converter based on kernel density estimation classifier[J]. Power System Technology, 2019, 43(6): 2204-2210(in Chinese).
13. Pu Tianjiao, Li Ye, Chen Naishi, et al. Key technology and research framework for optimal operation control of hybrid AC/DC system based on power electronic transformer[J]. Power System Technology, 2018, 42(9): 2752-2759(in Chinese).
14. Sun Qiu, Fan Chunju, Xu Feng, et al. Research on characteristics of pole-to-ground fault in mv power electronic transformer DC port[J]. Power System Technology, 2018, 42(11): 3616-3622(in Chinese).
15. WAN Shuting1, WEI Jiaoling1, LÜ Pengrui1, JIANG Danyu2, DOU Longjiang1, MA Zhiqin2 1. Hebei Key Laboratory of Electric Machinery Health Maintenance & Failure Prevention (North China Electric Power University), Baoding 071003, Hebei Province, China; 2. Electric Power Research Institute of Guangdong Power Grid Co., Ltd., Guangzhou 510080, Guangdong Province, China)
16. WAN Shuting1, WEI Jiaoling1, LÜ Pengrui1, JIANG Danyu2, DOU Longjiang1, MA Zhiqin2 (1. Hebei Key Laboratory of Electric Machinery Health Maintenance & Failure Prevention (North China Electric Power University), Baoding 071003, Hebei Province, China; 2. Electric Power Research Institute of Guangdong Power Grid Co., Ltd., Guangzhou 510080, Guangdong Province, China)

AUTHOR INDEX

Bai Lu............44
Bai Peixian............242, 251
Cao Chang............138, 151
Chai Yongming............130
Chang Xincai............1
Chao Han............159
Chen Bin............195, 199
Chen Jiandong............234
Chen Qingwei............217, 238
Chen Xiaoqi............44
Dong Bin............126, 130, 134
Dou Longjiang............251
Feng Deluan............32
Feng Xingxing............138, 151
Fu Jia............75
Gao Minghong............209
Gao Zhenkui............167
Ge Shan............24
Geng Yuxiong............229
Hu Xiaoguang............189
Huang Guo............199
Huang Guoqing............217
Huang Hanting............28
Huang Shanyue............55
Huang Yao............195
Jia Chunyang............5, 9
Jia Yixuan............242
Jinzhou Dai............159
Lai Lin............183
Li ChengYu............44
Li Jinbao............183
Li Maohua............189
Li Qinghua............189
Li Wei............234
Li Xiuwen............221
Li Yantao............44
Li Zhao............234
Li Zheng............195
Li Zhijie............5, 20
Liao Hailin............195, 199
Lin Zihao............61
Liu Haijun............134
Liu Hongfei............14
Liu Min............138, 151, 234
Ma Jun............229
Ma Xinning............173
Niu Junlong............9
Pan Zhuo............203
Peng Xu............159
Qi Yuhao............14
Shang Jingshi............221
Su Zhigang............199
Sun Mengxuan............5
Sun Yongxiu............20
Tan Xudong............1
Tang Shengwei............9
Wang Chunsheng............14
Wang Duo............111
Wang Haibao............1
Wang Peng............234
Wang Qingyong............221
Wang Shurui............82
Wang Xiaoai............229
Wang Yongtao............229
Wu Baizhen............117
Wu Dangdang............14
Wu Hao............225
Xiao Qunxiong............1
Xie Jingyi............126
Xie Yiling............251
Xiong Kangwei............32
Yan Yijun............5
Yang Kun............32
Yang Wu............159
Yao Yao............159
Yao Yaqi............104
Yu Meilin............93
Yu Tongzhu............14
Yuan Peizhi............69
Zhang Fen............44
Zhang Hao............234
Zhang Lixi............49
Zhang Panpan............217, 238
Zhang Xiaolei............44
Zhen Shisen............225
Zheng Chao............32
Zheng Kang............163
Zheng Yitian............229
Zhou Haijun............44
Zhou Mengmeng............44
Zhou Siyang............238
Zhou Yanan............130
Zhu Binrong............189, 195, 199